Contents

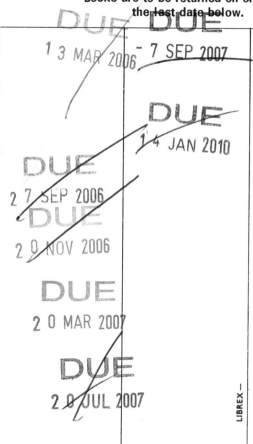

Handbook of Biodegradable Polymers

Editor: Catia Bastioli

Rapra Technology Limited

Shawbury, Shrewsbury, Shropshire, SY4 4NR, United Kingdom
Telephone: +44 (0)1939 250383 Fax: +44 (0)1939 251118
http://www.rapra.net

First Published in 2005 by

Rapra Technology Limited

Shawbury, Shrewsbury, Shropshire, SY4 4NR, UK

©2005, Rapra Technology Limited

A catalogue record for this book is available from the British Library.

Every effort has been made to contact copyright holders of any material reproduced within
the text and the authors and publishers apologise if any have been overlooked.

ISBN: 1-85957-389-4

Typeset, printed and bound by Rapra Technology Limited
Cover printed by The Printing House, Crewe, UK

14 Biodegradable Polymers and the Optimisation of Models for Source Separation and Composting of Municipal Solid Waste

Preface

The present volume reviews the most important achievements, the programs and approaches of institutions, private sector and universities to develop bioplastics and explores their potential utility. The volume covers: the most relevant bioplastics from renewable and non renewable origin and the present business situation; a review of the main studies on the environmental impact of bioplastics and a critical analysis of the methodologies involved; the potential of new areas such as biocatalysis in the development of new bioplastics. It also takes into consideration aspects related to the biodegradation of bioplastics in different environments and the related standards and case studies showing their use in helping to solve specific solid waste problems.

The demand for biodegradable polymers has steadily grown over the last ten years at an annual rate of between 20 and 30%. The market share, however, is very modest accounting for less than 0.1% of the total plastics market.

The limited growth of bioplastics in the last years can be explained by the few products available in the market, the performances sometimes not being fully satisfactory, the high prices, the limited legislative attention, the fact that biodegradability is a functional property not immediately perceived by the final users, requiring significant communication efforts.

However, the opportunity to utilise renewable resources in the production of some of these polymers and to reduce dependency on foreign petroleum resources with the exploitation of new functional properties in comparison with traditional plastics, could become a significant added benefit and accelerate the future growth. Besides biodegradability the technical developments made in the research process, could have significant advantages for the final consumers and contribute to the solution of technical, economical and environmental issues in specific market areas.

Renewable raw materials (RRM) as industrial feedstock for the manufacture of chemical substances and products, such as oils from oilseed crops, starch from cereals and potatoes, and cellulose from straw and wood, have recently received attention from policy makers. By employing physical, chemical and biochemical processes these materials can be converted into chemical intermediates, polymers and speciality chemicals for which, to date, fossil fuels have traditionally been used as feedstock.

The development of products from RRM can be a significant contribution to sustainable development in view of the less energy involved in their production and the wider range of disposal option at lower environmental impact. Legislative attention able to properly address this issue could become a further incentive to the development of products from RRM and maximise the environmental, social and industrial benefits.

The success of such highly innovative products is linked to the achievement of high quality standards. In the field of bioplastics, quality mainly means environmental quality. Standardisation Committees at national and international level have been working for many years in the definition of standard test methods to assure the biodegradability and full environmental compatibility of the new bio-plastics. Standards such as the European EN 13432 on the compostability of packaging (CEN TC261SC4WG2) and other related norms at international level are now in place, whereas standards on biodegradation of bioplastics in soil are still under discussion.

The quality of the bioplastic products is assured not only by the control of the biodegradability parameters but also by the assessment of a real functionality. A biodegradable product is useless if it does not perform as a traditional product or better in terms of mechanical resistance, duration, etc. For this reason the commitment of producers of bioplastics in the creation of a quality network able to guarantee the quality of the product in all the steps of the life cycle becomes very relevant.

The elaboration and the diffusion of a best practice in the field of organic waste collection, where the use of biodegradable compostable bags is a tool to improve the quality of the system, for example, has permitted to thousands of municipalities all over Europe to implement the proposed model. The cooperation with public bodies is also a key factor in the success of bioplastics, because the topics under discussion are strictly related to public interest, such as safety, environment, and health.

Today the bioplastics available in the market at different levels of development are mainly carbohydrate-based materials. Starch can be either physically modified and used alone or in combination with other polymers, or it can be used as a substrate for fermentation for the production of polyhydroxyalkanoates or lactic acid, transformed into poly lactic acid (PLA) through standard polymerisation processes. Also vegetable oil based polymers are under development.

It can be roughly estimated that for 1 kg of bioplastics 1 to 2 kg of corn or 5 to 10 kg of potatoes is needed. A potential of 500,000 ton/year of bioplastics therefore should require

50,000 - 100,000 hectares of land. The 2010 scenario of a replacement of 10% of plastics by bioplastics requiring 5 to 10 million tons/year of corn and, consequently, 1 million hectares of land, is perfectly compatible with the European set aside program.

The increasing use of bioplastics can open to entirely new generations of materials with new performances in comparison with traditional plastics. The possibility offered by physically modified starch to create functionalised nanoparticles able to modify the properties of natural and synthetic rubbers and of other synthetic polymers, the natural high barrier to oxygen of starch, and its derivatives and their high permeability to water vapour, already offer a range of completely new solutions to the plastic Industry.

Moreover the characteristic of bioplastics to recycle carbon dioxide and/or to biodegrade minimising the risks of pollution can offer significant environmental and social benefits in a wide range of disposal options such as sewage sludge water treatment plants, composting, and incineration.

The use of RRM, however, is not by itself a guarantee for a low environmental impact. Aspects such as the production processes, the technical performance and the weight of each final product and its disposal options have to be carefully considered along all the steps of product life. The engineering of bio-based materials for specific applications using Life Cycle Analysis (LCA) in a cradle to grave approach is therefore a critical aspect. Instruments like LCA should be used to clearly point out, which benefits the use of this new class of materials can offer. First results for bioplastics suggest an optimistic outlook.

Bioplastics are generally biodegradable according to the EN 13432 standard and, besides other disposal options, can be organically recycled through composting. Such characteristics may represent a significant advantage in sectors like packaging when composting infrastructures are available.

On the basis of the mandate M200/ref.3 of the EU-Commission the standard for biodegradable/compostable packaging EN 13432 was elaborated and published as harmonised standard in the Official Journal of the EU.

The development of biodegradable bioplastics is strictly linked to the parallel growth of composting infrastructures. The organic waste, and bioplastics belonging to this waste fraction, is recycled through composting. The diffusion of the composting technology is a prerequisite for the development of bioplastics even if other disposal options such as incineration or disposal in sewage water treatment may be possible.

The option of the organic recycling offered by the network of composting infrastructures has permitted in recent years the start-up of this new sector of industry, mainly focused on polymers from renewable raw materials and other biodegradable materials.

Properly addressing this issue by the next Biowaste Directive in Europe would be a further incentive to the development of products from RRM, with the aim of maximising environmental, economic, and social benefits. Bioplastics can also develop in sectors, such as in agricultural mulches, where retrieval of plastics is difficult or costly and bioplastics are perceived as a best fit.

In order to combine environment and economy and make the renewable resources a sustainable business option and not only a sustainable environmental option further efforts must be done.

The involvement of up-stream players that is the farmers and their associations is a very important prerequisite. In agriculture, new agronomical approaches and the development of new genotypes for non-food applications should be taken into consideration. Agriculture with lower environmental impact and lower costs is an important factor in the development of new bio-based products.

Efforts must also be made at the industrial level in order to develop less expensive and higher performance products and low impact technologies.

The involvement of the specific stakeholders can be achieved if a communication programme is launched and operated in parallel with the industrial activities. The success of the project is very much linked to the diffusion of a new environmental awareness, at all levels: politicians, public administrators, associations, citizens, etc., all must be reached by specific communications.

This, in turn, must give rise to specific legislative actions in order to quantify the social and environmental benefits linked to the non-food use of agricultural and natural raw materials and to the bioconversion of waste materials into industrial products.

ACKNOWLEDGMENTS

I would like to thank all of the contributors to this volume for their time and their efforts that went into producing this book, and also for their patience.

I would like to express my sincere appreciation to my Editor, Frances Powers of Rapra Technology Ltd and to her assistant Claire Griffiths for the high quality support during all stages of production of this book. Thanks are also extended to the Graphic Designer, Sandra Hall for typesetting the book and the design of the cover and to John Holmes for producing the Index. I would also like to thank Annalisa Righetti for the handling of the considerable correspondence associated with the book and the organisation of the reviewing process and Elisa Sonzini for her secretarial support. Without their help none of this would have happened.

Finally I dedicate this volume to the memory of Dr Luigi Marini, who has devoted, together with myself, a significant part of his life to study and promote bioplastics and is no longer with us.

Catia Bastioli
December 2004

1 Biodegradability of Polymers – Mechanisms and Evaluation Methods

Maarten van der Zee

1.1 Introduction

This chapter presents an overview of the current knowledge on the biodegradability of polymeric materials. The focus is in particular on the biodegradation of materials under environmental conditions. *In vivo* degradation of polymers used in biomedical applications is not covered because it is considered outside the scope of this handbook and has been extensively reviewed elsewhere [1, 2]. Nevertheless it is important to realise that the degradation of polymers in the human body is also often referred to as biodegradation.

A number of different aspects of assessing the potential, the rate, and the degree of biodegradation of polymeric materials are discussed. The mechanisms of polymer degradation and erosion receive attention, the major enzyme systems involved in biodegradation reactions are described, and factors affecting enzymic and non-enzymic degradation are briefly addressed. Particular attention is given to the ways for measuring biodegradation, including complete mineralisation to gasses (such as carbon dioxide and methane), water and possibly microbial biomass. Finally, some general conclusions about the biodegradability of polymeric materials are presented.

1.2 Background

There is a worldwide research effort to develop biodegradable polymers for agricultural applications or as a waste management option for polymers in the environment. Until recently, most of the efforts were synthesis oriented, and not much attention was paid to the identification of environmental requirements for, and testing of, biodegradable polymers. Consequently, many unsubstantiated claims to biodegradability were made, and this has damaged the general acceptance.

An important factor is that the term biodegradation has not been applied consistently. In the medical field of sutures, bone reconstruction and drug delivery, the term biodegradation has been used to indicate hydrolysis [3]. On the other hand, for environmentally degradable plastics, the term biodegradation may mean fragmentation, loss of mechanical properties, or sometimes degradation through the action of living organisms [4]. Deterioration or loss in physical integrity is also often mistaken for biodegradation [5]. Nevertheless, it is essential to have a universally acceptable definition of biodegradability to avoid confusion as to where biodegradable polymers can be used in agriculture or fit into the overall plan of polymer waste management. Many groups and organisations have endeavoured to clearly define the terms 'degradation', 'biodegradation', and 'biodegradability'. But there are several reasons why establishing a single definition among the international community has not been straightforward, including:

- The variability of an intended definition given the different environments in which the material is to be introduced and its related impact on those environments.

- The differences of opinion with respect to the scientific approach or reference points used for determining biodegradability.

- The divergence of opinion concerning the policy implications of various definitions.

- Challenges posed by language differences around the world.

As a result, many different definitions have officially been adopted, depending on the background of the defining organisation and their particular interests. However, of more practical importance are the criteria for calling a material 'biodegradable'. A demonstrated potential of a material to biodegrade does not say anything about the time frame in which this occurs, nor the ultimate degree of degradation. The complexity of this issue is illustrated by the following common examples.

Low density polyethylene has been shown to biodegrade slowly to carbon dioxide (0.35% in 2.5 years) [6] and according to some definitions can thus be called a biodegradable polymer. However, the degradation process is so slow in comparison with the application rate, that accumulation in the environment will occur. The same applies for polyolefin-starch blends which rapidly loose strength, disintegrate, and visually disappear if exposed to microorganisms [7-9]. This is due to utilisation of the starch component, but the polyolefin fraction will nevertheless persist in the environment. Can these materials be called 'biodegradable'?

1.3 Defining Biodegradability

In 1992, an international workshop on biodegradability was organised to bring together experts from around the world to achieve areas of agreement on definitions, standards and testing methodologies. Participants came from manufacturers, legislative authorities, testing laboratories, environmentalists and standardisation organisations in Europe, USA and Japan. Since this fruitful meeting, there is a general agreement concerning the following key points [10].

- For all practical purposes of applying a definition, material manufactured to be biodegradable must relate to a specific disposal pathway such as composting, sewage treatment, denitrification, or anaerobic sludge treatment.

- The rate of degradation of a material manufactured to be biodegradable has to be consistent with the disposal method and other components of the pathway into which it is introduced, such that accumulation is controlled.

- The ultimate end products of aerobic biodegradation of a material manufactured to be biodegradable are carbon dioxide, water and minerals and that the intermediate products include biomass and humic materials. (Anaerobic biodegradation was discussed in less detail by the participants).

- Materials must biodegrade safely and not negatively impact on the disposal process or the use of the end product of the disposal.

As a result, specified periods of time, specific disposal pathways, and standard test methodologies were incorporated into definitions. Standardisation organisations such as CEN, International Standards Organisation (ISO), and American Society for Testing and Materials (ASTM) were consequently encouraged to rapidly develop standard biodegradation tests so these could be determined. Society further demanded non-debatable criteria for the evaluation of the suitability of polymeric materials for disposal in specific waste streams such as composting or anaerobic digestion. Biodegradability is usually just one of the essential criteria, besides ecotoxicity, effects on waste treatment processes, etc. The standards and certification procedures resulting from these discussions are presented in Chapter 5 of this handbook.

In the following paragraphs, biodegradation of polymeric materials is looked upon from the chemical perspective. The chemistry of the key degradation process, is represented by Equations 2.1 and 2.2, where $C_{POLYMER}$ represents either a polymer or a fragment from any of the degradation processes defined earlier. For simplicity here, the polymer or fragment is considered to be composed only of carbon, hydrogen and oxygen; other elements may, of course, be incorporated in the polymer, and these would appear in an oxidised or reduced form after biodegradation depending on whether the conditions are aerobic or anaerobic, respectively.

Aerobic biodegradation:

$$C_{POLYMER} + O_2 \rightarrow CO_2 + H_2O + C_{RESIDUE} + C_{BIOMASS} \qquad (1.1)$$

Anaerobic biodegradation:

$$C_{POLYMER} \rightarrow CO_2 + CH_4 + H_2O + C_{RESIDUE} + C_{BIOMASS} \qquad (1.2)$$

Complete biodegradation occurs when no residue remains, and complete mineralisation is established when the original substrate, ($C_{POLYMER}$ in this example), is completely converted into gaseous products and salts. However, mineralisation is a very slow process under natural conditions because some of the polymer undergoing biodegradation will initially be turned into biomass [11]. Therefore, complete biodegradation and not mineralisation is the measurable goal when assessing removal from the environment.

1.4 Mechanisms of Polymer Degradation

When working with biodegradable materials, the obvious question is why some polymers biodegrade and others do not. To understand this, one needs to know about the mechanisms through which polymeric materials are biodegraded. Although biodegradation is usually defined as degradation caused by biological activity (especially enzymic action), it will usually occur simultaneously with - and is sometimes even initiated by - abiotic degradation such as photodegradation and simple hydrolysis. The following sections give a brief introduction into the most important mechanisms of polymer degradation.

1.4.1 Non-biological Degradation of Polymers

A great number of polymers are subject to hydrolysis, such as polyesters, polyanhydrides, polyamides, polycarbonates, polyurethanes, polyureas, polyacetals, and polyorthoesters. Different mechanisms of hydrolysis have been extensively reviewed; not only for backbone hydrolysis, but also for hydrolysis of pendant groups [12, 13]. The necessary elements for a wide range of catalysis, such as acids and bases, cations, nucleophiles and micellar and phase transfer agents, are usually present in most environments. In contrast to enzymic degradation where a material is degraded gradually from the surface inwards (primarily because macromolecular enzymes cannot diffuse into the interior of the material), chemical hydrolysis of a solid material can take place throughout its cross section except for very hydrophobic polymers.

Important features affecting chemical polymer degradation and erosion include:

(a) the type of chemical bond,

(b) the pH,

(c) the temperature,

(d) the copolymer composition, and

(e) water uptake (hydrophilicity). These features will not be discussed here, but have been covered in detail by [3].

1.4.2 Biological Degradation of Polymers

Polymers represent major constituents of the living cells which are most important for the metabolism (enzyme proteins, storage compounds), the genetic information (nucleic acids), and the structure (cell wall constituents, proteins) of cells [14]. These polymers have to be degraded inside cells in order to be available for environmental changes and to other organisms upon cell lysis. It is therefore not surprising that organisms, during many millions of years of adaptation, have developed various mechanisms to degrade naturally occurring polymers. For the many different new synthetic polymers that have found their way into the environment only in the last fifty years, however, these mechanisms may not as yet have been developed.

There are many different degradation mechanisms that combine synergistically in nature to degrade polymers. Microbiological degradation can take place through the action of enzymes or by products (such as acids and peroxides) secreted by microorganisms (bacteria, yeasts, fungi, etc). Also macro-organisms can eat and, sometimes, digest polymers and cause mechanical, chemical or enzymic ageing [15, 16].

Two key steps occur in the microbial polymer degradation process: first, a depolymerisation or chain cleavage step, and second, mineralisation. The first step normally occurs outside the organism due to the size of the polymer chain and the insoluble nature of many of the polymers. Extracellular enzymes are responsible for this step, acting either endo (random cleavage on the internal linkages of the polymer chains) or exo (sequential cleavage on the terminal monomer units in the main chain).

Once sufficiently small size oligomeric or monomeric fragments are formed, they are transported into the cell where they are mineralised. At this stage the cell usually derives metabolic energy from the mineralisation process. The products of this process, apart from adenosine triphosphate (ATP), are gasses, (e.g., CO_2, CH_4, N_2, H_2), water, salts and minerals, and biomass. Many variations of this general view of the biodegradation process can occur, depending on the polymer, the organisms, and the environment. Nevertheless, there will always be, at one stage or another, the involvement of enzymes.

Enzymes are the biological catalysts, which can induce enormous (10^8 - 10^{20} fold) increases in reaction rates in an environment otherwise unfavourable for chemical reactions. All enzymes are proteins, i.e., polypeptides with a complex three-dimensional structure, ranging in molecular weight from several thousand to several million g/mol. The enzyme activity is closely related to the conformational structure, which creates certain regions at the surface forming an active site. At the active site the interaction between enzyme and substrate takes place, leading to the chemical reaction, eventually giving a particular product. Some enzymes contain regions with absolute specificity for a given substrate while others can recognise a series of substrates. For optimal activity most enzymes must associate with cofactors, which can be of inorganic, (e.g., metal ions), or organic origin (such as coenzyme A, ATP and vitamins like riboflavin and biotin) [14].

Different enzymes can have different mechanisms of catalysis. Some enzymes change the substrate through some free radical mechanism while others follow alternative chemical routes. The enormous number of different enzymes (each catalysing its own unique reaction on groups of substrates or on very specific chemical bonds; in some cases acting complementarily, in others synergetically) makes it impossible to cover all enzymes within the limitations of this review. Therefore, the overview in this chapter will be restricted to some typical examples of polymer degradation through enzymic hydrolysis and enzymic oxidation.

1.4.2.1 Enzymic Hydrolysis

Glycosidic bonds, as well as peptide bonds and most ester bonds (e.g., in proteins, nucleic acids, polysaccharides, and polyhydroxyalkanoic acids), are cleaved by hydrolysis. A number of different enzymes are involved, depending of the type of bond to be hydrolysed: proteases, esterases, and glucoside hydrolases.

• **Proteases**

Proteolytic enzymes (proteases) catalyse the hydrolysis of peptide (amide) bonds and sometimes the related hydrolysis of ester linkages. Proteases are divided into four groups on the basis of their mechanism of action:

(a) the serine proteases,

(b) the cysteine proteases,

(c) the metal containing proteases and

(d) the aspartic proteases.

The names indicate one of the key catalytic groups in the active site. They have been reviewed in detail by Whitaker [17].

a) Serine endopeptidases

The *serine endopeptidases* include the chymotrypsin family (EC 3.4.21.1), trypsin (EC 3.4.21.4), elastase (EC 3.4.21.37), thrombin (EC 3.4.21.5), subtilisin (EC 3.4.21.62) and α-lytic proteases (EC 3.4.21.12). The enzymes are all endopeptidases. The substrate specificities of the individual members of this group are often quite different, which is attributed to different structures of the binding pockets.

b) Cysteine endopeptidases

The group of *cysteine endopeptidases* (also called sulfhydryl proteases or thiol proteases) include the higher plant enzymes papain (EC 3.4.22.2) and ficin (EC 3.4.22.3), but also numerous microbial proteolytic enzymes such as *Streptococcus* cysteine proteinase (EC 3.4.22.10). The enzymes have a rather broad substrate specificity, and specifically recognise aromatic substituents. The specificity is for the second amino acid from the peptide bond to be cleaved.

c) Metal containing proteases

It is of interest to note that almost all the proteolytic enzymes that belong to this group are exopeptidases. Some enzymes require zinc (Zn^{2+}) whereas others require manganese ions (Mn^{2+}) for their hydrolytic activity. Whether the metal ion, which appears to be divalent in all cases, performs a similar function in all of the enzymes in this group is not known.

d) Aspartic endopeptidases

The group name *aspartic endopeptidases* indicates that the carboxyl groups of two aspartic acid residues are the catalytic groups in the active site. The best studied of this group of enzymes is pepsin (EC 3.4.23.1), but chymosin (EC 3.4.23.4) and a large number of microbial proteases also belong to this group. Pepsin has a preference for hydrolysis at the aromatic amino acid residues.

- **Esterases (EC 3.1)**

Perhaps no other group of enzymes is so widely distributed in nature, and most tissues contain a great variety of enzymes with esterase activity. As a group, esterases are involved in the splitting of ester linkages by the addition of water as the second substrate. The alcohol may be monohydric or polyhydric, aliphatic or aromatic. The acid may be an organic or inorganic acid. The esterases may be subdivided into several groups primarily on the basis of specificity for the acid involved in the ester substrate: (a) carboxylic ester hydrolases (EC 3.1.1), (b) thiol ester hydrolases (EC 3.1.2), (c) phosphoric monoester hydrolases (EC 3.1.3), (d) phosphoric diester hydrolases (EC 3.1.4) and (e) sulfuric ester hydrolases (EC 3.1.6).

Of special interest is one group of carboxylic acid hydrolases; the lipases (EC 3.1.1.3). Lipases catalyse the hydrolysis of triglycerides into diglycerides, monoglycerides, glycerol and fatty acids. Some lipases are also capable of hydrolysing polyesters to monomeric or oligomeric products which can be taken up by microbial cells and metabolised further by other esterases [18].

a) Lipases

Lipases act only at a lipid-water interface and have very little activity on soluble substrates [17]. For extracellular lipases to become active at an oil/water interface, it has been suggested that the lipase's 'hydrophobic head' is bound to the oil droplet by hydrophobic interactions, while the enzyme's active site aligns with, and binds to the substrate molecule [19]. A similar mechanism could be valid for lipase activity on polyester surfaces.

Although a great deal of work has been done on the substrate specificity of lipases, almost nothing has been done in a definite way in elucidating its mechanism of action. This is due largely to the difficulty of studying the kinetics of an enzyme in a heterogeneous system (solid/liquid, emulsion). It is suggested that the mechanism of hydrolysis of the ester bonds of lipids resembles the mechanism of hydrolysis of peptide bonds by the serine proteases. Most likely an acyl-enzyme intermediate is formed. The acyl-enzyme intermediate is then hydrolysed by water to give the free fatty acid and regenerate the enzyme or the acyl group can be transferred to a nucleophilic group, such as the -OH of glycerol to give transesterification.

- **Glycosidases**

Glycoside hydrolases cleave the glycosidic bond in polysaccharides like starch, inulin, cellulose and their derivatives. The most important types are (a) the amylases (EC 3.2.1.1 and EC 3.2.1.2), which act on starch and derived polysaccharides to hydrolyse the α-1,4 and/or α-1,6 glucoside linkages, and (b) cellulase (EC 3.2.1.4), which act on β-1,4 glucoside linkages in cellulose and derived polymers.

a) Amylases

α-Amylase (EC 3.2.1.1) is widely distributed in microorganisms and hydrolyses α-1,4-glucoside linkages in starch randomly while maintaining the configuration about the C(1) position of the glycone. It is capable of bypassing the branching points. The α-1,6 glucosidic linkages in amylopectin are not hydrolysed by α-amylases. Its action on starch, therefore, causes the formation of linear and branched oligosaccharides of varying length.

β-Amylase (EC 3.2.1.2) is an exo-enzyme, occurring in a few microbes. It hydrolyses α-1,4-glucoside linkages from the non-reducing end of polysaccharides. It causes the inversion of the anomeric configuration about the C(1) position of the liberated maltose from α to β. β-Amylase is incapable of bypassing α-1,6 linkages of branched substrates like amylopectin. In some cases microorganisms form similar exo-acting enzymes but instead of maltose other oligosaccharides with defined size are formed. The reducing ends of oligomeric products like maltotetraose and maltohexaose, however, have the α-configuration [20].

Glucoamylase (EC 3.2.1.3) is also called amyloglucosidase or γ-amylase and is a typical enzyme of fungi. It is an exo-splitting enzyme which attacks α-1,4- and α-1,6-glucoside linkages of α-glucans from the non-reducing side. The branching points, however, are hydrolysed at a very slow rate. The action of the enzyme liberates one molecule of β-D-glucose at a time causing the complete conversion of polysaccharides to glucose. The enzyme has a preference for large substrates.

α-Glucosidase (EC 3.2.1.20) attacks the α-1,4- and/or α-1,6-glucoside linkages from the non-reducing ends in short saccharides which are normally formed by the action of other amylolytic enzymes. Unlike the glucoamylase action, the α-glucosidase action liberates glucose with an α-configuration. α-Glucosidase appears to be the final enzyme involved in the metabolism of starch. The enzyme is usually present with other amylolytic enzymes and is either extracellular, cell bound or intracellular.

Pullulanase (EC 3.2.1.41) is a typical bacterial endo-splitting enzyme that hydrolyses the α-1,6-glucosidic bonds of pullulan, amylopectin and their α- and β-limit dextrins. Pullulanase requires that each of the two chains of amylopectin linked by an α-1,6-glucosidic bond contain at least two adjacent α-1,4-linked glucose units.

b) Cellulases

Cellulase (an endoglucanase) (EC 3.2.1.4) is inactive against crystalline celluloses such as cotton and Avicel, but it hydrolyses amorphous celluloses (including amorphous regions of crystalline celluloses) and soluble substrates such as carboxyethyl cellulose and hydroxyethyl cellulose. Endoglucanase activity is characterised by random cleavage of β-glucosidic bonds.

Cellobiohydrolases (EC 3.2.1.91) are exo-splitting enzymes which degrade amorphous cellulose by consecutive removal of cellobiose from the non-reducing ends of the substrate. Endoglucanases and cellobiohydrolases act synergetically in the hydrolysis of crystalline cellulose.

Exoglucohydrolases catalyse the removal of glucose units from the non-reducing end of cellodextrins; the rate of hydrolysis decreases as the chain length of the substrate decreases.

β-Glucosidases (EC 3.2.1.21), like exoglucohydrolases, catalyse the removal of glucose units from the non-reducing end, but do not act on polymeric materials. The reaction rate increases as the chain length of the substrate decreases.

1.4.2.2 Enzymic Oxidation

Biological oxidation is catalysed by a large group of enzymes called oxidoreductases (EC 1; 611 enzymes total). Oxidation or reduction of a substrate can occur in a number of ways, as is shown in **Table 1.1**, where the distinction is made on the basis of electron acceptor (B, O_2 or H_2O_2) and products formed.

By far the largest number of oxidoreductases belong to the type *(1)* reactions in which enzymes catalyse the oxidation of the substrate by removal of hydrogens and/or electrons through participation of an acceptor B, such as NAD^+, $NADP^+$, ferricytochrome, and so on. In type *(2)* and *(3)* reactions, molecular oxygen is involved, and are thus only observed under aerobic conditions. A cofactor is involved in the mechanism of catalysis, which is regenerated when the cycle of events is completed. The cofactor therefore does not appear in the overall equation, as opposed to the cofactor in type *(1)* reactions which is used up and can only be regenerated by an other enzyme system.

Reactions *(4)* to *(7)* involve the oxidation of substrate by incorporation of one or more oxygen atoms in the substrate. The reactions are distinguished on the basis of the source of oxygen atoms, which can be water, H_2O_2 or O_2. The enzymes that incorporate O_2 are called 'oxygenases' since the reactions are similar to those known to occur by chemical and photochemical processes. Oxygenases can be subdivided in two classes. Monooxygenases catalyse the insertion of a single atom of oxygen in the substrate as a hydroxyl group and require a second reduced substrate which simultaneously undergoes oxidation. Usually this is NADH or NADPH. Dioxygenases catalyse the insertion of the whole oxygen molecule into the substrate. Sometimes the product is a dihydroxy derivative but more often the oxygen atoms are incorporated as a carboxyl or hydroperoxide grouping.

Table 1.1 Oxidative enzymes in biological systems		
Reaction formula		
$AH_2 + B \rightarrow A + BH_2$		(1)
$AH_2 + O_2 \rightarrow A + H_2O_2$	(H_2O_2-forming oxidases)	(2)
$AH_2 + \frac{1}{2} O_2 \rightarrow A + H_2O$	(H_2O-forming oxidases)	(3)
$A + H_2O + B \rightarrow AO + BH_2$		(4)
$A + H_2O_2 \rightarrow AO + H_2O$		(5)
$A + O_2 + BH_2 \rightarrow AO + B + H_2O$	(monooxygenases)	(6)
$A + O_2 \rightarrow AO_2$	(dioxygenases)	(7)

Examples of oxidative polymer degradation

An example of oxidative degradation of polymers is presented by White and co-workers [21] for the biodegradation of water-soluble poly(ethylene glycols) (PEG). PEG-dehydrogenase (EC 1.1.99.20), aldehyde-oxidising enzymes and ether cleaving enzymes were considered to operate in sequence to catalyse the oxidation of the terminal R-O-CH$_2$-CH$_2$OH group, via the aldehyde and carboxyl function to R-O-CHOH-COOH. The chain length of the PEG is subsequently shortened by two CH$_2$-units by the liberation of 2-hydroxyacetic acid. Other mechanisms observed for PEG degradation by anaerobic microorganisms include hydroxyl-shifts from the terminal carbon to the ether-linked carbon (analogous to the diolhydratase reaction) followed by a rapid dissociation of the resulting hemiacetal to acetaldehyde and a shortened PEG.

Aerobic biodegradation of lignin also is an oxidative process mediated by the extracellular enzyme lignin peroxidase (EC 1.11.1.14) in the presence of H$_2$O$_2$ [22]. Lignin peroxidase is a non-specific oxidative enzyme produced by a number of species of aerobic fungi, especially white-rot fungi, and a few aerobic bacteria, such as actinomycete species. Microorganisms which produce lignin peroxidase are generally also able to produce the H$_2$O$_2$ required. The mechanism of lignin peroxidase activity is considered to involve the formation of substrate radical intermediates. Such radicals might invade the lignin molecule and be the immediate effectors of its degradation [23]. Manganese peroxidases (EC 1.11.1.13) have been defined as a second class of oxidising enzymes. These oxidise Mn(II) to Mn(III) and it is proposed that such Mn(III), chelated to organic acids, function as an active radical that can mediate oxidative depolymerisation of lignin [24].

1.5 Measuring Biodegradation of Polymers

As can be imagined from the various mechanisms described above, biodegradation does not only depend on the chemistry of the polymer, but also on the presence of the biological systems involved in the process. When investigating the biodegradability of a material, the effect of the environment cannot be neglected. Microbial activity, and hence biodegradation, is influenced by:

- the presence of microorganisms

- the availability of oxygen

- the amount of available water

- the temperature

- the chemical environment (pH, electrolytes, etc.)

In order to simplify the overall picture, the environments in which biodegradation occurs are basically divided in two environments: *(a)* aerobic (with oxygen available) and *(b)* anaerobic (no oxygen present). These two, can in turn be subdivided into *(1)* aquatic and *(2)* high solids environments. **Figure 1.1** presents schematically the different environments, with examples in which biodegradation may occur [25, 26].

The high solids environments will be the most relevant for measuring biodegradation of polymeric materials, since they represent the conditions during biological municipal solid waste treatment, such as composting or anaerobic digestion (biogasification). However, possible applications of biodegradable materials other than in packaging and consumer products, e.g., in fishing nets at sea, or undesirable exposure in the environment due to littering, explain the necessity of aquatic biodegradation tests.

Numerous ways for the experimental assessment of polymer biodegradability have been described in the scientific literature. Because of slightly different definitions or interpretations of the term 'biodegradability', the different approaches are therefore not equivalent in terms of information they provide or the practical significance. Since the typical exposure environment involves incubation of a polymer substrate with microorganisms or enzymes, only a limited number of measurements are possible: those pertaining to the substrates, to the microorganisms, or to the reaction products. Four common approaches available for studying biodegradation processes have been reviewed in detail by Andrady [11]:

- Monitoring microbial growth

- Monitoring the depletion of substrates

- Monitoring reaction products

- Monitoring changes in substrate properties

	aquatic	high solids
aerobic	• aerobic waste water treatment plants • surface waters, e.g., lakes and rivers • marine environments	• surface soils • organic waste composting plants • littering
anaerobic	• anaerobic waste water treatment plants • rumen of herbivores	• deep sea sediments • anaerobic sludge • anaerobic digestion/ biogasification • landfill

Figure 1.1 *Schematic classification of different biodegradation environments for polymers*

In the following sections, different test methods for the assessment of polymer biodegradability are presented. Measurements are usually based on one of the four approaches given previously, but combinations also occur. Before choosing an assay to simulate environmental effects in an accelerated manner, it is critical to consider the closeness of fit that the assay will provide between substrate, microorganisms or enzymes, and the application or environment in which biodegradation should take place [27].

1.5.1 Enzyme Assays

1.5.1.1 Principle

In enzyme assays, the polymer substrate is added to a buffered or pH-controlled system, containing one or several types of purified enzymes. These assays are very useful in examining the kinetics of depolymerisation, or oligomer or monomer release from a polymer chain under different assay conditions. The method is very rapid (minutes to hours) and can give quantitative information. However, mineralisation rates cannot be determined with enzyme assays.

1.5.1.2 Applications

The type of enzyme to be used, and quantification of degradation, will depend on the polymer being screened. For example, Mochizuki and co-workers [28] studied the effects of draw ratio of polycaprolactone (PCL) fibres on enzymic hydrolysis by lipase (EC 3.1.1.3). Degradability of PCL fibres was monitored by dissolved organic carbon (DOC) formation and weight loss. Similar systems with lipases have been used for studying the hydrolysis of broad ranges of aliphatic polyesters [29-34], copolyesters with aromatic segments [30, 35-37], and copolyesteramides [38-39]. Other enzymes such as α-chymotrypsin (EC 3.4.21.1) and α-trypsin (EC 3.4.21.4) have also been applied for these polymers [40-41]. Biodegradability of poly(vinyl alcohol) segments with respect to block length and stereo chemical configuration has been studied using isolated poly(vinyl alcohol)-dehydrogenase (EC 1.1.99.23) [42]. Cellulolytic enzymes have been used to study the biodegradability of cellulose ester derivatives as a function of degree of substitution and the substituent size [43]. Similar work has been performed with starch esters using amylolytic enzymes such as α-amylase (EC 3.2.1.1), β-amylase EC 3.2.1.2), and glucan 1,4-α-glucosidase (EC 3.2.1.3) [44]. Enzymic methods have also been used to study the biodegradability of starch plastics or packaging materials containing cellulose [45-50].

1.5.1.3 Drawbacks

Caution must be used in extrapolating enzyme assays as a screening tool for different polymers since the enzymes have been paired to only one polymer. The initially selected enzymes may show significantly reduced activity towards modified polymers or different materials, even though more suitable enzymes may exist in the environment. Caution must also be used if the enzymes are not purified or appropriately stabilised or stored, since inhibitors and loss of enzyme activity can occur [27].

1.5.2 Plate Tests

1.5.2.1 Principle

Plate tests were initially developed to assess the resistance of plastics to microbial degradation. Several methods have been standardised by standardisation organisations such as the ASTM and the International Organisation for Standardisation (ISO) [51-53]. They are now also used to see if a polymeric material will support growth [27, 54]. The principle of the method involves placing the test material on the surface of a mineral salts agar in a petri dish containing no additional carbon source. The test material and agar surface are sprayed with a standardised mixed inoculum of known bacteria and/or fungi. The test material is examined after a predetermined incubation period at constant temperature for the amount of growth on its surface and a rating is given.

1.5.2.2 Applications

Potts [55] used the method in his screening of 31 commercially available polymers for biodegradability. Other studies, where the growth of either mixed or pure cultures of microorganisms, is taken to be indicative for biodegradation, have been reported [4]. The validity of this type of test, and the use of visual assessment alone has been questioned by Seal and Pantke [56] for all plastics. They recommended that mechanical properties should be assessed to support visual observations. Microscopic examination of the surface can also give additional information.

A variation of the plate test, is the 'clear zone' technique [57] sometimes used to screen polymers for biodegradability. A fine suspension of polymer is placed in an agar gel as the sole carbon source, and the test inoculum is placed in wells bored in the agar. After incubation, a clear zone around the well, detected visually, or instrumentally is indicative of utilisation of the polymer. The method has for example been used in the case of starch plastics [58], various polyesters [59-61], and polyurethanes [62].

14

1.5.2.3 Drawbacks

A positive result in an agar plate test indicates that an organism can grow on the substrate, but does not mean that the polymer is biodegradable, since growth may be on contaminants, on plasticisers present, on oligomeric fractions still present in the polymer, and so on. Therefore, these tests should be treated with caution when extrapolating the data to field situation.

1.5.3 Respiration Tests

1.5.3.1 Principle

Aerobic microbial activity is typically characterised by the utilisation of oxygen. Aerobic biodegradation requires oxygen for the oxidation of compounds to its mineral constituents, such as CO_2, H_2O, SO_2, P_2O_5, etc. The amount of oxygen utilised during incubation, also called the biochemical (or biological) oxygen demand (BOD) is therefore a measure of the degree of biodegradation. Several test methods are based on measurement of the BOD, often expressed as a percentage of the theoretical oxygen demand (TOD) of the compound. The TOD, which is the theoretical amount of oxygen necessary for completely oxidising a substrate to its mineral constituents, can be calculated by considering the elemental composition and the stoichiometry of oxidation [11, 63-66] or based on experimental determination of the chemical oxygen demand (COD) [11, 67].

1.5.3.2 Applications

The closed bottle BOD tests were designed to determine the biodegradability of detergents [65, 68]. These have stringent conditions due to the low level of inoculum (in the order of 10^5 microorganisms/l) and the limited amount of test substance that can be added (normally between 2 and 4 mg/l). These limitations originate from the practical requirement that the oxygen demand should be not more than half the maximum dissolved oxygen level in water at the temperature of the test, to avoid the generation of anaerobic conditions during incubation.

For non-soluble materials, such as polymers, less stringent conditions are necessary and alternative ways for measuring BOD were developed. Two-phase (semi) closed bottle tests provide a higher oxygen content in the flasks and permit a higher inoculum level. Higher test concentrations are also possible, encouraging higher accuracy with directly weighing in of samples. Alternatively the oxygen demand can be determined by periodically measuring the oxygen concentration in the aquatic phase by opening the flasks [64, 69-70], by measuring the change in volume or pressure in incubation flasks containing CO_2-absorbing agents [63, 71-72], or by measuring the quantity of oxygen produced (electrolytically) to maintain constant gas volume/pressure in specialised respirometers [63, 66, 69, 71].

1.5.3.3 Suitability

BOD tests are relatively simple to perform and sensitive, and are therefore often used as screening tests. However, the measurement of oxygen consumption is a non-specific, indirect measure for biodegradation, and it is not suitable for determining anaerobic degradation. The requirement for test materials to be the sole carbon/energy source for microorganisms in the incubation media, eliminates the use of oxygen measurements in complex natural environments.

1.5.4 Gas (CO₂ or CH₄) Evolution Tests

1.5.4.1 Principle

The evolution of carbon dioxide or methane from a substrate represents a direct parameter for mineralisation. Therefore, gas evolution tests can be important tools in the determination of biodegradability of polymeric materials. A number of well known test methods have been standardised for aerobic biodegradation, such as the (modified) Sturm test [72-76] and the laboratory controlled composting test [77-79], as well as for anaerobic biodegradation, such as the anaerobic sludge test [80, 81] and the anaerobic digestion test [82]. Although the principle of these test methods are the same, they may differ in medium composition, inoculum, the way substrates are introduced, and in the technique for measuring gas evolution.

1.5.4.2 Applications

Anaerobic tests generally follow biodegradation by measuring the increase in pressure and/or volume due to gas evolution, usually in combination with gas chromatographic analysis of the gas phase [83, 84]. Most aerobic standard tests apply continuous aeration; the exit stream of air can be directly analysed continuously using a carbon dioxide monitor (usually infrared detectors) or titrimetrically after sorption in dilute alkali. The cumulative amount of carbon dioxide generated, expressed as a percentage of the theoretically expected value for total conversion to CO_2, is a measure for the extent of mineralisation achieved.

16

A value of 60% carbon conversion to CO_2, achieved within 28 days, is generally taken to indicate ready degradability. Taking into account that in this system there will also be incorporation of carbon into biomass formation (growth) the 60% value for CO_2 implies almost complete degradation. While this criterion is meant for water soluble substrates, it is probably applicable to very finely divided moderately-degradable polymeric materials as well [11]. Nevertheless, most standards for determining biodegradability of plastics consider a maximum test duration of six months.

Besides the continuously aerated systems, described previously, several static respirometers have been described. Bartha and Yabannavar [85] describe a two flask system; one flask, containing a mixture of soil and the substrate, is connected to another chamber holding a quantity of carbon dioxide sorbent. Care must be taken to ensure that enough oxygen is available in the flask for biodegradation. Nevertheless, this experimental set-up and modified versions thereof have been successfully applied in the assessment of biodegradability of polymer films and food packaging materials [86-88].

The percentage of carbon converted to biomass instead of carbon dioxide depends on the type of polymer and the phase of degradation. Therefore, it has been suggested to use the complete carbon balance to determine the degree of degradation [89]. This implies, that besides the detection of gaseous carbon, also the amount of carbon in soluble and solid products needs to be determined. Soluble products, oligomers of different molecular size, intermediates and proteins secreted from microbial cells can be measured as COD or as DOC. Solid products, biomass, and polymer remnants require a combination of procedures to separate and detect different fractions. The protein content of the insoluble fraction is usually determined to estimate the amount of carbon converted to biomass, using the assumptions that dry biomass consists of 50% protein, and that the carbon content of dry biomass is 50% [89-91].

1.5.4.3 Suitability

Gas evolution tests are popular test methods because they are relatively simple to perform and sensitive. A direct measure for mineralisation is determined, and water-soluble or insoluble polymers can be tested as films, powders or objects. Furthermore, the test conditions and inoculum can be adjusted to fit the application or environment in which biodegradation should take place. Aquatic synthetic media are usually used, but also natural sea water [92, 93] or soil samples [85, 87, 88, 94] can be applied as biodegradation environments. A prerequisite for these media is that the background CO_2-evolution is limited, which excludes the application of real composting conditions. Biodegradation under composting conditions is therefore measured using an inoculum derived from matured compost with low respiration activity [77, 78, 95, 96].

A drawback of using complex degradation environments such as mature compost is that simultaneous characterisation of intermediate degradation products of determination of the carbon balance is difficult due to the presence of a great number of interfering compounds. To overcome this, an alternative test has been developed based on an inoculated mineral bed based matrix [97, 98].

1.5.5 Radioactively Labelled Polymers

1.5.5.1 Principle and Applications

Some materials tend to degrade very slowly under stringent test conditions without an additional source of carbon. However, if readily available sources of carbon are added, it becomes impossible to tell how much of the evolved carbon dioxide can be attributed to decomposition of the plastic. The incorporation of radioactive ^{14}C in synthetic polymers gives a means of distinguishing between CO_2 or CH_4 produced by the metabolism of the polymer, and that generated by other carbon sources in the test environment. By comparison of the amount of radioactive $^{14}CO_2$ or $^{14}CH_4$ to the original radioactivity of the labelled polymer, it is possible to determine the percent by weight of carbon in the polymer which was mineralised during the duration of the exposure [55]. Collection of radioactively labelled gasses or low molecular weight products can also provide extremely sensitive and reproducible methods to assess the degradation of polymers with low susceptibility to enzymes, such as polyethylene [6, 99] and cellulose acetates [100, 101].

1.5.5.2 Drawbacks

Problems with handling the radioactively labelled materials and their disposal are issues on the down side to this method. In addition, in some cases it is difficult to synthesise the target polymer with the radioactive labels in the appropriate locations, with representative molecular weights, or with representative morphological characteristics.

1.5.6 Laboratory-scale Simulated Accelerating Environments

1.5.6.1 Principle

Biodegradation of a polymer material is usually associated with changes in the physical, chemical and mechanical properties of the material. It is indeed these changes, rather than the chemical reactions, which make the biodegradation process so interesting from an application point of view. These useful properties might be measured as a function of the duration of exposure to a biotic medium, to follow the consequences of the biodegradation process on material properties. The biotic media can be specifically designed in a laboratory scale to mimic natural systems but with a maximum control of variables such as temperature, pH, microbial community, mechanical agitation and supply of oxygen. Regulating these variables improves the reproducibility and may accelerate the degradation process. Laboratory simulations can also be used for the assessment of long-term effects from continuous dosing on the activity and the environment of the disposal system [54].

1.5.6.2 Applications

The OECD Coupled Unit test [102] simulates an activated sludge sewage treatment system, but its application for polymers would be difficult as DOC is the parameter used to assess biodegradability. Krupp and Jewell [103] described well controlled anaerobic and aerobic aquatic bioreactors to study degradation of a range of commercially available polymer films. A relatively low loading rate of the semi-continuous reactors and a long retention time were maintained to maximise the efficiency of biodegradation. Experimental set-ups have also been designed to simulate marine environments [104], soil burial conditions [104-106], composting environments [107-110], and landfill conditions [111] at laboratory scale, with controlled parameters such as temperature and moisture level, and a synthetic waste, to provide a standardised basis for comparing the degradation kinetics of films.

A wide choice of material properties can be followed during the degradation process. However, it is important to select one which is relevant to the end-use of the polymer material or provides fundamental information about the degradation process. Weight loss is a parameter frequently followed because it clearly demonstrates the disintegration of a biodegradable product [112-114]. Tensile properties are also often monitored, due to the interest in the use of biodegradable plastics in packaging applications [58, 115, 116]. In those polymers where the biodegradation involves a random scission of the macromolecular chains, a decrease in the average molecular weight and a general broadening of the molecular weight distribution provide initial evidence of a breakdown process [85, 117-118]. However, no significant changes in material characteristics may be observed in recovered material if the mechanism of biodegradation involves bioerosion, i.e., enzymic or hydrolytic cleavage at the surface. Visual examination of the surface with various microscopic techniques can also give information on the biodegradation process [119-122]. Likewise, chemical and/or physical changes in the polymer may be followed by (combinations of) specific techniques such as infrared [8, 123] or UV spectroscopy [84, 124], nuclear magnetic resonance measurements [118-125], X-ray diffractometry [126, 127], and differential scanning calorimetry [128, 129].

1.5.6.3 Drawbacks

An inherent drawback in the use of mechanical properties, weight loss, molecular weight, or any other property which relies on the macromolecular nature of the substrate is that in spite of their sensitivity, these can only address the early stages of the biodegradation process. Furthermore, these parameters can give no information on the extent of mineralisation. Especially in material blends or copolymers, the hydrolysis of one component can cause significant disintegration (and thus loss of weight and tensile properties) whereas other components may persist in the environment, even in disintegrated form [11]. Blends of starch, poly(3-hydroxy butyrate) or poly(ε-caprolactone) with polyolefins are examples of such systems [9, 47, 130].

1.5.7 Natural Environments – Field Trials

Exposures in natural environments provide the best true measure of the environmental fate of a polymer, because these tests include a diversity of organisms and achieve a desirable natural closeness of fit between the substrate, microbial agent and the environment. However, the results of that exposure are only relevant to the specific environment studied, which is likely to differ substantially from many other environments. An additional problem is the time scale for this method, since the degradation process, depending on the environment, may be very slow (months to years) [27]. Moreover, little information on the degradation process can be gained other than the real time required for weight loss or total disintegration.

Nevertheless, field trials in natural environments are still used to extrapolate results acquired in laboratory tests to biodegradation behaviour under realistic outdoor conditions [119, 131]. Recent German regulations for the assessment of compostability of plastics even impose exposure of the product to a full scale industrial composting process to ensure that total disintegration will occur in real-life waste-processing [132].

1.6 Factors Affecting Biodegradability

The previous paragraphs have shown the importance of the environment on the rate and degree of biodegradation of polymeric substances. The other key aspects determining biodegradability are related to the chemical composition of the polymer. Of course the polymer chemistry governs the chemical and physical properties of the material and its interaction with the (biological) environment, which in turn affect the materials compatibility with particular degradation mechanisms. Many attempts have been made to correlate polymer structure to biodegradability. However, this proved to be challenging and so far only few general relationships between structure and biodegradability have been formulated. In many cases complex interplays between some of the different factors occur simultaneously, often creating difficulty in sorting out primary effects and correlations. Some of the general factors affecting biodegradability are listed below, but it should be considered that many exceptions to the 'rules' have also been reported.

The accessibility of the polymer to water-borne enzyme systems is vitally important because the first step in the biodegradation of plastics usually involves the action of extracellular enzymes which break down the polymer into products small enough to be assimilated. Therefore, the physical state of the plastic and the surface offered for attack, are important factors. Biodegradability is usually also affected by the hydrophilic nature (wettability) and the crystallinity of the polymer. A semicrystalline nature tends to limit the accessibility, effectively confining the degradation to the amorphous regions of the polymer. However, contradictory results have been reported. For example, highly crystalline starch materials and bacterial polyesters are rapidly hydrolysed.

The chemical properties that are important include *(a)* the chemical linkages in the polymer backbone, *(b)* the pendant groups, their position and their chemical activity, and *(c)* end-groups and their chemical activity. Linkages involving hetero atoms, such as ester and amide (or peptide) bonds are considered susceptible to enzymic degradation. However, this is not the case for polyamides, aromatic polyesters, and many other polymers containing hetero atoms in the main chain. The stereochemistry of the monomer units in the polymer chain also influences biodegradation rates, since an inherent property of many enzymes is their stereochemical selectivity. This stereoselectivity may nonetheless not be observed when inocula with a broad spectrum of microorganisms are used instead of enzyme solutions with high stereospecificity.

The molecular weight distribution of the polymer can have a dramatic effect on rates of depolymerisation. This effect has been demonstrated for a number of polymers, where a critical lower limit must be present before the process will start. The molecular origin for this effect is still subject to speculation, and has been attributed to a range of causes such as changes in enzyme accessibility, chain flexibility, fits with active sites, crystallinity, or other aspects of morphology.

Interactions with other polymers (blends) also affect the biodegradation properties. These additional materials may act as barriers to prevent migration of microorganisms, enzymes, moisture or oxygen into the polymer domains of interest. The susceptibility of a biodegradable polymer to microbial attack is sometimes decreased by grafting it onto a non-biodegradable polymer or by crosslinking. On the other hand, in the literature it has sometimes been suggested that combining a non-biodegradable polymer with one that is biodegradable, or grafting a biodegradable polymer onto a non-biodegradable backbone polymer may result in a biodegradable system. Whether the non-biodegradable component is in fact utilised and mineralised, however, is usually disregarded.

1.7 Conclusions

The overview presented previously makes it clear that there is no such thing as a single optimal method for determining the biodegradation of polymeric materials. First of all, biodegradation of a material is not only determined by the chemical composition and corresponding physical properties; the degradation environment in which the material is exposed also affects the rate and degree of biodegradation. Furthermore, the method or test to be used depends on what information is requested.

One should realise that biodegradability is usually not of interest by itself. It is often just one aspect of health and environmental safety issues or integrated waste management concepts. It is fairly obvious but often neglected that one should always consider why a particular polymeric material should be (or not be) biodegradable when contemplating how to assess its biodegradability. After all, it is the intended application of the material that governs the most suitable testing environment, the parameters to be measured during exposure, and the corresponding limit values. For example, investigating whether biodegradation of a plastic material designed for food packaging could facilitate undesired growth of (pathogenic) microorganisms requires a completely different approach from investigating whether its waste can be discarded via composting, (i.e., whether it degrades sufficiently rapid to be compatible with existing biowaste composting facilities).

In most cases, it will not be sufficient to ascertain macroscopic changes, such as weight loss and disintegration, or growth of microorganisms, because these observations may originate from biodegradation of just one of separate components. The ultimate fate of all individual components and degradation products must be included in the investigations. This implies that it is essential that both the polymeric materials and also intermediate degradation products have to be well characterised in order to understand the degradation process. For a good number of biodegradable materials this means that a lot of work still needs to be done.

References

1. T. Hayashi, *Progress in Polymer Science*, 1994, **19**, 663.

2. D.F. Williams and S.P. Zhong, *International Biodeterioration and Biodegradation*, 1994, **34**, 95.

3. A. Göpferich, *Biomaterials*, 1996, **17**, 103.

4. A-C. Albertsson and S. Karlsson in *Degradable Materials: Perspectives, Issues and Opportunities*, Eds., S.A. Barenberg, J.L. Brash, R. Narayan and A.E. Redpath, CRC Press, Boca Raton, FL, USA, 1990, 263.

5. A.C. Palmisano and C.A. Pettigrew, *Bioscience,* 1992, **42**, 680.

6. A-C. Albertsson and B. Rånby, *Journal of Applied Polymer Science: Applied Polymer Symposia,* 1979, **35**, 423.

7. R.G. Austin in *Degradable Materials: Perspectives, Issues and Opportunities,* Eds., S.A. Barenberg, J.L. Brash, R. Narayan and A.E. Redpath, CRC Press, Boca Raton, FL, USA, 1990, 209.

8. S.M. Goheen and R.P. Wool, *Journal of Applied Polymer Science,* 1991, **42**, 2691.

9. V.T. Breslin, *Journal of Environmental Polymer Degradation,* 1993, **1**, 127.

10. *Towards Common Ground - Meeting Summary of the International Workshop on Biodegradability,* Annapolis, MD, USA, 1992.

11. A.L. Andrady, *Journal of Macromolecular Science C,* 1994, **34**, 25.

12. T. St. Pierre and E. Chiellini, *Journal of Bioactive and Compatible Polymers,* 1986, **1**, 467.

13. T. St. Pierre and E. Chiellini, *Journal of Bioactive and Compatible Polymers,* 1987, **2**, 4.

14. L. Stryer, *Biochemistry,* 2nd Edition, W.H. Freeman and Company, San Francisco, CA, USA, 1981.

15. T.A. Anderson, R. Tsao and J.R. Coats, *Journal of Environmental Polymer Degradation,* 1993, **1**, 301.

16. P.J. Whitney, C.H. Swaffield and A.J. Graffam, *International Biodeterioration and Biodegradation,* 1993, **31**, 179.

17. J.R. Whitaker, *Principles of Enzymology for the Food Sciences,* 2nd Edition, Marcel Dekker Inc., New York, NY, USA, 1994.

18. A. Schirmer, C. Matz and D. Jendrossek, *Canadian Journal of Microbiology,* 1995, **41**, Supplement 1, 170.

19. C. Ratledge in *Biochemistry of Microbial Degradation,* Ed., C. Ratledge, Kluwer Academic Publishers, Dordrecht, The Netherlands, 1994, Chapter 4.

20. G. Antranikian in *Microbial Degradation of Natural Products,* Ed., G. Winkelmann, VCH, Weinheim, Germany, 1992, 27.

21. G.F. White, N.J. Russell and E.C. Tidswell, *Microbiological Reviews*, 1996, **60**, 216.

22. X.G. Tong and P.L. McCarthy in *Methane from Community Wastes*, Ed., R. Isaacson, Elsevier Applied Science, London, UK, 1991, 61.

23. P. Broda, *Biodegradation*, 1992, **3**, 219.

24. J.K. Glenn, L. Akileswaran and M.H. Gold, *Archives of Biochemistry and Biophysics*, 1986, **251**, 688.

25. M. Van der Zee, J.H. Stoutjesdijk, P.A.A.W. Van der Heijden and D. De Wit, *Journal of Environmental Polymer Degradation*, 1995, **3**, 235.

26. G. Eggink, M. Van der Zee and L. Sijtsma, *International edition of the IOP on Environmental Biotechnology*, IOP Milieubiotechnologie, The Hague, The Netherlands, 1995, p.7.

27. J.M. Mayer and D.L. Kaplan in *Biodegradable Polymers and Packaging*, Eds., C. Ching, D.L. Kaplan and E.L. Thomas, Technomic Publishing Co. Inc., Lancaster, PA, USA, 1993, p.233.

28. M. Mochizuki, M. Hirano, Y. Kanmuri, K. Kudo and Y. Tokiwa, *Journal of Applied Polymer Science*, 1995, **55**, 289.

29. Y. Tokiwa and T. Suzuki, *Journal of Applied Polymer Science*, 1981, **26**, 441.

30. Y. Tokiwa, T. Suzuki and K. Takeda, *Agricultural and Biological Chemistry*, 1986, **50**, 1323.

31. I. Arvanitoyannis, A. Nakayama, N. Kawasaki and N. Yamamoto, *Polymer*, 1995, **36**, 2271.

32. A. Nakayama, N. Kawasaki, I. Arvanitoyannis, J. Iyoda and N. Yamamoto, *Polymer*, 1995, **36**, 1295.

33. T. Walter, J. Augusta, R-J. Müller, H. Widdecke and J. Klein, *Enzyme and Microbial Technology*, 1995, **17**, 218.

34. M. Nagata, T. Kiyotsukuri, H. Ibuki, N. Tsutsumi and W. Sakai, *Reactive and Functional Polymers*, 1996, **30**, 165.

35. H.S. Jun, B.O. Kim, Y.C. Kim, H.N. Chang and S.I. Woo, *Journal of Environmental Polymer Degradation*, 1994, **2**, 9.

36. E. Chiellini, A. Corti, A. Giovannini, P. Narducci, A.M. Paparella and R. Solaro, *Journal of Environmental Polymer Degradation*, 1996, **4**, 37.

37. M. Nagata, T. Kiyotsukuri, S. Minami, N. Tsutsumi and W. Sakai, *Polymer International*, 1996, **39**, 83.

38. M. Nagata and T. Kiyotsukuri, *European Polymer Journal*, 1994, **30**, 1277.

39. M. Nagata, *Macromolecular Rapid Communications*, 1996, **17**, 583.

40. I. Arvanitoyannis, E. Nikolaou and N. Yamamoto, *Polymer*, 1994, **35**, 4678.

41. I. Arvanitoyannis, E. Nikolaou and N. Yamamoto, *Macromolecular Chemistry and Physics*, 1995, **196**, 1129.

42. S. Matsumura, Y. Shimura, K. Toshima, M. Tsuji and T. Hatanaka, *Macromolecular Chemistry and Physics*, 1995, **196**, 3437.

43. W.G. Glasser, B.K. McCartney and G. Samaranayake, *Biotechnology Progress*, 1994, **10**, 214.

44. C. Rivard, L. Moens, K. Roberts, J. Brigham and S. Kelley, *Enzyme and Microbial Technology*, 1995, **17**, 848.

45. A.A. Strantz and E.A. Zottola, *Journal of Food Protection*, 1992, **55**, 736.

46. V. Coma, Y. Couturier, B. Pascat, G. Bureau, J.L. Cuq and S. Guilbert, *Enzyme and Microbial Technology*, 1995, **17**, 524.

47. S.H. Imam, S.H. Gordon, A. Burgess-Cassler and R.V. Greene, *Journal of Environmental Polymer Degradation*, 1995, **3**, 107.

48. S.H. Imam, S.H. Gordon, R.L. Shogren and R.V. Greene, *Journal of Environmental Polymer Degradation*, 1995, **3**, 205.

49. M. Vikman, M. Itävaara and K. Poutanen, *Journal of Macromolecular Science A*, 1995, **32**, 863.

50. M. Vikman, M. Itävaara and K. Poutanen, *Journal of Environmental Polymer Degradation*, 1995, **3**, 23.

51. ASTM G21-96 (2002), *Standard Practice for Determining Resistance of Synthetic Polymeric Materials to Fungi*, 2002.

52. ASTM G22-76 (1996), *Standard Practice for Determining Resistance of Plastics to Bacteria*, 1996. (Withdrawn 2001)

53. ISO 846, *Plastics - Evaluation of the Action of Microorganisms*, 1997.

54. K.J. Seal in *Chemistry and Technology of Biodegradable Polymers*, Ed., G.J.L. Griffin, Blackie Academic and Professional, London, UK, 1994, 116.

55. J.E. Potts in *Aspects of Degradation and Stabilization of Polymers*, Ed., H.H.G. Jellinek, Elsevier Scientific Publishing Co., Amsterdam, The Netherlands, 1978, 617.

56. K.J. Seal and M. Pantke, *Material und Organismen*, 1986, **21**, 151.

57. F.P. Delafield, M. Doudoroff, N.J. Palleroni, C.J. Lusty and R. Contopoulos, *Journal of Bacteriology*, 1965, **90**, 1455.

58. J.M. Gould, S.H. Gordon, L.B. Dexter and C.L. Swanson in *Agricultural and Synthetic Polymers: Biodegradability and Utilization*, Eds., J.E. Glass and G. Swift, American Chemical Society, Washington, DC, USA, ACS Symposium Series No. 433, 1990, 65.

59. J. Augusta, R-J. Müller and H. Widdecke, *Applied Microbiology and Biotechnology*, 1993, **39**, 673.

60. H. Nishida and Y. Tokiwa, *Chemistry Letters*, 1994, **3**, 421.

61. H. Nishida and Y. Tokiwa, *Chemistry Letters*, 1994, **7**, 1293.

62. J.R. Crabbe, J.R. Campbell, L. Thompson, S.L. Walz and W.W. Schultz, *International Biodeterioration and Biodegradation*, 1994, **33**, 103.

63. ISO 9408, *Water Quality - Evaluation of Ultimate Aerobic Biodegradability of Organic Compounds in Aqueous Medium by Determination of Oxygen*, 1999.

64. ISO 10708, *Water Quality - Evaluation in an Aqueous Medium of the Ultimate Aerobic Biodegradability of Organic Compounds - Determination of Biochemical Oxygen Demand in a Two-phase Closed Bottle Test*, 1997.

65. OECD 301D, *Ready Biodegradability: Closed Bottle Test*, Guidelines for Testing of Chemicals, Organization for Economic Cooperation and Development (OECD), Paris, France, 1993.

66. OECD 302C, *Inherent biodegradability: Modified MITI Test (II)*, 1993.

67. ISO 6060, *Water Quality - Determination of the Chemical Oxygen Demand*, 1989.

68. ISO 10707, *Water Quality - Evaluation in an Aqueous Medium of the 'Ultimate' Aerobic Biodegradability of Organic Compounds - Method by Analysis of Biochemical Oxygen Demand (Closed Bottle Test)*, 1997.

69. ISO 14851, *Determination Of The Ultimate Aerobic Biodegradability Of Plastic Materials In An Aqueous Medium - Method By Measuring The Oxygen Demand In A Closed Respirometer*, 2004.

70. EN 14048, *Packaging - Determination Of The Ultimate Aerobic Biodegradability Of Packaging Materials In An Aqueous Medium - Method By Measuring The Oxygen Demand In A Closed Respirometer*, 2003.

71. OECD 301F, *Ready Biodegradability, Manometric respirometry test, Guidelines for Testing of Chemicals*, Organization for Economic Cooperation and Development (OECD), Paris, France, 1993.

72. L. Tilstra and D. Johnsonbaugh, *Journal of Environmental Polymer Degradation*, 1993, **1**, 247.

73. ISO 9439, *Water Quality - Evaluation Of Ultimate Aerobic Biodegradability Of Organic Compounds In Aqueous Medium - Carbon Dioxide Evolution Test*, 2000.

74. ISO 14852, *Determination Of The Ultimate Aerobic Biodegradability Of Plastic Materials In An Aqueous Medium - Method By Analysis Of Evolved Carbon Dioxide*, 1999.

75. EN 14047, *Packaging - Determination of the Ultiate Aerobic Biodegradability of Packaging Materials in an Aqueous Medium - Method by Analysis of Evolved Carbon Dioxide*, 2003.

76. OECD 301B, Ready biodegradability: Modified Sturm test, Guidelines for Testing of Chemicals, Organization for Economic Cooperation and Development (OECD), Paris, France, 1993.

77. ASTM D5338-98 (2003), *Standard Test Method for Determining Aerobic Biodegradation of Plastic Materials Under Controlled Composting Conditions*, 2003.

78. ISO 14855, *Determination Of The Ultimate Aerobic Biodegradability And Disintegration Of Plastic Materials Under Controlled Composting Conditions - Method By Analysis Of Evolved Carbon Dioxide*, 2004.

79. EN 14046, *Packaging - Evaluation of the Ultimate Aerobic Biodegradability and Disintegration of Packaging Materials Under Controlled Composting Conditions - Method by Analysis of Released Carbon Dioxide*, 2003.

80. ASTM D5210-92 (2000), *Standard Test Method for Determining the Anaerobic Biodegradation of Plastic Materials in the Presence of Municipal Sewage Sludge*, 2000.

81. ISO 11734, *Water Quality - Evaluation Of The 'Ultimate' Anaerobic Biodegradability Of Organic Compounds In Digested Sludge - Method By Measurement Of The Biogas Production*, 1998.

82. ASTM D5511-02, *Standard Test Method for Determining Anaerobic Biodegradation of Plastic Materials Under High-Solids Anaerobic-Digestion Conditions*, 2002.

83. M. Day, K. Shaw and J.D. Cooney, *Journal of Environmental Polymer Degradation*, 1994, **2**, 121.

84. P. Puechner, W.-R. Mueller and D. Bardtke, *Journal of Environmental Polymer Degradation*, 1995, **3**, 133.

85. R. Bartha and A. Yabannavar, *Proceedings of the Fourth International Workshop on Biodegradable Plastics and Polymers and Fourth Annual Meeting of the Bio-Environmentally Degradable Polymer Society*, Durham, NH, USA, 1995, p28.

86. A.L. Andrady, J.E. Pegram and Y. Tropsha, *Journal of Environmental Polymer Degradation*, 1993, **1**, 171.

87. A. Yabannavar and R. Bartha, *Soil Biology and Biochemistry*, 1993, **25**, 1469.

88. A.V. Yabannavar and R. Bartha, *Applied Environmental Microbiology*, 1994, **60**, 3608.

89. S. Urstadt, J. Augusta, R-J. Müller and W-D. Deckwer, *Journal of Environmental Polymer Degradation*, 1995, **3**, 121.

90. M. Itävaara and M. Vikman, *Chemosphere*, 1995, **31**, 4359.

91. B. Spitzer, C. Mende, M. Menner and T. Luck, *Journal of Environmental Polymer Degradation*, 1996, **4**, 157.

92. A.L. Allen, J.M. Mayer, R. Stote and D.L. Kaplan, *Journal of Environmental Polymer Degradation*, 1994, **2**, 237.

93. R. Courtes, A. Bahlaoui, A. Rambaud, F. Deschamps, E. Sunde and E. Dutriex, *Ecotoxicology and Environmental Safety*, 1995, **31**, 142.

94. P. Barak, Y. Coquet, T.R. Halbach and J.A.E. Molina, *Journal of Environmental Quality*, 1991, **20**, 173.

95. U. Pagga, D.B. Beimborn, J. Boelens and B. De Wilde, *Chemosphere*, 1995, **31**, 4475.

96. U. Pagga, D.B. Beimborn and M. Yamamoto, *Journal of Environmental Polymer Degradation*, 1996, **4**, 173.

97. M. Tosin, F. Degli-Innocenti and C. Bastioli, *Journal of Enviromental Polymer Degradation*, 1998, **6**, 79.

98. G. Bellia, M. Tosin, G. Floridi and F. Degli-Innocenti, *Polymer Degradation and Stability*, 1999, **66**, 65.

99. A-C. Albertsson, C. Barenstedt and S. Karlsson, *Journal of Environmental Polymer Degradation*, 1993, **1**, 241.

100. R.J. Komarek, R.M. Gardner, C.M. Buchanan and S. Gedon, *Journal of Applied Polymer Science*, 1993, **50**, 1739.

101. C.M. Buchanan, D. Dorschel, R.M. Gardner, R.J. Komarek, A.J. Matosky, A.W. White and M.D. Wood, *Journal of Enviromental Polymer Degradation*, 1996, **4**, 179.

102. OECD 303A, *Simulation test - Aerobic sewage treatment: Activated Sludge Units, Guidelines for Testing of Chemicals*, Organization for Economic Cooperation and Development (OECD), Paris, France, 2001.

103. L.R. Krupp and W.J. Jewell, *Environmental Science and Technology*, 1992, **26**, 193.

104. D.L. Kaplan, J.M. Mayer, M. Greenberger, R. Gross and S. McCarthy, *Polymer Degradation and Stability*, 1994, **45**, 165.

105. R. Dale and D.J., Squirrell, *International Biodeterioration and Biodegradation*, 1990, **26**, 355.

106. K.J. Seal and M. Pantke, *Material und Organismen,* 1990, **25**, 87.

107. R.M. Gardner, C.M. Buchanan, R. Komarek, D. Dorschel, C. Boggs and A.W. White, *Journal of Applied Polymer Science,* 1994, **52**, 1477.

108. C.M. Buchanan, D.D. Dorschel, R.M. Gardner, R.J. Komarek and A.W. White, *Journal of Macromolecular Science A,* 1995, **32**, 683.

109. R.A. Gross, J-D. Gu, D. Eberiel and S.P. McCarthy, *Journal of Macromolecular Science A,* 1995, **32**, 613.

110. EN 14045, *Packaging - Evaluation of the Disintegration of Packaging Materials in Practical Oriented Tests under Defined Compositions,* 2003.

111. G.P. Smith, B. Press, D. Eberiel, S.P. McCarthy, R.A. Gross and D.L. Kaplan, *Polymeric Materials Science and Engineering,* 1990, **63**, 862.

112. V. Coma, Y. Couturier, B. Pascat, G. Bureau, S. Guilbert and J.L. Cuq, *Packaging Technology and Science,* 1994, **7**, 27.

113. C.M. Buchanan, C.N. Boggs, D. Dorschel, R.M. Gardner, R.J. Komarek, T.L. Watterson and A.W. White, *Journal of Environmental Polymer Degradation,* 1995, **3**, 1.

114. D. Goldberg, *Journal of Environmental Polymer Degradation,* 1995, **3**, 61.

115. G. Iannotti, N. Fair, M. Tempesta, H. Neibling, F.H. Hsieh and M. Mueller in *Degradable Materials: Perspectives, Issues and Opportunities,* Eds., S.A. Barenberg, J.L. Brash, R. Narayan and A.E. Redpath, CRC Press, Boca Raton, FL, USA, 1990, 425.

116. J. Mergaert, A. Webb, C. Anderson, A. Wouters and J. Swings, *Applied Environmental Microbiology,* 1993, **59**, 3233.

117. L. Tilstra and D. Johnsonbaugh, *Journal of Environmental Polymer Degradation,* 1993, **1**, 257.

118. D.S.G. Hu and H.J. Liu, *Journal of Applied Polymer Science,* 1994, **51**, 473.

119. H.B. Greizerstein, J.A. Syracuse and P.J. Kostyniak, *Polymer Degradation and Stability,* 1993, **39**, 251.

120. L.V. Lopez-Llorca and M.F. Colom Valiente, *Micron,* 1993, **24**, 457.

121. H. Nishida and Y. Tokiwa, *Journal of Environmental Polymer Degradation,* 1993, **1**, 227.

122. C. Bastioli, A. Cerutti, I. Guanella, G.C. Romano and M. Tosin, *Journal of Environmental Polymer Degradation,* 1995, **3**, 81.

123. M.J. Kay, R.W. McCabe and L.H.G. Morton, *International Biodeterioration and Biodegradation,* 1993, **31**, 209.

124. N.S. Allen, M. Edge, M. Mohammadian and K. Jones, *Polymer Degradation and Stability,* 1994, **43**, 229.

125. A. Löfgren and A-C. Albertsson, *Journal of Applied Polymer Science,* 1994, **52**, 1327.

126. A-C. Albertsson and S. Karlsson, *Macromolecular Symposia,* 1995, **98**, 797.

127. J. Schurz, P. Zipper and J. Lenz, *Journal of Macromolecular Science A,* 1993, **30**, 603.

128. A-C. Albertsson, C. Barenstedt and S. Karlsson, *Journal of Applied Polymer Science,* 1994, **51**, 1097.

129. J.P. Santerre, R.S. Labow, D.G. Duguay, D. Erfle and G.A. Adams, *Journal of Biomedical Materials Research,* 1994, **28**, 1187.

130. A. Iwamoto and Y. Tokiwa, *Polymer Degradation and Stability,* 1994, **45**, 205.

131. K.K. Leonas, M.A. Cole and X-Y. Xiao, *Journal of Environmental Polymer Degradation,* 1994, **2**, 253.

132. DIN V 54900, *Testing of the Compostability of Plastics,* 1998.

2 Biodegradation Behaviour of Polymers in Liquid Environments

Rolf-Joachim Müller

2.1 Introduction

According to many definitions [1], biodegradation of plastics is usually primarily induced by the action of various microorganisms, although often non-biotic effects such as irradiation, thermal degradation or chemical hydrolysis contribute to the degradation process. The activity of microorganisms is closely connected to the presence of water. The supply of nutrients to the microorganisms and the transportation of excreted enzymes and metabolic products takes place by diffusion in an aqueous environment surrounding the cells. (Thus, it can be said that an aqueous environment is actually the natural one for a microbe). However, in environments regarded as non-liquid such as soil, compost or surfaces of solids, microbes can also be active as far as a certain aqueous micro-environment allows transportation processes necessary for the biological activity. For example in soil, microbial life takes place in the thin water-films located between the particles or in water-filled cavities in the soil components. A soil-humidity of around 50-60% is optimal for aerobic biological processes, where the humidity is given as percentage of the maximum water holding capacity, which takes into account also structural elements of the soil (actually it reflects the filling of the cavities in the material).

Although water is a basic component of the microbial world, many organisms need or prefer the contact to a solid matrix. For example many fungi exhibit a better growth on surfaces than in agitated liquids, which can, besides other effects, be attributed to the sensitivity of the fungal mycelium to mechanical forces. The differences of the optimal living conditions of the different microorganisms cause the presence of very special microbial communities in the various environments and thus, lead to specific degradation behaviours of substances (which are acting as energy and/or nutrient sources).

Talking about biodegradation of plastics in a liquid environment usually means the natural degradation in sweet water (lakes, rivers), in a marine environment or in aerobic and anaerobic sludges (waste water treatment). However, many degradation studies with plastics in laboratories were performed in defined synthetic or in complex liquid nutrient broths and this also can be regarded as a degradation in a liquid environment. While from studies in real natural aqueous systems, information about the behaviour of biodegradable plastics in a distinct natural environment can be obtained, laboratory studies with special aqueous media are used for fundamental studies of biodegradation processes or to optimise the evaluation of the intrinsic biodegradability of a plastic, e.g., in norms. Special liquid media provide defined and optimal living conditions for many organisms and thus, in many cases increase degradation rates leading to reduced test durations. Furthermore, analytical procedures to characterise the degradation process or to detect degradation intermediates are facilitated in a homogeneous and well defined liquid medium.

According to the situation described previously this chapter covers the biodegradation in real liquid environments as well as in especially designed laboratory test systems and also reviews the role of aqueous test environments in national and international standards in evaluating the biodegradability of plastics.

2.2 Degradation in Real Liquid Environments

In most cases biodegradability is a property which is related to the behaviour of the plastic articles after they become waste. Biodegradation of plastics in landfills was discussed in the earlier stage of the development of such materials to reduce the waste-volume and thus, save deposit-space. Nowadays, composting as an alternative waste treatment system to landfilling, incineration or recycling has come to be a major interest. However, biodegradability of plastics can also contribute to the property profile of a product during its application. In the agricultural field, mulching films made from biodegradable plastics are now being tested or controlled release formulations with fertilisers or agrochemicals are being developed. In this context degradation testing of plastics in soil has been intensified during the last years. Generally speaking, it can be stated, that most of the investigations on biodegradable plastics in the past focused on solid environments such as landfills, compost or soil.

However, for the fate of biodegradable plastics in liquid environments, not only the aspect of avoiding waste has to be considered, but also biodegradability as a novel property for special applications of plastics is an important consideration.

The prevention of marine environment pollution, for instance, is regulated by the MARPOL Treaty. This international convention prohibits the disposal of any plastics waste in the oceans, e.g., from ships or from offshore platforms. The International convention generated activities to check if biodegradable plastics used as an alternative to conventional polymers are suitable to be degraded in a marine environment [2]. A further problem exists from littering, where plastic items are washed away to the sea by rivers or blown by wind from the shores and can cause the death of numerous marine animals [3].

Especially in Japan, the Fisheries Agency is active in developing fishery equipment, (e.g., fishery nets), which are biodegradable and do not cause permanent harm to sea life when lost during fishing [4].

Beside these aerobic environments, biodegradability of plastics has also to be considered under anaerobic conditions. Especially with the collection and biological treatment of green waste from households (kitchens), anaerobic digestion (anaerobic composting) becomes more and more important, especially in some European countries. In addition to this waste management aspect, the introduction of biodegradable plastics to natural anaerobic environments, (e.g., sediments in lakes, rivers or oceans), may occur and therefore the biodegradation behaviour of plastics in the absence of oxygen is of practical interest, too.

Most of the investigations reported in the literature concerning the biodegradation in natural liquid environments consider natural polyhydroxyalkanoates (PHA) such as polyhydroxybutyrate (PHB) or the copolyester containing valerate units (PHBV). These biodegradable materials were of outstanding interest in the past, however, nowadays the commercial relevance of these materials is only limited. For commercially important biodegradable plastics, mainly synthetic polyesters, not much data about degradation in natural liquid environments are available.

2.2.1 Degradation in Sweet Water and Marine Environment

2.2.1.1 Poly(hydroxyalkanoates)

Doi and co-workers [5] exposed poly(hydroxybutyrate-*co*-hydroxyvalerate) (PHBV) with different copolymer compositions to sea water (1.5 m depth) at temperatures between 14 °C and 25 °C (depending on the season). There was no clear influence of the degradation rate on the hydroxyvalerate (HV) content of the copolymer detectable. Erosion rates (removal of polymer material from each surface of the film sample) were in the order of magnitude of 2.5 μm/week (at approximately 22 °C). A significant influence of the temperature could be found for the degradation of poly(3-hydroxybutyrate-*co*-4-hydroxybutyrate (3HB-*co*-4HB) polymers. Increasing the temperature from approximately 14 °C to 24 °C nearly doubled the degradation rate of the polymers (erosion rate at 24 °C approximately 3.8 μm/week). Imam and co-workers [2] tested the degradation of PHBV (12 mol% HV) and PHBV/starch-blends in tropical costal waters (in baskets at 0.5 m depth, temperature 25-32 °C) and stated for both materials (approximately 500 μm sheets) a significant weight loss. While pure PHBV degraded quite slow (10-40% weight loss within 400 days), the starch-blends were totally disintegrated within less than 150 days.

From these data an erosion rate of about 0.4-1.7 µm/week can be estimated for PHB and > 11 µm/week for PHBV.

The degradation of a PHBV (Biopol) in sweet water at a depth from 20 to 85 m was investigated by Brandl and Püchner [6] at temperatures ranging only from 6 to 8 °C. Despite the low temperatures and the reduced oxygen concentration in the deeper water layers 17 µm films of PHBV (8 mol% HV) were totally disintegrated within 254 days. Erosion data on PHBV bottles demonstrated, that the degradation rate significantly decreased with increasing water depth, although even at a distance of 85 m from the surface a clear biological degradation could be observed.

2.2.1.2 Synthetic Polyesters

Beside the work on natural PHA - polyesters, degradation experiments in sea water with synthetic polymers such as poly(ε-caprocalcone) (PCL) and modified polyethylene are reported in the literature, too.

Rutkowska and co-workers reported a complete defragmentation of PCL samples in sea water (Baltic sea) at temperatures between 9 °C and 21 °C [7] within 8 weeks. Temperature was stated to be a major influence factor for the degradation. For PCL, chemical hydrolysis and enzymic surface erosion are responsible in parallel for the polymer degradation. The same research group found for poly(ester urethanes) a significant weight loss in sea water (Baltic sea) within 12 months, while a poly(ether urethane) was not biologically attacked under the same experimental conditions [8].

Polyolefins such as polyethylene and polypropylene are usually not accessible to a direct microbial attack. For such polymers biological degradability is achieved by addition of starch, pro-oxidant additives or photo-sensitive components. Starch as natural polymer can be degraded by microorganisms and enhances the defragmentation of the polyolefins (if the starch is accessible to the microbes). The additives increase the initial reduction of polymer chain length by chemical processes to form short chain length oligomers which should finally be metabolised by microorganisms.

However, no significant changes in material properties nor any reliable weight loss of different modified polyethylenes and polypropylenes could be observed by Gonsalves and co-workers [9, 10] at sea water exposure (1-9 m depth, temperatures 13-30 °C) for 5 to 12 weeks. The primary (chemical) degradation depends on the exposure temperature and at the quite low temperatures in sea water, the reaction rate is probably too slow to observe any changes in the materials within the period of time investigated.

Photodegradable polyethylene proved to be degraded slower under sea water and freshwater floating conditions compared to environmental exposure to air [11]. The quite low temperatures (12-28 °C) and a shielding from sunlight by the water and biofouling were stated as reasons for the slower loss of physical properties in water. However, a disintegration of some samples could be observed within a period of time of 30 to 66 days. Similar observations were made by Leonas and Gorden in a laboratory simulation test [12].

A liquid environment with a high microbial activity is present in the activated sludge stage of a waste water treatment plant. Gilmore and co-workers tested the behaviour of different polymers in this environment [13]. Sheets of PHBV (500 µm; 26.5 mol% HV) were disintegrated within 60 days (at 22 °C), corresponding to an approximate erosion rate of 30 µm/week. At lower temperatures (12-19 °C) an erosion rate of approximately 6 µm/week was observed. Starch-filled polyolefins (without pro-oxidants) and blends of polyolefins with the degradable polyester PCL exhibited even in this active environment no hints of any biological attack (weight loss or changes in mechanical properties).

2.3 Degradation in Laboratory Tests Simulating Real Aquatic Environments

Field tests in real environments have a number of limitations and problems. Parameters such as temperature or water quality can vary during the test period and monitoring of the biodegradation process is usually limited to visual changes or at least to the determination of the weight loss of the samples. To overcome these deficiencies, controlled laboratory tests simulating natural (aquatic) environments are often used to investigate biodegradation processes.

2.3.1 Aerobic Liquid Environments

Investigations of Tsuji and Suzuyoshi [14] on PHB, PCL and polylactic acid (PLA) (films of 50 mm thickness) in a laboratory test with sea water at 25 °C resulted in erosion rates of 0.6 µm/week for PHB and 0.2 µm/week for PCL. These data are comparable to the findings in field tests. In contrast to both these polyesters, PLA did not show any significant weight loss in this experiment. This obviously can be attributed to the different degradation mechanisms. While PHB and PCL are primary attacked by enzymes (PHB depolymerases, lipases) at the surface, PLA is known to be at first mainly degraded by a non-enzymically catalysed hydrolysis mechanism, which is strongly temperature dependent. While, for instance, in compost (at temperatures up to 70 °C) PLA has been proved to be quite rapidly chemically depolymerised and then metabolised by microorganisms, this reaction mechanism is much slower at 25 °C, where PLA is in the glassy state below its glass transition temperature (T_g). Thus, it can be expected that PLA is, despite the presence of various polyester degrading microorganisms, only very slowly degraded in liquid environments such as sea water or sweet water.

A direct comparison of degradation rates in different liquid and non-liquid environments at different temperatures is given in a publication of Manna and Paul [15] using PHB as degradable polyester (250 µm sheets). It is quite surprisingly that no significant general differences in the degradation rate between liquid environments (fresh water, sewage sludge) and solid environments (compost, soil) at the same temperature could be observed (**Figure 2.1**).

A pronounced temperature dependence of the degradation is present, where in most cases highest weight losses were obtained at 30 °C (except with fresh water where a maximum degradation was observed at 40 °C). Compared to fresh water, microbial attack is somewhat higher in sewage sludge, an environment of high microbial activity. Erosion rates estimated from the weight loss data in fresh water are comparable to those presented previously for field tests (0.7 µm/week at 20 °C; 1.3 µm/week at 30 °C; 1.5 µm/week at 40 °C).

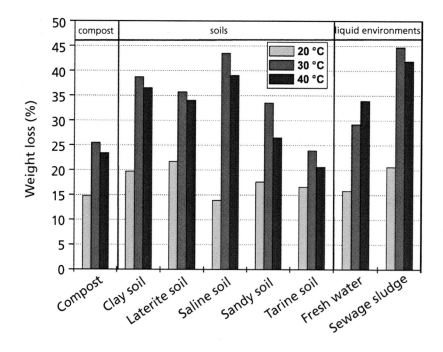

Figure 2.1 Weight loss of 250 µm sheets of PHB in different environments after 200 days of incubation

(Data from Manna and Paul [15])

A direct comparison of microbial degradation of PHBV (14 mol% HV; fibres of approximately 213 µm and 493 µm diameter) in sweet and sea water was performed by Ohura and co-workers [4] using natural water supplemented with some mineral salts to increase microbial activity. With this accelerated simulation test a complete degradation (weight loss measurement and determination of the biological oxygen demand - BOD) of the 213 µm fibres could be achieved within 2 weeks in sweet water (two sources) and within 4 weeks in sea water (two sources); also for the 493 µm fibres degradation in sweet water was almost twice as high as in sea water.

Some years before the same research group had investigated the degradation of a number of PHA-copolymers and a number of synthetic polyesters under similar test conditions in water from a river [16] (100 µm films at 25 °C, weight loss and BOD). Again quite rapid degradation was achieved with the accelerated test system used. A 100 µm film of P(3HB-*co*-3HV) (14% HV) was almost totally degraded within one week. The erosion rates obtained were in a range of 30 to 50 µm/week and much higher than observed in field tests under natural conditions.

However, the test system used allows the comparison of the biodegradation of different polymers in sweet water. PHBV-copolymer degradation rate increased with the copolyester composition up to a HV content of 14 mol%, and then decreased to nearly zero for P(3HB-*co*-3HV) (80% HV).

Copolyesters of 3HB with 4HB and a pure P(4HB) homopolyester were degraded completely within four weeks. For copolymers of 3HB with 3-hydroxypropionate (3HP) the copolyesters proved to be rapidly degradable while pure 3-HP was not attacked. Among a number of different synthetic aliphatic polyesters poly(ethylene succinate) degraded at approximately the same rate as the PHBV copolymers. Poly(ethylene adipate), poly(butylene adipate) and poly(butylene sebacate) exhibited a weight loss within four weeks of incubation, too, but degradation was slower than for the other materials. For poly(ethylene sebacate), poly(butylene succinate) and poly(hexylene succinate) no significant weight loss could be observed under the test conditions applied. Interestingly there is no correlation between the degradation rate and the melting points of the substances, a finding which has been made for enzymatic degradation tests [17, 18].

For synthetic polyesters the degradation behaviour in sea water compared to sweet water is different from that of the natural PHA [19]. While PHB and PHBV copolymers are quite rapidly degradable in both liquid environments, sea water and sweet water, synthetic polyesters seem to be less degradable in sea water. PCL (100 µm films) degrades in sweet water totally within approximately two weeks. In sea water (from the ocean) a BOD of 50-60% could be observed after only four weeks. In sea water from a bay, degradation is surprisingly high, reaching the maximum BOD level after only one week. Poly(ethylene succinate) in sweet water is more rapidly degradable than PCL, but exhibits no microbial degradation in sea water, from the ocean or from a bay. This behaviour can probably be attributed to the occurrence of the different microorganisms able to degrade the natural and synthetic polyesters. PHA are natural polymers and nature developed many organisms to degrade and utilise this carbon and energy source. Thus, PHA degraders are present in many environments. PCL is a synthetic polyester, but its structure resembles cutin, a natural polyester from plants. Thus, it is likely that many organisms do exist, that are able to degrade PCL. In contrast other synthetic polyesters have been shown to be degraded more or less by accident by microorganisms producing lipase-like enzymes with a broad substrate spectrum. The enzymes, and with this the corresponding organisms, have to fit to the particular polyester. It is suggested that for many synthetic polyesters the number of organisms able to attack these polymer structures is much smaller than for degraders of natural or natural-like polyesters. As a consequence, the probability of being degraded depends for the synthetic polyesters especially on the absolute number of organisms present in a particular environment. As general conclusion it can be supposed that the degradation of synthetic, non-natural polyesters is more dependent on the microbial population present in a distinct environment than is in the case of natural materials such as PHB.

A further specialised test system was used by Allen and co-workers [20] and Gonda and co-workers [21]. In these tests completely synthetic media inoculated with individual microbial strains or defined microbial consortia were used to investigate biodegradation of polymers.

Honda and Osawa [22] used a simulation test with a synthetic waste water inoculated with mud from a lake to investigate the behaviour of PCL for the denitrification of wastewater (at 25 °C). He found a remarkable degradation of the PCL plates used (erosion rates approximately 10-15 µm/week).

2.3.2 Anaerobic Liquid Environments

Compared to investigations of polymer degradation under aerobic conditions, very little information is available in the literature for degradation of plastics under anaerobic conditions. Again most of the investigations published are focused on PHA.

It can be expected that the degradation characteristics of polymers in an anaerobic environment are different from those observed in the presence of oxygen, since anaerobic microorganisms have a much more limited set of enzymes and thus, are more specialised with regard to substrates. Additionally, the energy benefit for the organisms is lower without having oxygen as an electron acceptor, resulting mostly in slowly growth of anaerobic microorganisms.

In 1992 Budwill and co-workers published a paper on the degradation of PHB and PHBV copolyesters in an anaerobic mineral medium inoculated with sewage sludge [23]. It could be shown that PHB powder as well as PHBV(13% HV and 20% HV) powder degraded almost completely in the laboratory simulation test (35 °C, degradation monitored via methane production) over a period of time of less than 3 weeks. No clear difference in degradation behaviour between the homopolyester PHB and the PHBV-copolyester was observed. In a later study the same authors extended the investigations to other conditions [24]. The degradation of PHB and PHBV with a microbial consortium from an anaerobic pond sediment at 15 °C was significantly slower than that with the sewage sludge at 35 °C (6 weeks lag-phase; complete degradation after 14 weeks). Again PHB and PHBV exhibited nearly the same degradation behaviour.

Anaerobic degradation of PHB and PHBV (8.4 mol% HV) in a mineral medium with sludge from a waste water plant of the sugar industry at 35 °C was tested by Reischwitz and co-workers [25]. In this case a significant degradation of the polyester powders (approximately 50 µm diameter) was observed also within three weeks, with no significant differences between PHB and PHBV.

Urmeneta found an almost complete degradation of PHBV powder (7 mol% HV) in a liquid anaerobic slurry from a sweet water sediment within six weeks (at 15 °C) up to an amount of 0.5 mg PHB/cm^3 sediment [26].

Shin and co-workers extended anaerobic biodegradation tests to other materials [27] (synthetic mineral medium inoculated with anaerobic sludge from a municipal waste treatment plant at 35 °C). They found a rapid degradation of PHBV (8 wt% HV) and cellophane films (50-75 µm) within three weeks, whereas no degradation (via biogas formation) was observed for the synthetic polyesters PLA and poly(butylene succinate). A corresponding finding was made by Gartiser and co-workers [28]. While a PHBV copolymer (60 µm film) was degraded at similar conditions to those used by Shin and co-workers in less than 3 weeks, no biogas formation could be observed for the synthetic polyester PCL over a period of 11 weeks. In this paper it could also be demonstrated that cellulose acetate polymers (degree of OH-substitution approximately 2.5) are in principle degraded under anaerobic conditions, however, the rate of metabolisation is significantly slower than that of PHA. A detailed investigation on the anaerobic degradability of various starch- and cellulose-esters is given by Rivard and co-workers [29]. The dependence of the anaerobic degradation rate on the degree of substitution and the kind of substituent is discussed.

An extensive investigation on the anaerobic degradability of a number of natural and synthetic polyesters was performed by Abou-Zeid [30]. Beside the poly(hydroxyalkanoates), PHB and PHBV (10 mol% HV) she tested PCL, the aliphatic homopolyester poly(butylene adipate) (SP4/6) and the copolyester poly(butylene adipate-co-butylene terephthalate) where about 40% of the diacid component consists of the aromatic terephthalic acid (BTA 40:60). Weight loss measurements of polymer films (40-74 µm thickness) in two different anaerobic sludges from waste water treatment plants and an anaerobic river sediment demonstrated that both natural PHA were rapidly eroded (up to 100% within 14 weeks at 35 °C) compared to the synthetic polyesters. While for PCL a low, but significant weight loss could be determined, SP4/6 and the aliphatic/aromatic copolyester BTA 40:60 exhibited no clear indication of a microbial attack under this test conditions (**Figure 2.2**).

Similar results were obtained by Abou-Zeid when monitoring the anaerobic biodegradation via biogas formation in a synthetic medium inoculated with anaerobic sludge (**Figure 2.3**). From these measurements it could clearly demonstrated that PHB, in contrast to aerobic conditions, degraded faster than the copolyester PHBV. Again the synthetic polyesters exhibited a significant slower anaerobic degradation rate. Finally, Abou-Zeid could show that synthetic aliphatic polyesters are in principle degraded, but for aliphatic-aromatic copolyesters of technical relevance (approximately 40 mol% aromatic component in the acids) no clear indication of an anaerobic attack could be found.

These observations may have some significance for the biological treatment of biodegradable plastics, as anaerobic digestion becomes more and more established beside aerobic composting. It cannot be supposed that synthetic polyesters will be significantly

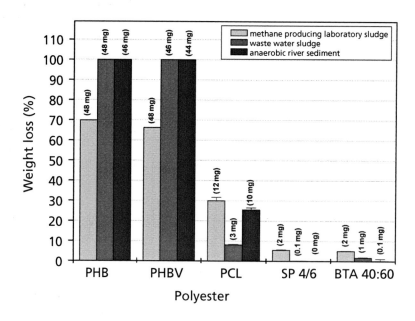

Figure 2.2 Weight loss of various polyesters in different anaerobic environments after 14 weeks at 35 °C. (Polyester films: diameter = 25 mm; surface area: 39.3 cm²; initial film weights = 39-49 mg; 3 films per test) [30]

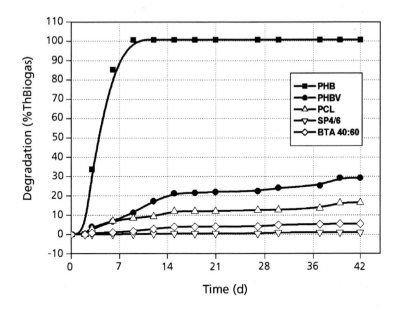

Figure 2.3 Time dependent mineralisation (percentage of the theoretical biogas volume) of various polyesters in anaerobic laboratory and waste water sludge at 37 °C over a period of 42 days. (Polyester films: Ø = 19 mm; surface area: 22.7 cm^2; initial film weights: 35-40 mg; 2 films per test) [30]

degraded in an anaerobic process with typical residence times of about three weeks. However, in most cases anaerobic digestion processes contain a final aerobic step for the stabilisation of the anaerobic compost. If the polyester materials disintegrate sufficiently in the anaerobic step and it will not disturb the technical process, the final biodegradation can take place during the aerobic stabilisation.

2.4 Degradation in Laboratory Tests with Optimised and Defined Liquid Media

For many investigations on biodegradable plastics reported in the literature, degradation tests in defined synthetic media, inoculated with mixed microbial populations or with individual strains, have been used. This kind of tests have significant advantages when investigating basic biological degradation processes of polymers. Due to the usage of defined, in most cases synthetic media, and the possibility to control the environmental parameters (such as temperature, pH, salinity, nutrient supply, etc.), these tests give better reproducible results than degradation tests under natural conditions. Compared to laboratory tests in a solid matrix, (e.g., soil burial test, controlled composting test), analytical procedures aiming on the analysis of intermediates or persistent residues are facilitated [31] in defined aqueous media.

The monitoring of the biodegradation process in such tests can be performed with various methods. Weight loss measurements of films or formed items are the easiest way, but do not necessarily prove the microbial metabolisation of the material. However, in combination with a detailed analysis of the intermediates and the residual polymer (also possible by quantitative chromatographic analysis when using fine polymer powders) useful information can be gained about the degradation mechanism and possible residual components, as demonstrated by Witt and co-workers for the degradation of the aliphatic-aromatic copolyester Ecoflex with a thermophilic actinomycete (*Thermobifidia fusca*), previously isolated from compost [32].

The methods used most often to measure the biodegradation process in laboratory tests with liquid media is to determine the consumption of oxygen, (e.g., Sapromat test) [33, 34], or the release of carbon dioxide (Sturm-test) caused by the metabolic activity of the microorganisms (respirometic tests) [35]. Due to the usually low amount of other carbon sources being present in addition to the polymer itself when using synthetic mineral media, only a fairly low background respiration has to be accounted for and the accuracy of the tests is usually good. These kind of tests have already been used for a long time for evaluating the degradability of diverse substances and chemicals in water, (e.g., in OECD guidelines, see **Table 2.1**) and now have been adapted to the application of non-water soluble polymeric materials. In particular the kind of analytical methods, especially for the determination of carbon dioxide, have been modified. The OECD guidelines describe the trapping of carbon dioxide in barium hydroxide solution in combination with manual titration. More sophisticated methods use the detection of oxygen and carbon dioxide concentration in the air stream (used for aeration) with infrared - detectors and paramagnetic oxygen - detectors. However, despite of the advantage of an automated and continuous measurement, there are also some disadvantages with these methods. The exact air flow has to be measured, the signals of the detectors must be stable for a quite long period of time and, if slow degradation processes have to be determined, the carbon dioxide-concentration or the drop in the oxygen-concentration is only very low. This increases the possibility of systematic errors during such long lasting experiments. Here other concepts, e.g., trapping carbon dioxide in a basic solution (approximately pH 11.5) with continuous titration or detection of the dissolved inorganic carbon [35] are useful alternatives. Other attempts to overcome the problems with carbon dioxide detection are based on non-continuously aerated, closed systems. Here either a sampling technique in combination with an infrared-gas-analyser [36] or a titration system [37] are applied. Another closed system with a discontinuous titration method is described by Solaro and co-workers [38]. Tests using small closed bottles as degradation reactors, determining the carbon dioxide in the head space [39] or the decrease in dissolved oxygen (closed bottle test) [40], are simple and quite insensitive to leakages, etc., but may cause problems due to the low amounts of material and inoculum used.

A crucial point in applying laboratory tests with synthetic liquid media is the source of microorganisms and the procedure of inoculation preparation. An optimum has to be achieved between the minimal input of external carbon into the synthetic medium (reducing the background O_2-consumption and CO_2-evolution) and the overall microbial activity in terms of number and diversity of microorganisms. Originally developed for evaluating the biodegradability of chemicals in waste water treatment plants, laboratory tests described in the OECD guidelines use aerobic sewage sludge as the source of microorganisms. As inoculum the complete sludge or also the particle-free supernatant solution of a sedimented sludge or a filtrate can be used, again there is a question of how much additional carbon source would be introduced into the system. It has been demonstrated, that the kind of pretreatment of the sewage sludge, (e.g., homogenisation), has a significant influence on the degradation of the polymers in the test [41]. However, since the predominant environments where biodegradable plastics are supposed to be degraded are compost or soil, attempts have been made to use extracts from soil or compost to simulate the microbial population in these environments, also in the liquid - phase degradation tests. These sources of inoculum are also included in current standard tests such as ISO 14851 [42] or ISO 14852 [43]. However, there has been some critical discussion about the sense of transferring microorganisms, which are adapted to life in a solid matrix, into a liquid environment. Fungi, for instance, often involved in polymer degradation in soils, do not show optimal growth conditions in a liquid medium and thus, will be under-represented in the aqueous tests. Van der Zee and co-workers discussed this in a paper on cellulose-acetate degradation [44] and found significant differences when comparing the degradation behaviour of cellulose-acetate in aquatic tests and a controlled composting test using mature compost as degradation matrix. However, beside differences in the microbial community other parameters such as the test temperature were also different (aquatic test at 20 °C; controlled composting test at 58 °C) in the tests compared by van der Zee and co-workers.

Despite the limitations of aqueous degradation tests discussed previously, these tests, usually performed in a carbon free, synthetic medium, have one important advantage - the possibility of establishing a reliable carbon balance. The polymer, as an energy and carbon source for the microorganisms, is not completely transformed into carbon dioxide, but a part of the polymer carbon is used to build new biomass or natural metabolic products other than carbon dioxide can be released into the medium; also this part of the polymer can be regarded as bio-degraded. In degradation tests in mature compost at about 60 °C (controlled composting test) very little biomass formation is observed and the carbon of the polymer is almost completely transformed into carbon dioxide. In aqueous media the fraction of carbon going into new biomass can be in the range of some 20% to 30% of the total carbon, and thus, taking carbon dioxide solely as a measure for biodegradation, usually underestimates the degree of biodegradation which is reached. Determination of the entire fate of the carbon from the polymer, (i.e., establishing a carbon balance), has been already discussed for ready biodegradability testing of low molecular weight chemicals [45]. This has been extensively investigated by Urstadt and co-workers for biodegradation of plastics [46]. In a system, where a water soluble substance is tested, the quantitative separation of the biomass from the medium does not usually constitute a problem, but if still residual, non-water-soluble material is present, mechanical techniques for the separation of the two solid fractions are in many cases not applicable and thus, other methods have to be used [47, 48]. Carbon balances have been included in currently developed standard test methods for evaluating the biodegradability of plastics, (e.g., ISO 14851 [42] or ISO 14852 [43]), and are also applied in non-standardised testing practice [49].

2.5 Standard Tests for Biodegradable Polymers Using Liquid Media

The evaluation of the biodegradability of low molecular weight chemicals has been an issue for many years, and a number of standard methods are available in this field [50] (**Table 2.1**).

However, for polymers the point of view was totally different in the past, since plastics had been optimised for many years to be as stable as possible against various environmental influences, among them biological attack. Thus, standard test methods dealing with the interaction of microorganisms with plastics focused at that time on unwanted changes of the material properties (mainly optical or mechanical properties) caused by biological action. Such processes were called biocorrosion. Standard test methods for biocorrosion of plastics were not really suitable to evaluate the biodegradability of plastics (meaning a metabolic conversion of the plastic material by microorganisms) although often used in the very beginning of the development of biodegradable plastics [51].

At the beginning of the 1990s the first attempts were made to establish norms to measure and evaluate biodegradation of non water soluble polymeric materials and first standards, often modifications of existing standards to assess biodegradability of low molecular weight substances, were published by the American Society for Testing and Materials (ASTM). While at the beginning environments where plastics were supposed to be degraded focused on marine environment and landfills, with the upcoming discussion about composting as an alternative method of treating biodegradable plastic waste, standardisation focused then on this topic. Nowadays degradation of plastics in soil is of major interest and standardisation bodies are now starting to establish evaluation schemes for this environment.

Table 2.1 Standard test methods for biodegradability of chemicals	
OECD Guidelines [54]	
301	Ready Biodegradability
301 A - 1992	DOC Die-Away Test
301 B - 1992	CO_2 Evolution Test
301 C - 1992	Modified MITI Test
301 D - 1992	Closed Bottle Test
301 E - 1992	Modified OECD Screening Test
301 F - 1992	Manometric Respirometry Test
302	Inherent Biodegradability
302 A - 1981	Modified SCAS Test
302 B - 1992	Zahn-Wellens Test
302 C - 1981	Modified MITI Test (II)
302 D - draft (2002)	Inherent biodegradability-Concawe test
303	Simulation Test
303 A - 2001	Aerobic Sewage Treatment: Activated Sludge Units
306 (2002)	Biodegradability in Seawater Aerobic mineralisation in surface water-simulation biodegradation test
310 draft (2002)	Ready biodegradability CO_2 in sealed vessels (Headspace test)
311 draft (2002)	Ready anaerobic biodegradability: Gas production from diluted anaerobic sewage sludge
ISO 7827 – 1994	Water quality - Evaluation in an aqueous medium of the 'ultimate' aerobic biodegradability of organic compounds - Method by analysis of dissolved organic carbon (DOC)
ISO 9439– 1999	Water quality - Evaluation of 'ultimate' aerobic biodegradability of organic compounds in an aqueous medium - Carbon dioxide evolution test
ISO 9408 - 1999	Water quality - Evaluation of 'ultimate' aerobic biodegradability of organic compounds in an aqueous medium by determination of oxygen
ISO 9887 - 1992	Water quality - Evaluation of the aerobic biodegradability of organic compounds in an aqueous medium - Semi-continuous activated sludge method (SCAS)
ISO 9888 - 1999	Water quality - Evaluation of the ultimate aerobic biodegradability of organic compounds in an aqueous medium - Static test (Zahn-Wellens method)
ISO 10634 - 1995	Water quality - Guidance for the preparation and treatment of poorly water-soluble organic compounds for the subsequent evaluation of their biodegradability in an aqueous medium

Table 2.1 Cont'd...	
ISO 10707 - 1994	Water quality - Evaluation in an aqueous medium of the 'ultimate' aerobic biodegradability of organic compounds - Method by analysis of biochemical oxygen demand (closed bottle test)
ISO 10708 - 1997	Water quality - Evaluation in an aqueous medium of the ultimate aerobic biodegradability of organic compounds - Determination of biochemical oxygen demand in a two-phase closed-bottle test
ISO 11733 - 1995	Water quality - Evaluation of the elimination and biodegradability of organic compounds in an aqueous medium - Activated sludge simulation test
ISO 11734 - 1995	Water quality - Evaluation of the 'ultimate' anaerobic biodegradability of organic compounds in digested sludge - Method by measurement of the biogas production
ISO 14592-1 - 2002	Water quality - Evaluation of the aerobic biodegradability of organic compounds at low concentrations - Part 1: Shake-flask batch test with surface water or surface water/ sediment suspension
ISO 14592-2 - 2002	Water quality - Evaluation of the aerobic biodegradability of organic compounds at low concentrations - Part 2: Continuous flow river model with attached biomass
ISO 14593 - 1999	Water quality - Evaluation of ultimate aerobic biodegradability of organic compounds in aqueous medium - Method by analysis of inorganic carbon in sealed vessels (CO_2 headspace test)
ISO TR 15462 - 1997	Water quality - Selection of tests for biodegradability
ISO 16221 - 2001	Water quality - Guidance for determination of biodegradability in the marine environment
EN ISO 7827 - 1995	Water quality - Evaluation in an aqueous medium of the 'ultimate' aerobic biodegradability of organic compounds - Method by analysis of dissolved organic carbon (DOC)
EN ISO 9439 - 2000	Water quality - Evaluation of ultimate aerobic biodegradability of organic compounds in aqueous medium - Carbon dioxide evolution test
EN ISO 9408 - 1999	Water quality - Evaluation of ultimate aerobic biodegradability of organic compounds in aqueous medium by determination of oxygen demand in a closed respirometer
EN ISO 9887 - 1994	Water quality - Evaluation of the aerobic biodegradability of organic compounds in an aqueous medium - Semi-continuous activated sludge method (SCAS)

Table 2.1 Cont'd...	
EN ISO 9888 - 1999	Water quality - Evaluation of ultimate aerobic biodegradability of organic compounds in aqueous medium - Static test (Zahn-Wellens method)
EN ISO 10634 - 1995	Water quality - Guidance for the preparation and treatment of poorly water-soluble organic compounds for the subsequent evaluation of their biodegradability in an aqueous medium
EN ISO 10707 - 1997	Water quality - Evaluation in an aqueous medium of the 'ultimate' aerobic biodegradability of organic compounds - Method by analysis of biochemical oxygen demand (closed bottle test)
EN ISO 11733 - 1998	Water quality - Evaluation of the elimination and biodegradability of organic compounds in an aqueous medium - Activated sludge simulation test
EN ISO 11734 - 1998	Water quality - Evaluation of the 'ultimate' anaerobic biodegradability of organic compounds in digested sludge - Method by measurement of the biogas production
DIN 38412 - 26 - 1994	German standard methods for the examination of water, waste water and sludge; bio-assays (Group L); surfactant biodegradation and elimination test for simualtion of municipal waste water treatment plants (L26)
DOC: *Dissolved Organic Carbon* MITI: *Ministry of International Trade and Industry, Japan* SCAS: *Semi-Continuous Activated Sludge*	

For biodegradation processes in liquid environments the standards established up to now can by structured as follows:

- Standards for laboratory test methods determining the intrinsic biodegradability of plastics

- Standards evaluating the biodegradability of plastics in a marine environment

- Standards evaluating the biodegradability of plastics in a waste water treatment (activated sludge)

- Standards evaluating the biodegradability of plastics in anaerobic sludges

A list of currently published standards is given in **Table 2.2**. While most of these standards are predominantly focused on how to measure the biodegradation in the specific

Table 2.2 National and international standards for biodegradable plastics	
ASTM D5210-92 (2000)	Standard test method for determining the anaerobic biodegradation of plastic materials in the presence of municipal sewage sludge
ASTM D5271-02	Standard test method for determining the aerobic biodegradation of plastic materials in an activated-sludge-wastewater-treatment system
ASTM D5511-02	Standard test method for determining anaerobic biodegradation of plastic materials under high-solids anaerobic-digestion conditions
ASTM D6340-98	Standard test methods for determining aerobic biodegradation of radiolabeled plastic materials in an aqueous or compost environment
ASTM D6691-01	Standard test method for determining aerobic biodegradation of plastic materials in the marine environment by a defined microbial consortium
ASTM D6692-01	Standard test method for determining the biodegradability of radiolabelled polymeric plastic materials in seawater
EN 13432 - 2000	Packaging - Requirements for packaging recoverable through composting and biodegradation - Test scheme and evaluation criteria for the final acceptance of packaging
DIN V 54900 - 1998	Testing of the Compostability of Plastics
ISO 14851 - 1999	Determination of the ultimate aerobic biodegradability of plastic materials in an aqueous medium - Method by measuring the oxygen demand in a closed respirometer
ISO 14852 -1999	Determination of the ultimate aerobic biodegradability of plastic materials in an aqueous medium - Method by analysis of evolved carbon dioxide
ISO/DIS 14853 - 1999	Determination of the ultimate anaerobic biodegradability of plastic materials in an aqueous system - Method by measurement of biogas production
ISO/DIS 15985 - 1999	Plastics - Determination of the ultimate anaerobic biodegradability and disintegration under high-solids anaerobic-digestion conditions - Method by analysis of released biogas
ISO/DIS 17556 - 2001	Plastics - Determination of the ultimate aerobic biodegradability in soil by measuring the oxygen demand in a respirometer or the amount of carbon dioxide evolved
JIS K6950 - 2000	Plastics - Testing method for aerobic biodegradability by activated sludge
JIS K6951 - 2000	Determination of the ultimate aerobic biodegradability of plastic materials in an aqueous medium - Method by analysis of evolved carbon dioxide

environment, some standards represent evaluation schemes, especially for biodegradable plastics in composting processes (ASTM D6002-96 [52], EN 13432 [53]) and also provide limit values and threshold levels for the evaluation of biodegradability. Generally all test schemes reflect the problems in measuring biodegradation processes in complex environments such as in biowaste during a composting process and thus, they forecast first to measure the intrinsic biodegradability of a plastic in defined laboratory tests and then to evaluate the disintegration behaviour under real composting conditions. In all schemes laboratory tests based on liquid media are allowed to prove the biodegradability. However, the requested threshold levels of 90% degradation (transformed carbon with respect to the carbon introduced), fixed in the evaluation schemes, requires the establishment of a carbon balance when using aqueous degradation tests, including also the biomass formed, into the calculation of the degree of degradation of the polymers, since, in most cases, more than 10% of the carbon from the polymer will be used to form new biomass instead of being transformed into carbon dioxide.

2.6 Summary

Generally any biological process is connected to the presence of water and thus, it could be stated that in principle all biological degradation takes place in a 'liquid environment'. However, in a macro-liquid environment such as in lakes, rivers, salt water or in special nutrient media in laboratory tests, biodegradation of plastics differs significantly from that in soil or in compost. This is connected on the one hand to differences in the kind and concentration of the microbial population, but also diffusion characteristics of enzymes or intermediates will play a role.

Compared to degradation in compost or in soil, the current interest in investigations of (non-soluble) plastics in aqueous environments is only limited. This is caused by the preferential application of biodegradable plastics as packaging materials (which are degraded in compost) or in agriculture (where degradation takes place in soil). However, in laboratory tests, evaluating the intrinsic biodegradability of plastics, tests in liquid media play an important role, since such test systems are comparable defined and reproducible due to the lack of a multiphase system.

References

1. A. Calmon-Decriaud, V. Bellon-Maurel and F. Silvestre, *Block Copolymers – Polyelectrolytes – Biodegradation*, Advances in Polymer Science, 1998, **135**, 207.

2. S.H. Imam, S.H. Gordon, R.L. Shogren, T.R. Tosteson, N.S. Govind and R.V. Greene, *Applied and Environmental Microbiology*, 1999, **65**, 2, 431.

3.   A.T. Pruter, *Marine Pollution Bulletin*, 1987, **18**, 305.

4.   T. Ohura, Y. Aoyagi, K-I. Takagi, Y. Yoshida, K-I. Kasuya and Y. Doi, *Polymer Degradation and Stability*, 1999, **63**, 1, 23

5.   Y. Doi, Y. Kanesawa, N. Tanahashi and Y. Kumagai, *Polymer Degradation and Stability*, 1992, **36**, 2, 173.

6.   H. Brandl and P. Püchner, *Biodegradation*, 1992, **2**, 237.

7.   M. Rutkowska, M. Jastrzebska and H. Janik, *Reactive and Functional Polymers*, 1989, **38**, 1, 27.

8.   M. Rutkowska, K. Krasowska, A. Heimowska, I. Steinka and H. Janik, *Polymer*, 2002, **76**, 2, 233.

9.   K.E. Gonsalves, S.H. Patel and X. Chen, *New Polymeric Materials*, 1990, **2**, 2, 175.

10. K.E. Gonsalves, S.H. Patel and X. Chen, *Journal of Applied Polymer Science*, 1991, **43**, 2, 405.

11. A.L. Andrady, J.E. Pegram and Y. Song, *Journal of Environmental Polymer Degradation*, 1993, **1**, 2, 117.

12. K.K. Leonas and R.W. Gorden, *Journal of Environmental Polymer Degradation*, 1993, **1**, 1, 45.

13. D.F. Gilmore, S. Antoun, R.W. Lenz and R.C Fuller, *Journal of Environmental Polymer Degradation*, 1993, **1**, 4, 269.

14. H. Tsuji and K. Suzuyoshi, *Polymer Degradation and Stability*, 2002, **75**, 2, 347.

15. A. Manna and A.K. Paul, *Biodegradation*, 2000, **11**, 5, 323.

16. Y. Doi, K-I. Kasuya, H. Abe, N. Koyama, S-I. Ishiwatari, K. Takagi and Y. Yoshida, *Polymer Degradation and Stability*, 1996, **51**, 3, 281.

17. Y. Tokiwa, T. Ando, T. Suzuki and T. Takeda, *Proceedings of the ACS Division of Polymeric Materials: Science and Engineering*, 1990, **62**, 988.

18. E. Marten, *Korrelationen Zwischen der Struktur und der Enzymatischen Hydrolyse von Polyestern*, Technical University Braunschweig, Germany, 2000. In German. [PhD Thesis] http://opus.tu-bs.de/opus/volltexte/2000/136.

19. K-I. Kasuya, K-I. Takagi, S-I. Ishiwatari, Y. Yoshida and Y. Doi, *Polymer Degradation and Stability*, 1998, **59**, 1-3, 327.

20. A.L. Allen, J. Mayer, R. Stote and D.L. Kaplan, *Journal of Environmental Polymer Degradation*, 1994, **2**, 4, 237.

21. K.E. Gonda, D. Jendrossek and H.P. Molitoris, *Hydrobiologia*, 2000, **426**,1, 173.

22. Y. Honda and Z. Osawa, *Polymer*, 2002, **76**, 2, 321.

23. K. Budwill, P.M. Fedorak and W.J. Page, *Applied and Environmental Microbiology*, 1992, **58**, 4, 1398.

24. K. Budwill, P.M. Fedorak and W.J. Page, *Journal of Environmental Polymer Degradation*, 1996, **4**, 2, 91.

25. A. Reischwitz, E. Stoppok and K. Buchholz, *Biodegradation*, 1998, **8**, 5, 313.

26. J. Urmeneta, J. Mas-Castella and R. Guerrero, *Applied and Environmental Microbiology*, 1995, **61**, 5, 2046.

27. P.K. Shin, M.H. Kim and J.M. Kim, *Journal of Environmental Polymer Degradation*, 1997, **5**, 1, 33.

28. S. Gartiser, M. Wallrabenstein and G. Stiene, *Journal of Environmental Polymer Degradation*, 1998, **6**, 3, 159.

29. C. Rivard, L. Moens, K. Roberts, J. Brigham and S. Kelley, *Enzyme and Microbial Technology*, 1995, **17**, 9, 848.

30. D. M. Abou-Zeid, *Anaerobic Biodegradation of Natural and Synthetic Polyesters*, Technical University Braunschweig, Germany, 2000. In English. [PhD thesis] http://opus.tu-bs.de/opus/volltexte/2001/246

31. M. Itävaara and M. Vikman, *Journal of Environmental Polymer Degradation*, 1996, **4**, 1, 29.

32. U. Witt, T. Einig, M. Yamamoto, I. Kleeberg, W-D. Deckwer and R-J. Müller, *Chemosphere*, 2001, **44**, 2, 289.

33. P. Püchner, W-R. Müller and D. Bardtke, *Journal of Environmental Polymer Degradation*, 1995, **3**, 3, 133.

34. J. Hoffmann, I. Reznicekova, S. Vanökova and J. Kupcec, *International Biodeterioration and Biodegradation*, 1997, **39**, 4, 327.

35. U. Pagga, A. Schäfer, R-J. Müller and M. Pantke, *Chemosphere*, 2001, **42**, 3, 319.

36. A. Calmon, L. Dusserre-Bresson, V. Bellon-Maurel, P. Feuilloley and F. Silvestre, *Chemosphere*, 2000, **41**, 5, 645

37. W-R. Müller, *LaborPraxis*, 1999, September, **94**.

38. R. Solaro, A. Corti and E. Chiellini, *Journal of Environmental Polymer Degradation*, 1998, **6**, 4, 203.

39. M. Itävaara and M. Vikman, *Chemosphere*, 1995, **31**, 11/12, 4359.

40. K. Richterich, H. Berger and J. Steber, *Chemosphere*, 1998, **37**, 2, 319.

41. F. Degli-Innocenti, *Labelling Biodegradable Products*, unpublished data of EC-project SMT4-CT97-2167.

42. ISO 14851, *Determination of the Ultimate Aerobic Biodegradability of Plastic Materials in an Aqueous Medium - Method by Measuring the Oxygen Demand in a Closed Respirometer*, 1999.

43. ISO 14852, *Determination of the Ultimate Aerobic Biodegradability of Plastic Materials in an Aqueous Medium - Method by Analysis of Evolved Carbon Dioxide*, 1999.

44. M. van der Zee, J.H. Stoutjesdijk, H. Feil and J. Feijen, *Chemosphere*, 1998, **36**, 3, 461.

45. P. Kuenemann, A. De Morsier and P. Vasseur, *Chemosphere*, 1992, **24**, 1, 63.

46. S. Urstadt, J. Augusta, R-J. Müller and W-D. Deckwer, *Journal of Environmental Polymer Degradation*, 1995, **3**, 3, 121.

47. A. Serandio and P. Püchner, *Gas, Wasser Abwasser*, 1993, **134**, 8, 482.

48. B. Spitzer, C. Mende, M. Menner and T. Luck, *Journal of Environmental Polymer Degradation*, 1996, **4**, 3, 157.

49. H.R. Stapert, *Environmentally Degradable Polyesters, Poly(ester-amide)s and Poly(ester-urethane)s*, 1998, University of Twente, The Netherlands. [PhD Thesis]

50. U. Pagga, *Chemosphere*, 1997, **35**, 12, 2953.

51. J. Augusta, R-J. Müller and H. Widdecke, *Chemie Ingenieur Technik*, 1992, **64**, 5, 410.

52. D6002-96 (2002) e1, *Standard Guide for Assessing the Compostability of Environmentally Degradable Plastics*, 2001

53. EN 13432, *Packaging - Requirements for Packaging Recoverable Through Composting and Biodegradation - Test Scheme and Evaluation Criteria for the Final Acceptance of Packaging*, 2000.

54. *OECD Guidelines for Testing of Chemicals*, Organisation for Economic Co-operation and Development, Paris, France, 1993.

3 Biodegradation Behaviour of Polymers in the Soil

Francesco Degli Innocenti

3.1 Introduction

3.1.1 Biodegradable Polymers and the Environment

A tremendous amount of work has been done at international level during the last decade to study the behaviour of the biodegradable polymers when exposed to different environments. However, looking at the scientific literature published in the 1990s, it appears that most of the work was focused on biodegradation under composting conditions [1-6], while other environments were neglected. The standardisation groups established both in Europe: CEN TC261 SC4 WG2 (European Standardisation Technical Committee on Packaging) and in the USA: ASTM D20.96 (ASTM subcommittee on Environmentally Degradable Plastics and Biobased Products) were mainly interested in defining the compostability of plastics, that is, the set of features plastic products must have in order to be safely recycled into compost. The reason for this preference was linked to the concurrent development of a new solid waste management policy, which aimed at reducing the use of landfilling to a minimum by the promotion of recycling. In Europe, the European Directive on Packaging and Packaging Waste (94/62/EC) declared that biological treatment (composting and biogasification) of packaging was a form of recycling [7]. Consequently, criteria and standard test methods were needed in order to verify the compatibility of plastics with composting and this stimulated research and standardisation.

3.1.2 Biodegradable Polymers and Soil

Several products made with biodegradable polymers are not made to be disposed of via composting at the end of their commercial life but rather to end up directly in soil. The biodegradable plastics used in agriculture are intended to biodegrade in soil. Since the agricultural soil is the medium for the production of food for humans and farm animals, the absence of negative effects linked to the in situ disposal of plastics and the absence of residue build-up are matters of concern. The definition of standard test methods and specific criteria to verify biodegradability and absence of eco-toxic effects in soil are nowadays required to clarify all these issues and launch the marketing of safe biodegradable polymers in agriculture.

3.2 How Polymers Reach Soil

Polymers can be applied into the soil intentionally or unintentionally. This classification is important because the environmental conditions can be different in one case or in the other. In practice, there are two main routes through which the biodegradable plastics terminate their life in soil: agriculture and littering.

Table 3.1 summarises the different modes of delivery. An object left on the ground will be exposed to several climatic factors (such as sunlight, temperature, rainfall, wind, and animals), while an object directly buried in soil will be protected from these factors but, on the other hand, will be exposed to the soil microbial populations. Furthermore, forest soil (where littering mainly occurs) is very different from agricultural soil.

3.2.1 Intentional Delivery

3.2.1.1 Through Compost

Compost is normally added to the agricultural soil as a fertiliser to add organic matter. Compost can contain residues of packaging and disposable items made with compostable plastics. The criteria of compostability established at international level requires a full disintegration of the compostable packaging within one composting cycle. From a practical viewpoint this means that the packaging is susceptible to be reduced into less than 2 mm particles, in less than three months of composting. Therefore, large, visible remains of the original packaging should not normally be present in the final mature compost. On the other hand, small plastic pieces could be still present in the compost due to incomplete degradation in the composting phase. The fate of these plastic particles is to be spread on soil together with the compost and to complete the mineralisation process in this environment.

3.2.1.2 Through Farming

There is an increasing interest towards the application of biodegradable polymers to replace the conventional polymers applied more and more in modern intensive agricultural

Table 3.1 Typical entry routes of polymeric materials into the soil				
Route of entry	Type of soil	Typical dimensions of the material	Environment	Main environmental factors
Compost application	Agricultural soil	Disintegrated, partly biodegraded material	Underground	Microbial
Littering	Forest soil; terrain along motorways, etc.	Bulky	Surface	Sunlight Fauna Microbial
Mulching	Agricultural soil	Pieces (after tillage)	First phase: surface Second phase: underground	Sunlight and heating (during use) Microbial (after tillage)
Other farming procedures	Agricultural soil	Small items used in agriculture strings, clips, etc.)	Mostly on surface. Pots are buried.	Sunlight Fauna Microbial (after burial)

techniques, (i.e., mulch films, drip irrigation tubes, string, clips, pots, etc.) [8]. Nowadays, biodegradable polymers have been effectively tested in many applications: mulch films, tunnel films, string, nets, clips, planting/flower pots, plant containers, controlled release of pesticides, herbicides, fertiliser, and pipelines for mulch. Mulch films consumption (conventional plastics) has greatly increased in the last decade: from 370,000 tonnes in 1991 to 540,000 tonnes in 1999 [9]. A current estimation of the European Plastic Converters for application of traditional plastics for agriculture in Europe is of about 700,000 tonnes a year [10].

The advantage of replacing conventional plastics with their biodegradable counterparts is due to economical and environmental reasons. In general, items made of traditional polymers must be removed after their use, (i.e., mulching films) or they are just left on the ground (pheromone traps). The removal and disposal of traditional plastics can be very expensive and difficult to perform, and in most European countries correct disposal is compulsory. Conventional plastics are expected to be collected and incinerated with energy recovery, or recycled. Uncontrolled incineration, or mechanical tillage of plastic residues in the field have high environmental impact but are unfortunately quite common practices in agriculture. Environmental effects of these practices are: air and field pollution, visual pollution, and accumulation of plastics in soil.

Using biodegradable mulch films, both recovery and final disposal are avoided, because the films are ploughed under after use and are expected to biodegrade *in situ*. Obviously, in order to assure a commercial success of the biodegradable products, the degradation time must be compatible with the application (mainly with the crop cycle). A degradation that is too fast is not acceptable because it can affect the performance of the product (for example, the early degradation of mulch film allows the growth of weeds). A degradation rate that is too slow, on the other hand, is also not acceptable, since plastic residues could interfere with root development.

3.2.2 Unintentional Delivery: Littering

The other way biodegradable plastics can be exposed to the soil environment is through littering. The use of biodegradable materials should not encourage littering. The biodegradability of a packaging should not be an excuse or a justification for littering in nature. The environmental burden of a massive littering of packaging would be very serious, no matter if biodegradable or traditional materials are spread. This can be already verified with paper. Paper is biodegradable and tends to fall apart if wet. Nevertheless, paper napkins, paper packaging, newspaper, etc., can last for a long time in the nature before disappearing. Therefore, no statements on the biodegradability of littered packaging in the nature should be allowed, and no commercial campaign should be based on the possibility of littering. On the other hand, considering that even in the most disciplined population there are always some careless people, biodegradability is undoubtedly a positive feature to solve the problem of littering. Therefore, a laboratory approach to verify the time of degradation after littering could also be developed. The results should not be used for commercial purposes but, rather, for a more comprehensive evaluation of the real environmental benefits of biodegradable plastics.

Plastic objects (such as bags, picnic cutlery, food packaging) are left or thrown on the ground. The fate of these objects is generally not to be buried. The typical environment is forests or terrain along motorways.

3.3 The Soil Environment

Soil varies widely from place to place. As a matter of fact, soil scientists have set up classification systems in which soil is considered to be composed of a large number of individual soils. The term 'soil' is a collective term for all the soils just as 'vegetation' is used to designate all plants [11]. The soil environment is affected by several uncontrolled parameters. The temperature (which is dependent on the regional climate and the seasonal fluctuations), the soil water content [dependent on rainfall (a climatic factor) and irrigation (if and when applied) and, also, influenced by the soil water holding capacity], the chemical composition (mineral compounds and organic matter), geographical factors and the pH. All these factors, joined together in different combinations, create different environments and strongly affect the soil ecology. As a consequence the microbiology and the biodegradation activity can change from soil to soil and from season to season.

The definition of the environmental parameters to be considered when planning a test system of soil biodegradation is the first dilemma encountered by researchers. This problem is less critical when defining biodegradability under composting conditions, because the variability of the composting environment is low. The composting environment is a rather homogeneous ecological niche and can be considered as a consistent micro-cosmos. This is due to the fact that compost is the result of an industrial process. Any composting manager, in any latitude, will impose similar conditions to the composting plant, in spite of different engineering regimes, in order to reach the same purpose: a fast conversion of the acidic, fermenting waste into a stabilised, earth-smelling, marketable compost. To obtain this result, the right combination of parameters (such as the carbon:nitrogen ratio, water content, porosity, ventilation) must be set at the beginning of the process and controlled during the reaction to assure a reliable conversion. These parameters favour the development of a microbial population which will display the same activity and will carry on the same functions. Therefore, the assessment of biodegradability is facilitated by this rather constant, homogeneous, 'standardised' environment. The rate and the final level of biodegradation of a given polymer will not be substantially different from a composting plant to another, because in any case a basically similar environment will be assured. On the other hand, the environmental factors in soil can be very different from one location to another and consequently the rate of degradation can be different to [12, 13]. Therefore, when studying biodegradation in soil, characterisation of the environmental factors can be important to correlate the biodegradation behaviour to a specific soil.

The environmental factors active in soil can be divided into two main classes: surface (ground) factors and underground factors. This classification is linked to two phases which typically characterise the life of a biodegradable item located in soil:

- A first phase on the surface, under the action of sun and other climatic factors.

- A second phase underground, buried in contact with active micro-organisms.

Usually the first phase is the functional phase: the object must satisfy some functional requirement, for example, the mulch film must control growth of weeds. If degradation happens during this phase, it will be considered a negative factor. The second phase corresponds to the disposal phase, when the item must disappear and be recycled through natural processes. In this phase, fast and complete degradation is a positive factor.

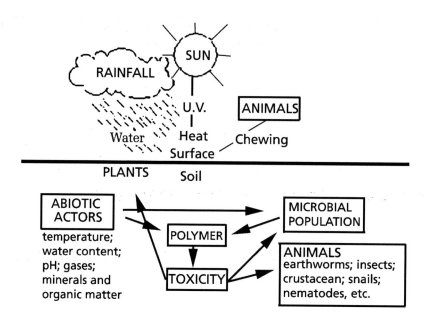

Figure 3.1 Environmental factors to which a plastic product is subjected at the surface and underground, when buried and possible interactions with living organisms

A schematic description of the factors affecting the biodegradation of a plastic item in soil is shown in **Figure 3.1**.

3.3.1 Surface Factors

The main environmental factors active at the surface and their possible effects on polymer degradation are summarised in **Table 3.2** and described in the following paragraphs.

3.3.1.1 Sunlight: the Effect of UV Irradiation

A plastic item left, (i.e., littering) or applied, (i.e., mulching) on soil is exposed to sunlight. This chapter does not address the photo-degradability of polymers (which is another branch in the science of polymer degradability) but it is mainly focused on the biodegradation behaviour of polymers in soil. Still, sunlight is an important environmental factor which can have an effect on the degradability and durability of biodegradable polymers. The typical effect of UV irradiation is to promote photochemical reactions causing oxidation and decrease the molecular weight (MW) of the polymer [14]. This, in turn, causes a

Table 3.2 Environmental factors active at soil surface and their possible effects on polymer degradation			
Environmental Factor	Main effect	Direct consequence on polymer	Biotic effects
Sunlight (UV)	Induction of photochemical reactions.	MW reduction (brittleness). Crosslinking (could impair biodegradability).	Germicidal effect. Reduction of microbial population on irradiated surface.
Sunlight (heat)	Local increase of temperature. Induction of chemical reactions.	Melting (lesions) MW reduction (brittleness).	Faster growth rate. Activation of temperature resistant (thermophilic) microbes.
Rainfall and irrigation	Increase of water activity.	Hydrolysis: MW reduction (brittleness). Leaching of plasticisers (brittleness).	Microbial growth and biodegradation can begin.
Macro-organisms	Gnawing.	brittleness caused by physical action.	Increment of exposed polymer surface can increase the biodegradation rate.

decrease of mechanical properties and possibly an increase of biodegradability. On the other hand, cross-reactions can form networks resistant to biodegradation. In any case, it must be born in mind that in most applications plastic items are only partially exposed to sunlight, while some other parts are directly buried and not exposed to sunlight at all. Therefore, the possible effect of sunlight on biodegradability is limited to the irradiated parts. A typical example is mulch film (see **Figure 3.2**).

3.3.1.2 Sunlight: the Effect of Heat

Sunlight causes local increase of temperature, especially in the case of black coloured mulch films and high irradiation levels, (i.e., temperate regions). High temperature can cause melting and lesions of mulch films. Furthermore, abiotic degradation processes can also be primed by temperature, leading to a decrease of molecular weight and causing brittleness [15]. This can cause a local increase of biodegradability. Also in this case, since many parts are not exposed to heat, as they are buried or sheltered, the effect of heat on biodegradation can only be partial and general conclusions should not be made.

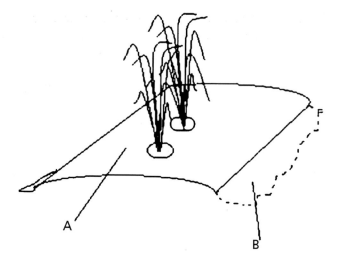

Figure 3.2 Scheme of a mulch film. Part A is exposed to the sunlight while the edges (B) are buried and exposed to microbial action. After crop harvest the part exposed to the sunlight is generally buried

3.3.1.3 Rainfall and Irrigation

The biodegradation process cannot start if the biodegradable material is dry (water activity below 0.7-0.8) and soil-free (low microbial contamination). Rainfall wets the plastic object and dirties it with splashes of soil. Rainfall can thus induce a transient microbial growth on the item surface, limited by nutrient availability, (for example, the nitrogen content of the soil) and stopped by water evaporation caused by sun irradiation. Therefore, rainfall is the cause of a temporary and limited biodegradation process of biodegradable items on the soil surface. Likewise, irrigation practices can strongly increase the biodegradation rate. Other effects caused by the presence of water are: abiotic hydrolysis and leaching of additives such as plasticisers, with decrease of mechanical properties, leading to a brittleness of the material. Recent experiments have shown that humidity can be suppressive of thermodegradation (Tosin and Degli Innocenti, unpublished results).

3.3.1.4 Macro-organism Degradation

Invertebrates and insects such as crickets, slugs, and snails can consume plastics as food. Mechanical damage is caused by the gnawing activities of termites, insects and rodents [16]. Plastics susceptible to this type of biodeterioration are usually protected with insect or rodent repellents [16]. Macro-organism degradation occurs in three stages: (a) mastication (chewing) (b) digestion (c) exocorporeal degradation. Mastication results in considerable deterioration of the physical and chemical structure of the polymers. Digestion by macro-organisms removes the digestible components by enzymic, mechanical and chemical action. Exocorporeal degradation involves the fate of non-digested faecal material and orally contacted pieces of polymer [17]. It has been reported that insects, attracted by some of the constituents of biodegradable polymers, (i.e., starch), have caused deterioration and brittleness by chewing films and producing small holes [18, 19]. For agricultural applications, insects or small animals could cause problems and this should be verified and, if needed, controlled to avoid early damage of the product. Even non-biodegradable polymers can show signs of insect damage [20]. It has been shown that the common soil isopod Armadilidium vulgare could ingest tritium labelled PE (³HPE) and ³HPE + starch blend disks. However, while the disks containing 10% starch were completely consumed, the 100% 3HPE disks were only partially consumed [21]. It has also been thought that macro-biodegradation could be enhanced by increasing the degree of attraction to the woodlouse thus elevating this ubiquitous creature to the status of potential plastic litter scavenger [22]. According to this viewpoint the macro-biological attack can be considered a beneficial component of the natural cycle, but it needs to be properly controlled. Also mites, collembolas and nematodes have been found in biodegradable plastic sheets buried in soil [23].

3.3.2 Underground Factors

In this chapter the term 'soil' will be used to indicate the topsoil, namely, the upper layers of a soil profile. Topsoils generally have a darker colour, due to accumulation of organic matter, and are the zone of root development, containing nutrients and water available to plants and microorganisms. Most biodegradation processes occur in the topsoil.

The main environmental factors active in the topsoil and their possible effects on polymer degradation are summarised in **Table 3.3** and described in the following paragraphs, differentiating between physical properties, chemical-physical properties and biological properties.

3.3.2.1 Structural Properties of Soil

The soil environment contains solids, liquids and gases. These phases vary in their composition. The arrangement of solids and their interactions with water and air define the soil physical properties. The main physical properties of soils are: particles size (defined by the soil texture or size distribution of soil particles) and soil aggregates (defined by the soil structure).

Table 3.3 Environmental factors active in soil and their possible effects on polymer degradation			
Soil factor	Main effect	Direct consequence on polymer	Biotic effects
Texture and soil structure	Determine porosity	Harsh texture can increase abrasion (mechanical degradation).	Porosity controls water and air circulation (see below).
Heat	Temperature change	Temperature controls rate of abiotic degradation, (i.e., hydrolysis), and mobility of polymeric chain (bio-availability).	Temperature controls the microbial population (living and active species in soil), growth rate of each single species, and enzymic activity.
Soil composition (mineral)	Determines the cation exchange capacity (CEC)	Contact between polymer and clayey soils can be difficult. Clay could have a catalytic role in polymer degradation.	High CEC assures higher levels of mineral nutrients (NH_4^+, K^+, Mg^{++}, Ca^{++}) which can otherwise become limiting factors.
Soil organic matter (SOM)	Source of nutrients CEC Better soil structure	A good soil structure allows a better contact between soil and polymer and higher gas diffusion.	SOM assures a healthy and active microbial population.
Water	Water activity (a_w)	Water induces hydrolysis ($\rightarrow$MW reduction). Leaching of plasticisers (brittleness).	a_w controls microbial growth and thus biodegradation. Too much water can cause anaerobic conditions and be negative.
Acid/ alkaline compounds	pH	Can induce hydrolysis ($\rightarrow$MW reduction).	The pH controls the microbial population (living and active species in soil), growth rate of each single species, and enzymic activity.
Air	Determines the O_2 and CO_2 content	Oxygen is needed for abiotic oxidation reactions leading to decrease of MW.	Air (O_2 - CO_2) controls the microbial population (living and active species in soil) growth rate of each single species.

3.3.2.1.1 Texture

Soil texture is defined by the particle size distribution, which is the most important physical property of soil. The mineral part of soil is classified as 'sand', 'silt' or 'clay' according to the particle size. The proportions of sand, silt, and clay determine the soil texture class. Clays are the smallest particles in soil (diameter < 2 µm); silts are larger (from 2 µm to 50 µm); sands are coarse (diameter from 0.05 mm to 2 mm). The term clay when applied to the texture refers to size; it should not be confused with the term clays used in paragraph 3.3.2.2.2.

3.3.2.1.2 Soil Structure

The soil particles, held together by chemical and physical forces in stable aggregates, form the soil structure. The aggregates may be characterised by their size, shape and surface roughness, even though the size has the most relevance. It is important to note the difference between soil texture and soil structure. The first cannot be easily subjected to modification by agricultural practices. On the other hand, physical changes due to agricultural practices, such as ploughing, cultivating, draining and fertilising (mainly organic fertilisation) as well as compression of soil due to transit on the land of agricultural machines, wetting or drying can strongly affect the structure. The size distribution of aggregates influences the amount of water that enters a soil, gas diffusion at the soil surface, heat transfer and soil porosity.

All these factors are very important for growth of microorganisms and biodegradation. A sandy, granular soil will have a relatively free gas diffusion. On the other hand, a clay, blocky (hard, difficult to plough)soil will be poorly aerated.

As a consequence, in the former soil, strictly aerobic microorganisms such as fungi (very active in biodegradation) can develop, while in the latter soil facultative or microaerofilic aerobes will develop. The microbial population found in a given soil will in turn, influence the biodegradation activity.

3.3.2.2 Physical Chemistry of Soil

3.3.2.2.1 Soil Temperature

Soil temperature is a relevant physical factor and has important effects on the biological and chemical processes taking place in the soil. Microbial growth and enzymic processes, in particular, will be strongly affected by the temperature as a consequence of the Arrhenius equation.

Temperature can also directly affect the polymer. For example, the rate of abiotic degradation processes such as hydrolysis, is controlled by temperature [15]. Furthermore, the mobility of the polymeric chains is related to the environmental temperature. This in turn affects the bio-availability of the polymer because an higher mobility will facilitate the contact between the susceptible chemical bonds and the enzymic active sites [24].

3.3.2.2.2 Soil Minerals and Cation Exchange Capacity

Clay minerals are soil secondary minerals derived from the weathering of rocks. Clays have a net negative charge at the surface. Cations are attracted by clay particles. This feature is referred to as the cation exchange capacity (CEC). The CEC of a soil is a measure of the quantity of cations that can be held by a given soil, against the forces of leaching. The more clay (and organic matter; see next paragraph) a soil contains, the higher the CEC. How does clay content affect biodegradation? The nature and content of clays determine the physical state (texture) of soil under different water regimes. The physical state determines the degree of polymer-soil contact and, therefore, the biodegradation process. For example, clayey soils form clumps which make it difficult to mix plastic items and soil together; furthermore air diffusion within the clumps is very limited. This in turn makes degradation difficult (unpublished results). Degradation in very clayey soils is therefore impaired by physical constraints.

On the other hand, the CEC of a soil, a factor controlled by clay and organic matter content, is important because it affects the availability of nutrients needed for a balanced microbial growth and a fast biodegradation process. A high CEC is associated with fertile soils, because many cations such as NH_4^+, K^+, Mg^{++}, Ca^{++} are important nutrients for living organisms and for the efficiency of the biodegradation processes.

It has been postulated that the presence of clay in soil promotes degradation of polymers. The hydrolysis would be catalysed by surface Bronsted and Lewis acidities associated with clay minerals [25]. This intriguing hypothesis, which has been developed to explain the behaviour of a specific class of polymers (silicone polymers), could also be extended to other classes of carbon-based polymers.

3.3.2.2.3 Soil Organic Matter

Soil organic matter (SOM) is formed by partially decomposed and partially re-synthesised plant and animal residues (lignin). SOM is important for two main reasons: as a nutrient reservoir and as a soil structure improver. Generally the majority of soils (including most agricultural soils), have a relatively poor SOM content, ranging from 0.5-10%. Despite the minor contribution to the total mass of minerals, SOM has a crucial role for soil fertility and exerts a profound influence on biodegradability. It contains all the essential nutrients, released during the process of decomposition (mineralisation): organic carbon compounds, nitrogen, phosphorus and sulfur. The availability of macronutrients is essential to get a fast biodegradation rate since the macronutrients can become limited. SOM, together with microorganisms (especially fungi), is involved in binding small soil particles into larger particles, with good air diffusion. Furthermore, SOM can directly affect water retention because of its ability to absorb up to 20 times its mass of water. In slightly acidic to alkaline soils, organic matter can act as a buffer in the maintenance of acceptable soil pH conditions. The high charge characteristics of a SOM (due to the humic fraction) enhance the CEC of a soil.

The addition of organic matter (10% of compost) has been shown to accelerate the rate of degradation although not changing the pattern of degradation [23].

3.3.2.2.4 Water

Water is essential for micro-organism growth, it is the solvent of soil solutions, and it occupies pore spaces competitively with soil gases. Water in a soil can be measured as water content, i.e., the amount of water present in a defined soil mass and it is expressed on a percentage basis (grams of water in 100 grams of soil). The water content can be measured by drying a soil sample at 105 °C and measuring the mass loss, which is then ascribed to evaporated water. An important parameter is water activity because it controls microbial growth. Water activity (a_w) is the ratio of the water vapour pressure in the soil system to the water vapour pressure of pure water.

$$a_w = P_{soil} / P_{water}$$

Microbial growth is possible in the range of a_w between 1 and 0.6, depending on the species. Most bacteria need an a_w higher than 0.98. Fungi are less sensitive and are able to grow at lower a_w (i.e., 0.8). Osmotolerant fungi are able to grow down to an a_w of 0.6 [26].

3.3.2.2.5 pH

Microorganisms are markedly affected by the environmental pH. An increasing soil acidity generally reduces the development of bacteria and on the other hand favours the development of fungi. Due to this there will be less nitrogen fixation and therefore the rate of soil mineralisation could, as a consequence, decrease. A study performed in our laboratory has shown that degradation of biodegradable polymers in acidic forest soils is rather slowed down. The same soils, if the pH is brought to neutrality with the addition of $CaCO_3$ become very active (Guerrini, Tosin, Degli Innocenti, unpublished results).

3.3.2.2.6 Gas Content

As has been discussed before, the gas content of a soil is proportional to the water content, because these two phases compete for the same pores. Therefore, the O_2 content of a soil decreases (and CO_2 increases) with increasing water content, as a consequence of soil respiration. The smaller the grains of a soil and therefore the finer its porosity, the slower will be the gas exchange within the soil. Anaerobic conditions are established under flooded conditions while a lower water content is conducive to aerobic conditions. Hardly any aerobic degradation of substances can be found in a water saturated soil [27]. Aerobic conditions are generally preferable for a fast biodegradation of plastics, even if exceptions do exist. The most notable example is the faster biodegradation of the poly hydroxy-butyrate-valerate under flooded anaerobic conditions [23].

3.3.2.3 Biological Properties of Soil

The living organisms of the soil are in the main, responsible for the continuous synthesis and degradation processes of SOM: they carry out essential environmental functions and they contribute to soil fertility through several biochemical reactions that improve soil structure and transform organic matter into nutrients necessary for life.

The specific populations inhabiting soils are dependent upon many factors [28]. The climate and the resulting vegetation significantly influence which organisms prevail. The soil factors, discussed in the previous paragraphs, such as temperature, acidity and moisture are also factors that govern the activity of organisms living in the soil. For these reasons, it is not easy to predict the number, kinds, and activities of organisms that one might expect to find in a given soil. But there are few generalisations that might be made. For example, compared to virgin areas, cultivated fields generally have lower numbers and weight of soil organisms. This is a consequence of the low SOM present in agricultural soils.

There are over 200 identified bacterial genera and a single soil sample may have over 4,000 genetically distinct bacteria [28]. The greatest population is located in the topsoil, a few millimetres below surface, since conditions of temperature, moisture, aeration, and food are more favourable. The solar radiation reduces the distribution of bacteria on the surface. Deeper in the soil the bacteria are then controlled by the nutrient availability, water content, pH, O_2 and CO_2 content, and temperature.

Fungi are the dominant organisms in soil, both in terms of processes and biomass. Fungi are active in the decomposition and mineralisation of several complex compounds such as cellulose, lignin and chitin [29, 30]. Fungi are mainly active in acid forest soils, but also play an important role in the other soils [31]. They are not able to oxidise and fix nitrogen.

Actinomycetes are fungus-like filamentous bacteria (Eubacteria) and they are especially numerous in soils high in humus, where the acidity is not too high. They have some characteristics typical of fungi such as hyphal growth form and production of extracellular enzymes. Actinomycetes have a very important role as soil decomposers; they are able to metabolise the SOM, such as cellulose, chitin, and phospholipids, transforming them into nutrients.

From a practical viewpoint, an important parameter is the soil metabolic activity. A simple method to assess the overall activity is by measuring the rate of endogenous respiration of a soil. The specific activity, namely the ability of a soil to degrade a specific polymer or substance, is also of great interest for practical reasons. Using agar plates containing the polymer of interest as the only carbon source, it is possible to isolate colonies that grow on the polymer. Nishida and Tokiwa found that polyhydroxybutyrate (PHB) and poly(ε-caprolactone) (PCL) degrading (depolymerising) microorganisms are distributed in many kinds of sources, including landfill leachate, compost, sewage sludge, forest soil, farm soil, paddy field soil, weed field soil, roadside sand and pond sediment [32].

This type of analysis can be performed for any polymer of interest, when an emulsion [32] or a fine powder [33] of the polymer can be used to prepare selective agar plates. This approach can be of great help to determine the microbial activity of a specific field and predict the biodegradation of the polymer.

3.4 Degradability of Polymers in Soil

3.4.1 The Standardisation Approach

Words such as biodegradable and biodegradability have no practical meaning unless the environment, the timeframe, the 'context' are specified. In the long-term, any polymer will possibly degrade. Even the traditional polymers, universally known as recalcitrant, can possibly undergo a biodegradation process after very long environmental exposure and, therefore, be claimed as biodegradable. Clearly, the biodegradation behaviour of traditional polymers such as polyethylene (PE) is exciting from a scientific viewpoint but insignificant from a practical viewpoint. The relevance of biodegradation is linked to waste management. The waste management is based on two rates: production and removal. The rate of production of plastic and packaging waste, which is, nowadays, very high, must be balanced by a similar disposal rate. In order to make biodegradability of polymers a real advantage rather than just a scientific oddity, it must have an impact on society and waste management.

Standardisation working groups are nowadays asked to define the criteria of acceptability of polymers expected to biodegrade in soil. Standardisation is therefore expected to provide definitions useful for the current society and capable of reassuring the users of biodegradable plastics, regulators, politicians, etc., about the suitability of this new class of products. Generally speaking a standard is successful only if all the stakeholders accept it and agree on a common vision of the problem.

Looking at the discussions held lately by the different standardisation working groups and experts, there are two starting points which seem to meet a common consensus and can therefore be considered as a basis on which definitions and test methods can be built on:

1. Test methods and procedures able to generate reliable, quantitative, and reproducible experimental results shall be used to measure biodegradability. This is important to allow the transparency of the evaluation process and to avoid claims based on qualitative data.

2. Criteria and requirements should be formulated so as to prevent the accumulation of man-made materials in soil, and ecotoxic effects.

The former starting point is a typical requirement of standardisation, which aims at unifying methods. The latter is a social need, because it is based on the requirements set by the different stakeholders involved in the use of biodegradable plastics in soil, namely the farmers, the public authority and public opinion.

Farmers need biodegradable plastic tools (mulch films, string, etc.), which can help in farming, as long as these are free from substances which can interfere with the agricultural production either in the short- or in the long-term. Soil, after the crop cycle, must be free not only from visible contaminants, such as plastic residues, but also from recalcitrant xenobiotics, (i.e., foreign compounds in biological systems), produced during degradation of mulch film. In short: field productivity should not be altered by the continuous application of biodegradable plastics. Therefore, in order to satisfy the requirements of the farmers, the biodegradation of plastics in soil should be fast and complete.

Absence of ecotoxicity and total biodegradability are the most important properties for public administrators and legislators, especially after the recent cases of the 'mad cow disease' and the presence of dioxins in chickens. Build up of biodegradable plastics, spread in the agricultural soils year after year, would not be acceptable.

Public opinion nowadays is also very sensitive about environmental problems, and would not be very keen on biodegradable 'green' plastics which do not biodegrade efficiently in soil.

The biodegradable plastics industry is willing to build a sustainable market in the agriculture sector but one single negative occurrence could destroy the credibility of the whole sector. Clear and qualified rules are therefore also important for industry to avoid an uncontrolled, 'short-term results' oriented market.

3.4.2 Test Methods and Criteria

Two main standardisation issues can be identified. The two missions are different and they should be developed separately in order to avoid misunderstanding.

1. Biodegradability and Environmental Compatibility of Polymers for Soil Applications. Focus is on the environmental effects of biodegradable polymers in soil. In order to prevent accumulation of non-biodegradable polymeric residues in soil, the inherent biodegradability must be assessed using standard test methods. Agricultural productivity and the environment should not be disturbed by eco-toxic substances generated by the biodegradation of the plastic material.

2. Durability of Products. Standard test methods are also necessary to predict the 'durability' of plastic products made with biodegradable polymers when in use, in order to verify if they can resist the severe environmental factors found during life cycle. Durability is of commercial interest and test methods are required to classify the products' performances.

It is important to note that biodegradability and durability are two different properties. The first is a property of polymers while the second is a property of a product. A product can be optimal for agricultural applications, offering the required commercial life and then a fast 'disappearance' and still not be environmentally compatible because it is not biodegradable or is unsafe. Conversley, a polymer shown to be compatible with the soil environment, could turn out not to be suitable for a given application because it is not stable under environmental conditions, or too persistent (because, for example, it is converted into mulch films that are too thick).

3.4.2.1 Biodegradability and Environmental Compatibility of Polymers for Soil Applications

First, the meaning of two terms frequently used erroneously as synonyms (even by the experts) must be clarified: biodegradability and biodegradation.

Biodegradability refers to a potentiality, (i.e., the ability to be degraded by biological agents).

Biodegradation refers to a process, happening under certain conditions, in a given time, with results which can be measured.

The inherent biodegradability of a polymer is inferred by studying a real biodegradation process under specific laboratory conditions and, from the test results, the conclusion that the polymer is biodegradable, (i.e., it can be biodegraded) can be drawn.

It must be noted that a fully biodegradable polymer can show a very limited biodegradation if environmental conditions are not suitable. In the previous paragraphs it has been clarified how soil can be affected by several parameters. A dry season, a cold temperature, an acidic soil, a limitation in nitrogen, etc., can affect the degradation rate of a polymer in a manner which is difficult to predict for each field, or region, or season. Only through repeated field trials performed in the area of interest, can one get sufficient knowledge about the specific behaviour of a given material in that area.

Biodegradability, as a general property (inherent biodegradability), is determined in the laboratory, by measuring the degree of biodegradation of the polymer when exposed to a microbial population. The CO_2 evolution or the O_2 consumption are measured and the level of conversion of the organic carbon into inorganic carbon is determined. Strictly speaking, this is a measure of mineralisation, which is the oxidation of the organic carbon of the polymer into CO_2 as a consequence of the microbial respiration. Several respirometric test methods are available nowadays to measure the inherent biodegradability of plastics. In principle, it is preferable to adopt a test method which reproduces the conditions of the environment of interest. So, for example, the evaluation of biodegradability under composting conditions is measured preferably in test systems devised to simulate the composting environment such as ISO 14855 [34]. Accordingly, in order to assess the biodegradability of plastic materials in soil, it is preferable to use a test system where the following conditions are met: temperature in the mesophilic range, a mesophilic microbial inoculum, aerobic conditions, and solid state.

A simple system for monitoring the consumption of oxygen by soil is the one described by Miles and Doucette [35]. The system was devised to follow the persistence and the biological effects of hydrocarbons in soil. It can nevertheless be used for testing polymers. Anderson [36] described several methods: a simple system for determination of oxygen consumption; an automated system for determination of oxygen consumption (the Sapromat); a simple system for determination of carbon dioxide production and a system based on radiolabelled substrates. Another interesting respirometric test apparatus which seems very appealing for its simplicity was described by Bartha and Pramer [37]. Nowadays, an International Standard test method is available, ISO 17556 [38]. The test material is mixed with soil to determine the mineralisation rate by measuring the biochemical oxygen demand or the amount of CO_2 evolved. A natural soil, collected from the surface layer of fields and/or forest, is used. A further standard test method based on soil is described in the American Standard ASTM-D5988 [39]. The test is performed using desiccators, available in most laboratories. A mixture of soil and test material (or compost containing test material after composting) is placed at the bottom of vessels, on the top of soil a perforated plate is laid and onto it a beaker containing KOH or $Ba(OH)_2$ is placed to trap the CO_2 evolved during the biodegradation process. A report indicates that the use of $Ba(OH)_2$ should be avoided because it is unsuitable for trapping CO_2 under static conditions [40]. The test soil can be a laboratory mixture of equal parts of sandy top soil, composted manure or natural soil. It can be also a mix of a natural soil and mature compost in the ratio 25:1. An interesting test system has been proposed to increase the reliability of the respirometric test methods. In order to decrease the amount of soil to a minimum it is proposed to use Perlite [41]. This is a chemically inert aluminosilicate largely used in horticultural applications as a component of growing substrates. The purpose is to reduce the amount of CO_2 produced by the soil itself compared to the investigated samples and therefore to maximise the signal-to-noise ratio. Compost has been also used as a solid matrix instead of soil, at room temperature [42].

Aquatic tests, such as the ISO 14851 [43] and ISO 14852 [44] can also be applied for demonstrating the inherent biodegradability of a polymer. The test temperature should be restricted to the mesophilic range (room temperature). The aquatic tests are considered the only reliable methods for performing carbon balance and characterisation to show complete degradation and also for the detection of potential metabolites (J. Fritz, personal communication). Albertsson [45] used soil as an inoculum of the aquatic test: 10 grams of garden soil (wet weight) were used to inoculate 250 cm^3 of a liquid culture medium applied in a radiolabelling respirometric technique. Radiolabelling respirometric techniques have also been applied in a soil-based test method [46]. Soil-water suspensions have also been used as media to test biodegradability by Suvorova and co-workers [47] and by Calmon-Decriaud and co-workers [48]. Sawada [49] found that the rate of degradation of biodegradable polymers in field tests is consistent with the results found in a laboratory test method based on the OECD Modified MITI Test [50] using activated sludge and measuring oxygen under aerobic conditions. In this very comprehensive study, soil burial tests were performed in 18 different locations in Japan and in one in the USA.

A terrarium for biodegradation of ^{14}C-labelled polymers was described by Guillet and co-workers [51, 52].

In order to perform a final mass balance, recovery from soil of undigested polymeric residues with an organic solvent extraction procedure can be performed [53]. This approach is based on the measurement of the polymer disappearance from soil. A polymer-specific solvent has to be used and a specific analytical method has to be set up [54]. Solvent extraction procedures and manual retrieval were used by Yabannavar and Bartha [55]. The manual retrieval was necessary because of unsatisfactory results obtained with extractions.

When recovering samples from soil, especially if outdoors, great care must be taken to withdraw a statistically representative sample [56].

3.4.2.2 Evaluation of Durability

A mulch film is subjected to strong environmental stresses. The possibility of predicting the effect of the combined environmental factors is extremely relevant for commercial success of plastic products. Needless to say, a mulch film destroyed before the end of the cultivation cycle can seriously impair the commercial yield of a crop. Any negative effect on the commercial yield will not be accepted by farmers. It is therefore important to have reliable test methods to predict durability. Furthermore, a plastic film which after use remains intact on the field for too long, can also be a practical problem for farmers, by preventing the use of the field for a second crop cycle. Durability is therefore a double performance issue: durability can be a problem during plant growth if it is scarce while it is a problem after harvest if too prolonged.

The environmental factors which influence the mechanical properties of the plastic products are typically due to sunlight (UV irradiation and heat), and/or to biodegradation of the buried parts. A typical example is the mulch film which is in part exposed to the sunlight while the lateral parts are buried to fix the whole film to the soil (**Figure 3.2**).

A possible test scheme for the assessment of durability of plastic products in soil is the following. The product is exposed to the surface factors (UV and heat from sun irradiation) to check durability at surface. In parallel the product is directly buried in soil to simulate the behaviour of the parts not exposed to sun (**Figure 3.2**). The films exposed to UV can then be buried to complete characterisation. The test results can be used to estimate the durability of products. Obviously substantiation of the laboratory results with field trials is needed. The same test approach can be used to define the corresponding problem of product durability after commercial life. The product, after crop harvesting is discarded in the field, generally buried, and it is supposed to disintegrate in a relatively short time. It is important to know that a given plastic product will disappear, and not cause visual pollution and impair root development or agricultural practices. The assessment of durability can be useful for predicting both performance in use and 'disappearance' of plastics after use in soil.

3.4.2.2.1 Soil Burial Test Methods

Soil burial test methods have been established and standardised for testing resistance of plastics to micro-organisms. The methods were originally used on plastics coming in contact with the ground, for example, construction materials and coated tents. The aim was to assess their resistance in soil, rather than their degradability. However, resistance and degradability are two complementary aspects of the same problem and a method devised for testing resistance can be applied for testing degradability as well. The test material is buried under laboratory or field conditions. Visual assessment of exhumed materials is carried out and mass loss and tensile strength measurements are also performed.

Soil burial tests are used to give an indication of the duration of the test material in a given soil under given conditions. They can be performed outdoors or indoors.

Outdoor Soil Burial

In theory the outdoor testing is expected to give the most faithful indications of 'real world' performance. However, field experiments are more difficult to perform than laboratory experiments and must be carefully designed. The exposure conditions are not controlled: temperature, rainfall, humidity and sunlight vary from day to day throughout the year and from year to year. The soil burial locations can also be disturbed by wildlife or even human activities, if the area is not restricted.

The choice of location can affect the test results. Characterisation and use of an habitual testing site is important in order to improve reproducibility and compare different test materials. It is also important to keep records of the environmental conditions during all the testing.

Generally speaking, outdoor experiments are advisable whenever the fate of a polymer in a given field or a region has to be predicted with precision. They are less suitable for general statements because of the difficulty to easily reproduce the experimental conditions.

Typical analysis performed after burial is the evaluation of the mass loss [49]. An analysis methodology based on numerical vision has been also developed [57]. Mechanical properties [49], molecular weight evaluation [54], IR spectroscopy [18] and electron microscopy [58] have been applied to characterise polymeric samples after degradation in soil.

A problem which can be encountered in outdoors testing is the interference of animals, which can damage the samples. To solve this problem, a fence of slatted plastic can be constructed about one meter beyond the plot boundary to keep out wildlife [56].

An example of equipment used to perform outdoor burial experiment is described by Goheen and Wool [18].

The plastic samples can be buried in perforated boxes which are then buried in soil. The perforation allows the samples to be attacked by microorganisms and keeps the soil moist [59].

Another possibility is to fasten the specimens on the surface of the ground, to cover it lightly with soil, and finally to protect the area with a net [60]. The following method was used in a very comprehensive test in Japan [49]. A mass of soil is removed from the surface down to approximately 10 cm and then is screened to remove stones, etc. Half of the resulting soil is put back into the hole and its surface is mildly levelled. The area of burying is divided according to the scheme of assessment periods and the test specimens are arranged according to a randomised block design. The space between the test specimens is about 5 cm between the rows and 10 cm between the columns. The remaining soil is then put back to cover the specimen (at about 5 cm in the soil).

It is also possible to run tests outdoors using containers filled up with soil. This makes the recovery of the samples easier. A possible example of this approach is to perform the soil burial test in plastic flower pots (60 x 20 x 20 cm) placed outdoors [61].

A typical method used for outdoor soil burial tests consists of closing the specimens in pockets prepared using a polypropylene (PP) net (**Figure 3.3**). The pocket (A) has the purpose of protecting the specimen (B) during recovery to avoid loss of fragments due to mechanical stress. Furthermore, a string (C) tied to the pocket and left unburied above the surface, will help to identify the burial site and to retrieve the sample. The mesh of the net should be large enough to allow contact of the specimen with soil but, at the same time, small enough to decrease the risk of loosing pieces during exhumation. A suitable mesh is about 4-6 mm. The pocket with the specimens should be inclined, as shown in **Figure 3.3**, to decrease the load caused by water in case of rainfall. A mark, such as a coloured label (D), should be used to identify, after recovery, the specimen. The specimen in the soil is subjected to a gradient of different local environmental conditions (oxygen, carbon dioxide, water content) by the different depth of burial. In the case of non-homogeneous degradation of the specimen, it is important to know the original orientation of the specimen in the soil [48].

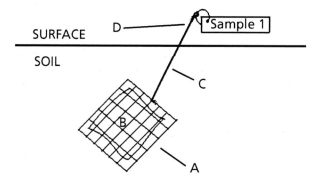

Figure 3.3 Simple device for outdoor soil burial test. A = protective net; B = plastic specimen; C = wire D = label for retrieval and identification

Indoor Soil Burial Tests

Practical reasons, together with the need to assure reproducibility have forced scientists to develop and use mainly laboratory soil burial tests. At laboratory level the environmental conditions are controlled and the management of the experiment very simple. Therefore the statements drawn are more general, reproducible and reliable than the results obtained outdoors. On the other hand, storage of moist soils at room temperature causes a loss of microbial biomass and a decrease in the general degradation potential of soil [27, 36].

An international test method applied to perform soil burial test in the laboratory is EN ISO 846:1997 [62]. An example of an 'Indoor Soil Box' is described by Goheen and Wool [18]. In order to maintain suitable moisture, plastic trays containing soil were covered with a mesh net and then with a thick paper moistened with tap water [63].

Abiotic control can be used as a negative control using sterile soil. This was obtained by heating up to 500 g of soil in an oven at 125 °C for six hours. Then the water lost during sterilisation was restored by using a 0.02 wt% aqueous solution of sodium azide (NaN_3) and thoroughly mixing [63].

Nylon meshes have been used to enclose separate test films before burial [64].

A soil burial test was developed by the American Association of Textile Chemists and Colorists (AATCC 30-1999) for testing fabric specimens [65]. The test method requires a viability control. The soil bed used as a matrix should be considered satisfactory if an untreated fabric loses its mechanical properties after seven days exposure. Recommended types of soil are garden and naturally fertile topsoils, composts and non-sterile greenhouse potting soils. An equal blend of good topsoils, well rotted and shredded manure, and coarse sand should be used. It is considered that these matrices usually have proper physical characteristics, along with an organic content sufficient to ensure a high degree of microbial activity and the presence of active organisms. The optimum moisture content is fixed at about 30% moisture of the dry weight. The air-dried soil bed is placed in trays, boxes or suitable containers and brought to the optimum moisture content by gradual addition of water accompanied by mixing to avoid water stagnation. After 24 h, the soil is sieved through a 6.4 mm mesh screen. The soil moisture content must be kept constant and the temperature maintained at 28 °C.

Experiments were performed using poly hydroxy-butyrate-valerate films as a test polymer to determine the optimum ratio of composted cow manure, topsoil and sand as well as moisture contents to maximise degradation rates. The results indicated that the micro-organisms from the composted cow manure were more active in degradation than micro-organisms from the topsoil with 25-50% manure being optimal. The optimal degradation rate in a 1:1:1 sand:topsoil:composted manure mixture was obtained using the 93.75% of the maximum moisture held [66].

3.4.2.2.2 Methods to Determine Environmental Ageing

The products for agriculture made with biodegradable polymers are exposed in a first phase to surface factors: UV radiation, heat (from sun irradiation), water (rainfall and irrigation), and mechanical stresses (wind, trampling, blown sand, rain, wave action, vehicular traffic, etc.). It is important to know how the products resist these factors because the first phase is the functional phase. Any premature damage could negatively affect their functionality and cause decreases in crop yield, (e.g., mulch film). It is known that exposure to physical-chemical factors can lead to significant degradation processes. For instance, the presence of water at a certain pH range can cause hydrolytic degradation of the polymeric chain [67, 68]. Therefore, to better simulate the life cycle of the plastic products, the products could be subjected to a weathering phase [69, 70]. A simulation of weathering effects can be performed alternating cycles of sunlight, humidity, and condensation with an accelerated weathering tester [71]. As an alternative, the samples can be exposed outdoors to sunlight and rainfall. The specimen should be kept in contact with soil to allow the colonisation by the microbial populations. Doing this the specimen is under the action of UV-radiation, humidity and biodegradation. A practical approach is, for example, to lay a film on soil or on grass. To prevent it from flying away under the wind action, a large mesh net should be placed on top of the film and tightly fixed as a cover. This procedure allows testing of the degradation time of a plastic item after littering. Alternatively, the film can be exposed to atmospheric conditions for a given time and then buried. This can be done to test the degradability of mulching film and takes into account all the different factors active outdoors, such as: chemical-physical degradation occurring during application and biodegradation after tillage. Yabannavar and Bartha exposed films to sunlight for periods of 6 or 12 weeks before burying in soil, following ASTM D1436-97 Standard [72] recommended practice for outdoor weathering of plastics using 45°-angle wooden racks, facing south [55]. Similarly in Thailand, plastic sheets were mounted on racks and exposed to natural solar radiation. Temperature, humidity, radiation, and rainfall were recorded during the experiment [59]. Similarly outdoor ageing tests were performed by fixing the polymer samples to a wooden board which is then placed at 45° facing south on the roof of a two-storey building. Samples for comparison were also placed on the ground where temperature and humidity are the same as on the roof. In this way it was possible to see what effects were due to environmental conditions (sunlight, temperature and humidity) and which were induced by soil-driven biodegradation [73].

A metallic frame rack used to expose samples attached to row to weathering agents is described by Ho and co-workers [69].

3.5 Effects of Biodegradable Polymers on Soil Living Organisms

Polymers are very long molecules not directly available to the living cells and therefore generally harmless. However, low molecular weight additives can be toxic. Furthermore, degradation changes the chemistry of polymers and, consequently, a plastic material can be safe before biodegradation, but may become toxic during degradation. An incomplete biodegradation can create intermediates, (i.e., low molecular weight molecules), which accumulate in the surrounding soil, temporarily or permanently. These degradation intermediates can be monomers, or oligomers, or metabolic derivatives and can interact with the living organisms. It is, therefore, important to assess the possible ecotoxic effects of the polymers introduced into soil [74]. It is not the purpose of this chapter to address the issue of the ecotoxicity of biodegradable polymers, which is covered by another chapter of the book. Here the focus is strictly on the ecotoxicity of biodegradable polymers in soil and to give some suggestions of possible test methods.

3.5.1 Performing the Assessment: Transient and Permanent Effects

Molecules with toxic activity can be reasonably expected at four different stages:

1. Just when the polymer is introduced into the soil, because of the migration of toxic low molecular weight additives present in the polymer. The additives can cause a temporary or permanent toxic effect, according to their chemical stability.

2. After ageing. During use the polymer is exposed to environmental factors such as sunlight which can degrade the polymer into low molecular weight eco-toxic intermediates.

3. During biodegradation. Ecotoxic molecules can be produced and released and the effect can be temporary or permanent, according to the lifespan of the toxic molecules. The temporary toxic effect of biodegradation on plant development is a well known phase occurring during the biodegradation process of any organic natural substrate. As a matter of fact, immature compost is not suitable for direct use in agriculture because of its phytotoxicity. This effect is mainly due to the microbial respiration induced by the biodegradation of the organic matter which impairs the respiration of roots. The negative effect of fresh compost disappears to become a beneficial effect as soon as the compost reaches the right degree of maturation [75]. Likewise, the addition to the soil of a huge amount of organic matter could depress the plant growth. Great care must be devoted in order not to mistake temporary phytotoxicity caused by the biodegradation process for a real permanent ecotoxicity.

4. After the biodegradation of the polymer, permanent toxic residues and intermediates can accumulate in soil.

A cost-effective solution for the detection of possible ecotoxic effects is to focus on permanent activities. A scheme which is currently being discussed by the standardisation groups involves burying the polymer in soil for 3-6 months. In this period the polymer is expected to pass through the first transient phytotoxic phase and undergo substantial degradation. The degree of degradation can be checked by controlling the weight loss of specimens. After full disintegration, the soil is assessed for ecotoxicity in comparison with a reference soil, where a reference material has been degraded in parallel, (i.e., cellulose). The assessment of the toxic activity present in the soil after degradation is very informative because at this moment all the possible stable toxic molecules will have been produced and accumulated. The ecotoxic activity assessed in this case is the cumulative sum of stable toxic molecules released during the different stages. It is, therefore, possible to verify with just one sampling the presence of permanent ecotoxic activities, independently from the moment of production (either during stage 1, 2, 3, or 4; see previously).

3.5.2 Test Material Concentration

In order to have a good chance to detect possible negative effects, it is advisable to apply high initial concentrations of the polymer under study. A 'high' concentration is considered to be one which is at least 1-2 orders of magnitude greater than the normal dose used in real applications. To better clarify this point: a 30 µm thick mulch film with a density of 1.1 is buried in a soil, with an apparent density of 1. Considering that the film is distributed in a 20 cm soil layer, the concentration of the mulch film in soil after tillage is 0.0165%. One to two orders of magnitude in this case is 0.16-1.6% therefore 1% represents a high concentration, suitable to test mulch films. The possibility of detecting toxic effects are maximised by using high concentrations of test material. If no toxic effect is detected applying the test material at such an high concentration, then the environmental risk at normal doses is negligible.

3.5.3 Preparation of the Soil Sample Ready for Ecotoxicity Testing

The test material is mixed with a known amount of soil and placed in a plastic box. The test is performed at room temperature with water content adjusted to a fixed value. The material is periodically inspected to verify the degree of disintegration. Under these conditions an accelerated disintegration process is expected because of the controlled environmental conditions. A sample of a reference soil, obtained in parallel with a reference material, should be used as a negative control. In outdoor tests great care must be taken to mark the area and the depth of burial. A practical system is based on plastic sheets fixed into the soil as a barrier of perimeter, used to delimitate the burial area.

3.5.4 Test Methods

3.5.4.1 Animal Toxicity

The inhabitants of the soil include: the nematodes (the most numerous multicellular animals on Earth), oligochaetes (earthworms), arthropods (crustacea, arachnids, insects), and gastropods (snails). The ecotoxicity tests which have been mostly used to test the effect of biodegradable polymers on solid substrates (such as compost and soil) are the Earthworm Acute Toxicity Test, ISO 11268-1 [76] and the Daphnia (a widespread crustacean) Acute Immobilisation and Reproduction Test [77].

3.5.4.2 Plant Toxicity

The assessment of the effects on plant growth is of most importance, for evident reasons, (i.e., for agricultural applications). The test method mostly applied is the OECD Guideline for Testing of Chemicals 208 - 'Terrestrial Plants, Growth Tests' [78].

3.5.4.3 Microbial Toxicity

Microbial toxicity tests are useful, inexpensive and fast methods. They are based on the reaction of a single species (or a mixture of microbes) to molecules. Measurable changes in terms of behaviour, metabolism, number, or growth rate of the investigated species are related to a toxic effect. Among these tests, bioluminescence tests based on the change in light emission by luminescent organisms (such as *Photobacterium phosphoreum* or *Vibrio fischeri*) when exposed to toxic molecules. A decrease in light emission is the response to serious damage to the metabolism of the bacterial cell. In particular the flash test was developed to evaluate toxic effect on solid and coloured samples [79, 80].

3.6 Biodegradability of Materials in Soil: A Survey of the Literature

Several biodegradable polymers have been already tested in soil. The polymer scientists very frequently have used the soil environment as the first (and frequently unique) test to screen new biodegradable materials produced from laboratory scale reactions. The soil burial test has been considered a very easy and fast test method to verify biodegradability. Everybody can easily get some soil and burying a piece of material is not a complicated technique. On the other hand, the description of the test conditions is frequently absent or not very detailed, making it difficult to evaluate the real meaning of the information obtained. In other cases, conclusions on biodegradability in soil were drawn from studies of fungal attack [81]. One of the few systematic studies of soil biodegradability was performed in Japan in the early 1990s [17]. The conclusion of this study was that polymers showed different rates of degradation according to the location of the experiment and the specific characteristics of each polymer.

Comparison of the degradation data obtained under different environmental conditions revealed that composting allows high levels of biodegradation in short times. Results similar to composting can be obtained in soil burial experiments only after much longer exposure times [82].

Table 3.4 gives the degradation behaviour of a specific polymer, with a very short description of the study performed, the results obtained, and the literature reference. Interesting enough, most of the literature produced in the past has originated from studies aimed at showing the biodegradability in soil of PE and other traditional plastics, solely or in combination with other polymers and additives. The great mass of data obtained show the substantial recalcitrance of traditional polymers to biodegradation in soil. On the other hand, polyesters such as polyhydroxybutyrate (PHB) and it's copolymers, polycaprolactone (PCL), polybutylene succinate adipate, show degrees of mineralisation which, on one hand, suggest a substantial biodegradability in soil and, on the other hand, indicate a rather high variability. For other materials (polylactic acid, polyvinyl alcohol (PVOH)) the data collected are insufficient to draw a conclusion about the effective biodegradability in soil.

The conclusion which arises from the available literature is that a substantial effort must still be spent in order to establish a unified methodological approach to generate reliable and reproducible data. This will be the challenge, in our opinion, of research and standardisation for the next few years in the field of biodegradable polymers.

Acknowledgements

Many thanks to Sara Guerrini, Giulia Gregori, and Gaetano Bellia for their key help in finding and classifying the literature. Many thanks to Sara Guerrini for reading the manuscript. Last but not least, many thanks also to my family for tolerating the spoiling of some week-ends.

Table 3.4 Degradation in soil of different materials. The name of the material, either chemical and/or commercial name (as reported in the original paper), the degradation behaviour, and the bibliographic reference are shown		
Material	Degradation	Refs.
Bioflex (starch-based blends; Biotec, Germany)	58% ML in 40 d	[83]
Bionolle 3000 (Showa Highpolymer)	100% CO_2 in 45 d	[84]
Cellophane	100% ML in 2 y	[57]
Cellulose	53% CO_2 in 158 d	[85]
Cellulose	91% CO_2 in 55 d	[86]
Cellulose	50% CO_2 in 74 d	[87]
Cellulose (bacterial)	100% ML in 9 m	[49]
Cellulose (regenerated) coated with polyurethane (PU)/chitosan	~50% ML in ~30 d	[64]
Cellulose (regenerated) coated with PU/elaeostearin	~30% ML in ~30 d	[64]
Cellulose (regenerated) coated with PU/nitrocellulose	~70% ML in ~30 d	[64]
Cellulose (Sigma Chemicals)	48% CO_2 in 130 d	[41]
Cellulose acetate DS = 2.5/starch blend	~20% ML in 90 d	[88]
Cellulose, regenerated	~90% ML in ~30 d	[64]
Cellulose/vinyl acetate/methyl acrylate copolymers	Δ DSC, DMA	[89]
Copolyamides/starch blends	→ starch content	[47]
Ecolene (degradable PE) + starch (USI Far East, Taiwan)	Resistance as a mulch	[83]
Ecolyte (protodegraded)	2% $^{14}CO_2$ in 230 d	[51]
Ecostar - plus (LDPE + pro-oxidant photosensitiser dye + starch)	~1-2% CO_2 in 44 d	[13]
Ethylene acrylic acid	Δ FTIR, TG, SEM, tensile properties	[90]
Ethylene/vinylacetate copolymers + 28% starch	→ vinyl acetate content	[91]
Gelatin/(50%)phenol formaldehyde	4.6% ML in 2 w	[92]
Gelatin, crosslinked	100% ML in 6 d	[63]

Table 3.4 Continued ...		
Material	Degradation	Refs.
Green choice (PE + starch; Zn Sang Co., Taiwan)	11% ML in 40 d	[83]
Jute, coated	→ soil water content Δ tensile properties	[71]
Lignin: alkali lignin	22% CO_2 in 202 d	[86]
Lignin: hydrolytic lignin	0% CO_2 in 202 d	[86]
Mater Bi (Novamont, Italy)	20% CO_2 in 130 d	[41]
Mater Bi (starch-based blend; Novamont, Italy)	12-72% ML in 55 d in different soils	[12]
Methyl methacrylate-butadiene-styrene copolymer	Δ FTIR, TG, SEM, tensile properties	[90]
Nylon 6,6	Δ IR, tensile properties	[93]
Paper	100% ML in 2 y	[57]
Paper: Brown Kraft	Half life: 2 w	[94]
Paper: Brown Kraft coated by polymerised linseed oil	Half life: 7 w	[94]
Paper Brown Kraft coated by polymerised soybean oil	Half life: 4 w	[94]
Phenol formaldehyde	0% ML in 2 w	[92]
Plant polymer/traditional polymers blends	Δ soil burial and SEM analysis	[95]
Poly(1,1bis[5-(methoxycarbonyl)-2furyl]ethane)	0% ML in 360 d	[96]
Poly(1,3-propandiol-terephthalic-adipic acid)	→ monomers molar ratio	[97]
Poly(1,3-propandiol-terephthalic-sebacic acid)	→ monomers molar ratio	[97]
Poly(1,4:3,6-dianhydro-D-glucitol-1,1bis[5-(methoxycarbonyl)-2furyl]ethane)	<10% ML in 100 d	[98]
Poly(1,4:3,6-dianhydro-D-glucitol-sebacic acid)	100% ML in 30 d	[96]
Poly(2,6-dimethylphenylene oxide)	Δ solvent recovery ~85% after 40 d	[54]
Poly(2-methylphenylene oxide)	Δ solvent recovery ~47% after 40 d	[54]
Poly(2-pyrrolidone)	~20-100% ML in 120 d → soil	[67]

Table 3.4 Continued ...		
Material	Degradation	Refs.
Poly(8-oxa-6-azabicyclo[3.2.1]octan-7-one	~0-15% ML in 180 d → soil	[67]
Poly(ethylene terephthalate)/(15%) poly(ε-caprolactone)	0.8% ML in 8 m	[82]
Poly(ethylene terephthalate)/(20%) poly(ε-caprolactone)	0.4% ML in 8 m	[82]
Poly(lactic-glycolic acid) copolymer	Δ soil micro-organism isolation	[99]
Poly(*m*-cresol)	11% CO_2 in 202 d	[86]
Poly(methyl-glutamate)	10% ML in 24 m	[49]
Poly(methyl-glutamate)	Resistant	[23]
Poly(*p*-ethylphenol)	5% CO_2 in 202 d	[86]
Poly(*p*-phenylphenol)	64% CO_2 in 202 d	[86]
Poly(succinic acid-1,4-butandiol)-*co*-(succinic acid-1,4-cyclohexanedimethanol)	Brittle after 4 w	[100]
Poly(vinyloxyacetate)	Δ Bacterial growth test	[101]
Poly(ε-caprolactam)	0% ML in 180 d	[67]
Poly[(ethylenehexandioate)-*co*-(ethyleneterephthalate)]	→ monomers molecular ratio	[102]
Poly[(tetramethylenehexandioate)-*co*-(tetramethyleneterephthalate)]	→ monomers molecular ratio	[102]
Poly[(trimethylenedecandioate)-*co*-(trimethyleneterephthalate)]	→ monomers molecular ratio	[102]
Poly[(trimethylenehexandioate)-*co*-(trimethyleneterephthalate)]	→ monomers molecular ratio	[102]
Polyacrylate	~1% ^{14}C-CO_2 in 76 d with white rot fungus	[46]
Polyacrylate/polyacrylamide copolimer	~7% ^{14}C-CO_2 in 76 d with white rot fungus	[46]
Polyacrylic acid/sodium alginate network	Δ Fungal resistance test	[81]
Polybutylene succinate (Bionolle 1010)	60% ML in 10 m	[103]
Polybutylene succinate (Bionolle 1010) + compatibiliser (Modic 15%)	100% ML in 7 m	[103]

Table 3.4 Continued ...		
Material	Degradation	Refs.
Polybutylene succinate (Bionolle)	Δ SEM	[58]
Polybutylene succinate adipate (Bionolle 3001)	~70% CO_2 in 1 y	[104]
Polybutylene succinate adipate (Bionolle 3001)/(30%) starch	~70% CO_2 in 60 d	[104]
Polybutylene succinate (Bionolle) foam	8.5% ML in 4 m	[105]
Polybutylene succinate (Bionolle) irradiated	6.5% ML in 4 m	[105]
Polybutylene terephthalate (Ultradur)	→ monomers molecular ratio	[102]
PCL	Degradable	[23]
PCL	101% CO_2 in 9 m	[40]
PCL	32% ML in 24 m	[49]
PCL	~98% ML 2 y	[57]
PCL	Δ SEM	[58]
PCL	44% ML in 8 m	[82]
PCL	50% O_2 with soil micro-organisms in 350 h	[106]
PCL	95% ML in 12 m	[107]
PCL (Union Carbide)	20% CO_2 in 130 d	[41]
PCL (Union Carbide)	Δ Soil micro-organisms	[32]
PCL + starch blend (Bioplastics; Michigan State University USA)	48% ML in 40 d	[83]
PCL irradiated by γ rays	60% ML in 6 m	[61]
Polydimethylsiloxane	→ dryness of soil	[25]
Polydimethylsiloxane (Dow Corning)	→ climatic conditions	[56]
Polyester amide	75% ML in 10 w	[108]
Polyester amide/cotton fibre	85% ML in 10 w	[108]
Polyester amide/flax fibre	80% ML in 10 w	[108]

Table 3.4 Continued ...		
Material	Degradation	Refs.
Polyesteramide	20% O_2 with soil micro-organisms in 600 h	[106]
PE	Δ 6 y soil burial. Damage from ants and roots	[20]
PE	0% ML in 2 y	[57]
PE	Δ IR analysis after 10 y	[109]
PE cellulose + additives	40% ML in 9 m	[110]
PE – starch composite	Δ SEM analysis	[17]
PE + alginate	15% ML in 9 m	[110]
PE + alginate + additives	14-18% ML in 9 m	[110]
PE + cellulose	22-35% ML in 9 m	[110]
PE + chitosan	12-16% ML in 9 m	[110]
PE + starch + additives	0% ML in 2 y	[57]
PE + starch + oxidant additives	0% ML in 2 y	[57]
Polyethylene 2,6-naphthalene dicarboxylate	No degradation by soil micro-organisms after 50 d.	[111]
High density polyethylene (HDPE)	Resistant to biodegradation	[23]
HDPE	0% ML in 24 m	[49]
HDPE	Δ IR, tensile properties	[93]
HDPE ^{14}C-labelled	0.4% [14]CO_2 in 800 d	[45]
HDPE/polypropylene blend + Bioefect 72000	Δ DSC, DMA	[112]
Polyethylene HD/polypropilene blend + MaterBi AF05H	Δ DSC, DMA	[112]
HDPE/polypropylene blend + starch	Δ DSC, DMA	[112]
HDPE/polypropylene/additives blends	Brittle	[113]

Table 3.4 Continued ...		
Material	Degradation	Refs.
Low-density polyethylene (LDPE)	Δ IR, tensile properties	[93]
LDPE	MW decrease after 32-37 y	[114]
LDPE	~0.5-1.5% 14CO$_2$ in 10 y	[115]
LDPE	Holes after 32 y	[116]
LDPE + UV sensitisers	~1-4% ^{14}CO$_2$ in 10 y	[115]
LDPE + (18%) starch	~1-2% CO$_2$ in 44 d	[13]
LDPE + (50%) octanoated starch	2.75% ML in 6 m	[117]
LDPE + starch + prooxidant (ADM)	~1-2% CO$_2$ in 44 d	[13]
LDPE + starch blends	→ starch content	[18]
LDPE + starch blends	Δ SEM, FTIR, tensile properties	[118]
LDPE + starch blends	→ starch content	[90]
PE photodegradable after photoexposure	~3.5-5% CO$_2$ in 12 w	[55]
Polyethylene terephthalate (PET)	0.4% ML in 8 m	[82]
PET	Δ IR, tensile properties	[93]
PET	No degradation by soil micro-organisms after 50 d	[111]
PET (Ecolyte)	~15% ^{14}CO$_2$ in 2 y	[51]
PE/oxidised PE/starch blend	<1% ^{3}H released in 2 y	[21]
PE/starch blends	Degradation limited to starch	[17]
PHB	97% CO$_2$ in 55 d	[86]
PHB	95% CO$_2$ in 92 d	[40]
PHB	~97% ML in 2 y	[57]
PHB	Erosion rate = 5 μm/week at 25 °C	[119]

Table 3.4 Continued ...		
Material	Degradation	Refs.
PHB	100% ML in 10 w	[120]
PHB (ICI)	→ Soil and temperature	[15]
PHB (ICI)	Δ soil micro-organism	[32]
PHB (ICI, UK)	4-99% ML in 25 d in different soils	[12]
PHB-*co*-10%-3-HV) (ICI)	→ Soil and temperature	[15]
PHB-*co*-4-hydroxybutyrate)	100% ML in 2 w	[120]
PHB-HV (Biopol)	20% ML in 6 m	[121]
PHB-HV (Biopol)/ethylene vinyl acetate blends	→ Biopol content	[122]
PHB-HV (Biopol)/purified cellulose blend (70:30)	23% ML in 6 m	[121]
PHB-HV copolymer	~50% CO_2 in 44 d	[13]
PHB-HV copolymer	Degradable	[23]
PHB-HV copolymer	58% ML in 24 m	[49]
PHB-HV copolymer	~72% ML in 2 y	[57]
PHB-HV copolymer (Aldrich)	35% CO_2 in 130 d	[41]
Polylactic acid (PLA)	~4-84 ML in 2 y	[57]
PLA	Δ tensile properties and GPC	[69]
PLA	Δ GPC	[70]
PLA	Δ Soil micro-organisms isolation	[99]
PLA (Cargill)	14% CO_2 in 45 d	[84]
PLA + Bionolle 3000	→ The molar ratio of the blends	[84]
PP	3% CO_2 in 12 w	[55]
PP	Δ IR, tensile properties	[93]
PP	Δ DSC, DMA	[112]
PP + cellulose + additives	24-29% ML in 9 m	[110]

Table 3.4 Continued ...		
Material	**Degradation**	**Refs.**
PP + sodium alginate additives	18% ML in 9 m	[110]
PP-*co*-ethylene	0% ML in 10 m	[123]
PP-*co*-ethylene/polybutylene succinate (Bionolle 1010) electron beam irradiated blends	→ Pre-treatment	[103]
PP-*co*-ethylene/Bionolle blend (1:1) + compatibiliser (Modic 15%)	60% ML in 10 m	[123]
PP-*co*-ethylene/Bionolle blend (1:3) + compatibiliser (Modic 15%)	100% ML in 7 m	[123]
PP-*co*-ethylene/Bionolle blend (3:1) + compatibiliser (Modic 15%)	0% ML in 10 m	[123]
PP-*co*-ethylene + compatibiliser (Modic 15%)	0% ML in 10 m	[123]
Polystyrene (PS)	Δ Tensile properties ~23% after 6 m	[59]
PS	No degradation after 32 y	[115]
PS + starch blends (85:15)	~50% tensile properties after 6 m	[59]
PU + molasses blends	Δ DSC, TG	[124]
PU from coffee grounds	4-9% ML in 9 m	[60]
PU from molasses	15% ML in 12 m	[60]
PVOH	Resistant to biodegradation	[23]
PVOH	8% ML in 24 m	[49]
PVOH	8% CO_2 in 158 d	[85]
PVOH	~ 8% CO_2 in 74 d	[125]
PVOH	~5% ML in 150 d	[126]
PVOH	9% CO_2 74 d	[87]
PVOH (Hoechst)	8% CO_2 in 130 d	[41]
PVOH + chitin-graft-poly(2-methyl-2-oxazoline)	~70% ML in 150 d	[126]
PVOH + waste gelatin	32% CO_2 in 30 d	[85]
PVOH (Idroplast)	~10% CO_2 in 80 d	[41]
PVOH + chitin-graft-poly(2-ethyl-2-oxazoline)	~50% ML in 150 d	[126]

Table 3.4 Continued ...		
Material	Degradation	Refs.
PVOH + partially deacetylated chitin	~80% ML in 150 d	[126]
PVOH + sugar cane bagasse blend	23% CO_2 in 158 d	[85]
Polyvinylchloride (PVC) + additives	29% CO_2 in 12 w	[55]
PVC	No degradation after 32 y	[116]
Rubber (Neoprene)	Δ SEM	[95]
Rubber, nitrile	Δ SEM	[95]
Rubber, natural	Δ SEM	[95]
Rubber, natural + plant polymer blend	Δ SEM	[95]
Rubber, nitrile + plant polymer blend	Δ SEM	[95]
Sky-Green (polyester made of succinic acid, adipic acid, butanediol, ethylene glycol)	6-77% ML in 55 d in different soils	[12]
Starch	0.5 kg CO_2/kg C in 30 d	[13]
Starch (octanoated)	5% ML in 56 d	[117]
Starch (7.7%) + Polyethylene + additives	12.7% CO_2 in 12 weeks	[55]
Sugar cane bagasse	35% in 158 d	[85]
Tetrahydropyrane-based polymers	Δ MW decrease	[68]
Urea formaldehyde resin	No degradation after 32 y	[116]
Wood: *Cryptomeria japonica*	8% ML in 12 m	[60]
Wood: *Fagus sieboldi*	50% ML in 12 m	[60]

ADM: Archer Daniels Midland Co.
DSC: Differential Scanning Calorimetry
DMA: Dynamic Mechanical Analysis
SEM: Scanning Electron Microscopy
LDPE: low-density polyethylene
FTIR: Fourier-transform IR Spectroscopy
TG: Thermogravimetric analysis
GPC: Gel Permeation Chromatography

ML: mass loss
DS: degrees of substitution
IR: infra-red spectroscopy
→ : function of
Δ: studied by means of
HV: Hydroxyvalerate

References

1. U. Pagga, D.B. Beimborn, J. Boelens and B. De Wilde, *Chemosphere*, 1995, **31**, 11-12, 4475.

2. M. Tosin, F. Degli Innocenti and C. Bastioli, *Journal of Environmental Polymer Degradation*, 1996, **4**, 1, 55.

3. F. Degli Innocenti and C. Bastioli, *Journal of Environmental Polymer Degradation*, 1997, **5**, 4, 83.

4. M. Tosin, F. Degli Innocenti and C. Bastioli, *Journal of Environmental Polymer Degradation*, 1998, **6**, 2, 79.

5. F. Degli Innocenti, M. Tosin and C. Bastioli, *Journal of Environmental Polymer Degradation*, 1998, **6**, 4, 197.

6. G. Bellia, M. Tosin and F. Degli Innocenti, *Polymer Degradation and Stability*, 2000, **69**, 1, 113.

7. *Official Journal of the European Communities*, L365, 31.12.1994, p.10.

8. *Proceedings of the International Symposium on Biodegradable Materials and Natural Fibre Composites in Agriculture and Horticulture*, Hanover, Germany, 2002.

9. J-P. Jouet, *Plasticulture*, 2001, **120**, 108.

10. *European Plastics Converters, Plastics for Agriculture*, http://www.eupc.org/markets/agri.html.

11. N.C. Brady, *The Nature and Properties of Soils*, Macmillan Publishing Company, New York, NY, USA, 1984.

12. M-N. Kim, A-R. Lee, J-S. Yoon and I-J. Chin, *European Polymer Journal*, 2000, 36, 8, 1677.

13. P. Barak, Y. Coquet, T. R. Halbach and J.A.E. Molina, *Journal of Environmental Quality*, 1991, **20**, 173.

14. G. Scott, *Trends in Polymer Science*, 1997, 5, 11, 361.

15. J. Mergaert, A. Webb, C. Anderson, A. Wouters and J. Swings, *Applied and Environmental Microbiology*, 1993, **59**, 10, 3233.

16. H.O.W. Eggins, J. Mills, A. Holt and G. Scott in *Microbial Aspects of Pollution*, Eds., G. Sykes and F.A. Skinner, Academic Press, New York, NY, USA, 1971, p.267.

17. R.P. Wool, D. Raghavan, G.C. Wagner and S. Billieux, *Journal of Applied Polymer Science*, 2000, **77**, 8, 1643.

18. S.M. Goheen and R.P. Wool, *Journal of Applied Polymer Science*, 1991, **42**, 10, 2691.

19. R.P. Wool in *Biodegradable Plastics and Polymers*, Eds., Y. Doi and K. Fukuda, Elsevier Science, Amsterdam, The Netherlands, 1994, p.250.

20. A.V. Vasil'ev, *International Polymer Science and Technology*, 1989, **16**, 11, T/84.

21. T.A. Anderson, D.M. Scherubel, R. Tsao, A.W. Schwabacher and J.R. Coats, *Journal of Environmental Polymer Degradation*, 1997, 5, 2, 119.

22. G.J.L. Griffin, *Pure and Applied Chemistry*, 1980, **52**, 399.

23. M. Kimura, K. Toyota, M. Iwatsuki and H. Sawada in *Biodegradable Plastics and Polymers*, Eds., Y. Doi and K. Fukuda, Elsevier Science, Amsterdam, The Netherlands, 1994, p.92.

24. R-J. Müller, E. Marten and W-D. Deckwer in *Biorelated Polymers: Sustainable Polymer Science and Technology*, Eds., E. Chiellini, H. Gil, G. Braunegg, J. Buchert, P. Gatenholm and M. Van der Zee, Plenum Publishers, New York, NY, USA, 2001, p.303.

25. U.B. Singh, S.C. Gupta, G.N. Flerchinger, J.F. Moncrief, R.G. Lehmann, N.J. Fendinger, S.J. Traina and T.J. Logan, *Environmental Science and Technology*, 2000, **34**, 2, 266.

26. A. Balows, H.G. Trüper, M. Dworkin, W. Harder, K-H. Schleifer, *The Prokaryotes*, 2nd Edition, Springer-Verlag, Berlin, Germany, 1992.

27. J.B. Wesnigk, M. Keskin, W. Jonas, K Frigge and G. Rheinheimer in *The Handbook of Environmental Chemistry, Part K, Volume 2, Biodegradation and Persistence: Reactions and Processes*, Ed., B. Beek, Springer-Verlag, Berlin, Germany, 2000, p.253.

28. P. Morgan and R.J. Watkinson, *Critical Reviews in Biotechnology*, 1989, 8, 4, 305.

29. V. Torsvik, J. Goksoryl, F.L. Daae, R. Sorheim, J. Michalsen and K. Salte in *Beyond the Biomass: Compositional and Functional Analysis of Soil Microbial Communities*, Eds., K. Ritz, J. Dighton and K.E. Giller, Wiley, London, UK, 1994, p.39.

30. R.C. Cooke and A.D.M. Rayner, *Ecology of Saprotrophic Fungi*, Longman, London, UK, 1984.

31. A.D.M. Rayner and L. Boddy, *Fungal Decomposition of Wood: Its Biology and Ecology*, John Wiley, Chichester, UK, 1988.

32. H. Nishida and Y. Tokiwa, *Journal of Environmental Polymer Degradation*, 1993, **1**, 3, 227.

33. F. Degli-Innocenti, G. Goglino, G. Bellia, M. Tosin, P. Monciardini, L. Cavaletti in *Microbiology of Composting*, Ed., H. Insam, N. Riddech and S. Klammer, Springer-Verlag, Berlin, Germany, 2002, p.273.

34. ISO 14855, *Determination of the Ultimate Aerobic Biodegradability and Disintegration of Plastic Materials under Controlled Composting Conditions - Method by Analysis of Evolved Carbon Dioxide*, 1999.

35. R.A. Miles and W.J. Doucette, *Chemosphere*, 2001, **45**, 6-7, 1085.

36. J.P.E. Anderson in *Biotechnology and Biodegradation*, Eds., D. Kamely, A. Chakrabarty and G.S. Omenn, Gulf Publishing Company, Houston, TX, USA, 1989, p.129.

37. R. Bartha and D. Pramer, *Soil Science*, 1965, **100**, 1, 71.

38. ISO 17556, *Plastics - Determination of the Ultimate Aerobic Biodegradability in Soil by Measuring the Oxygen Demand in a Respirometer or the Amount of Carbon Dioxide Evolved*, 2003.

39. ASTM D5988, *Standard Test Method for Determining Aerobic Biodegradation in Soil of Plastic Materials or Residual Plastic Materials after Composting*, 2003.

40. A. Modelli, B. Calcagno and M. Scandola, *Journal of Environmental Polymer Degradation*, 1999, **7**, 2, 109.

41. R. Solaro, A. Corti and E. Chiellini, *Journal of Environmental Polymer Degradation*, 1998, **6**, 4, 203.

42. L. Chen, S.H. Imam, S.H. Gordon and R.V. Greene, *Journal of Environmental Polymer Degradation*, 1997, **5**, 2, 111.

43. ISO 14851, *Determination of the Ultimate Aerobic Biodegradability of Plastic Materials in an Aqueous Medium – Method by Measuring the Oxygen demand in a Closed Respirometer*, 1999.

44. ISO 14852, *Determination of the Ultimate Aerobic Biodegradability of Plastic Materials in an Aqueous Medium – Method by Analysis of Evolved Carbon Dioxide*, 1999.

45. A-C. Albertsson, *Journal of Applied Polymer Science*, 1978, **22**, 12, 3419.

46. J.D. Stahl, M.D. Cameron, J. Haselbach and S.D. Aust, *Environmental Science and Pollution Research*, 2000, **7**, 2, 83.

47. A.I. Suvorova, I.S. Tujkova and E.I. Trufanova, *Macromolecular Symposia*, 1999, **144**, 331.

48. A. Calmon-Decriaud, V. Bellon-Maurel and F. Silvestre, *Advances in Polymer Science*, 1997, **135**, 207.

49. H. Sawada in *Biodegradable Plastics and Polymers*, Eds., Y. Doi and K. Fukuda, Elsevier Science, Amsterdam, The Netherlands, 1994, p.298.

50. *OECD Guidelines for the Testing of Chemicals, Method 301C: Degradation and Accumulation: Modified MITI Test (I)*, Organisation for Economic Co-operation and Development Paris, France, 2003.

51. J.E. Guillet, H.X. Huber and J.A. Scott, *Journal of Macromolecular Science A*, 1995, **32**, 4, 823.

52. J.E. Guillet, H.X. Huber and J.A. Scott in *Biodegradable Polymers and Plastics*, Ed., M. Vert, J. Feijen, A. Albertsson, G. Scott and E. Chiellini, Royal Society of Chemistry, Cambridge, UK, 1992, 55.

53. G. Bellia, M. Tosin, G. Floridi and F. Degli-Innocenti, *Polymer Degradation and Stability*, 1999, **66**, 1, 65.

54. L-X. Li, E.A. Grulke and P.J. Oriel, *Journal of Applied Polymer Science*, 1993, **48**, 6, 1081.

55. A.V. Yabannavar and R. Bartha, *Applied and Environmental Microbiology*, 1994, **60**, 10, 3608.

56. R.G. Lehmann, J.R. Miller and G.E. Kozerski, *Chemosphere*, 2000, **41**, 5, 743.

57. A. Calmon, S. Guillaume, V. Bellon-Maurel, P. Feuilloley and F. Silvestre, *Journal of Environmental Polymer Degradation*, 1999, 7, 3, 157.

58. E. Ikada, *Journal of Environmental Polymer Degradation*, 1999, 7, 4, 197.

59. S. Kiatkamjornwong, M. Sonsuk, S. Wittayapichet, P. Prasassarakich and P-C. Vejjanukroh, *Polymer Degradation and Stability*, 1999, **66**, 3, 323.

60. H. Hatakeyama, S. Hirose, T. Hatakeyama, K. Nakamura, K. Kobashigawa and N. Morohoshi, *Journal of Macromolecular Science A*, 1995, **32**, 4, 743.

61. D. Darwis, H. Mitomo and F. Yoshii, *Polymer Degradation and Stability*, 1999, **65**, 2, 279.

62. EN ISO 846, *Plastics - Evaluation of the Action of Microorganisms*, 1997.

63. P.G. Dalev, R.D. Patil, J.E. Mark, E. Vassileva and S. Fakirov, *Journal of Applied Polymer Science*, 2000, 78, 71, 1341.

64. L. Zhang, J. Zhou, J. Huang, P. Gong, Q. Zhou, L. Zheng and Y. Du, *Industrial and Engineering Chemistry Research*, 1999, **38**, 11, 4284.

65. AATC Test Method 30 – 1999, *Antifungal Activity, Assessment on Textile Materials: Mildew and Rot Resistance of Textile Materials*, 1999.

66. R. Stote, J. McCassie, J.Mayer, A.Shupe and K.Dixon in Proceedings of the *Third Annual Meeting of the Bio/Environmentally Degradable Polymer Society*, Boston, MA, USA, 1994. p.48.

67. K. Hashimoto, T. Hamano and M. Okada, *Journal of Applied Polymer Science*, 1994, **54**, 10, 1579.

68. M. Okada, S. Ito, K. Aoi and M. Atsumi, *Journal of Applied Polymer Science*, 1994, **51**, 6, 1045.

69. K-L.G. Ho, A.L. Pometto III, P.N. Hinz, A. Gadea-Rivas, J.A. Briceno and A. Rojas, *Journal of Environmental Polymer Degradation*, 1999, 7, 4, 167.

70. K-L.G. Ho, A.L. Pometto III, A. Gadea-Rivas, J.A. Briceno and A. Rojas, *Journal of Environmental Polymer Degradation*, 1999, 7, 4, 173.

71. M.A. Khan, M.K. Uddin, M.N. Islam and K.M. Idriss Ali, *Journal of Applied Polymer Science*, 1995, **58**, 1, 31.

72. ASTM D1436-97, *Standard Test Methods for Application of Emulsion Floor Polishes to Substrates for Testing Purposes*, 2002.

73. M.A. Khan, K.M.I. Ali, F. Yoshii, K. Makuuchi, *Angewandte Makromolekulare Chemie*, 1999, **272**, 1, 94.

74. R. Truhaut, *Ecotoxicology Environmental Safety*, 1977, **1**, 151.

75. F. Zucconi, M. de Bertoldi in *Compost: Production, Quality and Use*, Eds., M. de Bertoldi, M.P. Ferranti, P. L'Hermite and F. Zucconi, Elsevier Applied Science, London, UK, 1987, p.30.

76. ISO 11268-1, *Soil Quality - Effect of Pollutants on Earthworms (Eisenia fetida) - Part 1: Determination of Acute Toxicity using Artificial Soil Substrate*, 1993.

77. *OECD Guidelines for the Testing of Chemicals, Method 202, Effects on Biotic Systems: Daphnia sp., Acute Immobilisation Test and Reproduction Test*, Organisation for Economic Co-operation and Development, Paris, France, 2003.

78. *OECD Guideline for Testing of Chemicals, Method 208, Effects on Biotic Systems: Terrestrial Plants, Growth Test*, Organisation for Economic Co-operation and Development, Paris, France, 2003.

79. J. Lannalainen, R. Juvonen, K. Vaajasaari and M. Karp, *Chemosphere*, 38, 5, 1069.

80. F. Degli-Innocenti, G. Bellia, M. Tosin, A. Kapanen and M. Itavaara, *Polymer Degradation and Stability*, 2001, **73**, 2, 101.

81. S.H. Yuk, S.H. Cho, B.C. Shin and H.B. Lee, *European Polymer Journal*, 1996, **32**, 1, 101.

82. E. Chiellini, A. Corti, A. Giovannini, P. Narducci, A.M. Paparella and R. Solaro, *Journal of Environmental Polymer Degradation*, 1996, **4**, 1, 37.

83. S-R. Yang and C.H. Wu, *Macromolecular Symposia*, 1999, **144**, 101.

84. S.P. McCarthy, A. Ranganthan and W. Ma, *Macromolecular Symposia*, 1999, **144**, 63.

85. E. Chiellini, P. Cinelli, A. Corti, E.R. Kenawy, E.G. Fernandes and R. Solaro, *Macromolecular Symposia*, 2000, **152**, 83.

86. R. Farrell, M. Ayyagari, J. Akkara and D. Kaplan, *Journal of Environmental Polymer Degradation*, 1998, **6**, 3, 115.

87. E. Chiellini, A. Corti and R. Solaro, *Polymer Degradation and Stability*, 1999, **64**, 2, 305.

88. J.M. Mayer, G.R. Elion, C.M. Buchanan, B.K. Sullivan, S.D. Pratt and D.L. Kaplan, *Journal of Macromolecular Science A*, 1995, **32**, 4, 775.

89. C. Flaqué, L. Contat Rodrigo and A. Ribes-Greus, *Journal of Applied Polymer Science*, 2000, **76**, 3, 326.

90. D. Bikiaris, J. Prinos and C. Panayiotou, *Polymer Degradation and Stability*, 1997, **58**, 1-2, 215.

91. G.J.L. Griffin and H. Mivetchi in Proceedings of the *3rd International Biodegradation Congress*, Rhode Island, NY, USA, 1975, p.807

92. T.H. Goswami and M.M. Maiti, *Polymer Degradation and Stability*, 1998, **61**, 2, 355.

93. G. Colin, J.D. Cooney, D.J. Carlsson and D.M. Wiles, *Journal of Applied Polymer Science*, 1981, **26**, 509.

94. R.L. Randal, *Journal of Sustainable Agriculture*, 2000, **16**, 4, 33.

95. S.N. Ghosh and S. Maiti, *European Polymer Journal*, 1998, **34**, 5-6, 849.

96. M. Okada, K. Tachikawa and K. Aoi, *Journal of Applied Polymer Science*, 1999, **74**, 14, 3342.

97. U. Witt, R-J. Müller, and W-D. Deckwer, *Journal of Macromolecular Science A*, 1995, **32**, 4, 851.

98. M. Okada, K. Tachikawa and K. Aoi, *Journal of Polymer Science*, 1997, **35**, 13, 2729.

99. A. Torres, S.M. Li, S. Roussos and M. Vert, *Applied and Environmental Microbiology*, 1996, **62**, 7, 2393.

100. I.K. Jung, K. Hee Lee, I-J. Chin, J. San Yoon, M. Nam Kim, *Journal of Applied Polymer Science*, 1999, **72**, 4, 553.

101. S. Matsumura, J. Takahashi, S. Maeda and S. Yoshikawa, *Die Makromolecular Chemie - Rapid Communications*, 1988, **9**, 1, 1.

102. U. Witt, M. Yamamoto, U. Seeliger, R-J. Muller and V. Warzelhan, *Angewandte Chemie International Edition*, 1999, **38**, 10, 1438.

103. Zainuddin, M.T. Razzak, F. Yoshii and K. Makuuchi, *Polymer Degradation and Stability*, 1999, **63**, 2, 311.

104. J.A. Ratto, P.J. Stenhouse, M. Auerbach, J. Mitchell and R. Farrell, *Polymer*, 1999, **40**, 24, 6777.

105. K. Bahari, H. Mitomo, T. Enjoji, F. Yoshii and K. Makuuchi, *Polymer Degradation and Stability*, 1998, **62**, 3, 551.

106. C. David, I. Dupret and C. Lefèvre, *Macromolecular Symposia*, 1999, **144**, 141.

107. J.E. Potts, R.A. Clendinning, W.B. Ackart and W.D. Niegish, *Polymer Science and Technology*, 1973, **3**, 61.

108. L. Jiang and G. Hinrichsen, *Angewandte Makromolekulare Chemie*, 1999, **268**, 1, 18.

109. A-C. Albertsson, S.O. Andersson and S. Kalsson, *Polymer Degradation and Stability*, 1987, **18**, 1, 73.

110. M. Ratajska, S. Boryniec, A. Wilczek and M. Szadkowski, *Fibres and Textiles in Eastern Europe*, 1998, **6**, 3, 41.

111. C. Lefevre, C. Mathieu, A. Tidjani, I. Dupret, C. Vander Wauven, W. De Winter and C. David, *Polymer Degradation and Stability*, 1999, **64**, 1, 9.

112. L. Contat-Rodrigo and A. Ribes-Greus, *Journal of Applied Polymer Science*, 2000, **78**, 1707.

113. L. Contat-Rodrigo and A. Ribes-Greus, *Macromolecular Symposia*, 1999, **144**, 153.

114. Y. Ohtake, T. Kobayashi, H. Asabe, N. Murakami and K. Ono, *Journal of Applied Polymer Science*, 1998, **70**, 9, 1643.

115. A-C. Albertsson and S. Karlsson, *Journal of Applied Polymer Science*, 1988, 35, 5, 1289.

116. Y. Otake, T. Kobayashi, H. Asabe, N. Murakami and K. Ono, *Journal of Applied Polymer Science*, 1995, 56, 13, 1789.

117. D. Bikiaris E. Pavlidou, J. Prinos, J. Aburto, I. Alric, E. Borredon and C. Panayiotou, *Polymer Degradation and Stability*, 1988, 60, 2-3, 437.

118. P.K. Sastry, D. Satyanarayana and D.V. Mohan Rao, *Journal of Applied Polymer Science*, 1998, 70, 11, 2251.

119. Y. Doi, *Macromolecular Symposia*, 1995, 98, 585.

120. Y. Doi, A. Segawa and M. Kunioka, *Polymer Communications*, 1989, 30, 6, 169.

121. M. Avella, G. La Rota, E. Martuscelli, M. Raimo, P. Sadocco, G. Elegir and R. Riva, *Journal of Materials Science*, 2000, 35, 4, 829.

122. F. Gassner and A.J. Owen, *Polymer*, 1992, 33, 12, 2508.

123. Zainuddin, M. Thabrani Razzak, F. Yoshii and K. Makuuchi, *Journal of Applied Polymer Science*, 1999, 72, 1283.

124. T. Hatakeyama, T. Tokashiki and H. Hatakeyama, *Macromolecular Symposia*, 1998, 130, 139.

125. E. Chiellini, A. Corti, S. D'Antone and R. Solaro, *Macromolecular Symposia*, 1999, 144, 127.

126. A. Takasu, K. Aoi, M. Tsuchiya and M. Okada, *Journal of Applied Polymer Science*, 1999, 73, 7, 1171.

4 Ecotoxicological Aspects in the Biodegradation Process of Polymers

Johann Fritz

This chapter contains an overview of the direct environmental impact, of biodegradable polymers. The theme is complex because different types of ecosystems are involved and must be considered separately. Therefore it is the intention to keep all explanations short. In that sense the very basics of organic waste recovery, ecotoxicology and soil and sediment ecology are summarised in a few words, but references to a selection of specialised textbooks are given.

The structure of the following sections should allow the reader to become more familiar with the theme step by step. The sections are:

- a very general overview about the need for ecotoxicity testing including theoretical derivations for potential environmental influences,

- a short introduction to the science of ecotoxicology with a list of commonly used methods,

- special requirements needed when testing polymers and environmental samples,

- an overview about currently available research results, and

- a summary and some impressions about further research needs.

4.1 The Need of Ecotoxicity Analysis for Biodegradable Materials

When plastics are used for throw-away products (such as packaging for example) or for products with a limited lifetime (bird nets, plant foils, grass nails or sapling plant pots to name a few) they may end up sooner or later as waste or litter. Since all those conventional polymers have been optimised to be stable against microbial attacks and to withstand moisture, light and atmospheric oxygen, such waste and litter is almost inert to environmental attack.

Biodegradable materials on the other hand are designed to fulfil the specific needs of an application they are intended for and to become mineralised by microorganisms present in the environment or at a treatment facility for organic waste. Therefore biodegradable polymers will have a strong interaction with the ecosystem. They will become feed stock for the autochthone microorganisms and degradation residues and metabolites may be produced or enriched at that location.

Figure 4.1 gives an overview of the possible pathways of biodegradable materials. The most significant deviation is between biodegradation due to organic recovery of biowaste and biodegradation in the environment. The first is a combined thermo- and mesophilic process in presence of a dense population of microorganisms that are supported by watering and aeration. The second, the same in terrestrial or aquatic environments, is a biodegradation at meso- or psychrophilic conditions achieved by a less dense population of microorganisms and without active support. Therefore both types of biodegradation should be described separately. When designing an artificial biodegradable polymer where the degradation process will occur should be considered – this should preferably be at a waste treatment facility or in the environment.

Incineration and landfill as additional treatment techniques for residual waste are mentioned for completeness. The content of pollutants and harmful substances in residual waste is almost always determined by waste fractions other than biodegradable materials. Therefore the established rules for incineration and landfill will cover biodegradable polymers well; additional considerations of ecotoxicological impacts are not needed.

4.1.1 Standards and Regulations for Testing of Biodegradable Polymers

During the last few years some national and international standards, such as DIN 54900 [1], ASTM D6002:1996 [2] and EN 13432 [3], have been published. The intended goal of them all is to provide producers and users as well as authorities with test schemes and quality criteria (pass levels) for biodegradable materials. The three standards are different in detail but have the same basic four-step test scheme:

1. Estimation of the possibility of biodegradability based on the chemical composition (polymer structure) and absence of intentionally added components which are known to be or are under suspect of becoming toxic or harmful to the environment (for example heavy metals)

2. Determination of the degradability caused by microbial activity and quantification by either oxygen demand, carbon dioxide release or methane production considering the time needed for full mineralisation

3. Determination of the disintegration under real or simulated composting or anaerobic digestion conditions and quantification by gravimetric determination of a sieve residue

4. Investigation of the quality of the compost resulting from the material disintegration test by analysis of chemical and physical parameters and by determination of ecotoxicological effects to at least higher plants

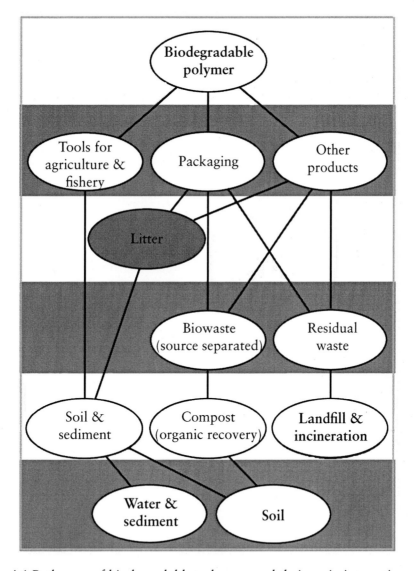

Figure 4.1 Pathways of biodegradable polymers and their main interactions with different ecosystems related to the intended use

The standards discussed do consider a utilisation of biodegradable materials by regular waste processing and do not deal with procedures and criteria for degradation in soil or aquatic ecosystems. Nevertheless, they are trend setting by defining the final compost as product. Every product has a market, a value and a circle of consumers and therefore needs quality criteria for regulation of both the price and the potential uses. All three standards contain the requirement that any introduction of man-made polymers into the established organic waste recovery must not negatively influence the quality of the final compost.

It is easy to extend that philosophy to natural terrestrial and aquatic ecosystems although the specific standards are currently not finished: Man-made polymers must not have any negative influence on the environment to which they are applied. It should be acknowledged that any local ecosystem is in a natural balance that is worth protecting. It is expected that the standards in progress will follow that philosophy.

During the past decades, conventional agriculture has caused an almost global contamination of groundwater with residues of fertilisers (especially nitrate) and pesticides. The protection of water resources, not necessarily drinking water alone, is one more aspect to be considered when biodegradable polymers are released to the environment in vast amounts. Finally national laws and regulations for drinking water have to be followed.

Natural ecosystems which are not of commercial interest are the least protected in all the previously mentioned examples, because only a very few standards and regulations could be applied to them. Nevertheless, they should be remembered when environmental effects of human intervention on the nature are discussed.

4.1.2 Detection of the Influences on an Ecosystem Caused by the Biodegradation of Polymers

As already mentioned, compost is a product under governmental quality regulation. It seems to be an easy task when declaring an unwanted effect caused by the degradation of polymers as any change of the quality relevant parameters. Since the composition of compost is dependent upon the composition of the original biowaste, all of the analysed chemical or biological parameters of the final product will vary over a broad range. For determining the influence of biodegradable materials two test batches will be needed, one with the test polymer added and another without any additions. The physical and chemical parameters as well as ecotoxicological effects can then be compared between those two compost products.

Nevertheless, the interpretation of differences between the two batches could be very difficult. How should a change in the pH-value (for example from 7.9 to 8.2), or in the plant available fractions of nutrients in relation to their total content be interpreted? A more readily applied result could be obtained from bioassays, since any decrease of the germination rate of the seed or any reduction of plant growth (production of green plant biomass) could easily be declared as a loss of compost quality, independent from the chemical composition.

Table 4.1 gives an overview of the environmentally relevant quality criteria defined in the three standards. It is important to mention that the avoidance of quality losses and negative impacts to small ecosystems starts with the material design. Known toxic substances and others, which are suspected to become harmful to the environment during or after the degradation process, must not be used. The bioassays included are a safety net to detect all those stable biodegradation residues and metabolites which may be formed by microorganisms and which were not present in the original polymer.

No standards are available to deal with the changes in the quality of soil and aquatic ecosystems caused by biodegradation of polymers. Again, the intentions of the standards for organic recovery should be extrapolated to those environments too. Since the chemical compositions of different soils as well as of aquatic ecosystems and their sediments could be very different from each other, the definition of acceptable changes will cause many more problems than for compost. For those ecosystems the application of biotests will be the most important impact control for biodegradable polymers.

Table 4.1 Quality criteria of the steps 2 and 4 defined in the analysis schemes for biodegradable materials		
Standard	Material design	Criteria after composting
DIN 54900 (1997) [1]	Maximum 50% minerals Maximum 30% of heavy metals allowed in compost	RAL-criteria [4] for chemical composition Ecotoxicity tests with summer barley
ASTM D6002 (1996) [2]	No special requirements	National US standards for chemical composition Ecotoxicity tests with three plant species, earthworms and rotifers
EN 13432 (2001) [3]	No toxic or harmful substances Maximum 50% minerals Maximum 50% of heavy metals allowed in compost	European national standards for chemical composition Ecotoxicity tests with two plant species
RAL: Reichs-Ausschuss für Lieferbedingungen		

4.1.3 Potential Influences of Polymers After Composting

The positive effects of the application of compost in agriculture are described in numerous publications, for example Allison [5], Voelker [6], Gottschall [7] and Hartl [8]. The most common effect is the fertilisation by mineral nutrition elements. Further Sekhon and Meelu [9] mention the supporting effect of organic matter on the micro-flora and on physical properties of less fertile soil.

Potential hazards caused by the introduction of toxic components (especially heavy metals) with contaminated compost are revealed very often. These typical impurities are covered by the analytical quality control of most of the national regulations in Europe and will lead to a classification as second or third quality and to a limited use of the compost. These national standards are dealing with well-known contaminants that may derive from typical biowaste and are focussing on heavy metals and a handful of halogenated or aromatic hydrocarbons. The inclusion of bioassays with higher plants in some standards is more to determine the maturity of the compost than with the appearance of ecotoxic effects caused by anything other than the chemicals being determined.

The collection of new, artificial, biodegradable materials together with the traditional biowaste and their composting (or anaerobic digestion) includes new risks of the introduction or generation of not known and therefore not analysed substances. Hope-Simpson [10] demonstrated for the first time that residues from composting of coated paper could be toxic to plants and make it impossible to use such a compost in agriculture. The reason was an enrichment of the nutrient element boron to a toxic level.

Insam [11] gives a very comprehensive overview about accepted test methods for investigations of biodegradable packaging for their suitability for various established composting processes. The methods are focussed on the determination of the degradability but do consider biotests as routine quality control. Also Pagga [12] does consider biotests as necessary quality control for compost batches containing degraded artificial packaging polymers.

The term compost quality should not be limited to physical and chemical parameters. While such analysis could describe the contents of nutrients and the presence of a small number of selected pollutants, the appearance of unidentified metabolites and residues could be detected more reliably by the application of biotests.

The increased cost for the additional analysis, before the introduction of materials on the market, will be rewarded by the confidence of compost operators and compost users. Mandatory biotests are needed during the phase of material development and are not necessary as an additional routine quality control of each batch of compost.

The inclusion of mandatory ecotoxicity tests in the most relevant standards is already realised (**Table 4.1**). The extent of the investigations may differ between the standards, but they share the same intention: the detection of negative influences, which are not covered by the routine chemical analysis. Because of lack of practical experience, especially about how to deal with complex matrices like compost in conventional bioassays, not many mandatory methods are currently listed in the German and in the European standard. Nevertheless, an option to include more or maybe specially adapted methods is kept in the EN 13432 [3].

4.1.4 Potential Influences of Polymers During and After Biodegradation in Soil and Sediment

It is a justified claim that artificial materials should not inhibit the growth and crop yield of agricultural plants. That is valid for a short-term view as well as for a longer-term evaluation. Negative effects should neither appear during the same vegetation period that a biodegradable material is applied nor at the following years. The long-term observation is necessary, because repeated applications of biodegradable polymers may lead to an accumulation of potentially ecotoxic substances which are below the no effect concentration level (NOEC) after one single application.

In the standards dealing with organic recovery of biodegradable materials the necessity of ecotoxicological investigations is clearly stated. Not only because of the possible formation of unknown metabolites but also because of the behaviour of additives in polymers (for example conventional softeners) which are already know to be problematic.

This does not change if instead of a thermophilic composting process the slower, meso- or psychrophilic biodegradation in soil, in aquatic environments or in their sediments is observed. As is well known, the thermally initiated hydrolysis step of ester or ether bonds could be crucial for following the uptake of the built oligo- and monomers into the cells. Also for degradation at ambient temperature the first hydrolysis step must either be catalysed due to microbial activity or initiated by other physical and chemical forces (for example by sunlight or by oxidation in presence of air). The probability of the appearance of undegraded residues and of their further accumulation in soil and sediments is increased compared to a thermophilic composting process.

On the other hand in many other publications the positive effects of organic substances in soil are described. Harvest residues (green plants), organic fertiliser (stable manure) and other organic, biodegradable substances can contribute positively to the physical structure and will therefore indirectly increase the soil's fertility. Higher water holding capacity and elevated ion exchange capacity are the most often claimed causes for such improvements of the soil quality. Further a prospering soil micro-flora (applied with compost or grown due to biodegradation of plant residues) may induce disease resistance of plants. All those agricultural publications do consider only natural polymers, such as starch, cellulose, ligno-cellulose (wood), proteins and fats. The positive influences on plant growth, crop yield and quality are explained by the increased content of organic matter (humic substances) in the soil by several authors (Danneberg [13], Dick and McCoy [14], Gottschall [7], Hartl [8] and Knafl [15]). At least Sekhon and Meelu [9] do correlate the content of organic matter directly with the soil fertility.

Also very often negative effects are described, caused by the presence of biodegradable substances in soil and appearing during the time of plant growth. The most prominent is the formation of toxic fermentation by-products released in the early stages during the biodegradation of organic substances. This phenomenon is described by Lynch and co-workers [16] and Toussan and co-workers [17] and is well known in relation to the incorporation of crop residues in soil. The prime reason for reduced plant growth is the generally increased microbial activity, which may further lead to a drop in the pH-value and to an abnormal high oxygen demand, as described by Subba Rao [18] and Alloway and co-workers [19]. All these effects are of a temporary nature and will end soon after the biodegradation is completed. Other negative impacts are explained by the mobilisation of heavy metals, which are already present in the soil (or in the compost). While metals, which are bound to or are included in the mineral matrix, behave inertly in the ecosystem, the mobile and therefore bioavailable fractions can cause serious harm to plants and animals and can accumulate in the food chain (excerpt from Förstner and co-workers [20], Scrudato and co-workers [21] and Suffet and MacCarthy [22]).

The effects and influences caused by the deposition of communal residual and specific industrial wastes are well described in the literature. More literature is available detailing the ecosystem responses to known organic pollutants (mostly pesticides and polycyclic aromatic hydrocarbons) or their residues and metabolites during biodegradation. Almost no literature is available concerning the interaction between the biodegradation of organic substances, the appearance and the related mechanisms of non-reversible ecotoxicological effects. Related empirical and research results dealing with material of other than biogenic origin are missing in the literature. If such artificial polymers are completely biodegradable it could be assumed that they might not cause effects which are basically different from those of plant residues.

Although agricultural production plants are at the centre of interest when discussing ecotoxicological effects, other soil and water organisms should be included as well for an extension to a broad ecological assessment. Coleman and Crossley [23] claim that many commercially uninteresting organisms are an essential part of soil and determine its long time fertility.

4.2 A Short Introduction to Ecotoxicology

Several textbooks deal with principles and with applied aspects of the relatively young science of ecotoxicology, for example Calow [24], Fendt [25], Forbes and Forbes [26] and Landis and Yu [27] to name a few. It does not make sense to repeat all the basics and details here. Nevertheless, a very short summary of the most important facts is presented to give the reader an impression of the subject.

4.2.1 Theory of Dose-Response Relationships

Every living organism keeps in its cell or in its body several hundred or thousands of chemical substances in a steady state. Whenever an influence from outside disturbs that steady state, the organism endeavours to reach the balance again as soon as possible. If the disturbance is the presence of a toxic substance either the production of the inhibited enzyme is increased or new chemicals or enzymes are produced to deactivate the disturbing substance or to keep its effect to a minimum. However, for the establishment of the physiological balance the organism needs energy and nutrient resources that are available in limited amounts in most natural environments.

The intensity of a stress (its dose) and the intensity of the counter-reaction of the organism (the response) follow a relationship that is significant for a species and significant for the type of stress. A theoretical example of a dose-response relationship is given in **Figure 4.2**. Such relationships could be drawn up experimentally by applying a stress in known doses to selected test organisms and by the measurement of a significant reaction. This could be the cell growth, biomass production, the production of specific enzymes or metabolites or simply the survival of the organisms.

The graphical evaluation made from doses and responses will give a relationship curve from which some key parameters can be derived. These are important standardised key values, which are commonly used for hazard evaluation and risk assessment of chemicals. A toxicity data collection for many chemicals can be found in Rippen [28].

4.2.2 Test Design in Ecotoxicology

4.2.2.1 Investigation Level

Molecular biological (enzymes, proteins, genes), single species, multispecies and ecosystem level tests offer different insights into the behaviour and effects of pollutants. Single

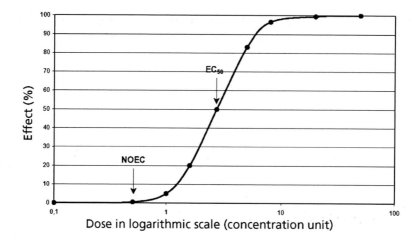

Figure 4.2 Graphical representation of a dose-response relationship with some key parameters marked: EC_{50} is the concentration at which half of the maximal effect is observed; NOEC is the maximum concentration at which no significant effect is observed

species test are most commonly used because they are generally simpler to perform and interpret than higher level tests and they provide more practically relevant information than molecular biological tests. Multispecies, community and ecosystem level tests permit the detection of indirect pollutant effects that are the result of species interactions, as well as direct toxicity effects (for details see Calow [24]). Molecular biological tests have recently increased in popularity since suitable analytical methods have become available to detect either mutations in single DNA strains or to identify genes which become activated in a stress situation.

4.2.2.2 Length of the Exposure Period

Tests are distinguished mainly as acute (short-term) or chronic (long-term). The duration of the exposition is defined relative to the test organism's life cycle. Acute biotests detect immediate responses following a chemical reaction at one or more points in the organism's metabolism, negatively influencing its physiological fitness. Chronic tests detect influences, which include subtle effects such as elongated generation time, reduced life span or changes in behaviour or morphology.

It should be considered that a complex sample could undergo a significant change due to microbial activity during the runtime of a bioassay. Rarely it will be possible to document changes in the composition by conventional chemical analysis. Nevertheless, some representative parameters, such as pH value, conductivity, colour and total carbon content could be determined and reported.

4.2.2.3 End Points

A further criterion is the type of the measured endpoint. Both lethal and sublethal endpoints are frequently investigated in ecotoxicological studies. Toxicity tests are often designed to measure the exposure concentrations at which 50% of the test population is killed (LC_{50}) or exhibits a defined sublethal effect, (e.g., growth rate, expressed as EC_{50}). Screening tests are performed very often to get an impression about the ecotoxic properties of an unknown sample. The end-point may not be defined at the start of the test but the organisms are observed carefully not only for lethal effects.

4.2.3 Toxicity Tests and Bioassays

Hill and co-workers [29] give some well formulated definitions:

A **toxicity test** is an experiment in which organisms of a single species are exposed in the laboratory to a clean natural sample (soil, sediment or water) which has been spiked with a known chemical or mixture of chemicals, generally at a range of concentrations. The purpose is to measure the degree of response associated with specific concentrations of the chemical(s).

A **bioassay** is an experiment in which organisms of a single species are exposed in the laboratory to field samples or extracts of this, potentially containing one or more contaminants, with the aim of measuring possible biological effects of those contaminants.

The term **biotests** is also very often used in the literature. It could be used as universal expression if both, toxicity tests and bioassays are mentioned in one sentence or if a general explanation is formulated.

To start with the more simple relationship: The determination of the direct toxicity from undegraded materials will be set up as toxicity test. For the testing of degradation residues and metabolites deriving from the biodegradation of polymers' bioassays are generally used. Although well known materials are mixed with the matrix, the chemical composition of the sample may not be known precisely even after a short time of biodegradation.

To distinguish between toxicity tests and bioassays is not only of theoretical relevance. From the first, a clear dose-response graph could be drawn and effect concentrations

(EC_{10}, EC_{50}, EC_{100}) as well as the NOEC could be calculated. The evaluation of bioassays should preferably be limited to the expression of inhibition values of the original sample and to the calculation of dilution rates (G-values) at which a specific rate of inhibition, most commonly 10%, is reached.

4.2.4 Ecotoxicity Profile Analysis

An ecotoxicological profile analysis is a procedure developed to evaluate the environmental relevance of chemicals (commonly used for pesticides). The base for a risk assessment is a list of properties of the investigated chemical. That list includes concentration used, pattern of application, biodegradability, type and properties of metabolites, environmental dispersion and accumulation. Related results from laboratory tests are weighted and classified following very specific rules. Finally one relative index number is obtained which allows a simple comparison between ecological impacts of different chemicals (for details see Freitag and co-workers [30]).

The profile index, which is obtained from several selected single species biotests, could be an acceptable way to characterise the ecotoxicological relevance of a biodegradable material. Since the profile analysis was developed and is frequently used to characterise mainly agrochemicals some input parameters, such as concentration used and environmental dispersion need to be adapted. Nevertheless, similarities do exist and if organisms of different trophic levels are used for investigations on biodegradable materials the knowledge about their environmental impact will be significantly increased.

4.3 Recommendations and Standard Procedures for Biotests

In national and international standard method collections Deutsches Institut für Normung (DIN), International Organisation for Standardisation (ISO), Organisation for Economic Cooperation and Development (OECD), a wide range of biotests could be found. For the selection of a suitable test system potential target organisms, which are representative of the environment under investigation, should be selected. All possible situations should be considered, even migration of biodegradation intermediates from a terrestrial to an aquatic ecosystem.

In most cases a test for direct material toxicity to animals (birds, rats, rabbits, etc.), may not be of primary interest. Such interactions should be considered only, if a product is applied in a shape and colour that may be confused with feed by such animals and may be consumed by them in relevant amounts. The encapsulation of fertilisers or pesticides with biodegradable polymers may be such a case. On the other hand those active substances are most probably more harmful to animals than the polymer coating.

Not considering very special or exotic applications, an overview of suitable bioassays is given in **Table 4.2**. Aquatic test systems could be conducted from solid samples if an elutriate is made (for example according to DIN 38414-4 [31]). Liquid samples deriving from aqueous degradation tests should be analysed for aquatic ecotoxicity only. However, such samples could be used for the periodical watering of plants tests. The results should be interpreted very carefully, because metabolites and biodegradation residues could have a limited solubility in water or could be strongly adsorbed to the sample matrix.

In the following paragraphs the most suitable test methods are described in more detail and recent research results are added as far as possible. The list of methods may not be complete, since a lot more research and development are necessary to evaluate established methods for their use in complex matrices such as compost and soil.

Further it is a challenging task to differentiate between biotic and abiotic changes in an environment caused by one single application of biodegradable polymers. Coleman [23] says that even ecosystems that appear to be homogeneous do have a number of local inhomogenities (hot spots). These can be the specific support of microorganisms in the plant rhizosphere or the aggregation of bacteria around animal faeces or plant litter to name but two. A shift in the composition of the population (for example the propagation of some bacteria species) is the consequence. This may disturb an ecosystem but should not be mixed up with the effects of a toxic material component or biodegradation residue.

4.3.1 Bioassays with Higher Plants

Since plants are in most cases the primary target organisms for agricultural applications the methods and influencing parameters of plant biotests will be described in more detail than for all the other test organisms.

4.3.1.1 Test Set Up and General Conditions

The OECD method 208 [32] requires the application of at least three tests in parallel using three different plant species. That seems to be justified, since the grouped species are sensitive to different inhibition mechanisms. Such tests could be varied for the determination of the germination rate or of the plant growth (biomass production) or for both effects at the same time.

For the determination of the germination rate, a known number of seeds is put on top of the sample and watered properly. After the species specific germination time the number of young plants is counted and compared with the germination rate in the reference

Table 4.2 Overview about the most suitable bioassays for ecotoxicity testing of polymers during and after their biodegradation

Test organism	Sample types	Standard/literature	Comments
Higher plants, terrestrial (cress, millet, rape etc.)	compost and soil	OECD 208 [32]; ISO 11269-1 and ISO 11269-2 [33]	many species available
Higher plants, aquatic (*Lemna sp.*)	freshwater and sediments	OECD-draft [34]	currently not standardised
Fish (various species)	fresh- and seawater	OECD 203 [35], 204 [36], 210 [37]; ISO 7346 [38]; DIN 38412 L15 [39]	static or flow through design
Earthworm (*Eisenia foetida*)	soil and sediments	OECD 207 [40]; ISO 11268 [41]	not suitable in presence of digestible materials
Collembola (*Folsomia candita*)	compost and soil	ISO 11267 [42]	
Protozoa (ciliates)	soil	Berthold [43]	not standardised
Protozoa (*Colpoda mauposi*)	freshwater	DEV L10 [44]	designed for wastewater
Crustaceae (*Daphnia magna*)	freshwater	OECD 202 [45], 211 [46]; ISO 6341 [47], 10706 [48]; DIN EN ISO 5667-16 [49], 38412 L30 [50]	acute and chronic
Crustaceae (*Artemia sp.*)	seawater	ISO 14669 [51]	
Algae (*Scenedesmus subsp., Selensatrum cap., Chlorella sp.*)	freshwater	OECD 201 [52]; ISO 8692 [53]; DIN 38412 L33 [54]	test used very often
Algae (*Skeletonema costatum, Phaeodactylum tricornutum*)	seawater	ISO 10253 [55]; DIN 38412 L45 [56]	
Bacteria (*Pseudomonas putida*)	freshwater	ISO 10712 [57]	
Luminescent bacteria (*Vibrio fischeri, Photobacterium sp.*)	sea-water	ISO 11348 [58]	very short exposure time
Enzymic activity	soil and sediment	OECD 216 [59], 217 [60]; ISO 9509 [61]	measurement of N- and C-transformation
Various organisms and DNA	-	OECD 471 - 486 [62]	several mutagenity tests
Multispecies tests	all environments	Calow [24]	not standardised

substrate. The test time could be extended for two or three weeks (depending on the plant species) and the grown plants could then be harvested for determination of the biomass production. Similarly to the germination rate, the biomasses obtained from the samples are also compared with those from the reference substrate.

For the determination of the plant growth rate, young plants are pre-cultivated in a reference substrate and are then transferred to the prepared samples. That procedure requires some experience since the small roots must not be injured and the amounts of adhering reference substrate should be as small as possible. After a typical growth time the plants are harvested and the biomass produced is evaluated.

In all variants the samples (150 g to 250 g) are placed into trays made of polyethylene or glass. A thin layer of washed sand on the bottom can form a drain layer and a very thin layer on the top (spread carefully over the seed) avoids the drifting of the seed during the watering.

The reference matrix for plant tests should be chosen with care. When investigating the ecotoxic effects of biodegradable materials these are commonly mixed with compost or soil for the biodegradation test. The original matrix should be used as reference and for dilution of the samples. It is unavoidable that the test plants will grow differently in each compost batch and each type of soil. Standardised matrices, such as are given in the standard methods, should be avoided since those are optimised for toxicity tests of chemicals, added in defined amounts and not treated any further before the start of the biotests. For the calculation of ecotoxic effects deriving from the degradation process of polymer materials the difference to results from the same matrix without any additions is needed.

4.3.1.2 Special Test Conditions

The watering of the tests has a major impact on the plant growth and should be done carefully. Over the whole period of the plant growth the water content should be kept as constant as possible and at an optimised level. If the water content is too low, the ion transport from the roots to the leaves is inhibited or the plants will die from thirst. If the water content is too high, all the pores in the sample will be flooded and the oxygen transfer (normally by diffusion) will be interrupted. Anoxic or even anaerobic conditions will be the consequence. Again the plant growth will be inhibited or they will die. The water content should therefore be kept between 70% and 100% of the water holding capacity of the sample matrix. This fact is not mentioned in the standard methods. The influence of the water content on the plant growth in bioassays is shown in **Figure 4.3**. To keep the water content in the recommended range it is helpful to know the weight of each test tray calculated at 70% and 100% water saturation. During the test time whenever the lower weight is reached, water is added up to the weight of 100% saturation.

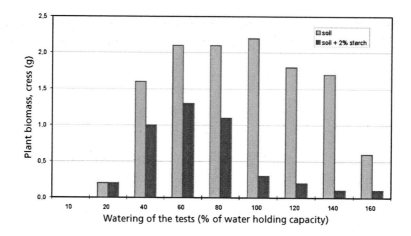

Figure 4.3 Influence of the amount of water used for the periodical watering on the plant growth in bioassays with standardised substrate alone and with 2% starch added. Derived from [64]

Current research results have shown that the response of the three species cress, millet and rape may be spread over a wide range using soil samples, which have degraded different biogenic and synthetic polymers. In our group we found that typical physical and chemical parameters determined from such soil samples did not correlate strictly with the observed inhibition of the plant growth. Some of the substances responsible for the effects may not be detected by a conventional soil analysis and remain unknown. Following Chen and Inbar [63] and interpreting results obtained by our group [64] the most probable causes may be:

- decrease of the oxygen content in the soil pores by microbial consumption due to the degradation process and in the consequence a decrease of the redox-potential

- formation of fatty acids, especially formic, acetic and propionic acid as typical metabolic intermediates

- decrease of the pH value for the same reason as above

- formation of stable toxic metabolites and degradation residues which are in most cases very difficult to detect

- a shift in the composition of the microbial community and an enrichment of potentially plant pathogenic organisms (or organisms with allelopathic properties).

The first three causes for an inhibition of plant growth are temporal effects, directly connected to increased microbial activity during the degradation process and may end at the time the degradation is completed. The last two causes could be of temporal nature but could be present over a longer time.

4.3.2 Bioassays with Earthworms (Eisenia foetida)

According to the OECD method 207 [40] bioassays with earthworms could be applied as acute or as chronic tests. In both cases 10 or more animals with known weight are exposed to 300 g to 500 g of sample. For the acute toxicity the test duration is limited to two weeks and the number of survivors and their weight are determined and compared with the results obtained from reference samples. For the chronic test the animals are left in the vessels for up to two months (and are fed, of course), additionally the number of cocoons (reproduction) is evaluated.

For the analysis of biodegradable materials the acute test seems not to be suitable, especially when the polymers are digestible for the earthworms. Compost is an optimal nutrition source for the animals and should never be tested for earthworm toxicity. The weight gain from the feed effect may cover possible smaller inhibition responses. In recent research work Stacher [64, 65] achieved weight increases of up to 50% of the initial values using soil samples where biobased materials had been added.

It may be an interesting task for future research to look at the digestion residues of biodegradable polymers obtained from long-term biotests with the earthworms. The animal's conversion pathway may differ significantly from those of microorganisms and lead to other metabolites and residues.

4.3.3 Preparation of Elutriates for Aquatic Ecotoxicity Tests

For the preparation of aqueous elutriates from solid samples the standard method DIN 38414-4 [31] can be used. Ten parts of water are added to one part dry matter of the samples and the mixture is continuously shaken for 24 hours. A clear solution is obtained by centrifugation or filtration and can be used for aquatic ecotoxicity tests. If necessary a smaller amount of water can be used (for example a dilution of 1:5).

By the standard procedure only water soluble fractions of the original sample are obtained in the elutriate. Hydrophobic and even some soluble substances, which are adsorbed or immobilised by ion exchange bonds to the organic or inorganic matrix, will be retained in the solid residue. If substances with a moderate solubility are present, it may be that in every elution the same concentration (saturated solution) may be achieved.

It may not be of major importance in which proportion the elutriates are made from compost or soil samples. The same elution procedure should be used for all samples if the results are to be compared with or related to each other. The most suitable proportions are 1:5 and 1:10 (dry matter to water). Using the proportion 1:5 gives an increased chance of detecting small inhibition effects.

The elution method DIN 38414-4 [31] is disputed concerning its universal use for aquatic biotests from solid samples. Hund [66] has observed significant differences in the results between terrestrial and aquatic biotests made from the same samples. From our group recently obtained research results with biodegradable materials did show in almost all cases no or very small inhibition effects, exceptions had been rare, [67, 68]. From those few results no further serious predictions could be made about the suitability of the DIN 38414-4 elution procedure for the testing of biodegradable materials.

4.3.4 Bioassays with Algae

The standardised methods OECD 201 [52] and DIN 38412 L33 [54] could both be applied for aqueous samples and for elutriates from compost and soil. Generally the liquid sample is inoculated with a defined algal species in a specified concentration (for example 10^4 cells per cm^3 final volume) and a mixture of essential mineral nutrients is added from a concentrated stock solution. The growth of the algae is determined periodically over 72 hours. The growth curve is integrated and the relative numbers are compared between samples and a control. A reference, in most cases potassium dichromate, is used to validate the test.

4.3.4.1 Selection of Algal Species Based on their Physiological Properties

The most commonly used freshwater algae species are *Selenastrum capricornutum*, *Scenedesmus subspicatus* and *Chlorella sp.* and the marine algae *Skeletonema costatum* and *Phaeodactylum tricornutum*. They are well known for their physiological properties and for their sensitivity to a couple of toxic elements and organic substances. Their optimal growth conditions are known as well as their reproduction rate, which is a validity criterion for the standardised tests.

Algal species which are representative for a local ecosystem can be isolated as well and used for bioassays. Some basic properties should be determined and included in the final report. Local species have the advantage of being adapted to the environment under investigation and are further not inhibited by an unbalanced relationship of minerals or by the geogenic presence of some heavy metals. Kusel-Fetzmann [69] recommends that such local factors should be determined and reported anyway and that bioassays from real samples using standardised algae species may fail to produce interpretable results (this is discussed in the OECD-Report [70]). A procedure for the isolation of algae has been described by Kusel-Fetzmann [69] and has been extended by Fritz [64]. Numerous toxicity results using the standardised algae species are available from literature. Therefore it may be of advantage to use at least one of them in parallel with local species. The results will be easier to compare and will be easier to interpret. On the other hand, the effort of an isolation and characterisation may not be justified if only one sample is analysed. If the goal of a study is a comprehensive investigation of influences on an ecosystem it may be of advantage to isolate and use local species.

4.3.4.2 Influence on the Algae Growth

The autotrophic growth (reproduction) of algae is dependent on the availability of light, carbon dioxide and mineral nutrients. Under standardised test conditions the mineral nutrients are the limiting factor determining the maximum growth rate. If an elutriate from compost or soil contains higher concentrations of the major minerals (nitrogen, potassium and phosphorus) the growth of the algae may be increased when compared to the control test. Such a growth support may compensate for minor inhibition effects deriving from toxic substances in the elutriate and can make it impossible to detect them.

The growth of the algae should be limited either by the availability of carbon dioxide or by the light intensity or by the physiological maximum for cell reproduction. If the concentration of mineral nutrients in the test medium is high enough, so that additional minerals from samples do not increase the growth, any inhibition caused by the presence of toxic substances may become effective. But it should be considered that the mineral salt concentration in the samples can be very high. The optimum concentration could be exceeded even without addition of further nutrients and the algae may not grow optimally. If this is suspected an analysis should be made prior to the biotest.

Carbon dioxide is one more factor to be considered because its availability in the test is dependent on the pH-value of the sample. The lower the pH, the lower is the solubility of carbon dioxide. Especially elutriates from compost and soil can have very different pH values. It does make sense to adjust the pH values of all test vessels to the same level as the control.

Compost elutriates are almost always a brown colour deriving from dissolved humic substances. The colour of soil elutriates is usually light, mostly yellow ochre or pale brown. The light absorption in some compost samples could exceed an extinction value of 10 cm^{-1} at wavelengths between 400 and 500 nm. For comparison: the two absorption wavelengths of chlorophyll are at 435 and at 485 nm. Therefore deeply coloured samples may decrease the illumination intensity in the test vessels and reduce the amount of light energy. Although no toxic substances are present, the growth of the algae will be reduced.

The relationship between colour intensity (measured as light extinction at 485 nm) and algae growth can be seen from **Figure 4.4**. Deeply coloured samples with extinction values above 1 cm⁻¹ should not be used for algae tests. Since it is impossible to remove the colour from elutriate samples without changing their composition the colour effect is unavoidable. Such samples can be tested only if dilutions are made until the colour is pale enough. However, the comparability of the results with other less intensively coloured samples is limited. No general solution for that problem can be offered currently.

4.3.5 Bioassays with Luminescent Bacteria

Standardised test methods, such as ISO 11348 [58], could be used for aqueous samples and elutriates. Light emitting marine bacteria, such as *Vibrio fischeri* or *Photobacterium sp.*, are used. A defined bacterial inoculum is added to the sample solutions and the change of the intensity of the bioluminescence is measured over a period of 30 minutes. Ready to use test kits, for example LumisTox (Dr. Lange) or ToxAlert (Merck) are available and do comply with all the requirements defined in the standard methods.

Since the organisms are of marine origin the biotest should preferably be used for samples deriving from marine environments, such as sea water or marine sediment. Nevertheless, the tests are commonly used for freshwater samples as well (sewage or waste water treatment

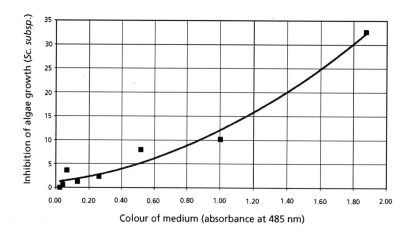

Figure 4.4 Algal growth in non-toxic but coloured mineral medium. The colour content is expressed as light extinction at 485 nm. The spectrum of the ink used was very similar to that of humic substances typically found in compost [64]

effluent), because of the short contact time they are detecting universal cytotoxic effects mainly. Sodium chloride must be added to freshwater samples and elutriates to reach a minimum salinity of 2%.

Again the colour of compost elutriates may absorb the emitted light of the bacteria and lead to improper results.

Currently no results are available from the literature for the systematic testing of biodegradable polymers which give any further helpful comments.

4.3.6 Bioassays with Daphnia

The freshwater micro-crustacean *Daphnia magna* is the most often used organism for the standardised methods OECD 202 [45] and DIN 38412-30 [50]. For a short-term test, five animals of a defined physiological state are added to about 25 cm^3 of an aqueous sample or its dilutions. The survival of the Daphnia is observed over a period of 48 hours and compared with the survivors of the control test. For a long-term test the same set up is used but the animals are fed with algae and exposed over a period of up to four weeks. The number of young Daphnia (due to reproduction) is counted and compared with those of the control test. Defined concentrations of potassium chromate are used as a positive reference to validate the required sensitivity of the animals.

Currently no research results are available for systematic investigations concerning specific influences of typical soluble substances in compost samples. Looking at the results from our own research [64] it is to be expected, that dissolved humic substances in high concentrations may inhibit Daphnia in a short-term test. Since those elutriates have contained remarkable concentrations of some heavy metals too, no generally valid interpretation could be given.

4.3.7 Evaluation of Bioassay Results Obtained from Samples of Complex Composition

Most of the standardised ecotoxicity methods do recommend the use of a synthetic or other well-known and fully defined control. That control is needed to detect the behaviour (growth, survival) of the test organisms in the absence of toxic or harmful agents. Those results are the reference values, defined as zero percent inhibition or 100% vitality and are the basis for the calculation of inhibition from the samples.

Complex samples, such as compost, soil and sediment could either inhibit or support the growth or reproduction rate of the test organisms compared to a synthetic control. The reasons for inhibition could be some physical properties, such as water holding capacity or particle size or even pH-value. The reasons for increased life activity could be organic (earthworm) or mineral (plants, algae) nutrients. In such cases it is very difficult to detect any chemical inhibition besides supporting effects if real samples, which have degraded polymers, are compared to synthetic controls.

It is therefore essential to use controls of exactly the same composition as the samples. That could be achieved, if a bigger amount of a natural matrix (equal if compost, soil or sediment) than needed for the biodegradation experiments is collected. One part is separated and treated in exactly the same way as the biodegradation experiments except that no polymers are added. Bioassays are then conducted with the samples and with the separate control at the same time.

Some examples (taken from [64]) may demonstrate the necessity for separate matrix blind tests:

- The biomass production of millet grown in a soil that has degraded wood (sawdust) compared to a synthetic reference (standardised culture substrate) was 67%. This could be expressed as 33% inhibition. But the soil sample without any additions has produced not more than 52% of plant biomass compared to the standard substrate. In relation to the untreated soil the net effect of the wood biodegradation should be seen as an increase of 28% of biomass, which is a positive effect and not an inhibition.

- The analysis of aqueous samples from a Sturm test (OECD 301B) [71] degrading industrial softeners (in this case tributyl-acetyl-citrate) has resulted in an algae growth of 117% compared to a pure mineral medium (as recommended in the OECD method number 201 [52]). No inhibition would have been the official result. The mineral medium, which has been used in the biodegradation tests, gave an algal growth rate of 169% compared to the biotest control. Such a support could be explained by the higher concentrations of mineral nutrients in the degradation test medium. Comparing the initial Sturm medium with the residuals after degradation of the softener, a net inhibition effect of 31% (or 69% relative biomass production) remains. An inhibition effect which should be considered.

As a summary, the inclusion of a control which is identical to the sample or at least representative in its chemical composition and in its physical properties is strongly recommended for all investigations about degradation intermediates and residues.

4.3.8 Testing of Sediments

Hill and co-workers [29] give a comprehensive overview of the special problems and needs for toxicity testing of sediment samples. The arguments do not need to be repeated here in detail, but it should be remembered that sediment is in permanent interaction with the water body above it. Distribution of chemicals, sedimentation of solid components and inhomogenities in the sediment layer (gradients of particle size, oxygen and biological activity) all need to be considered when designing a test and taking samples for laboratory analysis. No adjusted method is currently available for dealing with these special prerequisites. Therefore planning and running combined biodegradation and ecotoxicological tests in sediment should be done very carefully.

4.4 Special Prerequisites to be Considered when Applying Bioassays for Biodegradable Polymers

Some special situations, which may influence the result of a bioassay, are already listed in the test descriptions. To extend that list, more information about possible chemical and physical changes of the degradation matrix because of microbial activity has been collected. It appears to be of most importance to identify probable impacts on the result of a bioassay caused by reasons other than the presence of toxic residues or metabolites.

4.4.1 Nutrients in the Sample

Two controversial effects may occur; both initiated due to the microbial degradation activity:

Nutrient consumption in soil and aquatic environments (mainly nitrogen, sulfur, phosphorus, potassium and magnesium) which are essential for both the degrading microorganisms and plants or algae which should grow in the environment (see **Figures 4.5** and **4.6** for examples).

Release of elements from the polymer material (mainly nitrogen) which can act as a fertiliser for algae and plants (as nitrate) as well as be toxic (as ammonium).

The presence of nutrients and any change of their concentration will influence the growth of higher plants and algae in bioassays. It is a prerequisite to know the chemical composition of the test material. To analyse the degradation matrix for the main nutrients is always helpful for the interpretation of biotest results and for building relationships to the controls.

The concentration of ammonium can reach critical values in compost, since the nitrification process (oxidation of ammonium to nitrate) is inhibited at thermophilic conditions. It is even possible that the composting process itself will be disturbed or break down if raw

125

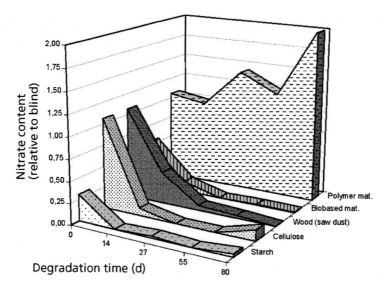

Figure 4.5 Course of the concentration changes of nitrate in agricultural soil as it appeared during the degradation of six biodegradable materials. The values are presented as changes relatively to the control experiments (control = 1). The initial concentration was 648 mg NO_3^-/kg dry matter. Derived from [67]

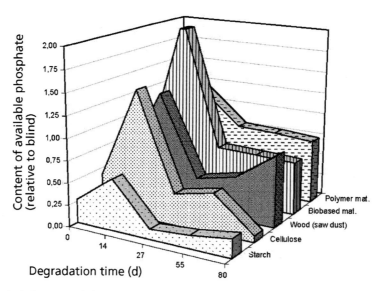

Figure 4.6 Course of the concentration changes of plant available phosphate in agricultural soil as it appeared during the degradation of six biodegradable materials. The values are presented as changes relatively to the control experiments (control = 1). The initial concentration was 262 mg PO_4^{3-}/kg dry matter. Derived from [67]

material with high nitrogen content (proteins, amides, etc.), is used. As a consequence a higher amount of undegraded organic material will be present in the samples. During the runtime of a bioassay with such a disturbed compost sample the microbial community can become active again and interact in an unpredictable way with the test organisms.

4.4.2 Biodegradation Intermediates

The microorganism flora in soil and sediment is highly diverse in variety; further aerobic as well as anaerobic species will be present at the same time. Depending on the conditions in the degradation experiment either typical aerobic or anaerobic degradation intermediates will be produced. The most critical of them are short chain fatty acids, with the main components acetic and propionic acid. Those and others will be released as intermediates of anaerobic fermentation pathways during the starting phase when the concentration of the substrate is high and will be metabolised later. The free acids are highly toxic to several organisms, to bacteria as well as to plants. If bioassays are conducted from samples containing such organic acids a strong inhibition effect will be observed in the beginning but will disappear in the course of the degradation experiment. An example is shown in **Figure 4.7**.

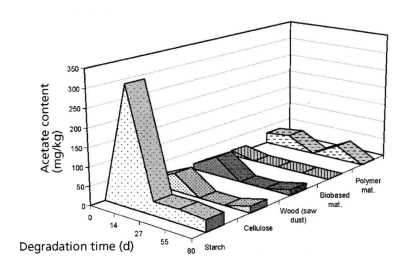

Figure 4.7 Course of the concentrations of acetate in agricultural soil as it appeared during the degradation of six biodegradable materials. Analysed from elutriates made with 0.01 M HCl. Derived from [67]

4.4.3 Diversity of the Microorganism Population

Coleman [23] states that the composition of the microorganism community in soil (but it should be valid for sediments too) changes if organic material is introduced (leave litter, crop residues or organic fertilisers). Depending on the chemical nature of the introduced material some species are supported and others will be suppressed. That is a natural phenomenon, which happens several times or at least once a year in every natural ecosystem.

If biodegradable materials are introduced to an ecosystem nothing different should happen. But consider that any shift of the population may activate certain microorganisms, which can be pathogenic to crop plants or soil animals. That whole paragraph should be seen as theoretical derivation as far as no specific literature is available dealing with such effects caused by synthetic polymers. It was intended to mention the possibility but the appearance should be confirmed or excluded either in ecosystem level tests or in real scale experiments.

4.4.4 Humic Substances

During every biodegradation process humic substances are formed, which are similar for compost, soil and aquatic environments. Suffet and MacCarthy [22] state that these are highly complex organic molecules composed of aliphatic and aromatic hydrocarbons, sugars, terpenes and other polar organic components containing elements like nitrogen, sulfur, oxygen and phosphorus. Not to go into any detail about formation, transformation and degradation of humic substances, they may interact in several ways with other components of the matrix.

Following the recommendations of Odum [72] and Korte [73] possible interactions (synergism, antagonism) between different classes of substances should be considered when applying bioassays. The presence of humic substances in environmental samples may seriously change the properties (degree of toxicity) of organic and inorganic pollutants. Rippen [28] summarises the possible influences of humic substances on the toxicity of organic pollutants against the algae *Selenastrum capricornutum* as follows (partly reformulated and extended with claims from Lee and co-workers [74]):

- Dissolved humic substances may adsorb on the cell surface and force or inhibit the uptake into the cell by changing the membrane permeability (similar to Munari [75]).

- Dissolved humic substances cause changes in the chemical structure of organic chemicals (mainly by radical reactions) forming reaction products with higher or lower toxicity (also mentioned at Suffet and MacCarthy [22]).

- Dissolved humic substances may capture the organic pollutant by chemical bonding and therefore lower its concentration or remove it from the biosystem (decrease its bioavailability).

A similar overview but focussing on the interactions between humic substances and heavy metals was found by Munari [75] and Götz [76]. The two main arguments can be used as an example to reveal the complexity of the theme:

- Dissolved humic substances can form chemical complexes with heavy metals (especially copper, lead, nickel and zinc) which derive from the soil matrix and had been originally immobilised. In the form of water soluble complexes those heavy metals are emitted into ground- and surface-water.

- Dissolved humic substances may detoxify already bioavailable heavy metals by formation of stable complexes preventing uptake into the cells.

Suffet and MacCarthy [22] give the most comprehensive overview about appearance, chemistry and properties of humic substances. They discuss influences on waste, on the drinking water preparation, on several interactions with heavy metals, on detergents and organic pollutants, reactions with light (UV-radiation) and ion exchange effects.

Humic substances deriving from aquatic and terrestrial environments are not identical in their composition and structure but very similar which is mentioned by Suffet and MacCarthy [22] as well as by Chen and Inbar [63]. Such complex chemical interactions with ecological consequences may therefore appear in all ecosystems.

Compost contains humic substances in very high concentrations. Their appearance should be given more attention since they are one of the few analytically measurable parameters and are additionally an indicator for compost maturity, as claimed by Chen and Inbar [63].

4.4.5 Evaluation of Test Results and Limits of Bioassays

If different bioassays are applied from one sample, widely varying results may be the outcome. That is a known but unsolved problem that does appear in the testing of chemicals or pesticides as well. In each single case it is to question if such variations may be caused by errors or mistakes in the practical handling of the bioassays or if they do express the different sensitivity of the used organisms. Errors and mistakes will include the effects caused by properties of a complex sample other than chemical toxicity as described previously.

Steinberg and co-workers [77] claim that with a series of single species assays a set of toxic potentials will be gained but their extrapolation to the level of an ecosystem will not be possible. The recommendation of Korte [73] to summarise single test results into an ecotoxicological profile could be set for the analysis of biodegradation effects as well.

From a set of standardised biotest results a relative index could be calculated representing something like an 'average' ecotoxic hazard. This has been common practice for the risk evaluation of chemicals and pesticides for a long time [78]. Nevertheless, some difficulties arise since the importance (the relative weight) of each single test result is not the same for different matrices and application scenarios. An example may demonstrate this: Hund [66] has found the earthworm test to be one of the most sensitive for testing of chemicals in soil. But Fritz found that the sensitivity of earthworms to be almost zero in presence of digestible substances (residues) after biodegradation tests [64].

Much more research will be necessary to characterise potential negative effects to the environment deriving from polymers and their degradation products with the needed accuracy and raggedness. What is listed here should be seen as a first step.

4.5 Research Results for Ecotoxicity Testing of Biodegradable Polymers

Very few research results have been published about this theme because such investigations are novel. Even the interpretation of available results has to be done carefully since comparisons are not possible in many cases or relationships and deductions are based on data which had been generated for purposes other than the impact analysis of biodegradable polymers.

The following paragraphs should be seen from that point of view. Data for investigations on aquatic (especially marine) environments and their sediments are not published. Nevertheless, I have tried to collect and summarise the current knowledge.

4.5.1 The Relationship Between Chemical Structure, Biodegradation Pathways and Formation of Potentially Ecotoxic Metabolites

Van der Zee [79] gives a comprehensive overview about the relationship between biodegradability and the chemical structure of several biopolymers and synthetic materials. The work does not deal directly with potential hazards of degradation residues and metabolites to the environment but gives some insight into degradation processes and prerequisites for an inherent biodegradability. The results could be extrapolated to a prediction about the appearance of possible degradation residues and their accumulation in the environment.

Scott [80] went one step further and claims relationships between the chemical composition of a polymer material and the need for deeper investigations about the environmental behaviour of degradation end products. He states that polymers containing elements other than carbon, hydrogen and oxygen should be analysed more extensively for the appearance of unwanted effects after the biodegradation process. Halogens or heavy metals, for example, which are often introduced with pigments, may form by-products during the degradation process, which are not acceptable for environmental reasons.

But even pure hydrocarbons can cause ecological problems if they have an incompatible chemical structure. The relationship between the biodegradability of aliphatic and aromatic hydrocarbons is addressed by Müller [81]. Biodegradability is not simply correlated with the degree of polymerisation; additional factors like availability of the polymeric bonds for extracellular bacterial enzymes and the content of aromatic monomers do have major influence. Further the distribution of aromatic monomers in the polymeric chain does determine the degree of biodegradability and therefore indirectly the appearance of probably toxic residues.

4.5.2 Ecotoxicity of the Polymers

Dang and co-workers [82, 83], discuss the measurement of toxic effects of biodegradable polymers. Using a cell culture test system the authors demonstrated the functional suitability of the method with four samples. The results have been determined mainly by the presence of leachable substances, effects of non soluble polymers have not been observed.

The work of Stacher [65] has demonstrated that a direct measurement of the plant toxicity of biodegradable materials is impossible. Initiated degradation processes in the test trays had caused a dramatically reduced plant growth. Those effects appeared even if natural polymers, such as starch or cellulose were used and even if the soil was sterilised before the test. Tests with not readily biodegradable polymers should be possible since the microorganism community is not activated that much. In such a case ecotoxic effects will most probably be related to an incompatibility of leachable components from the material. That outcome should be considered for the conduction of bioassays with plants and with other test species as well.

The earthworm (*Eisenia foetida*) is definitely not suitable for determining the ecotoxicity of materials that are digestible by the animals. The feeding effect will result in an increased growth, which may more than compensate potential inhibition effects [65].

Aquatic bioassays made with elutriates from polymeric materials are possible although limited to water-soluble components. The elutrition procedure should be designed properly to simulate the conditions at the natural environment of the application. Examples could be the continuous elutrition in aquatic environments or a periodic exposure at times of rainfall in terrestrial environments. However, neither a standardised nor an otherwise validated method is currently available for such investigations.

4.5.3 Ecotoxic Effects Appearing After Degradation in Compost or After Anaerobic Digestion

Both these processes for the organic recovery of waste are finally very similar although the degradation pathways are different in principle. At least both are technical processes in an artificial environment producing compost as a main product. The compost pile and the digestion sludge do not need an analysis for ecotoxic effects caused by the introduction of biodegradable polymers. Guidelines for the detection of disturbances in the processes are already covered in some of the test scheme standards. Marketable mature compost, which is used in high quantities in agriculture, is the sample to be analysed for ecotoxic effects.

The analytical detection of residues and metabolites had been possible from a laboratory degradation test using a mineral bed matrix. Tosin and co-workers [84] described such a test system and the detection of the metabolite, 4,4′-diaminodiphenylmethane (a known toxic substance) as result of the degradation of polyurethane caprolactone co-polymer. Especially the problems arising with high concentrations of organic substances in the matrix (humic substances) could be avoided. But it may be difficult to detect metabolites from a certain polymer when it is not known what to search for. The success of such methods and analysis procedures as a primary source of information may therefore not be assured. Nevertheless, such investigations could be helpful to discover the causes of ecotoxic effects already observed at other experiments.

The use of bioassays will give the most relevant information about the appearance of negative effects in compost. With all the limitations about test species and other known influences some data have been generated by Fritz and co-workers [68] analysing a set of commercially available polymers. Single results from several bioassays as well as their summary into an ecotoxicity profile can be the most proper data base to get an impression about the influence of degradation residues in the compost on a complex ecosystem. **Figure 4.8** gives a summary of the results.

4.5.4 Ecotoxic Effects Appearing During Degradation in Soil

From a theoretical point of view, inherently biodegradable polymers should not behave other than dead biomass (leave litter, wood, whole plants) which is a significant part of the natural carbon cycle. General effects of degradable substances on physical and chemical soil properties as well as on the soil ecology are described by Coleman [23]. They are almost always of positive nature distinct as a long time increase of productivity and soil fertility.

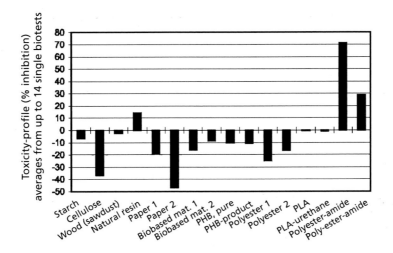

Figure 4.8 Average toxicity data obtained from laboratory composting tests in which natural and synthetic polymers have been degraded. Positive values are inhibitions; negative values are to understand as increased growth or biological activity. PHB: polyhydroxybutyrate; PLA: polylactic acid. Derived from [64]

The detection of ecotoxic effects deriving from degradation residues and metabolites is demonstrated with some examples in the **Figures 4.9** and **4.10**. Both, the initial suppression effect of readily degradable substances as well as the existence of additional ecotoxic effects could be demonstrated in those experiments. From **Figure 4.9** it can be seen that the addition of biopolymers improved the original soil quality after the biodegradation had been completed. In summary, neither the application of one single species test nor the measurement at only one time of the degradation experiment will be enough to differentiate between the both mentioned effects.

4.6 Conclusion

It is very difficult if not impossible to extrapolate the appearance of ecotoxic degradation metabolites or residues exclusively from the chemical structure of a polymer. Nevertheless, some basic guidelines concerning the presence (or absence) of heteroatoms and aromatic compounds in the polymer chain could be followed. The use of combined tests for biodegradability and ecotoxicity is strongly recommended.

Results from ecotoxicological investigations of biodegradable polymers after composting are rare. Biotest design and evaluation of results should focus on the detection of influences

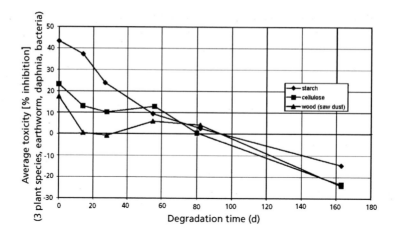

Figure 4.9 Course of the average ecotoxicity (seven single species bioassays) of three biopolymers during 160 days degradation in soil. Positive values are inhibitions and negative ones are supported growth or biological activity. Derived from [64]

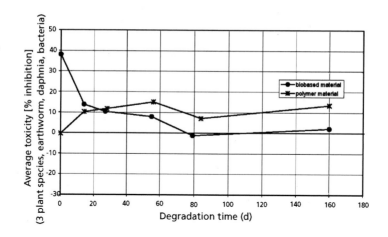

Figure 4.10 Course of the average ecotoxicity (seven single species bioassays) of two commercially available biodegradable materials during 160 days degradation in soil. Positive values are inhibitions and negative ones are supported growth or biological activity. Derived from [64]

on the compost quality that predestine its possible application and specific use. Special laboratory composting test systems, for example based on an inert mineral matrix, can be used if there is a demand for the analytical detection of toxic metabolites or residues.

Temporal inhibition of higher plants during the biodegradation of organic materials (harvest residues) in soil have been known empirically for a long time. Although they are not based on chemical toxicity of the material or its components they should be considered when bigger amounts of biodegradable polymers are applied in agriculture. Those temporal effects as well as an inhibition based on other reasons could be detected and distinguished from each other in laboratory degradation trials. The experiments should run for at least three months with samples taken periodically for chemical analysis and bioassays. If there is any doubt about the results obtained, field studies should also be used.

A crucial fact for all considerations and investigations about ecotoxicity of biodegradable polymers are synergistic and antagonistic effects between the polymer, the degradation intermediates, the residues, the formed metabolites and the matrix (degradation environment). One of the better described, although not fully understood, is the interaction of organic or inorganic pollutants with humic substances.

4.6.1 Consequences for Test Schemes for Investigations on Biodegradable Polymers

Probable ecotoxic effects arising from the biodegradation of polymers are rarely predictable from theoretical approaches. Furthermore the routine chemical analysis for quality control of compost and soil will not be suitable to detect all possible effects harmful for the environment. Bioassays will be necessary to supplement the other analyses and to complete the information about the environmental behaviour of biodegradable polymers. That need is already expressed in the inclusion of mandatory bioassays in the relevant standards for compostable products.

In the following section, I have attempted to formulate a recommendation for an extended test scheme based on the existing standards and on the rare research data available.

4.6.1.1 Materials Intended for Organic Recovery

Step 1: Analysis of the components for known toxic or harmful substances; search for polymers which do biodegrade slowly or are not biodegradable (theoretical approach from the product data sheet); search for hetero-atoms (chloride, nitrogen, phosphorus, and sulfur, etc.), and aromatic hydrocarbons (again a theoretical approach but analysis is possible).

Step 2: Determination of the biodegradability (already standardised).

Step 3: Determination of the material disintegration (standardised); compost batches which are intended for further analysis should run at least 90 days (independent from the factual time needed for complete biodegradation) and addition of a mesophilic maturation phase of at least three weeks.

Step 4: Compost analysis for conventional parameters determining any quality change compared to the blind and ecotoxicity test with at least two plant species; if negative influences are detected additional bioassays should be done using more plant species, daphnia, algae, luminescent bacteria or special local species.

4.6.1.2 Materials Intended for Applications in the Environment

Step 1: (as before), material analysis and search for critical contents.

Step 2: Determination of the biodegradability under the same conditions as predominating at the environment for the intended application; for example modified Sturm test for applications in aquatic ecosystems and soil degradation test for applications in agriculture.

Step 3: Determination of the material disintegration under relevant conditions (same as for step 2) with a batch size of at least 2 litres (liquid) or 20 kg (solid); artificial or natural sediments should be included at aquatic test systems.

Step 4: Determination of ecotoxic effects using bioassays compatible with the degradation environment; for example three plant species for soil applications or daphnia and two algae species for freshwater applications or marine algae and luminescent bacteria for marine applications.

4.6.2 Conclusion

Remarkable ecotoxic effects caused by the biodegradation of polymers have appeared rarely in all reviewed papers and reports. In general a very small hazard potential could be assumed. An extensive investigation, applied before the market introduction of a newly designed or composed material, may act as safety net to avoid unwanted negative effects on ecosystems. The additional cost of such analysis will be paid back by the confidence of the material users and waste treatment plant operators.

The currently available standards provide an analysis scheme and limit values for a general acceptance of biodegradable materials, which are intended for waste utilisation. Additional standards dealing with material applications in the environment will follow soon. All those guidelines should be understood as minimal requirements and could and will be extended in future revisions.

Hurlbert [85] states: 'Even a small reduction of the growth rate of an organism in a laboratory test, which may be interpreted as unimportant, may lead to the disappearance of a species in a natural ecosystem.' Although that statement sounds conservative it should be recognised before any introduction of artificial substances into nature.

The current development state of bioassays for compost and soil samples does not allow to differentiate between effects in the range of some single percents. More research will be needed to fully discover the complex chemical reactions involved in the biodegradation process of artificial polymers. In the meanwhile the material producers should spend much attention on the environmental behaviour of their products. The goal is to avoid any appearance of negative effects to the environment caused by a practical application in bigger scale. Such accidents could have precedence character and may hurt the image of the whole category of materials for years.

A broad use of biodegradable materials should replace conventional, not degradable and therefore waste, entailing plastics in all variants of possible applications. That new generation of products could be one of the many puzzle pieces to harmonise modern life style and protection of nature - if some essential requirements are fulfilled.

References

1. DIN 54900, Parts 1-4, *Testing of the Compostability of Plastics*, 1998.

2. ASTM D6002-96 (2002), *Standard Guide for Assessing the Compostability of Environmentally Degradable Plastics*, 2002.

3. EN 13432, *Packaging - Requirements for packaging recoverable through composting and biodegradation - Test scheme and evaluation criteria for the final acceptance of packaging*, 2000.

4. *Compost Quality Assurance According to the RAL-Quality Mark No. 251 (Kompost, Gütesicherung nach RAL Gütezeichen 251)*, Ed., Deutsches Institut für Gütesicherung und Kennzeichnung eV, Beuth, Berlin, Germany, 1998. [In German]

5. F.E. Allison, *Soil Organic Matter and its Role in Crop Production*, Elsevier Scientific Publications, Amsterdam, The Netherlands, 1973.

6. H. Voelker, *Influence of compost to the microbiological activity and to the humic substances in the soil (Einfluss von Kompost auf die Mikrobiogische Aktivitat und das Huminstoffsystem des Bodens)*, the University of Trier, Germany, 1990. [Ph.D. Thesis] [In German]

7. R. Gottschall, *Composting - Optimised Preparation and Use of Biogenic Materials in Organic Farming (Kompostierung - Optimale Aufbereitung und Organischer Materialien im Ökologischen Landbau)*, Verlag C.F. Müller, Karlsruhe, Germany, 1992. [In German]

8. W. Hartl, *Proceedings of the Austrian compost quality assurance symposium (Tagungsband Kompostgutesicherung in Österreich)*, Vienna, 1996. [In German]

9. G.S. Sekhon and O.P. Meelu, *Stressed Ecosystems and Sustainable Agriculture*, Ed., S.M. Virmani, J.C. Katyal, H. Eswaran and I.P. Abrol, Oxford and IBH Publishing Co. Pvt. Ltd., New Delhi, India, 1995, 231.

10. M. Hope-Simpson, *Composting of Waxed Corrugated and Post-Household Boxboard Container Waste*, Final Report of the ecological agriculture project of the Paper & Paperboard Packaging Environmental Council, McGill University, Quebec, Canada, 1992.

11. H. Insam, *Test Methods for Proving the Compostability of Packaging - Literature and Recommendations (Tests zur Feststellung der Kompostierbarkeit von Verpackungsmaterialien - Literaturübersicht und Empfehlungen)*, Ministry for Environment, Youth and Family, Vienna, Austria, 1994. [In German]

12. U. Pagga, *Journal of Environmental Polymer Degradation*, 1996, **4**, 3, 173.

13. O.H. Danneberg, *Formation of Humic Substances and Reactions of Nitrogen during the Composting of Maize Straw (Huminstoffaufbau und Stickstoffumbau Während der Rotte von Maisstroh)*, the University for Agriculture, Vienna, Austria, 1970. [PhD. Thesis] [In German]

14. W.A. Dick and E.L. McCoy in *Science and Engineering of Composting - Design, Environmental, Microbiological and Utilization Aspects*, Ed., H.A.J. Hoitink and H.M. Keener, Renaissance Publications, Worthington, MA, USA, 1993, 621.

15. S. Knafl, *Humic Substances in Soils of Conventional and Organic Farming (Huminstoffe in Konventionell und Organisch Biologisch Bewirtschafteten Böden)*, the University for Agriculture, Vienna, Austria, 1989. [Diploma Thesis] [In German]

16. J.M. Lynch, K.B. Gunn and L.M. Panting, *Plant and Soil*, 1980, **56**, 93.

17. T.A. Toussan, A.R. Weinhold, R.G. Linderman and Z.A. Patrick, *Phytopathology*, 1968, **58**, 1, 41.

138

18. N.S. Subba Rao, *Soil Microorganisms and Plant Growth*, Science Publishers Inc., Enfield, CT, USA, 1995.

19. B.J. Alloway and D.C. Ayres, *Pollutants in the Environment - Chemicals Principles for Evaluation of Air, Water and Soil Contamination (Schadstoffe in der Umwelt - chemische Grundlagen zur Beurteilung von Luft-, Wasser- und Bodenverunreinigungen)*, Spektrum Akademischer Verlag, Heidelberg, Germany, 1996. [In German]

20. U. Förstner, C. Colombi and R. Kistler, *Metals and their Compounds in the Environment: Occurrence, Analysis and Biological Relevance*, VCH, New York, NY, USA, 1991.

21. R.J. Scrudato, W. McDowell, R. Hinrichs and J. Pagano, *Proceedings of the International Conference of Heavy Metals in the Environment*, Genf, Switzerland, 1989.

22. *Aquatic Humic Substances: Influence on Fate and Treatment of Pollutants*, Eds., I.H. Suffet and P. MacCarthy, American Chemical Society, Washington, DC, USA, 1989.

23. D.C. Coleman and D.A. Crossley, *Fundamentals of Soil Ecology*, Academic Press Limited, London, UK, 1996.

24. P. Calow, *Handbook of Ecotoxicology*, Blackwell Science Ltd., Oxford, UK, 1993.

25. K. Fent, *Ecotoxicology - Environmental Chemistry, Toxicology, Ecology (Ökotoxikologie - Umweltchemie, Toxikologie, Ökologie)*, Georg Thieme Verlag, Stuttgart, Germany, 1998. [In German]

26. V.E. Forbes and T.L. Forbes, *Ecotoxicology in Theory and Practice*, Chapman & Hall, London, UK, 1994.

27. W.G. Landis and M-H. Yu, *Introduction to Environmental Toxicology: Impacts of Chemicals Upon Ecological Systems*, Lewis Publishers, Boca Raton, FL, USA, 1999.

28. G. Rippen, *Handbook of Environmental Chemicals (loose-leaf book) (Handbuch Umweltchemikalien)*, (Loseblatt-Ausgabe), Ecomed-Verlag, Landsberg/Lech, Germany, 1984-2001.

29. I.R. Hill, P. Matthiessen and F. Heimbach, *Guidance Document on Sediment Toxicity Tests and Bioassays for Freshwater and Marine Environments*, Society of Environmental Toxicology and Chemistry, Renesse, The Netherlands, 1993.

30. D. Freitag, H. Geyer, A. Kraus, R. Viswanathan, D. Kotzias, A. Attar, A. Klein and F. Korte, *Ecotoxicology and Environmental Safety*, 1982, 6, 60.

31. DIN 38414-4, *German Standard Methods for the Examination of Water, Waste Water and Sludge; Sludge and Sediments (Group S); Determination of Leachability by Water (S4)*, 1984.

32. OECD Test No.208, *Terrestrial Plants, Growth Test*, 1984.

33. ISO 11269-1 and 2, *Soil Quality - Determination of the Effects of Pollutants on Soil Flora*, 1995.

34. *Lemna Growth Inhibition Test*, Draft OECD test guideline, OECD, Paris, France, 1998.

35. OECD Test No.203, *Fish, Acute Toxicity Test*, 1992.

36. OECD Test No.204, *Fish, Prolonged Toxicity Test: 14-Day Study*, 1984.

37. OECD Test No.210, *Fish, Early-Life Stage Toxicity Test*, 1992.

38. ISO 7346 1-3, *Water Quality - Determination of the Acute Lethal Toxicity of Substances to a Freshwater Fish Brachydanio rerio Hamilton Buchanan (Teleostei, cprinidae)*, 1996.

39. DIN 38412-15, *German Standard Methods for the Examination of Water, Waste Water and Sludge; Bioasays (Group L): Determination of the Effects of Substances in Water on Fish - Test L15*, 1982.

40. OECD Test No.207, *Earthworm, Acute Toxicity Test*, 1984.

41. ISO 11268 1-3, *Soil Quality - Effects of Pollutants on Earthworms (Eisenia fetida)*, 1997.

42. ISO 11267, *Soil Quality - Inhibition of Reproduction of Collembola (Folsomia candida) by Soil Pollutants*, 1999.

43. A. Berthold and T. Jakl, *Journal of Soils and Sediments*, 2002, 2, 4, 179.

44. DEV L10, *Determination of the Inhibition of Toxic Wastewater to Protozoa*, 1968.

45. OECD Test No.202, *Daphnia sp. Acute Immobilisation Test and Reproduction Test*, 2004.

46. OECD Test No.211, *Daphnia magna Reproduction Test*, 1998.

47. ISO 6341, *Water Quality - Determination of the Inhibition of the Mobility of Daphnia magna Straus (Cladocera, Crustacea), Acute Toxicity Test*, 1998.

48. ISO 10706, *Water Quality - Determination of Long Term Toxicity of Substances to Daphnia magna Straus (Cladocera, Crustacea)*, 2000.

49. DIN EN ISO 5667-16, *Sampling of Water - Part 16: Guidance on Biotesting of Samples*, 1998.

50. DIN 38412-30, *German Standard Methods for Examination of Water, Waste Water and Sludge Bioassays (Group L); Determining the Tolerance of Daphnia to the Toxicity of Waste Water by Way of a Dilution Series (L30)*, 1989.

51. ISO 14669, *Water Quality - Determination of Acute Lethal Toxicity to Marine Copepods (Cladocera, Crustacea)*, 1999.

52. OECD Test No.201, *Alga, Growth Inhibition Test*, 1981.

53. ISO 8692, *Water Quality - Fresh Water Algal Growth Inhibition Test with Scenedesmus subspicatus and Selenastrum capricornutum*, 1989.

54. DIN 38412-33, *German Standard Methods for the Examination of Water, Waste Water and Sludge; Bioassays (Group L); Determining the Tolerance of Green Algae to the Toxicity of Waste Water (Scenedesmus Chlorophyll Fluorescence Test) by Way of Dilution Series (L33)*, 1991.

55. ISO 10253, *Water Quality - Marine Algal Growth Inhibition Test with Skeletonema costatum and Phaeodactylum tricornutum*, 1995.

56. DIN 38412, *Growth Inhibition Test using the Marine Algae Skeletonema costatum and Phaeodactylum tricornutum (L45)*, 1998.

57. ISO 10712, *Water Quality - Pseudomonas putida Growth Inhibition Test (Pseudomonas Cell Multiplication Inhibition Test)*, 1995.

58. ISO 11348 1-3, *Water Quality - Determination of the Inhibitory Effect of Water Samples on the Light Emission of Vibrio fischeri (Luminescent Bacteria Test)*, 1998.

59. OECD Test No.216, *Soil Microorganisms: Nitrogen Transformation Test*, 2000.

60. OECD Test No.217, *Soil Microorganisms: Carbon Transformation Test*, 2000.

61. ISO 9509, *Water Quality – Biological Methods – Method for Assessing the Inhibition of Nitrification of Activated Sludge Microorganisms by Chemicals and Waste Waters*, 1989.

62. OECD Test No.471 - 486, *Section 4: Genetic Toxicology Testing*, 1997.

63. Y. Chen and Y. Inbar, *Science and Engineering of Composting - Design, Environmental, Microbiological and Utilization Aspects*, Ed., H.A.J. Hoitink and H.M. Keener, Renaissance Publications, Worthington, MA, USA, 1993, 551.

64. J. Fritz, *Ecotoxicity of Biogenic Materials During and After their Biodegradation*, the University for Agriculture, Vienna, Austria, 1999. [Ph.D. Thesis]

65. C. Stacher, *Ecotoxicology of Biodegradable Materials*, the University for Agriculture, Vienna, Austria, 1998. [PhD Thesis]

66. K. Hund, *Development of Biological Test Systems for Identifying Soil (Entwicklung von Biologischen Testsystemen zur Kennzeichnung der Bodenqualität)*, Study No. 94-011, Environmental Protection Agency, Berlin, Germany, 1994. [In German]

67. J. Fritz, U. Link, C. Stacher and R. Braun, *Proceedings of the International ORBIT Conference on Biodegradable Polymers*, Wolfsburg, Germany, 2000.

68. J. Fritz, U. Link and R. Braun, *Starch/Stärke*, 2001, 53, 3-4, 105.

69. E. Kusel-Fetzmann, *Comparing Ecotoxicity Aanalysis of Sselected Pullutants using Algae as Indicator Organisms (Vergleichende Toxizitätsbestimmungen Ausgewählter Schadstoffe Mittels Algen als Indikatororganismen)*, Ministry for Agriculture and Forestry, Vienna, Austria, 1989. [In German]

70. *Report on the Assessment of Potential Environmental Effects of Chemicals - The Effects on Organisms Other than Man and on Ecosystems*, OECD, Paris, France, 1980.

71. OECO Test No. 301B, *Section 3: Degradation and Accumulation - CO_2 Evolution Test*, 1992.

72. E.P. Odum, *Fundamentals of Ecology (Grundlagen der Ökologie)*, Thieme-Verlag, Stuttgart, Germany, 1983. [In German]

73. F. Korte, *Textbook of Ecological Chemistry (Lehrbuch der Ökologischen Chemie)*, Thieme-Verlag, Stuttgart, Germany, 1992. [In German]

74. S.K. Lee, D. Freitag, C. Steinberg, A. Kettrup and Y.H. Kim, *Water Research*, 1993, **27**, 199.

75. M. Munari, *Evaluation of the Toxic Effect of the Herbicide Atrazine to Freshwater Green Algae Alone and in Combination with Humic Acid (Beurteilung der toxischen Wirkung des Herbizides Atrazin auf Süßwasser Grünalgen (Sel. cap.) alleine und in Kombination mit Huminsäuren)*, the University for Agriculture, Vienna, Austria, 1987. [PhD Thesis] [In German]

76. R. Götz, *Forum Städte-Hygiene*, 1978, **29**, 33.

77. C. Steinberg, J. Klein and R. Brüggemann, *Ecotoxicological Test Methods (Ökotoxikologische Testverfahren)*, Ecomed Verlag, Landsberg/Lech, Germany, 1995. [In German]

78. *Environmental Test Methods and Guidelines*, Environmental Protection Agency, http://www.epa.gov/epahome/Standards.html.

79. M. van der Zee, *Structure-Biodegradability Relationships of Polymeric Materials*, the University Twente, The Netherlands, 1997. [Ph.D. Thesis]

80. G. Scott, *Polymers and the Environment*, The Royal Society of Chemistry, Cambridge, UK, 1999.

81. R-J. Müller, *Mechanistic Studies on the Biodegradation of Polyester*, Ed., W. Sand, DECHEMA Monographs Vol. 133, Hamburg, Germany, 1996, 211.

82. M.H. Dang, F. Birchler, K. Ruffieux and E. Wintermantel, *Journal of Environmental Polymer Degradation*, 1996, 4, 3, 197.

83. M.H. Dang, F. Birchler and E. Wintermantel, *Journal of Environmental Polymer Degradation*, 1997, 5, 1, 49.

84. M. Tosin, F. Degli-Innocenti and C. Bastioli, *Journal of Environmental Polymer Degradation*, 1998, **6**, 2, 79.

85. S.H. Hurlbert, *Research Review*, 1975, 57, 82.

5 International and National Norms on Biodegradability and Certification Procedures

Bruno De Wilde

5.1 Introduction

In contrast to other novel materials or processes, normalisation of testing procedures, characteristics and requirements immediately played an important role in the development of biodegradable materials. An important reason for this is the fact that a significant benefit attributed to these materials, namely the biodegradation, cannot, or at best only with a lot of difficulty, be checked by the client himself. For many other materials the customer can easily check a novelty or an improvement compared to an earlier generation product and the need for independent verification is needed much less.

Further reasons for the importance of normalisation are the dubious or even outright false claims on biodegradability which have been made throughout the history of biodegradable materials up until today. This creates confusion and distrust and illustrates the need for objective and impartial judgement. In the early 1990s this distrust was illustrated in a publication by Greenpeace [1] in which degradable plastics were heavily criticised. An early milestone in these developments was a publication [2] in 1990 from a group of US State Attorneys General about responsible environmental advertising, in which the urgent need for 'Standards' was one of the main recommendations.

The development of national and international norms on biodegradability proved not to be an easy task as it necessitated the bringing together of biological processes, e.g., composting, in all their complexity on one hand and polymer chemistry on the other hand. Yet, in the last 10 years significant progress has been made and several standards and norms were developed as will be discussed further in this chapter.

The first function of a norm is to act as an arbitrator and make sure that uniform, unbiased and scientifically correct rules are used. However, norms also have a second, important function as it facilitates communication between producers, authorities and consumers (**Figure 5.1**). This is even more so when international norms can be established and national borders can more easily be crossed.

145

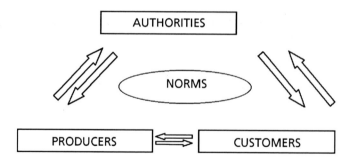

Figure 5.1 Norms as tool of communication

One of the pioneer organisations, which made serious efforts towards standardisation of biodegradation, is the The Organisation for Economic Co-operation and Development (OECD) which published a first series of standardised tests in 1981 with the *OECD Guidelines for the Testing of Chemicals* [3]. Besides biodegradation tests, these guidelines include various other test procedures such as physico-chemical properties, effects on biotic systems, bioaccumulation and health effects.

While the OECD tests form a good reference and the basic principles of biodegradation tests are similar for all environments and all test products, the OECD tests proved to be insufficient to answer all questions regarding biodegradable materials. First of all, they dealt with aquatic environments only and did not cover a composting environment. Yet, the major applications for biodegradable plastics are in products ending up in composting piles. It quickly became clear that a composting environment is a much more biologically aggressive environment compared to water. A second reason for the insufficiency of the OECD guidelines was related to the type of test item. The OECD biodegradation tests were developed for low molecular weight chemicals and proved to be less suited for high molecular weight materials such as polymers. Historically, one of the first products to be scrutinised for biodegradation was detergents and their components.

In the late 1980s and the early 1990s the American Society for Testing and Materials (ASTM) took the lead in the development of standards for biodegradable plastics and for some years worked hard on this topic. Some test procedures became definite norms by 1992-1994. After a few years a working group within Comité Européen de Normalisation – European Committee for Standardisation (CEN) and another one within Deutsches Institut für Normung – German Normalisation Institute (DIN) became focal points. A lot of brainstorming and development of norms and standards for biodegradable materials took place during several meetings per year. Whereas ASTM was the first group to develop some new test procedures, the European groups were the first to start considering and defining criteria and pass levels, more in particular on compostability. In 1998 a German pre-norm (DIN V 54900) [4] was published, followed in 2000 by a CEN norm on compostability (EN 13432) [5].

Parallel to these regional or national initiatives, efforts were also made on a global, worldwide level in order to develop international norms for biodegradable materials. Within International Organisation for Standardisation (ISO) a working group was created dealing with biodegradable plastics and various standard test methods in the meantime have been definitely adopted as ISO standards.

By the end of the 1990s, when standards and criteria were in place, the first certification systems on compostability were started, e.g., OK Compost, DIN-Certco. Through certification systems an independent, external organisation is making sure that the previously mentioned norms are used correctly. This should facilitate the market introduction of biodegradable materials even further.

5.2 Organisations for Standardisation

An overview of the most important normalisation institutes in the field of biodegradable materials is given in **Table 5.1**, including name, address, geographical spread and membership structure. Mostly these institutes work on a voluntary basis on the initiative of industry. In many instances however, the originally voluntary industry standards are subsequently taken over in legislation and other government documents.

Although most nations have their own standardisation body, a clear trend towards globalisation and internationalisation can be noticed. ISO standards are automatically transformed into national standards. The Vienna treaty makes sure there is an agreement between ISO and CEN standards. National initiatives are limited.

With regard to normalisation in the field of biodegradable plastics, some relevant working groups are:

*   ISO TC 61/SC 5/WG 22 – Plastics – Biodegradability

*   ASTM D 20.96 – Degradable Plastics

*   CEN TC 261/SC 4/WG 2 – Packaging – Degradability and Organic Recovery

*   CEN TC 249/WG 9 – Plastics – Degradability

*   DIN FNK-AA 103.3 – Biodegradable plastics
(Note: TC = technical committee, SC = subcommittee and WG = working group)

Table 5.1 Overview of normalisation institutes

Name (short)	Name (long)	Geographical spread	Address	Characteristics
ASTM	American Society for Testing and Materials	USA/Canada	100 Barr Harbor Drive West Conshohocken PA, 19428-2959 USA http://www.astm.org	Open, fee-based membership
CEN	Comité Européen de Normalisation (European Committee for Standardisation)	EU and EFTA countries and Czech Republic (EFTA = Iceland, Norway, Switzerland)	36, Rue de Stassart B-1050 Brussels Belgium http://www.cenorm.be	Limited membership through national standardisation bodies (delegation)
DIN	Deutsches Institut für Normung eV	Germany	Burggrafenstrasse, 6 D-10787 Berlin Germany http://www.din.de	Open, fee-based membership
ISO	International Organisation for Standardisation	Worldwide	1, rue de Varembleé Case Postal 56 CH-1211 Genève Switzerland http://www.iso.org	Limited membership through national standardisation bodies (delegation)
JIS	Japanese Institute for Standardisation	Japan	4-1-24 Akasaka Minato-ku Tokyo 107-8440 Japan http://www.jsa.or.jp	Depending on Ministry of Trade and Industry (MITI)
OECD	Organisation for Economic Co-operation and Development	OECD countries	2, rue André-Pascal F-75775 Paris Cédex 16, France http://webnet1.oecd.org	Limited membership through national OECD co-ordinator

EFTA: European Fair Trade Association

Typically, normalisation organisations are primarily organised by the different types of materials that are dealt with. In most cases biodegradability norms have been developed by working groups working in the material category of plastics and can therefore strictly speaking only be applied to plastics. An exception is made by the CEN TC 261 working group which comes under packaging and therefore not only including plastics but also paper and cardboard packaging, packaging from natural materials, etc.

Each standardisation organisation has its own set of regulations on how a norm is precisely developed and finally approved. However, in all cases it boils down to the same fundamental chronology. At first a proposal is introduced by a member and is then eventually approved as a working item if sufficiently supported. In a next phase the working group that is supposedly composed of experts, elaborates a 'committee draft' that it has to approve by consensus or by majority vote depending on the specific regulation. Ideally, at this stage a round-robin test is also performed or other scientific evidence is obtained to evaluate the reliability of the proposed test procedure. The committee draft is then sent to the members of the standardisation organisation and eventually outside observers for comments and/or approval by higher echelons. A distinction is made between editorial and technical comments. In a next step the comments are discussed in the working group, approved or disapproved and the necessary changes are made to the proposal. This new proposal, (e.g., DIS or draft international standard at ISO; prEN or preliminary European norm), is again sent to outside members for a new revision and commenting. The document can be sent between the working group and the outside members several times (mostly 1-3) until a sufficient number of people endorse the proposal. Finally, the norm is finished when it is officially published by the standardisation organisation. Subsequently, it is easily available from the international standardisation organisation or from the national mirror organisations.

An important note is that it is possible to review or update the standard after a few years. In some organisations this review is even automatically included in the procedures. Likewise there is the possibility of adapting standards to the latest developments on technical or normative level, e.g., streamlining a national norm with an international norm.

5.3 Norms

Both the rate and the maximum level of biodegradation of a specific material are very much determined by the environmental niche in which the material is to be disposed of. These environmental niches can differ with regard to:

- moisture content: ranging from water to high-solids

- oxygen availability: aerobic or anaerobic

- temperature: e.g., high in compost, ambient to low in soil and water

- concentration of microorganisms: e.g., high in wastewater treatment plant, low in open sea

- salt concentration

An overview of the different environmental niches in which a material can end its life is given in **Figure 5.2**. Because of the difference in biodegradation characteristics of a given material related to the environmental niche, it was necessary to develop different standards for test procedures and acceptance criteria for each niche individually.

Most norms and standards are dealing with a testing procedure, e.g., measurement of biodegradation by following a certain parameter in a given environment. In such cases, the main purpose of the norm is to harmonise test conditions, e.g., temperature, nutrients, pH, concentration of test substance, concentration of inoculum, etc. Some norms however are related to required properties for a certain characteristic and the necessary criteria and pass levels. Typical examples are the compostability norms, which comprise several aspects for which each specific criteria and pass levels are defined.

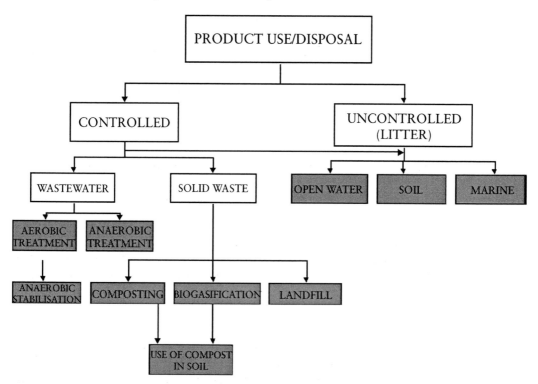

Figure 5.2 Overview different environmental niches for end-of-life

5.3.1 Aquatic, Aerobic Biodegradation Tests

5.3.1.1 Based on Carbon Conversion ('Sturm' Test)

One of the earliest and best known biodegradation tests is the aquatic, aerobic biodegradation test in which the conversion of carbon to CO_2 is measured and used to calculate the percentage of biodegradation. A schematic overview of the biodegradation reactions under aerobic, respectively anaerobic conditions is given in **Figure 5.3**. From Equation 5.1 in **Figure 5.3** it can be seen that the organic carbon in the test item or parent compound is primarily converted into CO_2. A minor part is converted into microbial carbon (the so-called biomass yield or $C_{biomass}$ in the equation). The biomass yield is typically between 10% and 40% depending on the substrate which is converted. The $C_{residual}$ consists of (partially) undegraded test item/parent compound but can also be metabolites (to be considered as in-between degradation products). Likewise it could be further split into $C_{polymer}$ and $C_{metabolite}$. In any case the $C_{residual}$ cannot be considered as being fully biodegraded.

Aerobic biodegradation:

$$C_{polymer} + O_2 \Rightarrow CO_2 + H_2O + C_{residual} + C_{biomass} \tag{5.1}$$

Anaerobic biodegradation:

$$C_{polymer} \Rightarrow CH_4 + CO_2 + C_{residual} + C_{biomass} \tag{5.2}$$

Figure 5.3 Equations for aerobic and anaerobic biodegradation

This test is also known under its more popular name of 'Sturm' test after the scientist who developed this test and wrote the first publication on it [6]. The test later became standardised at the OECD level (OECD #301B) [7] and ISO level (ISO 9439) [8]. Yet, these two test procedures were written for low-molecular-weight compounds to be tested at low concentrations and were less suitable for biodegradable polymers with a high molecular weight.

For this latter field of biodegradable materials such as bioplastics or packaging, the test procedure was slightly modified and 'officialised' under the form of ISO 14852 [9]. This norm was developed by the working group on biodegradable plastics ISO TC 61/SC 5/WG 22 and published in 1999.

- *ISO 14852 – Determination of the ultimate aerobic biodegradability of plastic materials in an aqueous medium – Method by analysis of evolved carbon dioxide*

Principle: The test item is placed in an aqueous mineral medium, spiked with inoculum and incubated under batch conditions (this is a single, 'one-shot' feeding). The test item is the sole source of organic carbon and energy. The mineral medium provides the necessary nutrients (nitrogen, phosphorus, potassium, macro- and micro-nutrients) and buffering capacity (avoiding noxious pH shifts). The inoculum can be either activated sewage sludge, compost eluate, soil eluate or a combination of two or three of these. The mixture is incubated at constant temperature and continuously stirred and aerated with carbon dioxide-free air. The temperature can be ambient (20-25 °C), mesophilic (30-40 °C) or thermophilic (50-60 °C). The duration of the test is not really specified. The test should be run until a 'plateau in activity' is reached. In practice a minimum duration is typically four weeks. The maximum duration, in contrast, is defined precisely in the norm and is six months. Carbon dioxide is trapped in an alkaline solution and quantified by titration or dissolved inorganic carbon (DIC) measurement. The percentage of biodegradation is determined by the amount of carbon in the test item that is converted to carbon dioxide. Depending on the frequency of carbon dioxide determinations the rate of biodegradation can also be established.

The ISO 14852 norm also explicitly mentions the possibility and a procedure for biomass determination. The procedure is based on determination of protein and some assumptions regarding protein and carbon content of biomass. The determination of biomass makes it possible to further complete the biodegradation equation (**Figure 5.3**, Equation 5.1).

The two similar but older norms, OECD 301B [7] first published in 1981 and ISO 9439 [8] first published in 1990 are different because for several aspects less flexibility is permitted. Temperature for example must be ambient (20-25 °C); inoculum must be activated sewage sludge pretreated according to a strict procedure or surface water. Also the possibility for biomass determination is not mentioned and the concentration of test item is lower.

Another norm, which is very similar, is ASTM D5209 [10] originally published in 1992 and comparable to ISO 9439. The ASTM norm is now being modified to be in line with ISO 14852. Two European norms are identical to the ISO norms: EN 29439 [11] to ISO 9439 and EN 14047 [12] to ISO 14852. The latter European norm however has expanded the field of application to packaging whereas ISO is limited in principle, to plastics only. In Japan the ISO 14852 method was transposed by the Japanese Institute for Standards (JIS) without changes into the JIS K 6951 [13].

5.3.1.2 Based on Oxygen Consumption ('MITI' Test)

Besides the 'Sturm' test another frequently cited aquatic, aerobic biodegradation test is the ISO 14851 [14], in parallel developed by the same working group ISO TC 61/SC 5/WG 22 and also published in 1999.

- *ISO 14851 – Determination of the ultimate aerobic biodegradability of plastic materials in an aqueous medium – Method by measuring the oxygen demand in a closed respirometer*

The principle of the test procedure is very similar to ISO 14852. The major and basically only difference is the parameter for measuring the biodegradation. Instead of carbon dioxide production, the oxygen consumption is measured (see Equation 5.1 in **Figure 5.3**). Further, the percentage of biodegradation is calculated by comparing the biological oxygen demand (BOD) to the chemical oxygen demand (COD), determined by chemical oxidation of the test item, or even better the theoretical oxygen demand (ThOD), calculated on the basis of the stochio-metrical formula.

Again, ISO 14851 had some predecessors, namely OECD 301C [15] and ISO 9408 [16]. Differences between the latter two procedures and ISO 14851 lie in flexibility of test conditions, source of inoculum and possibility for biomass determination. The test is often named after MITI (the Japanese Ministry of Trade and Environment) because they proposed the test to OECD. The OECD procedure prescribes the need to take inoculum from at least 10 (mostly aquatic) sources and make a mixed inoculum.

In North America a similar norm is ASTM D5271 [17] originally published in 1992. The ASTM norm has now been modified to be in line with ISO 14851. Also for this type of test, two European norms are identical to the ISO norms: EN 29408 [18] to ISO 9408 and EN 14048 [19] to ISO 14851. The latter European norm however has expanded the field of application to packaging compared to ISO being limited in principle to plastics only. In Japan the ISO 14851 was transposed without changes into the JIS K 6950 [20].

5.3.1.3 Other

Several other norms for aquatic, aerobic biodegradation tests have been published as well but are not frequently used for biodegradable polymers. Mostly the conditions of incubation (mineral medium, inoculum, temperature, concentration, etc.), are identical or at least similar to the Sturm and MITI methods. The differences lie in the parameters that are being measured and the method for calculating the biodegradation. The following different tests can be used:

- Dissolved organic carbon (DOC) die away test: OECD 301A [21], ISO 7827 [22], ASTM E1279-89 [23]. Based on the disappearance of DOC through biodegradation, likewise a prerequisite for this procedure is the solubility of the test item in water.

- Two-phase BOD test: ISO 10708 [24]. Based on oxygen consumption.

- Closed bottle test: OECD 301D [25], ISO 10707 [26]. Based on oxygen consumption, monitoring of decrease in dissolved oxygen.

- Zahn-Wellens test: OECD 302B [27] and ISO 9888 [28]. Monitoring of decrease of COD or DOC.

- Oil or lubricant biodegradation test: CEC-L-33-T-82 [29], ASTM D5864-00 [30], ASTM D6139-00 [31]. Monitoring of disappearance of parent compound.

Other aquatic, aerobic biodegradation tests are not operated under batch mode (single, one-shot feeding) but under semi-continuous or continuous conditions. In this case, the reactors are incubated for a long period and fed on regular intervals, e.g., once daily (semi-continuous) or on a continuous basis, e.g., by feeding with a pump. These tests have been designed for simulation of aquatic wastewater treatment plants, testing of products ending up in the wastewater stream, (e.g., detergents) and evaluation of long-term effects:

- Semi-continuous activated sludge (SCAS): OECD 302A [32], ISO 9887 [33], and ASTM E1625-94 (2001) [34].

- Continuous activated sludge (CAS) or coupled-units test: OECD 303A [35], ISO 11733 [36].

In both tests, monitoring of biodegradation is achieved through analysing of parent compound (COD or DOC) and its decrease.

5.3.2 Compost Biodegradation Tests

5.3.2.1 Controlled Composting Test

In the first years of the existence of biodegradable plastics, it became clear that the aerobic, aquatic biodegradation tests were not appropriate to evaluate the biodegradation of these polymers in composting. The environmental conditions are of course very different: high temperatures up to 60-65 °C in composting as opposed to ambient temperature in water, different moisture content, etc. An important difference is also the activity of fungi and actinomycetes. Whereas in water these organisms can be detected but are not really active, in compost they are dominantly present and very active. It has been known for a long time that fungi can degrade some materials much better and faster than bacteria. The biodegradation of lignin by white rot fungi is a well-known example [37, 38]. This observation led to the development of a novel biodegradation method in which a composting process was simulated as good as possible while still measuring the biodegradation based on carbon conversion very precisely [39, 40]. This test procedure became first standardised at ASTM level: ASTM D5338-92 [41]. Later, and after a few modifications it became adopted at ISO level in 1999 as ISO 14855 [42].

- *ISO 14855 – Determination of the ultimate aerobic biodegradability and disintegration of plastic materials under controlled composting conditions - Method by analysis of evolved carbon dioxide*

Principle: The test item is mixed with mature compost and incubated under batch conditions at 58 °C under optimum oxygen and moisture conditions. The mature compost acts at the same time as the carrier matrix, the source of microorganisms and the source of nutrients. The mixture is continuously aerated with carbon dioxide-free air. The exhaust air is analysed for carbon dioxide. A schematic layout of the test is given in **Figure 5.4.**

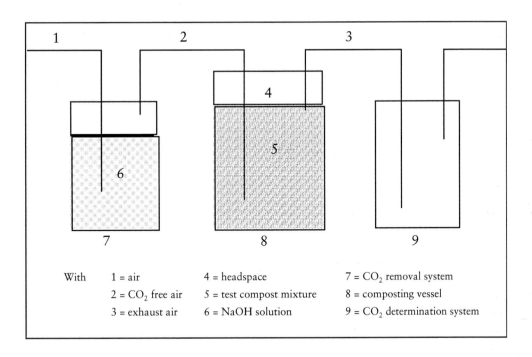

Figure 5.4 Schematic layout of controlled composting test

The maximum test duration is six months while a typical minimum duration is 45 days. The carbon dioxide produced during the composting is measured continuously or at regular intervals. After subtracting the background carbon dioxide production from the blank compost inoculum (without any extra carbon source addition), the percentage of biodegradation is determined by the net amount of carbon of the test item that is converted to carbon dioxide. Also the rate of biodegradation can be established. A positive reference control, cellulose, is tested in parallel to check the activity of the inoculum. Strict requirements are imposed on the results for cellulose to validate the test. The test item is preferably added in the form of fine powder. However, when film samples or formed products are added, the test procedure also allows an evaluation of the disintegration under composting conditions.

In **Figure 5.5** an example is given of a carbon dioxide biodegradation curve of the blank compost inoculum and the cellulose positive reference, each in three replicates. From the difference in carbon dioxide the net biodegradation can be calculated and graphically represented as in **Figure 5.6**.

The ISO procedure was adopted in a new edition of the ASTM standard, ASTM D5338-98e1 [41] and can also be found back in the European norm EN 14046 [43] which is expanding the field of application from plastics to all possible types of packaging materials. In Japan, the ISO 14855 was adopted without changes as JIS K 6953 [44].

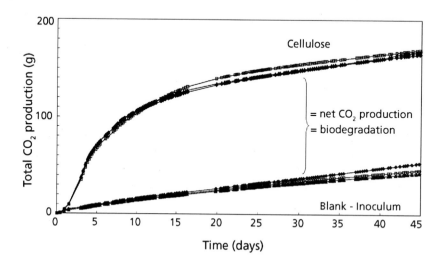

Figure 5.5 Carbon dioxide evolution curve

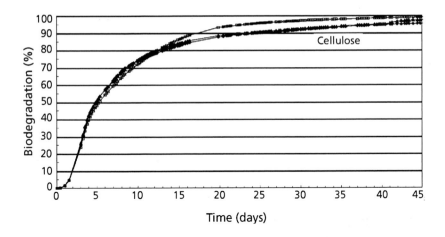

Figure 5.6 Biodegradation curve

5.3.2.2 Mineral Bed Composting Test

The two major types of tests for determining the biodegradation of plastics each show some specific advantages and specific disadvantages. The aquatic tests allow not only a measurement of biodegradation through carbon dioxide production or oxygen consumption but also the measurement of biomass and (dissolved) metabolites. However, the aquatic environment is a poor simulation of a composting environment and biodegradation is much less aggressive. In contrast, the controlled composting test is a much better simulation of the composting environment and shows a high fungal activity. Yet, because of the complex compost matrix a precise measurement of metabolites and biomass is not possible.

In an effort to combine the possibilities and advantages of both types of tests, a novel test method was developed in which the compost matrix is replaced by a mineral, inert medium [45, 46]. In this novel test procedure the carrier matrix consists of vermiculite which is a type of expanded clay mineral with an overall physical structure and water holding capacity behaviour very similar to compost. This medium is inoculated with a compost extract and brought to the right moisture content by the addition of a mineral medium, which at the same time is introducing the necessary nutrients. The physical structure and the compost eluate inoculum initiate a microbiological (aggressive) activity similar to the activity in a real composting pile. At the same time the inert matrix permits the extraction and determination of metabolites and biomass.

The absence of an extra carbon source also prevents the phenomenon of a 'priming effect' in a mineral bed test. The addition of a readily degradable and energy-rich test material in a conventional controlled composting test sometimes results in an extra stimulated activity of the compost matrix and a higher background production of carbon dioxide compared to the blank compost reactors. As a result the net carbon dioxide production in the test reactors is overestimated and a biodegradation percentage above 100% can be obtained. The absence of an extra carbon source under the form of compost in the mineral bed test prevents this extra activity or priming effect.

The novel method using the vermiculite, mineral inert matrix has been proposed at the ISO TC61/SC 5/WG 22 as an amendment to the ISO 14855 [42].

• *ISO 14855 Amendment 1: Use of a mineral bed instead of mature compost*

At the end of 2001 the proposal is at the stage of a draft amendment ('DAM 1'). Some further research and development is still needed but it is expected that the new method will be an official ISO standard within a short term.

5.3.2.3 Other Compost Biodegradation Tests

Another, more drastic approach to improve the precision of the measurement of CO_2 production and biodegradation and ascertain that effectively the CO_2 derived from a test material is determined, is by the use of ^{14}C radiolabelled test material. An aerobic aquatic test procedure as well as an aerobic, composting test procedure using such material in which the production of $^{14}CO_2$ is measured by absorption and liquid scintillation counting has been developed in the ASTM subcommittee D20.96 on degradable plastics and has been published as ASTM D6340-98 [47].

• *ASTM D6340-98 - Standard Test Methods for Determining Aerobic Biodegradation of Radiolabelled Plastic Materials in an Aqueous or Compost Environment*

Besides norms for biodegradation tests in which the biodegradation is measured directly, some exposure tests have also been standardised, e.g., ASTM D5509-96 [48] and ASTM D5512-96 [49]. In these tests plastic test items are exposed to composting conditions and these conditions are precisely defined. After a certain period of incubation a loss of property is measured, such as weight, molecular weight, tensile strength, tear resistance, etc. It may be clear that these parameters are only secondary results of biodegradation and are no proof of a complete biodegradation and mineralisation as shown in Equation 5.1 of **Figure 5.3**.

5.3.3 Compostability Norms

Because of the complex nature of many biodegradable materials such as plastics or packaging and also because of several aspects related to the composting process and to the compost as high-quality end product, the biodegradation of a test material alone is not sufficient to evaluate its overall compostability. As illustrated in **Figure 5.7** three basic conditions must be fulfilled:

1) *Biodegradation*: the complete breakdown to mineral endproducts (carbon dioxide, water) and biomass. This can be situated at the chemical, molecular level of a given test material and is therefore an inherent material characteristic. The physical form of the test material is irrelevant.

2) *Disintegration*: the degradation on a visual, physical level. The test item must physically fall apart and disintegrate into invisible particles. The physical form of the test item is essential and typically materials will be approved until a certain thickness (plastics) or a certain weight per surface (paper materials).

3) *Compost quality*: may not be negatively influenced by the addition of a biodegradable material.

Several compostability norms have been developed in different standardisation committees but in all norms these three basic requirements can be found back as the principal rationale. In some standards a fourth requirement has been added, saying that the addition of a compostable product may not hinder the normal operation of the composting process. Yet, a precise method or procedure to evaluate and approve this is missing.

5.3.3.1 EN 13432 – European Compostability Norm for Packaging Waste [5]

In 1994 the EU European Commission adopted the Packaging Waste Directive 94/62/EEC. This directive defines and imposes recovery and recycling targets for packaging waste and specifically mentions the possibility for 'organic recovery' through composting or biogasification. Yet, in 1994 no criteria were defined for a packaging material how to fulfil

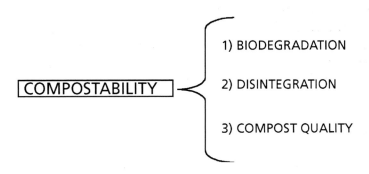

Figure 5.7 Basic requirements for compostability

the requirements for organic recovery. To solve this problem, a mandate (M 200) was given by the European Commission to CEN to develop a standard defining the requirements for organic recovery or in other words to develop a compostability norm for packaging waste. The mandate also asked for four other norms, namely with regard to reduction, reuse, (material) recycling and energy recovery (incineration) of packaging waste. Within CEN the task related to the compostability norm was assigned to CEN TC261/SC4/WG2. After several years of intensive discussions and development work a definite CEN norm was published in September 2000, EN 13432 [5].

*   *EN 13432 - Packaging – Requirements for packaging recoverable through composting and biodegradation – Test scheme and evaluation criteria for the final acceptance of packaging*

Principle: This norm specifies a scheme and combines different tests, criteria and pass levels which a packaging has to fulfil in order to be accepted as being compostable. Throughout the norm several references are made to other norms which are describing individual test procedures. In the scheme, which in an annex is further illustrated by a detailed flow chart, four major stages can be distinguished, each with specific requirements:

1) **Material characteristics:** identification of and information on different constituents; organic matter content (determined as volatile solids - must be at least 50%); heavy metals (maximum concentrations defined for 11 elements).

2) **Biodegradation:** preferably determined by ISO 14855 [42], alternatively determined by another suitable international standard such as ISO 14851 [14] or 14852 [9]. Pass level is 90% biodegradation in absolute terms or in relative terms compared to the positive reference, cellulose. Biodegradation is considered to be the sum of mineralisation and biomass formation. The maximum acceptable test duration is six months. Constituents below 1% must not be evaluated as long as the total of these 'irrelevant' constituents is below 5%. Also natural materials which are chemically not modified, must not be evaluated for biodegradation.

3) **Disintegration:** to be evaluated either in a pilot-scale composting test (ISO 16929) [50] or in a full-scale test. The test material is added in a concentration of 1% (on wet weight basis). At the end of a 12 week composting cycle a maximum of 10% of the original weight of the test material may be retrieved after sorting and careful manual selection in the >2 mm compost fraction.

4) **Compost quality:** Some physico-chemical parameters are determined and ecotoxicity tests performed to evaluate the quality of the compost. This is done by comparing a blank compost (obtained from organic waste to which no test material was added) to test compost (obtained from the same organic waste to which 10% of test material

was added at the start of the preceding pilot-scale composting test). Note: the pilot-scale composting test for measurement of disintegration can be combined with the pilot-scale test for compost preparation for ecotoxicity tests. If the compost is to be used for ecotoxicity tests, 9% of test material is added in the form of powder at start of the composting trial as well as 1% of the test material in its final form.

The compost is analysed for typical physico-chemical parameters such as pH, salt content, density, nitrogen, etc. The ecotoxicity tests include two plant tests in which the germination and the plant growth (biomass) are compared between the blank compost and the test compost. The test compost cannot show a significant negative difference with the blank compost.

About a year after its acceptance as a European norm, EN 13432 [5] was also endorsed by the European Commission. In 2001 the decision was taken to formally accept it as an *EU harmonised norm* [51]. It is interesting to note that of the five packaging waste norms required by the mandate M200 (reduction, reuse, material recycling, energetic recycling, organic recycling), three were not accepted, one was accepted with some remarks (reduction) and one was accepted without any changes (organic recycling). The upgrading to 'EU harmonised norm' status gives the norm a higher juridical value. If a packaging waste fulfils the requirements of EN 13432, it automatically also fulfils the requirements of the Packaging Waste Directive 94/62/EEC. Likewise the norm is not only a standard proposed by the CEN organisation (mainly industry) but also is the rule required by the European authorities.

5.3.3.2 DIN V 54900 [4]

Around 1992, when CEN started working on compostability, the German DIN standards organisation created a new working group, with as its main task, the development of a norm on compostability. Instead of being in the packaging group as a branch of materials, as within CEN, in Germany the group DIN FNK-AA 103.3 falls into the category of plastics. In September 1998 a document was published which however is not yet a final DIN norm but instead a Vornorm, which is like a kind of 'pre-norm' without the status of a full-fledged and definite norm. This means that changes are possible after a certain period and that a review is necessary after a few years. Nevertheless this document was the first publicly available norm on compostability containing precise criteria and pass levels. The norm is composed of three different parts:

- *DIN V 54900 – Testing of the compostability of plastics*

Part 1: Chemical Testing

Part 2: Testing of the complete biodegradability of plastics in laboratory tests

Part 3: Testing under practice relevant conditions and testing of the quality of the composts

In general the DIN V 54900 is quite similar to the EN 13432, certainly with regard to overall philosophy and approach. Yet, some distinct differences can also be noticed. Again, the norm is not only defining principles and test procedures but also specific criteria and pass levels.

The first part defines material characteristics: identification and information; organic matter content (must be at least 50%); heavy metals (maximum concentrations for seven elements, somewhat stricter than in EN 13432); in addition also the level of polychlorinated phenols (PCB) and dioxins and furanes must be determined although no maximum concentrations have been defined.

The second part on biodegradation is very similar to EN 13432, e.g., with regard to selected test methods, six-month period, irrelevant components below 1%, no need for testing natural materials. Differences are the required pass level (60% for homopolymers, 90% for heteropolymers; both in absolute terms), the acceptable total for irrelevant components (3%) and the requirement to test components between 1% and 10% separately.

The third part describes both disintegration and compost quality. For the evaluation of disintegration both a pilot-scale and a full-scale composting test are required. The pilot-scale test is comparable to ISO 16929 [50] and EN 13432 [5], the duration, concentration of test item and pass levels are exactly the same. The full-scale test must be executed in an installation of low technical level (no computer control, no forced aeration; category: Baumusterkategorie V according to the classification system of the German Composting Association [51]). The test material is introduced via net sample bags into the composting pile. The concentration of test material in the net sample bag is 1% and the pass level is less than 10% residue on weight basis in the >2 mm fraction (identical as for pilot-scale test).

The compost quality is evaluated via the standard compost quality analyses needed for certification by the Bundesgütegemeinschaft Kompost eV. This includes one plant germination and growth test, using summer barley, in a similar way as for EN 13432 [5].

In principle, the DIN prenorm should be discussed again at the working group on biodegradable plastics within a few years after its first publication in order to be transposed into a definite and full-fledged DIN norm. However, this will most probably not happen as the norm will probably be merged with or replaced by an ISO or EN norm.

5.3.3.3 ASTM D6002-96 [53] and D6400-99 [54]

Whereas in the early 1990s ASTM took the lead in developing biodegradation test methods specifically suited for plastics, the committee was a little hesitant to define criteria and pass levels for plastics to qualify as being compostable. This hesitation is illustrated by the publication of two different ASTM documents *ASTM D6002-96* and *ASTM D6400-99*.

The first document describes a standard guide without specific requirements and criteria to be met by compostable plastics. Instead it gives an overview of the various tests which are available in the field of biodegradable plastics, going from simple screening tests to field and full-scale assessment, together with some very basic and general requirements. These tests are structured in a tiered approach with Tier 1 consisting of rapid screening tests, Tier 2 lab- and pilot-scale composting assessment and Tier 3 field/full-scale assessment. Whereas this standard gives a good overview of the different tests which are available and how they relate to one another or fit within the overall picture, it does not give any precise indication or direction when a biodegradable plastic can be accepted as being compostable and when not.

The second standard however, ASTM D6400-99, is much more specific and precise with regard to the requirements which must be met by compostable plastics. In that sense, it is comparable to the EN 13432 [5] and DIN V 54900 [4] and as a matter of fact is also built around the same three basic requirements (biodegradation, disintegration and compost quality). Although the basic philosophy and rationale is similar to the European and German norm, the detailed figures and requirements show some differences.

The material characteristics are limited to criteria for heavy metals only (less severe than in EN and DIN norm), a minimum content of organic matter is not defined. For biodegradation only composting tests can be used (no aquatic biodegradation tests), the pass level is 60% for homopolymers and 90% for heteropolymers and products consisting of more than one polymer. These percentages must be seen as 'relative percentages' compared to the biodegradation of a known reference tested in the same test. The time period to reach these pass levels is 180 days for non-radiolabelled materials and 365 days for radiolabelled materials. The disintegration requirements are similar to those in the EN 13432 norm: a maximum of 10% of original dry weight may remain in the >2 mm fraction after a composting test. Finally, ecotoxicity tests include a cress germination test (without evaluation of plant growth) and another plant test in line with OECD #208 [55] involving both germination and growth.

5.3.4 Compost Disintegration Tests

A second necessary characteristic of compostable materials besides biodegradation, is the disintegration, to be determined on the level of a finished material or product. Typically, disintegration is very important when a maximum thickness of a compostable material has to be determined or when a multi-layer product of different individually compostable materials is fabricated. The disintegration has been evaluated in various tests, ranging from simple burial tests to labour-intensive full-scale tests. In the last years, the various developments and experiences have all been brought together and summarised into a procedure proposed at ISO/TC61/SC5/WG22 and which is now close to final adoption and acceptance, ISO FDIS 16929 [50].

- *ISO 16929, Plastics - Determination of the disintegration of plastic materials under defined composting conditions in a pilot-scale test.*

Principle: The test material is mixed with fresh biowaste in a precise concentration (1% for measurement of disintegration) and introduced into a pilot-scale composting bin (140 litres or larger). Composting will start spontaneously because of the presence of natural, ubiquitous microorganisms in the biowaste and will result in a temperature increase. The composting mass is regularly turned and mixed. Several parameters are followed and have to stay within certain limits to guarantee a proper and typical composting process, e.g., temperature, pH, moisture, gas composition, etc. After 12 weeks of composting, the test is terminated and the disintegration of the test item is evaluated by sieving over 10 and 2 mm. The compost obtained at the end of the test can also be used for chemical analyses and ecotoxicity tests.

Another method introduced at ISO is much simpler in execution and has to be considered as a preliminary screening test, ISO 20200 [56].

- *ISO 20200, Plastics - Determination of the disintegration of plastic materials under simulated composting conditions in a laboratory-scale test.*

Principle: In this test the sample is mixed with synthetic waste and introduced in small containers (5-20 litres) which are not actively aerated. Disintegration is evaluated by manual sorting and sieving over 10, 5 and 2 mm.

5.3.5 Soil Biodegradation Tests

Many potential applications for biodegradable plastics lie in direct use in soil. Examples are primarily agricultural products such as mulching film, clips, planting pots, etc., but also other products such as road constructions aids (for slopes), body bags, replacements for clay pigeons, etc. The biodegradation behaviour in soil is therefore an important question and cannot always be deducted from other biodegradation tests. In soil the temperature will not raise to thermophilic ranges like in composting. As some biodegradable plastics do need the 'thermal trigger' to start the first hydrolysis and the ensuing mineralisation, the biodegradation rate and percentage will be different between composting and soil conditions. On the other hand, in many instances soil will turn out to be a much more biologically aggressive environment than water because of the high fungal activity. In water fungi are present but do not exert a strong biodegradation activity. **Figure 5.8** gives an order of aggressiveness with regard to biodegradation for various environments. The marine environment is weaker than fresh water because of the much lower concentration of microorganisms.

COMPOST	>	SOIL	>	FRESH WATER	>	MARINE WATER
T + fungi + bacteria		fungi + bacteria		bacteria		dilute bacteria

Figure 5.8 Order of aggressiveness with regard to biodegradation for various environments

Various standard methods have been developed for the evaluation of biodegradation in soil. The first official test procedure was OECD 304 A [57] on inherent biodegradability in soil. ISO 11266 [58] on biodegradation of organic chemicals in soil was first published in 1994. The first norm for bioplastics was ASTM D5988-96 [59] which covers not only the biodegradation of plastics directly applied to soil but also the biodegradation of residual plastic materials after composting. Also at ISO level a method specifically for plastics is close to being finally adopted ISO 17556 [60].

* *ISO 17556, Plastics – Determination of the ultimate aerobic biodegradability in soil by measuring the oxygen demand in a respirometer or the amount of carbon dioxide evolved*

Principle: The test item is introduced into a selected soil and incubated under batch conditions at 20-25 °C under optimum oxygen and moisture conditions. The soil acts at the same time as the carrier matrix, the source of microorganisms and the source of nutrients. The maximum test duration is six months while a typical minimum duration is three-four months. Either oxygen consumption or carbon dioxide production is monitored and the net amount is calculated after subtracting the background activity of the soil. This is compared to the theoretical oxygen demand or carbon dioxide production (based on stochiometrical chemical formula) to calculate the percentage of biodegradation.

The previously-mentioned norms on soil biodegradability are only dealing with test methods for measuring the mineralisation but are not proposing any acceptance criteria and/or pass levels with regard to, for example, time frames and percentages to be achieved. Within the CEN organisation however, TC 249/WG 9, is trying to develop such criteria.

5.3.6 Aquatic, Anaerobic Biodegradation Tests

When oxygen is available in a specific environment, this is called an aerobic environment. When no oxygen is available, one talks about anaerobic conditions. Several anaerobic environments do exist, especially in places where oxygen is consumed or depleted more rapidly than it is replaced by diffusion. Examples include bottoms of rivers, canals and lakes with a lot of organic debris on the bottom; landfills; the rumen of herbivores, etc. Besides these 'natural' examples anaerobic conditions also exist in several man-controlled environments such as septic tanks, anaerobic wastewater treatment plants, sludge digesters or solid waste biogasification plants. These anaerobic environments show a high biological activity that can be quite different from aerobic conditions.

Through anaerobic biodegradation organic carbon is converted into biogas, a mixture of methane and carbon dioxide (see Equation 5.2 of **Figure 5.3**). Chemical substances or organic polymers can show a very different biodegradation pattern under anaerobic conditions compared to aerobic conditions. Likewise the need to develop separate anaerobic biodegradation tests was quickly recognised. These tests can be further subdivided into two major categories according to the moisture content: aquatic tests and high-solids or dry tests.

The aquatic, anaerobic biodegradation test was first published by the European Centre for Ecotoxicology and Toxicology of Chemicals as ECETOC Technical Report N° 28 [61]. Later, more or less the same procedure was adopted as ISO 11734 [62] in 1995. Another very similar norm is ASTM D5210-92 [63]. Within the field of bioplastics a new version with some minor modifications was developed by the ISO TC 61/SC 5/WG 22. This version, ISO/DIS 14853 [64] is close to final adoption as an international norm.

- *ISO FDIS 14853, Plastics – Determination of the ultimate anaerobic biodegradability in an aqueous system – Method by measurement of biogas production*

Principle: The test material is placed in an aqueous mineral medium, spiked with inoculum (anaerobic sludge) and incubated under batch conditions at mesophilic temperature (35 °C). The test material is the sole source of organic carbon and energy. The mineral medium provides the necessary nutrients and buffering capacity. Precautions are taken to keep the mineral medium and the reactor totally oxygen-free. The duration of the incubation is 60 days. Biodegradation is measured by following the biogas production (measured by volume displacement or pressure build-up) and the increase of DIC in the medium. The percentage of biodegradation is determined by the amount of carbon of the test material that is converted to carbon dioxide and methane. Depending on the frequency of biogas determinations the kinetics of biodegradation can also be established.

5.3.7 High-Solids, Anaerobic Biodegradation Tests

ISO/DIS 14853 [64] is representative for anaerobic wastewater treatment or for anaerobic sludge stabilisation, two systems which are always operated under aquatic conditions (moisture content > 95%) and at mesophilic temperatures. Yet, other commercial biogasification systems are being used which are working under much drier conditions (moisture content as low as 60%) and eventually also at higher thermophilic temperature (around 55 °C). These different conditions lead to different biodegradation characteristics and hence the need for a specific test procedure. For example the lower moisture content results in a much higher concentration of microorganisms and therefore a much higher biodegradation rate.

A new biodegradation test method for bioplastics was first developed by the ASTM and was accepted as ASTM D5511-94 [65]. Later, the same method was also introduced at ISO level as ISO/DIS 15985 [66].

- *ISO/DIS 15985, Plastics – Evaluation of the ultimate anaerobic biodegradability and disintegration under high-solids anaerobic-digestion conditions – Method by analysis of released biogas.*

Principle: A small amount of test material is added to a large amount of highly active inoculum that has been stabilised before the start of the biodegradation test. The inoculum consists of residue obtained directly from a high-solids biogasification unit or obtained after dewatering of anaerobic sludge. Optimal conditions for pH, nutrients, volatile fatty acids, etc., are provided and the mixture is left to ferment batch wise. The volume of biogas produced is measured and used to calculate the percentage of biodegradation based on carbon conversion.

- Landfill Simulation Tests

Another category of dry, anaerobic biodegradation tests are landfill simulation tests. These tests have primarily been developed in the USA where biologically active landfills represent a viable waste management option for the future. In Europe however, much less interest is shown for biodegradation characteristics in landfills, especially after the adoption in 1999 of the EU landfill directive, which is phasing out the disposal of biodegradable materials in landfills.

In landfill simulation tests the biological activity is much slower compared to the high-solids anaerobic digestion test due to the (much) lower concentration of microorganisms. Biodegradation is evaluated through loss of properties after exposure by ASTM D5525-94a [67], measurement of biogas production by ASTM D5526-94 [68] or monitoring of radiolabelled test materials (ASTM draft).

5.3.8 Marine Biodegradation Tests

A separate category of biodegradation tests, although receiving less attention in the field of bioplastics, is formed by the marine biodegradation tests. Yet, some promising applications of bioplastics are related to the marine environment, e.g., fishing lines, fishing nets, disposables on ships, etc.

OECD #306 [69] on biodegradability in seawater describes two test methods. The first one, the shake flask method, is based on the determination of loss of DOC and can therefore be rarely used for biodegradable plastics. The second one, the closed bottle test, is based on oxygen consumption and uses very low concentrations of test item (2 mg/l). Therefore, again, it is not really suited for biodegradable plastics.

Another test has been developed by the ASTM committee D20.96, ASTM D5437-93 [70].

- *ASTM D5437-93 Standard Practice for Weathering of Plastics under Marine Floating Exposure*

As the title clearly indicates, this norm only describes an exposure procedure in which bioplastic samples are placed in seawater. After certain time intervals, samples are retrieved and loss of property is evaluated. However, loss of property is a secondary parameter and is no proof of complete biodegradation or mineralisation as represented in Equation 5.1 of **Figure 5.3**.

Another ASTM method, ASTM D6691-01 [71] is determining the aerobic biodegradation of plastic materials in the marine environment by a defined microbial consortium. The latest development at ASTM is the inclusion of a marine variant in a new revision of the Sturm test, ASTM D5209 [10]. Yet, this project is still in development. As it looks now, it would be the first norm that determines the biodegradation of plastics under marine conditions by measuring directly the mineralisation and not a secondary parameter.

Within the CEN organisation, the TC/249/WG/9 on characterisation of degradability of plastics has included biodegradation of plastics under marine conditions as one of the working items. However, progress is very slow.

5.3.9 Other Biodegradation Tests

Several other tests have been used also to demonstrate or to evaluate biodegradation. Some of these tests were originally developed to verify resistance against biodeterioration or biofouling. Examples are ASTM G21-90 (resistance to fungi) [72], ASTM G22-76 (resistance to bacteria) [73], NF EN ISO 11721-1 [74] and NF EN ISO 846 [75]. These tests at most show a susceptibility to biological attack but are totally unsuited to demonstrate a far-reaching, let alone complete biodegradation or mineralisation.

A similar judgement can be given on a wide variation of tests which basically consist of some sort of immersion or burial in a given environment (soil, compost, surface water) followed by physical or chemical analyses. The most used analysis is weight loss. Yet, weight loss is dependent on disintegration and possibility for retrieval and therefore no proof of a complete mineralisation as mentioned in **Figure 5.3**. Other analyses include tensile strength, elongation, molecular weight, etc.

For a better understanding of the degradation mechanisms, tests have been reported in which specific microorganisms or enzymes have been used to evaluate the degradation of polymers or organic compounds. Some of these tests have even been normalised, e.g., ASTM D5247-92 [76], aerobic biodegradability by specific microorganisms. Nonetheless, these tests are mostly used for internal evaluation purposes and only very rarely for outside communication, marketing or certification purposes.

5.4 Certification

5.4.1 Introduction

The intensive work and development of international standards related to biodegradability and compostability of bioplastics and packaging materials is an important and vital element in the market development and breakthrough of these materials. Still, this is only the first step in the communication and build-up of credibility towards customers and authorities. The next step required is the formation of an independent and reliable certification system linked to a logo. The certification body is needed to evaluate the often complex information and make a correct judgement on the overall characteristics of a given material. In a way one could see it as standards being the theory and certification systems turning the theory into practice.

Two major reasons have added to the need for certification systems at an early stage. The first is the variety of aspects and factors related to composting and compostability. Parameters range from purely chemical analyses to evaluation of biological processes. Besides the direct customer of a producer also a third (authorities) and even a fourth party (composter) can be involved. This is the authorities defining the waste separation rules and the composting facility accepting and treating the compostable waste fraction. The second reason for the early need of certification is the complexity of the compostable products. Often these are packaging materials consisting of various (physical) components and (chemical) constituents. Information on the compostability is coming from different sources.

It must be noticed that most certification systems are specifically aimed at the evaluation of compostability. These are also the most successful certification systems. Other systems however are related to environmental fate and safety, e.g., the Japanese GreenPla), or to biodegradation in soil or water, e.g., OK Biodegradable).

Very long discussions have been held trying to develop a single, international certification and logo system for compostability valid for different countries in Europe, North America and the Far East. In spite of these efforts, national resentments proved to be too strong and various systems have emerged which are mostly limited to one country or at best a few countries. These systems are different from each other with regard to technical content (mostly slight differences in norms), geographical coverage, application and administration. The bioplastic industry however is still striving for a harmonisation of these systems and several memoranda of understanding have been signed between various certification bodies. The major goal is to get to a mutual recognition of certificates. A unified, worldwide system seems not really possible in the short-term.

5.4.2 Different Certification Systems

5.4.2.1 DIN-Certco

Probably the best-known and most used certification system is the DIN-Certco compostability certification scheme [77]. This system is managed by DIN-Certco which is a quality control organisation based in Berlin (Germany) and linked to DIN. Products that are certified can carry a compostability logo (see **Figure 5.9**). The property rights on the logo belong to Interessengemeinschaft Biologisch Abbaubare Werkstoffe, an industry association of bioplastic producers (IBAW); DIN-Certco however is responsible for the usage rights.

At the start in 1997 the system was based on the DIN V 54900 [4] prenorm on compostability of plastics. In addition to the norm a certification scheme was published in which some further (technical) rules for certification were specified. These can considered to be a kind of by-law. At the third revision of this certification scheme published in July 2001 [14], the EN 13432 [5] and ASTM D6400-99 [54] norms are also mentioned besides the DIN V 54900 as standards along which the compostability can be evaluated. On a few matters of conflict, e.g., heavy metals and test duration of radiolabelled biodegradation tests, the scheme is giving the ultimate guideline for the certification to follow.

Figure 5.9 DIN-Certco compostability logo

Testing must be done in test laboratories that are approved by DIN-Certco. The approval is based on the EN ISO/IEC 17025-1 standard [78] for quality control and assurance in test laboratories.

The applicant for the compostability logo must submit the completed forms to DIN-Certco. After a quick review for completeness, the dossier is further distributed amongst a certification committee for a definite evaluation and eventual agreement for certification. The certification committee is composed of representatives of different professional interest groups, e.g., waste management industry, compost quality organisation, retailers, farmers, bioplastic industry, environmental group, and university. After approval the material or product receives a certificate and is allowed to carry the compostability logo.

A distinction is made between polymeric materials, compostable materials, intermediates and additives on one hand and products and product ranges on the other hand. Whereas new, basic polymeric materials have to go through the complete testing programme, other categories or products only have to be submitted to a reduced testing programme, e.g., checking of disintegration only or just an administrative review to check if all constituents are compostable and the dimensions (thickness) are within the approved range.

If a packaging material is to be certified the content of it or product which is going to be packed in it, also needs to be evaluated on its suitability for composting.

For the initial approval a sample of the material or product must also be delivered for archiving and an infrared (IR) spectrum. The latter can considered being a kind of fingerprint analysis for identification. After certain time intervals, (e.g., once in the first year of certification), samples are retrieved from the market for conformity checks. These new samples are submitted to IR analyses, which are used to check the similarity between the retrieved material and the originally certified material.

More information on the DIN-Certco labelling and certification system can be found on their website [79]. In October 2004 about 45 plastics were certified as well as about 40 product families. The system is mainly used in Germany.

5.4.2.2 OK Compost

In Belgium, the OK Compost compostability certification and labelling system was launched in 1994 by the quality control organisation AIB-Vinçotte International (AVI). The initiative was stimulated by the request of local governments who wanted to use compostable biowaste collection bags. Instead of collection bins some local governments prefer plastic bags because it forms the basis of a tax collection system applying 'the polluter pays principle'. Only waste bags with the imprint of the city can be used.

The OK Compost label has played an important role in the development of the market for compostable biowaste collection bags, which is now well established in Belgium (several million bags sold per year). The biowaste collection bags are now state-of-the-art technology and functionality and true compostability has been proven.

Originally, the OK Compost system was based on the draft proposals for the European norm on compostability completed with some further technical specifications by AVI. Since the definite publication of EN 13432 [5], this norm is the principal guideline for the OK Compost system. Producers of basic materials or definite products must submit a dossier to AVI, which is making an evaluation and eventually granting the OK Compost certificate. The products can carry the OK Compost logo (see **Figure 5.10**). More information can be found on their website [80]. In October 2004 about 40 materials, products or additives are certified according to OK Compost.

Figure 5.10 OK Compost logo
Reproduced with kind permission of AIB Vinçotte

- ## OK Biodegradable

AVI has introduced a second certification and labelling system besides OK Compost. This is OK Biodegradable with further specifications according to whether the applications are for soil or water. These environmental conditions are different from composting and can show very different biodegradation characteristics as explained previously in the paragraphs on standards for biodegradation tests, e.g., it is perfectly possible that a biodegradable plastic needs a thermal (abiotic) trigger to start hydrolysis and continue biological mineralisation afterwards and is a perfect candidate for OK Compost certification. Yet, the same bioplastic will not start hydrolysis in a soil at ambient temperature and in this situation the OK Compost logo is of little use.

Extrapolating this reasoning further to certification, AVI thought it useful to start a separate OK Biodegradable logo (see **Figure 5.11**) guaranteeing biodegradation in the specified environment. Because international standards in this field have not yet been developed, the specifications and pass levels to be met are defined by a certification scheme prepared by AVI itself.

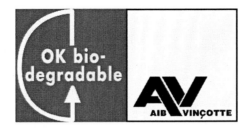

Figure 5.11 OK Biodegradable logo

Reproduced with kind permission of AIB-Vinçotte

5.4.2.3 BPI Logo

In the USA, a compostability certification and logo (see **Figure 5.12**) programme was started in 2000 by a joint effort of International Biodegradable Products Institute (BPI) and the US Composting Council (USCC). BPI is an industry organisation of bioplastic producers [81]. The USCC represents the interests of the composting industry. The certification program is based on the ASTM D6400 [54] and ASTM D6868 [82]. Applicants have to submit a dossier, which is reviewed by the Scientific Review Committee. In May 2001 the first series of products officially received the certificate.

Figure 5.12 BPI-USCC logo

Reproduced with kind permission of the BPI

5.4.2.4 GreenPla Certification System

In Japan a certification system has been started by the Biodegradable Plastics Society (BPS), an industry association on biodegradable plastics, which in many aspects is different from the European and American systems. The main focus of the system is biodegradability and environmental safety. Compatibility with a typical biological waste treatment system or disintegration within a specific time period is not an issue at this stage.

The material must be biodegradable which is defined by a minimum of 60% of mineralisation. A time frame to achieve this is not specified. Further criteria are related to maximum levels of heavy metals (same 11 metals as for EN 13432 [5]) and a minimum concentration of organic matter.

The major difference with the other certification systems lies in the need for toxicological safety data on the material itself (not to be confused with ecotoxicity tests after a preceding composting cycle). In the GreenPla system, the proof of toxicological safety is given by either oral acute toxicity tests with rats or environmental safety tests with algae, Daphnia or fish. Alternatively, the approval as a food additive is also sufficient.

More information on the Japanese GreenPla system can be found on the website [83]. A visual representation of the logo is given in **Figure 5.13**.

Figure 5.13 GreenPla logo

5.4.2.5 Other Certification and Logo Systems

In various other, mainly European, countries' initiatives have also been taken to launch compostability logo's.

In Finland, the Jätelaitosyhdistys (Finnish solid waste association, organisation co-ordinating activities of composting facilities) has launched the apple-logo (see **Figure 5.14**) for compostable products. The prerequisites for the logo are identical to EN 13432 [5]. A few products, mainly biowaste bags, have been certified.

Other initiatives have been started in Austria, The Netherlands (by the certification institute 'Stichting Milieukeur'), Spain and Italy but very little information is known about these systems. The relevance of these systems to market development seems rather limited until now. Nevertheless it is possible that these systems might play an important role in the near future.

Figure 5.14 Finnish compostability logo

References

1. A.G. Sadun, T.F. Webster and B. Commoner, *Breaking Down the Degradable Plastics Scam,* Greenpeace, Washington, DC, USA, 1990.

2. *The Green Report – Findings and Preliminary Recommendations for Responsible Environmental Advertising,* The Task Force (group of 10 US State Attorneys General), St. Paul, MN, USA, 1990.

3. *Guidelines for the Testing of Chemicals,* OECD, Paris, France, 1981.

4. DIN V 54900, *Testing of the Compostability of Plastics,* 1998.

5. EN 13432, *Packaging - Requirements for Packaging Recoverable Through Composting and Biodegradation - Test Scheme and Evaluation Criteria for the Final Acceptance of Packaging,* 2000.

6. R.N. Sturm, *Journal of American Oil Chemists' Society,* 1973, 50, 159.

7. OECD 301B, *Ready Biodegradability Test: CO_2 Evolution Test,* 1992.

8. ISO 9439, *Water Quality - Evaluation of Ultimate Aerobic Biodegradability of Organic Compounds in Aqueous Medium - Carbon Dioxide Evolution Test,* 2000.

9.   ISO 14852, *Determination of the Ultimate Aerobic Biodegradability of Plastic Materials in an Aqueous Medium – Method by Analysis of Evolved Carbon Dioxide*, 1999.

10. ASTM D5209-92, *Standard Test Method for Determining the Aerobic Biodegradation of Plastic Materials in the Presence of Municipal Sewage Sludge.* (Discontinued 2001)

11. EN 29439, *Water Quality - Evaluation in an Aqueous Medium of the 'Ultimate' Aerobic Biodegradability of Organic Compounds - Method by Analysis of Released Carbon Dioxide*, 1993.

12. EN 14047, *Packaging - Determination of the Ultimate Aerobic Biodegradability of Packaging Materials in an Aqueous Medium - Method by Analysis of Evolved Carbon Dioxide*, 2002.

13. JIS K6951, *Determination of the Ultimate Aerobic Biodegradability of Plastic Materials in an Aqueous Medium - Method by Analysis of Evolved Carbon Dioxide*, 2000.

14. ISO 14851, *Determination of the Ultimate Aerobic Biodegradability of Plastic Materials in an Aqueous Medium – Method by Measuring the Oxygen Demand in a Closed Respirometer*, 1999.

15. OECD 301C, *Ready Biodegradability Test, Modified MITI Test*, 1992.

16. ISO 9408, *Water Quality - Evaluation of Ultimate Aerobic Biodegradability of Organic Compounds in Aqueous Medium by Determination of Oxygen*, 1999.

17. ASTM D5271-02, *Standard Test Method for Determining the Aerobic Biodegradation of Plastic Materials in an Activated-Sludge-Wastewater-Treatment System*, 2002.

18. EN ISO 9408, *Water Quality - Evaluation of Ultimate Aerobic Biodegradability of Organic Compounds in Aqueous Medium By Determination of Oxygen Demand in a Closed Respirometer*, 1999.

19. EN 14048, *Packaging - Determination of the Ultimate Aerobic Biodegradability of Packaging Materials in an Aqueous Medium - Method by Measuring the Oxygen Demand in a Closed Respirometer*, 2002.

20. JIS K6950, *Plastics - Testing Method For Aerobic Biodegradability By Activated Sludge*, 2000.

21. OECD 301A, *Ready Biodegradability Test: DOC Die-Away Test*, 1992.

22. ISO 7827, *Water Quality - Evaluation in an Aqueous Medium of the 'Ultimate' Aerobic Biodegradability of Organic Compounds - Method by Analysis of Dissolved Organic Carbon (DOC)*, 1994.

23. ASTM E1279-89, *Standard Test Method for Biodegradation By a Shake-Flask Die-Away Method*, 2001.

24. ISO 10708, *Water Quality - Evaluation in an Aqueous Medium of The Ultimate Aerobic Biodegradability of Organic Compounds - Determination of Biochemical Oxygen Demand in a Two-Phase Closed Bottle Test*, 1997.

25. OECD 301D, *Ready Biodegradability Test: Closed Bottle Test*, 1992.

26. ISO 10707, *Water Quality - Evaluation in an Aqueous Medium of the 'Ultimate' Aerobic Biodegradability of Organic Compounds - Method by Analysis of Biochemical Oxygen Demand (Closed Bottle Test)*, 1994.

27. OECD 302B, *Inherent Biodegradability Test: Zahn-Wellens/EMPA Test*, 1992.

28. ISO 9888, *Water Quality - Evaluation of Ultimate Aerobic Biodegradability of Organic Compounds in Aqueous Medium - Static Test (Zahn-Wellens Method)*, 1999.

29. CEC L-33-A-93, *Biodegradability of Two-stroke Cycle Outboard Engine Oils in Water*, 1993.

30. ASTM D5864-00, *Standard Test Method for Determining Aerobic Aquatic Biodegradation of Lubricants or Their Components*, 2000.

31. ASTM D6139-00, *Standard Test Method for Determining the Aerobic Aquatic Biodegradation of Lubricants or Their Components Using the Gledhill Shake Flask*, 2000.

32. OECD 302A, *Inherent Biodegradability Test: Modified SCAS*, 1981.

33. ISO 9887, *Water Quality - Evaluation of the Aerobic Biodegradability of Organic Compounds in an Aqueous Medium - Semi-Continuous Activated Sludge Method (SCAS)*, 1992.

34. ASTM E1625-94, *Standard Test Method for Determining Biodegradability of Organic Chemicals in Semi-Continuous Activated Sludge (SCAS)*, 2001.

35. OECD 303A, *Simulation Test: Aerobic Sewage Treatment, A: Activated Sludge Units*, 2001.

36. ISO 11733, *Water Quality - Evaluation of the Elimination and Biodegradability of Organic Compounds in an Aqueous Medium - Activated Sludge Simulation Test*, 1998.

37. W. Zimmermann, *Journal of Biotechnology*, 1990, **13**, 119.

38. T.K. Kirk in *Microbial Degradation of Organic Compounds*, Ed., D.T. Gibson, Marcel Dekker, Inc., New York, NY, USA, 1984, 399.

39. B. De Wilde, L. De Baere and R. Tillinger, *Test Methods for Biodegradability and Compostability*, Mededelingen Faculteit Landbouw, University of Gent, Gent, Belgium, 58/4a, 1993 p.1621.

40. U. Pagga, D.B. Beimborn, J. Boelens and B. De Wilde, *Chemosphere*, 1995, **31**, 11/12, 4475.

41. ASTM D5338-98e1, *Standard Test Method for Determining Aerobic Biodegradation of Plastic Materials Under Controlled Composting Conditions*, 1998.

42. ISO 14855, *Determination of the Ultimate Aerobic Biodegradability and Disintegration of Plastic Materials under Controlled Composting Conditions - Method by Analysis of Evolved Carbon Dioxide*, 1999.

43. EN 14046, *Packaging - Evaluation of the Ultimate Aerobic Biodegradability and Disintegration of Packaging Materials Under Controlled Composting Conditions - Method by Analysis of Released Carbon Dioxide*, 2003.

44. JIS K6953, *Determination of the Ultimate Aerobic Biodegradability and Disintegration of Plastic Materials Under Controlled Composting Conditions - Method by Analysis of Evolved Carbon Dioxide*, 2000.

45. G. Bellia, M. Tosin, G. Floridi and F. Degli Innocenti, *Polymer Degradation and Stability*, 1999, **66**, 65.

46. B. Spitzer and M. Menner, *Dechema Monographs*, 1996, **133**, 681.

47. ASTM D6340-98, *Standard Test Methods for Determining Aerobic Biodegradation of Radiolabeled Plastic Materials in an Aqueous or Compost Environment*, 1998.

48. ASTM D5509-96, *Standard Practice for Exposing Plastics to a Simulated Compost Environment* (Discontinued 2002).

49. ASTM D5512-96, *Standard Practice for Exposing Plastics to a Simulated Compost Environment Using an Externally Heated Reactor* (Discontinued 2002)

50. ISO 16929, *Plastics - Determination of the degree of disintegration of plastic materials under defined composting conditions in a pilot-scale test*, 2002.

51. *Official Journal of the European Communities*, L190/21, 12.7.2001.

52. *Hygiene Baumusterprüf-systemen*, BGK Nr. 225, Bundesgütegemeinschaft Kompost eV, Cologne, Germany, 1996.

53. ASTM D6002-96 (2002)e1, *Standard Guide for Assessing the Compostability of Environmentally Degradable Plastics*, 2002.

54. ASTM D6400-99e1, *Standard Specification for Compostable Plastics*, 1999.

55. OECD 208, *Terrestrial Plants: Growth Test*, 1984.

56. ISO 20200, *Plastics - Determination of the Degree of Disintegration of Plastic Materials Under Simulated Composting Conditions in a Laboratory-Scale Test*, 2004.

57. OECD 304A, *Inherent Biodegradation in Soil*, 1981.

58. ISO 11266, *Soil Quality - Guidance on Laboratory Testing for Biodegradation of Organic Chemicals in Soil under Aerobic Conditions*, 1994.

59. ASTM D5988, *Standard Test Method for Determining Aerobic Biodegradation in Soil of Plastic Materials or Residual Plastic Materials After Composting*, 2003.

60. ISO 17556, *Plastics - Determination of the Ultimate Aerobic Biodegradability in Soil by Measuring the Oxygen Demand in a Respirometer or the Amount of Carbon Dioxide Evolved*, 2003.

61. *Evaluation of Anaerobic Biodegradation*, Technical Report No.28, ECETOC, Brussels, Belgium, 1988.

62. ISO 11734, *Water Quality - Evaluation of the 'Ultimate' Anaerobic Biodegradability of Organic Compounds in Digested Sludge - Method by Measurement of the Biogas Production*, 1995.

63. ASTM D5210-92, *Standard Test Method for Determining the Anaerobic Biodegradation of Plastic Materials in the Presence of Municipal Sewage Sludge*, 2000.

64. ISO/DIS 14853, *Determination of the Ultimate Anaerobic Biodegradability of Plastic Materials in an Aqueous System - Method by Measurement of Biogas Production*, 1999.

65. ASTM D5511-02, *Standard Test Method for Determining Anaerobic Biodegradation of Plastic Materials Under High-Solids Anaerobic-Digestion Conditions*, 2002.

66. ISO/DIS 15985, *Determination of the Ultimate Anaerobic Biodegradability of Plastic Materials in an Aqueous System - Method by Measurement of Biogas Production*, 1999.

67. ASTM D5525-94a, *Standard Practice for Exposing Plastics to a Simulated Active Landfill Environment* (Discontinued 2002)

68. ASTM D5526-94, *Standard Test Method for Determining Anaerobic Biodegradation of Plastic Materials Under Accelerated Landfill Conditions*, 2002.

69. OECD 306, *Biodegradation in Seawater*, 1992.

70. ASTM D5437-93, *Practice for Weathering of Plastics Under Marine Floating Exposure* (Discontinued 1999).

71. ASTM D6691-01, *Standard Test Method for Determining Aerobic Biodegradation of Plastic Materials in the Marine Environment by a Defined Microbial Consortium*, 2001.

72. ASTM G21-96, *Standard Practice for Determining Resistance of Synthetic Polymeric Materials to Fungi*, 2002.

73. ASTM G22-76 *Standard Practice for Determining Resistance of Plastics to Bacteria* (Withdrawn 2001).

74. NF EN ISO 11721-1, *Textiles - Determination de la Resistance aux Micro-Organismes des Textiles Contenant de la Cellulose - Essai d'enfouissement - Partie 1: Evaluation d'un Traitement d'imputrescibilite*, 2001.

75. NF EN ISO 846, *Plastics - Evaluation Of The Action Of Microorganisms*, 1997.

76. ASTM D5247-92, *Standard Test Method for Determining the Aerobic Biodegradability of Degradable Plastics by Specific Microorganisms* (Discontinued 2001)

77. *Certification Scheme – Products Made of Compostable Materials*, 3rd revision, DIN-CERTCO, Berlin, Germany, 2001.

78. EN ISO/IEC 17025, *General Requirements For The Competence Of Testing And Calibration Laboratories*, 2000.

79. DIN-CERTCO, *www.en.dincertco.de*, click on products and then environment

80. *OK Compost and OK Biodegradable Marks*, http://www.aib-vincotte.com/uploads/documents/086_03_E.pdf

81. *Biodegradable Products Institute*, www.bpiworld.org

82. ASTM D6868, *Standard Specification for Biodegradable Plastics Used as Coatings on Paper and Other Compostable Substrates*, 2003.

83. *Biodegradable Plastics Society*, www.bpsweb.net/02_english/index.htm

6 General Characteristics, Processability, Industrial Applications and Market Evolution of Biodegradable Polymers

Gregory M. Bohlmann

6.1 General Characteristics

Biodegradable polymers have the potential to be the solution to a range of environmental concerns associated with conventional, non-degradable polymers. Of primary concern is the solid waste problem associated with the decreasing availability of landfills around the world. Other concerns include the benefits of sustainable or renewable raw material sources rather than petrochemical sources and the issue of global warming caused by increasing amounts of carbon dioxide in the atmosphere.

Many options are being explored in public and private sectors to address these environmental concerns related to the use of polymers in society. Incineration is often used as a means of solid waste disposal, but concerns have grown over the environmental impact of incineration emissions. Mechanical recycling is beneficial for many polymer products, but has limitations especially concerning food packaging. Composting is well suited for the disposal of biodegradable materials, but infrastructure is lacking in most regions of the world.

The degradation of synthetic polymers has been investigated since their commercial introduction because nearly all plastics are affected by exposure to natural weathering forces such as sunlight, oxygen, water, and heat. Historically, most research has focused on developing stable and durable polymer structures that resist these forces. Modern plastics such as polyethylene (PE), polypropylene (PP), polystyrene (PS), polyethylene terephthalate (PET) and polyvinyl chloride (PVC) are strong, inexpensive, easily processible, and durable. Durability attributes have led to difficulties when these plastic materials enter the waste stream.

Most man-made polymers are resistant to biological degradation because their carbon components cannot be broken down by the enzymes of microorganisms. In addition, the hydrophobic character of plastics inhibits enzyme activity, and the low surface area of plastics along with their inherent high molecular weight (MW), further compounds their resistance to microbial attack [1]. In the past two decades, biodegradable polymers have been developed in the laboratory and commercialised that are designed to biologically degrade. The industry has been challenged to develop biodegradable polymers that are easily processible, have good performance properties and are cost competitive with conventional polymers. This chapter provides descriptions of the polymers commercialised to date, and also shows how successful industry has been in commercialising these products.

6.1.1 Polymer Biodegradation Mechanisms

Two key steps occur in the biodegradation of polymers [2]. First is a depolymerisation, or chain cleavage, step in which the long polymer chain is converted into smaller oligomeric fragments. Hydrolysis and/or oxidation may be responsible for this step. Extracellular enzymes may also be responsible, acting either endo (random cleavage on internal linkages of the polymer) or exo (sequential cleavage of the terminal monomer unit). This first step is important because large structural material, like macromolecules, cannot pass through the outer membranes of living cells [3].

The second step, known as mineralisation, occurs inside the cell where small-size oligomeric fragments are converted into biomass, minerals and salts, water, and gases such as CO_2 and CH_4. Biodegradability of polymers is described in more detail in Chapter 2.

Most of the methods that measure the extent of biodegradation are respirometric, which is primarily related to carbon dioxide evolution. Other methods include assessing the rate of MW loss; measuring the loss of polymer physical properties, (e.g., tensile strength per ASTM standard D3826-98 [4]); measuring the rate of increase of the microbial culture colony size contacting the material; and using classical oxygen uptake procedures, [e.g., biochemical oxygen demand (BOD)] and radioactive tracer techniques that use ^{14}C labelling. Certification procedures are described in Chapter 2.

Biodegradation implies the use of the plastic substrate as the carbon source for the microorganism metabolism. Biodegradation results in the production of CO_2 under aerobic environments or CH_4 under anaerobic environments, as well as humic materials. Humic material is an important component of the biodegradation process because it can enhance productivity of agricultural land. Thus composting polymeric materials is a biological recycling of the polymeric carbon. Composting is defined by Narayan as 'accelerated degradation of heterogeneous organic matter by a mixed microbial population in a moist, warm, aerobic environment under controlled conditions' [1]. A typical compost system supports a diverse microbial population in a moist aerobic environment in a temperature range of 40-70 °C [5].

Merely exposing a material, whether a natural or synthetic polymer, to a biologically active environment does not guarantee its biodegradation. Several factors are important in biodegradation, including macromolecule size, structure, and chemistry; microbial population and enzyme activity; and various environmental conditions such as darkness, high humidity, and adequate mineral and other organic nutrients, as well as temperature, pH, and oxygen requirements [2].

Provided that the appropriate environmental conditions are present, conventional plastics resist biodegradation primarily because of their molecular size, structure, and chemical composition. Potts and co-workers have conducted studies on the biodegradation of synthetic polymers and, in general, found that molecular weight is the most critical factor in the process [6]. For synthetic high molecular weight polymers, only aliphatic polyesters (polyether sulfone) and some aliphatic/aromatic copolyesters were found to be biodegradable. PE oligomers become biodegradable at MW below 500, although more rigorous testing is needed for confirmation. Polyvinyl alcohol (PVOH) is probably the only carbon chain synthetic polymer to be fully biodegradable, although recent studies indicate that PE can be slowly biodegraded by pretreatment with surfactants or an oxidation process [7].

6.1.2 Polymer Molecular Size, Structure and Chemical Composition

Biodegradation involves the actions of a microorganism's extracellular enzymes, which break down a polymer into products that are small enough to be assimilated, such microorganisms tend to attack the ends of large molecules, and the number of ends is inversely proportional to the MW. To make plastics degradable, it is necessary to break them down into very small particles with a large surface area and then to reduce their MW. Chain branching and crystallinity also inhibit these activities. Not only does a lower degree of polymerisation yield a higher concentration of chain end groups for attack by microorganisms, but it also discourages the formation of crystalline domains that are generally difficult to biodegrade [8].

The carbon chain backbones of synthetic polymers are difficult to cleave enzymically. Biodegradation may be enhanced if *N*-substituted amide links, ester links, or under some circumstances, ether links are present in the backbone. Biodegradation processes may also be inhibited by a variety of agents such as additives, impurities, and even intermediate products of degradation.

6.1.3 Biodegradable Polymer Classes

Three broad classes of commercially important biodegradable polymers are discussed in this chapter:

1. unmodified polymers that are naturally susceptible to microbial-enzyme attack,

2. synthetic polymers, primarily polyesters, and

3. naturally biodegradable polymers that have been modified with additives and fillers.

Naturally biodegradable polymers produced in nature are renewable. Some synthetic polymers are also renewable because they are made from renewable feedstocks, for example polylactic acid (PLA) is derived from agricultural feedstocks.

6.1.4 Naturally Biodegradable Polymers

Natural polymers are produced in nature by all living organisms. Biodegradation reactions are typically enzyme-catalysed and occur in aqueous media. Natural macromolecules containing hydrolysable linkages, such as protein, cellulose, and starch, are generally susceptible to biodegradation by the hydrolytic enzymes of microorganisms. Thus the hydrophilic/hydrophobic character of polymers greatly affects their biodegradability. It also has a great impact on their performance and durability in humid conditions. The major category of biodegradable polymers consists of those with hydrolysable linkages along the polymer chain backbone: polyesters, polyamides, polyureas, polyanhydrides, poly(amide-enamine)s, polyurethanes (PU), and polyphosphazene. Outside of natural fibres like wool and silk, polysaccharides such as starch are the most prevalent naturally biodegradable polymers in commercial use.

Starch, or plant nutrient material, is composed of two polysaccharides: α-amylose and amylopectin. The primary structure of α-amylose is linear because of the exclusive α (1-4) linkages between the D-glucose monomers (see **Figure 6.1**).

α-Amylose typically has a MW of 1.6×10^5 to 2.6×10^6 [8]. Amylopectin is branched because of the presence of α (1-6) linkages as well as the α (1-4) linkages, as shown in **Figure 6.2**.

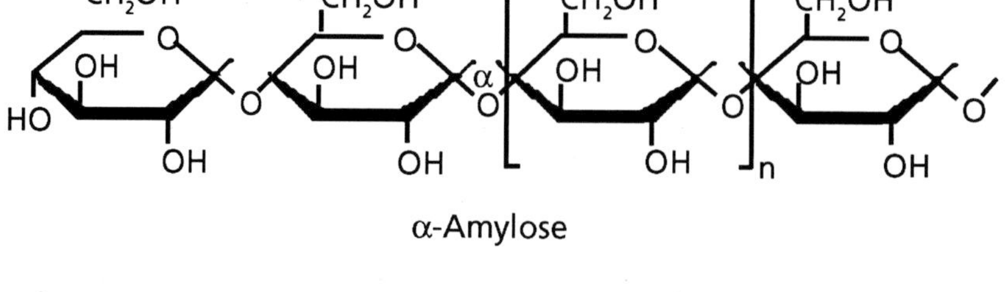

α-Amylose

Figure 6.1 α-Amylose

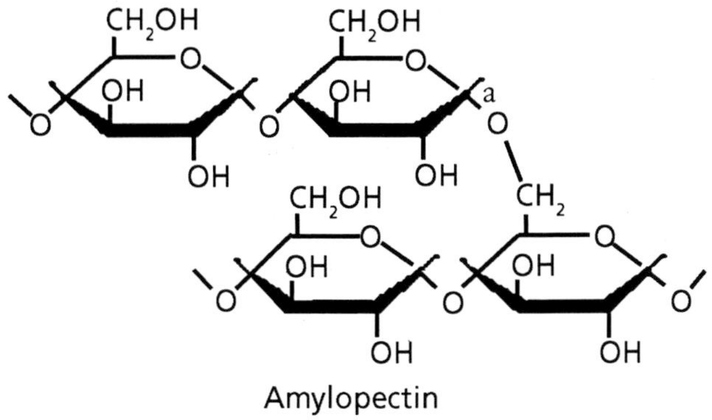

Amylopectin

Figure 6.2 Amylopectin

Amylopectin, the water-soluble portion of starch, has a molecular weight of 5×10^7 to 4×10^8 [9]. Normal corn starch is composed of 20-30% -amylose and 70-80% amylopectin [10].

All green plants make and store D-glucose in the form of starch granules. In the granular form, starch is quasicrystalline; i.e., it displays spherocrystalline patterns [10]. The sizes and shapes of granules are specific to the plant of origin. Upon heating, starch granules decompose before they melt. There is a process, however, known as destructurising, that modifies starch morphology so it can become a thermoplastic melt. In the destructurising process starch is heated under pressure above the melting and glass-transition (Tg) temperatures of its components so that they undergo endothermic transitions [11]. Thermoplastic starch alone can be processed as a conventional plastic; however its sensitivity to humidity makes it unsuitable for most applications [12].

The main use of thermoplastic starch alone is in foam applications. Starch-based foams have been found to be an effective alternative to PS foam in loose-fill protective packaging. Starch-based foams offer the advantage that they are readily biodegradable if they escape into the environment. They also offer superior antistatic properties. Starch-based foam, however, has some disadvantages: it is brittle and the density is higher than PS. Foamed, starch-based articles are prepared by heating starch in an extruder in the presence of water with subsequent extrusion.

One of the most crucial properties in packaging applications is bulk density. A comparison of bulk densities of expanded polystyrene loose-fill with starch-based foams is given in **Table 6.1.**

An important starch-based foam is Novamont's Mater-Bi® Class V grade material. The content of thermoplastic starch is more than 85% [12]. Novamont in Italy has patented

Table 6.1			
	Expanded polystyrene [13]	Novamont starch-based [14]	Free-flow packaging starch-based [15]
Bulk density, kg/m^3	4.0-4.8	5-13	4.8-32

low-density starch-based foams with bulk densities of 5-13 kg/m^3, which correspond to 19-31 kg/m^3 specific density [14]. The composition includes one or more thermoplastic polymers to give the molten mass high melt strength. The composition can also include a nucleating agent, lubricant, plasticisers, flame retardants, and rodent repellents. Foamed material is prepared in a two-step process. First a mixture of starch and thermoplastic starch is mixed in an extruder in the presence of water to obtain a plasticised matrix. Total water content of the resulting pellets is 5-20 wt%. The pellets are then foamed using a single screw extruder.

Several other companies have patented starch-based foams including Free-Flow Packaging in the United States [15], K&S Bio-Pack in Germany [16], and Nippon Gohsei in Japan [17]. The base mixture for Free-Flow Packaging's patented foam includes nonmodified starch, PVOH, proteinaceous grain meal, glycerin, vegetable oil, and glycerol monostearate. Nippon Gohsei's patented composition also includes a vinyl alcohol resin.

Aliphatic polyesters are perhaps some of the most easily biodegraded polymers found in nature. One reason for this is the effect that chain flexibility has on biodegradability. For degradation of polymers by enzymes, the polymer chain must be flexible enough to fit into the active site of the enzyme. This characteristic most likely accounts for the biodegradability of aliphatic polyesters, which are flexible, whereas wholly aromatic polyesters, which are more rigid, are generally considered bioinert.

Many types of microorganisms produce and store the aliphatic polyester, polyhydroxybutyrate (PHB), during sugar fermentation. Certain bacteria feeding on sugars enzymatically produce PHB that is stored as 'bacterial fat.' PHB can then be extracted from the bacteria, dried, formed as powder or conventional resin, and moulded into film or rigid forms. Through variation in nutrition of the microorganism, certain bacteria can make a range of copolymers based on hydroxybutyric and hydroxyvaleric acids, resulting in random copolymers called poly(hydroxybutyrate-co-hydroxyvalerate) (PHBV). The structure of the copolymer is shown in **Figure 6.3**.

Because microorganisms synthesise PHB and PHBV for use as carbon and energy reserve materials, it can be inferred that many microorganisms are capable of degrading and metabolising these polymers. The rate of degradation depends on a number of factors

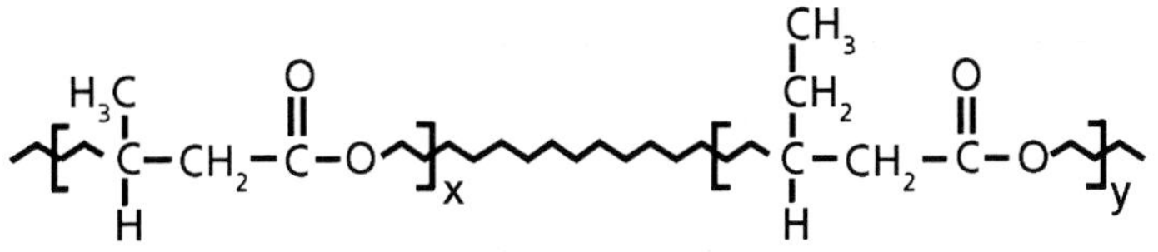

Poly(hydroxybutyrate-co-hydroxyvalerate)
(PHBV)

Figure 6.3 Structure of PHBV

such as environment, temperature, pH, oxygen concentration, surface area, molecular mass, and degree of crystallinity [18].

PHB is stiffer and more brittle than PHBV and its chemical properties differ. Table 6.2 lists some of the property differences. Its solvent resistance is inferior, but it has better natural resistance to UV weathering. PHB is also optically active and piezoelectric. The properties of PHBV depend on the valerate content. Incorporating the hydroxyl valerate (HV) monomer into the copolymer reduces the level of crystallinity and melting point, resulting in a decrease in stiffness but an increase in toughness or impact resistance.

More than a dozen organisations have patented technology relating to microbial production of polyhydroxyalkanoate (PHA) (emphasis on PHB, PHBV, and related polyesters). The bacterial production of PHB was first characterised in 1925 by Lemoigne at the Pasteur Institute in Paris and has since been extensively studied [20]. WR Grace in the United States patented PHB and produced small quantities for commercial evaluation in the late 1950s and early 1960s [21-23]. ICI in the United Kingdom continued evaluation of PHB

Table 6.2 Properties of PHB and PHBV			
Property	PHB	PHBV (10% HV)	PHBV (20% HV)
Melting point, °C	180	140	130
Tensile strength, MPa	40	25	20
Flexural modulus, GPa	3.5	1.2	0.8
Extension to break, %	8	20	50
Sources: [18, 19]			

in the 1970s and 1980s and commercialised BIOPOL™ polymers in 1981. The BIOPOL™ business and related technology were sold to Monsanto in the United States in 1996 and subsequently sold by Monsanto to Metabolix in the United States in 2001.

ICI was issued with a number of patents in the 1980s that describe the preparation of PHBV by cultivating *Ralstonia eutrophus* in a two-stage fermentation process [24-26]. The first stage is operated as a conventional fermentation with glucose as the carbon source and nutrient salts present for the nitrogen source. In the second stage propionic acid is added as an additional carbon source and the nitrogen nutrient is limited to induce the microorganism to produce 3-hydroxyvalerate units.

The PHBV polymer is accumulated as discrete granules within the cell cytoplasm and each granule is thought to be surrounded by a lipid and protein membrane [18]. Recovery of the polymer from within the cell may be accomplished by a variety of extraction routes.

The cost of making PHBV by microbial fermentation is very high; estimated to be over $9 per kilogram from a 2,000 ton per year plant [27]. Research is ongoing at several institutions such as Metabolix and Michigan State University to reduce the cost of PHA production by producing the polymer in plants. The cost of PHB produced by bacterial fermentation is substantially higher than that of other biomaterials such as starch or lipids that accumulate in many species of higher plants [28]. Using plants to make PHB is theoretically possible because acetyl-CoA and acetoacetyl-CoA, the precursors to PHB synthesis in *R. eutrophus*, are also found in plants and involved in the syntheses of a variety of compounds.

PHB production in plants has been demonstrated experimentally using a small weed in the mustard family [28]. Through genetic engineering a hybrid plant was obtained that expressed all bacterial and endogenous plant enzymes required for PHB synthesis. One problem observed is that the high level of acetoacetyl-CoA reductase activity in the plant caused a smaller plant size and a reduction in seed production. A possible solution maybe to use a species such as potato where the production of large quantities of starch in the root is not required for the viability of the seeds or plant. Procter & Gamble in the United States has patented a process for the recovery of the desired PHA poly(3-hyroxyvalerate-co-hydroxyhexanoate) from transgenic potatoes [29]. The company licenses PHA technology under the name Nodax™.

Researchers at the University of Warwick in the United Kingdom are exploring the use of genetically altered oilseed rape plants for making PHB. This research is being funded by the United Kingdom's Biotechnology Directorate. PHB is produced in the plant by using the yeast protein GAL4 (transcriptional activator) to regulate and coordinate the activation of genes in the oilseed rape plant [30].

Unanswered questions persist that influence the potential of PHA production in plants. Achieving control over the final composition of the polymer may be more difficult in plants than in bacterial fermentation. Isolation and quality of the purified polymer from plants is also a hurdle. A very important consideration in commercialisation is the level of PHA achieved in the plant. Monsanto has achieved PHA levels as high as 5% in plants [31]. The long-term goal is to produce a level of PHA comparable to the oil content in soybeans of 20% to achieve commercial viability. Monsanto is no longer actively researching PHA; Metabolix licensed its technology in 2001.

6.1.5 Synthetic Biodegradable Polymers

While natural polymers are produced by living organisms, synthetic biodegradable polymers are only produced by mankind. Biodegradation reactions are the same for both, i.e., typically enzyme-catalysed and occurring in aqueous media. The major category of synthetic biodegradable polymers consists of polyesters with hydrolysable linkages along the polymer chain backbone. PVOH is also reported to be biodegradable.

Interest in simple aliphatic polyesters such as PLA was pioneered by Carothers in the 1930s [32]. The susceptibility of these polymers to hydrolytic degradation led to DuPont discontinuing work in this area. The ability of the human body to degrade these materials led to medical applications in the 1970s; the simplest poly(α-hydroxyacid), polyglycolic acid (PGA), has been successfully used in degradable surgical sutures [33]. PGA is usually obtained by polymerising diglycolide with a tin catalyst. Similarly, PLA can be obtained from dilactide by stannous octoate catalysed, ring-opening polymerisation.

The structures of PGA and PLA are shown in **Figure 6.4**.

Biodegradable polyesters such as PLA, PGA, and their copolymers have been used widely in medicine and surgery for the controlled release of drugs, biodegradable surgical sutures, and implants for fixation of fractures, primarily because of their high biocompatibility [34]. In orthopaedic surgery, biodegradable fixation devices such as screws, plates, and

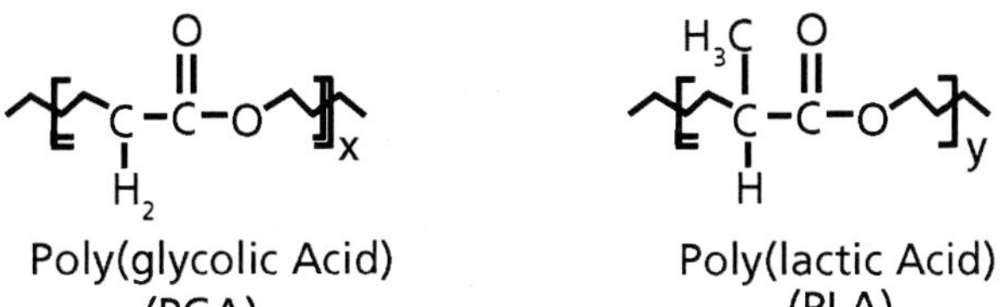

Poly(glycolic Acid)
(PGA)

Poly(lactic Acid)
(PLA)

Figure 6.4 Structures of PGA and PLA

pegs have the advantage of temporarily securing bone prostheses. The device is eventually absorbed by the body after bone tissue growth into the porous matrix structure of the prosthesis effectively affixes the implant to the bone.

An important aspect of the biodegradation of polyesters is the susceptibility of polyesters to hydrolytic degradation. Degradation proceeds by random hydrolytic chain scission of the ester linkages, eventually producing the monomeric hydroxyacid [35]. Two distinct stages in the degradation process have been identified. The first stage, which is nonenzymatic, is restricted to random hydrolytic cleavage of ester linkages. The second stage, which is also nonenzymatic, begins when the molecular weight of the polymer has decreased to the point that chain scission can produce an oligomer small enough to diffuse from the polymer bulk. Catastrophic loss of mechanical strength can occur during this second phase. The extent of degradation increases at higher degradation temperatures (when the temperature is between 40 and 60 °C) [4]. As the average MW approaches 10,000, microorganisms are able to digest the low MW lactic acid oligomers to produce carbon dioxide and water [36].

Recently, companies have been developing PLA materials for use in biodegradable applications such as film for compost bags or thermoformed food containers. At one time or another in the 1990s, five companies worldwide had commercial or semicommercial plants for making PLA. Because lactic acid is difficult to polymerise directly to high polymers in a single step on a commercial scale, most companies used a two-step process. Lactic acid is first oligomerised to a linear chain with a MW of less than 3,000 by removing water. The oligomer is then depolymerised to lactide, a cyclic dimer. This six-membered ring is purified and subjected to ring-opening polymerisation to produce a PLA with a MW of greater than 50,000-110,000. Companies in the United States holding patents for this process include Cargill [37, 38], Camelot Technologies [39] and Ecological Chemical Products [40]. Mitsui Chemical in Japan has developed a process for making high-molecular-weight PLA direct from lactic acid without the oligomerisation step [41].

Like other commercial biodegradable polymers, PLA is a high-cost material relative to conventional thermoplastics. Recent developments, particularly with regard to lactic acid sourcing, promise lower production costs in the future. The dominant producer in this field, Cargill Dow Polymers, has announced that PLA should be commercially available in the $1-2 per kilogram range from a world-scale plant that is scheduled to start-up in 2002 [42]. The upper end of that price range should be feasible as lactic acid prices fall with capacity expansions and process technology improvements according to the Process Economics Program [43].

According to Lunt, PLA polymers range from amorphous glassy polymers with a Tg of 60 °C to semi-crystalline/highly crystalline products with crystalline melting points of 130-180 °C [36]. Many of the basic properties of PLA lie between those of crystal

PS and PET. Selection of the PLA stereochemistry can have a major effect both on the polymer's properties, processability and biodegradability. Flexible PLA film can be made by incorporating a plasticiser.

The initial development efforts for large scale PLA applications has been in fibres. PLA is not necessarily biodegradable as a fibre due to its crystallinity. **Table 6.3** compares PLA fibre properties with those of PET and rayon, two materials that it may displace.

Another well-known biodegradable aliphatic polyester is poly(ε-caprolactone) (PCL). These polymers can be divided into two groups based on MW. Material with a MW of up to several thousand is a waxy solid or viscous liquid. These PCL are used as PU intermediates, reactive diluents for high solids coatings, and plasticisers for vinyl resins. The other type of PCL has a MW greater than 20,000 and is resinous with good mechanical strength. The primary worldwide PCL producers are Dow Chemical (formerly Union Carbide) in the United States, Solvay in Europe, and Daicel Chemical Industries in Japan.

PCL is generally prepared from the ring-opening polymerisation of ε-caprolactone. Union Carbide has patented stannous octanoate as a polymerisation initiator that can achieve MW as high as 100,000 [45]. The structure of PCL is shown in **Figure 6.5**.

Table 6.3 Fibre property comparison			
Fibre property	PET	PLA	Rayon
Specific gravity	1.39	1.25	1.52
T_m, °C	254-260	130-175	None
Tenacity, g/d	6.0	6.0	2.5
Elastic recovery at 5% strain	65	93	32
Moisture regain, %	0.2-0.4	0.4-0.6	11
Source: [44]			

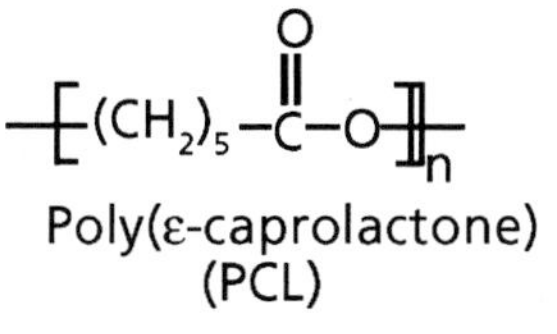

Poly(ε-caprolactone)
(PCL)

Figure 6.5 Structure of PCL

Because of its ease of polymerisation to high MW and its commercial availability, PCL has been the subject of a number of studies pertaining to its biodegradability [6]. Although PCL is an expensive polymer, it is used extensively in biodegradable applications typically as a starch blend. **Table 6.4** provides typical properties for PCL at three degrees of polymerisation.

Table 6.4 Typical properties of PCL			
Property	CAPA®640	CAPA®650	CAPA®680
Mean molecular weight	37,000	50,000	80,000
Melting point, °C	58-60	58-60	60-62
Tensile strength, kg/cm²	140	360	580
Elongation at break, %	660	800	900
Source: [46]			

PCL films exhibit mechanical properties similar to those of polyolefin films with stiffness in between low-density polyethylene (LDPE) and high-density polyethylene (HDPE). The following table compares selected properties of linear low-density polyethylene (LLDPE) with those of PCL slot cast film:

Table 6.5 Film properties of LLDPE and PCL		
Property	LLDPE	PCL
Film gauge	180	130
Tensile strength, MPa	40	20
Flexural modulus, GPa	3.5	0.8
Extension to break, %	8	50
Source: [47]		

Another commercialised type of biodegradable aliphatic polyester is succinate-based. Showa Highpolymer in Japan produces a family of aliphatic polyesters known as Bionolle®. Bionolle is produced from glycols and aliphatic dicarboxylic acids such as succinic acid or modified acids. The structures of two different types of succinate aliphatic polyesters, polybutylene succinate (PBSU) and polyethylene succinate (PESU) are shown in **Figure 6.6**.

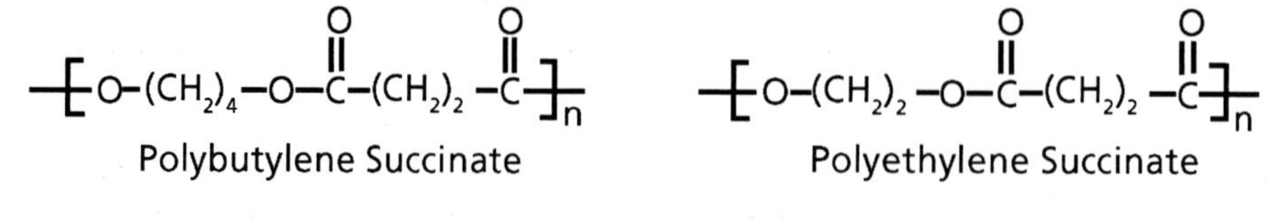

Polybutylene Succinate   Polyethylene Succinate

Figure 6.6 Structures of PBSU and PESU

Initial R&D work with succinate polyesters was successful only in producing polymers with MW less than 5,000. These polymers were weak and brittle. Beginning in 1980, Takiyama and co-workers began developing high MW aliphatic polyesters for pressure sensitive or thermosetting adhesives [48]. This experience proved useful in developing high MW succinate polyesters and in 1990 the means was discovered for making these polyesters with a number average MW in the range of 20,000-200,000. High MW polymers are prepared in two steps. First a hydroxy-terminated aliphatic polyester prepolymer is made by dehydration condensation of a dihydric alcohol, such as 1,4-butanediol, with succinic acid. The prepolymer is reacted with a diisocyanate chain extender to form a high MW succinate polyester [49].

Currently, PBSU and polybutylene succinate adipate copolymer are commercially available and PESU and polyethylene succinate adipate copolymer are under development but not yet commercial. **Table 6.6** compares the basic properties of commercial grades of Bionolle® with conventional polyolefins.

In contrast to most aliphatic polyesters, aromatic polyesters like PET provide excellent material properties [50]. To combine good material properties with biodegradability, aliphatic/aromatic copolyesters have been developed. Several major polyester producers in Europe and the United States have recently begun marketing aliphatic/aromatic copolyesters for biodegradable applications. BASF markets a product, Ecoflex®, which is a copolyester of butanediol, adipic acid, and dimethyl terephthalate. Eastman's Eastar Bio Copolyester 14766 is a similar aliphatic/aromatic copolyester. DuPont markets a modified PET known as Biomax®.

Table 6.6 Basic properties of Bionolle® pressed sheet					
Property	PBSU #1000	PBSU Co.#2000	PBSU Co. #3000	LDPE	HDPE
MFR at 190 °C, g/10 min	1.5-26	4.0	28	0.8	11
Melting point, °C	114	104	96	110	129
Yield strength, kg/cm^2	336-364	270	192	100	285
Elongation, %	560-323	710	807	700	300
Stiffness 103, kg/cm^2	5.6-6.6	4.2	3.3	1.8	12
Izod impact at 20 °C, kg-cm/cm	30-4.2	36	>40	>40	4
MFR: melt flow rate *Reprinted from Polymer Degradation and Stability, Volume 59, T. Fujimaki, Processability and Properties of Aliphatic Polyesters, 'BIONOLLE,' Synthesized by Polycondensation Reaction, 209-214, Copyright 1998, with permission from Elsevier Science*					

The structure of one type of aliphatic/aromatic copolyester is shown in **Figure 6.7**.

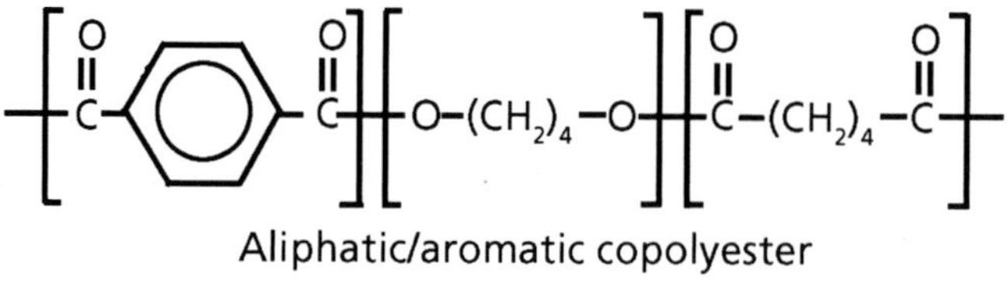

Aliphatic/aromatic copolyester

Figure 6.7

Copolymerisation of aliphatic monomers with aromatics such as terephthalic acid is one way to improve the performance properties of aliphatic polyesters. Questions have been raised, however, within the industry regarding the complete biodegradability of aliphatic/aromatic copolyesters because aromatic polyesters such as PET are resistant to microbial attack [3]. Researchers at the Gesellschaft für Biotechnologische Forschung in Germany have discovered that the biodegradability of these copolyesters is related to the length of the aromatic sequence [3]. Block copolyesters with relatively long aromatic sequences are not rapidly degraded by microorganisms. In the case of polybutylene terephthalate oligomers, oligomers with a length $n \geq 3$ show very little degradation over a period of several months, in contrast with aromatic sequences of $n = 1$ or $n = 2$ where degradation occurs within four weeks [51].

For some applications such as blown-film production, higher melt viscosities and hence higher MW are necessary [52]. These properties can be achieved by incorporating diisocyanates into the polymer chain as a chain extender. An important question is raised as to how chain extension influences biodegradability of the polymer. Studies done at Gesellschaft für Biotechnologische Forschung in Germany indicate that the biodegradation rates of chain-extended 1,4-butanediol/adipic acid/terephthalic acid copolyesters are the same as nonextended copolyesters in a compost environment [52].

Aliphatic/aromatic copolyesters may be prepared either as random copolymers or block copolymers. Random copolymers are more readily biodegraded than copolymers with long aromatic blocks. Generally, copolyesters with about 35-55 mol% aromatic component (in reference to the total amount of acid components) are in an optimal range that guarantees biodegradability and suitable mechanical and physical properties [51].

BASF has patent applications for biodegradable copolyesters that have an aromatic component [53, 54]. These polymers are prepared in a two-step process. First polytetramethylene adipate is prepared from 1,4-butanediol and adipic acid. Next this polymer is reacted with dimethyl terephthalate (DMT), 1,4-butanediol, pyromellitic dianhydride, and polyethylene glycol (MW of 600) in the presence of a titanium catalyst to obtain an aromatic polyether ester that is biodegradable. A sulfonate compound may also be incorporated into the polymer.

Eastman Chemical has patented three different families of linear, random aliphatic aromatic copolyesters [55]. The three families are composed of the following diols and diacids:

1. Diacids: Glutaric acid (30-65 mol%), diglycolic acid (0-10 mol%), terephthalic acid (TPA; 5-60 mol%).

   Diol: 1,4-butanediol (100 mol%).

2. Diacids: Succinic acid (30-85 mol%), diglycolic acid (0-10 mol%), TPA (5-60 mol%).

   Diol: 1,4-butanediol (100 mol%).

3. Diacids: Adipic acid (30-65 mol%), diglycolic acid (0-10 mol%), (25-60 mol%).

   Diol: 1,4-butanediol (100 mol%).

Copolyester mechanical properties are dependent on the content of the terephthalic acid in the copolymer. Tensile strength and behaviour of the elongation at break are examples. Tensile strength increases from 8 N/mm^2 to 12 N/mm^2 as the terephthalic acid composition increases from 31% to 39%. The material becomes stiffer with increasing aromatic composition. Elongation at break is constant at 500% with compositions up to 44% aromatic, but drops rapidly with higher aromatic composition [52].

Aliphatic/aromatic copolyesters have the potential to be lower priced than most biodegradable polymers. They can be produced from widely available, low priced monomers, e.g., adipic acid, butanediol and terephthalic acid. They can also be produced in existing polyester facilities, so they require little or no new capital investment. Witt and co-workers have indicated that prices of $2-4 per kilogram may be possible [52]. This is consistent with process economic evaluations undertaken by the Process Economics Program [27].

DuPont has patented aliphatic/aromatic copolyesters that contain sulfo groups [56-59]. Polyesters that are copolymerised with 5-sulfoisophthalic acid hydrolyse readily. The polyesters are reported to be biodegradable and can be processed at higher temperatures than other biodegradable polymers. These polyesters also offer the cost advantages mentioned earlier for aliphatic/aromatic copolyesters.

Bayer has developed polyesteramides that are non-aromatic. BAK 1095 is based on caprolactam, butanediol, and adipic acid and BAK 2195 is based on adipic acid and hexamethylenediamine [60]. DIN 54900 [61] tests indicate that BAK 1095 is completely biodegradable. These polymers were patented in 1995 [62]. Grigat and co-workers reports the modulus of stiffness for BAK 1095 to be 220 N/mm^2 and the tensile strain at break to be greater than 400% [60]. Bayer dropped its biodegradable polymers business in 2001.

Polyvinyl alcohol (PVOH) is a water soluble polymer prepared by the hydrolysis of polyvinyl acetate. The degree of solubility and biodegradability as well as other physical attributes can be controlled by varying the MW and the degree of hydrolysis of the polymer [63]. Polyvinyl acetate, if hydrolysed to less than 70%, is claimed to be nonbiodegradable under conditions similar to those that biodegrade the fully hydrolysed polymer [64].

The high degree of crystallinity of PVOH makes it impossible to process as a thermoplastic. Unplasticised PVOH thermally degrades at about 150 °C, but the crystalline melting point is 180-240 °C [65]. Attempting to thermally process unplasticised PVOH leads to release of water and the formation of conjugated double bonds. Consequently, PVOH film had to be produced by an expensive solution casting process. Recently, several companies have developed biodegradable PVOH that can be processed as a thermoplastic: Environmental Polymers Group (EPG) in the United Kingdom, Idroplast in Italy, Millenium Polymers in the United States and PVAX Polymers in Ireland [66].

Environmental Polymers Group has patented an extrusion process together with PVOH formulation technology to produce thermoplastic PVOH pellets which can be converted into film and sheet products [67]. EPG PVOH, which is typically 40-50% crystallinity, can be used to produce films with tensile and tear strengths superior to PE and PVC (see **Table 6.7**).

Table 6.7 Typical properties of EPG PVOH film				
Property	EPG PVOH	Cellophane	PVC	PE
Clarity (light transmitted), %	60-66	58-66	48-58	54-58
Water vapour transmission at 40 °C and 90% RH	1500-2000	1300-2000	120-180	35-180
Tear strength, Elmendorf Nm/m	147-834	2-4	39-78	29-98
Tensile strength, MN/m^2	44-64	55-131	20-76	17-19
Elongation at break, %	150-400	-	5-250	50-600

RH: relative humidity
Reprinted from Materials World, April 2000, with permission from N. Hodgkinson and M. Taylor

6.1.6 Modified Naturally Biodegradable Polymers

Carbohydrates are naturally occurring organic compounds that are related to simple sugars. They are extremely widespread in plants, accounting for as much as 80% of plant dry weight. Polysaccharides are biodegradable polymers made up of simple carbohydrate subunits, i.e., saccharides such as glucose. The linear polysaccharide, cellulose, is probably the single most abundant organic compound on earth and it is the chief structural component of plant cells. Enzymes (cellulases) that can catalyse the degradation of cellulose to glucose are common in microorganisms. Starch is the second most abundant polysaccharide. In animals it is the chief source of carbohydrates; in plants it is present in the form of small insoluble starch granules. Polysaccharides such as starch readily gelatinise in hot water to form a paste that can be cast into film. Such films, however, are sensitive to water and become brittle on drying. Less brittle films can be produced by combining polysaccharide with other materials such as plasticisers and synthetic polymers.

Since the 1970s, numerous attempts have been made to enhance the biodegradability of synthetic polymers by incorporating polysaccharide-derived materials [68-73]. The microstructure plays a fundamental role in determining the biodegradation rate of this class of products. The proposed sequence for biodegradation is:

1. The polysaccharide material is first consumed by microorganisms, a process that also increases the surface area of the synthetic polymer and weakens the polymer matrix.

2. The remaining synthetic polymers break into smaller fragments as a result of other environmental mechanisms, which are eventually small enough for assimilation by microorganisms. Reportedly even PE is biodegradable if the MW is below 500 [2].

The dominant commercial starch-based biodegradable polymers are marketed by Novamont in Italy under the name Mater-Bi®. This starch-based technology is unique because the modification goes beyond conventional compounding. In the Mater-Bi® technology, starch is destructurised by applying sufficient work and heat to almost completely destroy the crystallinity of amylose and amylopectine in the presence of macromolecules able to form a complex with amylose. Novamont produces several different classes of Mater-Bi®, all containing starch with different classes of synthetic components such as PCL [12]. The material obtained is available for making film and sheets, foams, and injection moulding.

The biodegradability of various Mater-Bi® classes does vary somewhat. The aerobic biodegradation rate of a Mater-Bi® Z class for film and sheet products compares favourably to biodegradation rates of pure cellulose under composting conditions. Although Mater-Bi® starch/EVOH copolymer products are not compostable, they do biodegrade in the soil.

Modified starch-based polymers are promoted as potential solutions to the current litter/municipal solid-waste management problems in as much as they serve two purposes:

1. Using polysaccharides that are naturally derived macromolecules from renewable sources reduces the amount of plastic needed from non-renewable petrochemical sources.

2. Incorporation of the polysaccharide enhances the overall biodegradability of the combined product.

Another argument for using agriculturally based additives is the potential reduction in farm subsidies it could allow. For example, use of corn starch may provide another use for the nation's corn production.

Proposed polysaccharide-derived materials as biodegradable fillers include a variety of starches, cellulose, lignin, sawdust, casein, mannitol, lactose, and other materials. These fillers have been tried in compositions of as much as 80% in a wide range of synthetic resins, including PE, PP, PS, ethylene-acrylic acid copolymers, PVC, and vinyl alcohol copolymers. Often additional additives such as fatty acids and processing aids are incorporated to improve the biodegradability of the finished product. Starch-based polymers are discussed further in Chapter 3 of this handbook.

Questions concerning the validity of polysaccharide-filled synthetic polymers focus on problems with processing, certain undesirable product properties, and efficacy of landfill biodegradation. Given the hydrophilic nature of starches, care must be taken to avoid exposing them to moisture during processing and storage. In most cases introduction of polysaccharides reduces product strength, and additional resin may be required to maintain product integrity. After the biodegradation of the polysaccharide portion, the surface area is increased, leaving a porous and weakened matrix. However, a nonbiodegradable polymer may remain. As noted previously, biodegradation depends on several factors, so even the polysaccharide portion may not degrade if the proper microorganisms and biodegradation conditions are not present.

The starch-based polymer compositions containing PCL have mechanical properties very similar to those of LDPE (see **Table 6.8**). They are primarily designed for film and layer applications such as compost bags.

6.2 Processability

In general for the purpose of processability classification, resins are classified as a thermoplastic or a thermoset depending on the effect of heat. Thermoplastics soften and

Table 6.8 Properties of Z grade mater-bi			
Property	ZF03U/A	Zl01U	LDPE
MFI, g/10 min	4-5.5	1.5	0.1-22
Strength at break, MPa	31	28	8-10
Elongation at break, %	886	780	150-600
Young's modulus, MPa	185	180	100-200
Tear strength			
Primer, N/mm	68	55	60
Propagation, N/mm	68	55	60

MFI: *melt flow index*
Reprinted from Polymer Degradation and Stability, 59, C. Bastioli, Properties and Applications of Mater-Bi Starch-Based Materials, 263-272, Copyright 1998, with permission from Elsevier Science

flow as liquids by the application of heat and pressure. When cooled, they solidify. These phase changes related to heating and cooling can be repeated with little or no detriment to the polymer's physical properties. Thermoplastics can be very quickly and efficiently processed into finished products by a variety of thermo-processing techniques. They can also be recycled easily. These are a few of the reasons that thermoforming and other thermo-processing techniques have displaced thermosets in many applications [74]. Thermosets crosslink upon heating and once moulded, cannot be reheated and moulded again [75]. Nearly all commercially important biodegradable polymers are thermoplastics with the main exception being PVOH, which acts as a thermoplastic only when modified,

Thermoplastics fall into two broad classes as determined by their morphology: amorphous and crystalline. Crystalline thermoplastics have melting and freezing points while amorphous ones do not have melting points [75]. For amorphous polymers crystallisation does not take place preventing the formation of ordered regions. A transition from a liquid (or rubbery state) to a solid (or glassy state) for an amorphous polymer is termed the glass transition, T_g.

Table 6.9 Melting points for conventional and biodegradable thermoplastics			
Conventional thermoplastics	Melting point, °C	Biodegradable thermoplastics	Melting point, °C
LDPE	110	PCL	60 [46]
HDPE	127	Succinate polyesters	96-114 [48]
PP	176	Copolyesters	79-137 [52]
Nylon 6	225	Meso PLA	130 [36]
Polyester 4GT	230	PHBV	130-140 [18]
Polyester 2GT	265	Polyesteramides	125-175 [60]
Nylon 6,6	265	100% L-PLA	180 [36]
Source for conventional thermoplastics: [75]			

6.2.1 Extrusion

Almost all thermoplastics are processed by extrusion at some stage of commercial manufacture. Many plastics are first extruded during polymer manufacture and compounding operations before reaching the final extrusion device used to make the fabricated products [75]. The two main types of extruders are single screw extruders and twin screw extruders. Single screw extruders are simpler, less expensive, and more widely used than twin screw extruders. Twin screw extruders are often used for difficult compounding applications, devolatilisation, and for extruding finished products from viscous polymers with limited heat stability (notably many biodegradable polymers) [76].

Starch-based polymers have been the most studied class of biodegradable polymers for their extrusion characteristics. Extrusion processing plays a large role in establishing the polymer's properties. Starch can be made thermoplastic using technology very similar to extrusion cooking [12]. Starch exists as granular beads of about 15-100 μm in diameter that can be compounded with another synthetic polymer as a filler [77]. However, under special heat and shear conditions during extrusion it can be transformed into an amorphous thermoplastic by a process known as destructurising.

Starch can be destructurised in the presence of more hydrophobic polymers such as aliphatic polyesters [78]. Aliphatic polyesters with low melting points are difficult to process by conventional techniques such as film-blowing and blow-moulding. Films made from PCL are tacky as extruded and have low melt strength over 130 °C. Also, the slow crystallisation of this polymer causes the properties to change with time. Blending starch with aliphatic polyesters improves their processability and biodegradability [79].

Addition of starch has a nucleating effect, which increases the rate of crystallisation [78]. The rheology of starch/PCL blends depends on the extent of starch granule destruction and the formation of thermoplastic starch during extrusion. Ko reports that increasing shear and heat intensities can reduce the melt viscosity, but enhance the extrudate-swell properties of the composite [80].

Starch/aliphatic polyester compositions are prepared by blending a starch-based component and an aliphatic polyester in a corotating, intermeshing twin screw extruder, the most commonly used equipment today for plastifying starch [81]. The corotating, self-cleaning screws on these machines prevent caking and charring of cooked starch [82]. Temperature and pressure conditions are such that the starch is destructurised and the composition forms a thermoplastic melt. The resulting material has an interpenetrated or partially interpenetrated structure. According to a Novamont patent, preparation of the blends involves several steps [83]:

- Forming thermoplastic starch in an extruder by mixing starch with EVOH and plasticiser (this step is optional).

- Swelling the thermoplastic starch and aliphatic polyester with additional plasticiser and water in the first stage of an extruder.

- Shearing the mixture in the extruder.

- Degassing the melt under vacuum to a water content of 1.5-5 wt%.

- Cooling the product in a water bath or in air.

Other companies also have patented starch/aliphatic polyester blends. Metraplast in Germany has a patent application for compositions of latex, starch, PHB and/or cellulose powder [84]. The materials are mixed and plasticised in a screw-type extruder. The plasticised mass can be injected directly into a mould or extruded in the form of a strand to make granulates. Nihon Shokuhin Kako in Japan has patented compositions of gelatinised fat- or oil-treated starch and aliphatic polyesters [85]. The composition is made by mixing the treated starch with the aliphatic polyester in the presence of water or water and a plasticiser. Gelatinisation of the starch generally refers to heating the starch in the presence of water, causing the starch to lose all crystallinity and increase its viscosity.

Researchers at Michigan State University have patented the use of aliphatic polyester-grafted starch as a compatibiliser between starch and aliphatic polyesters such as PCL [86]. These grafted compatibilisers provide enhanced interfacial adhesion between the starch and polyester phases. Compositions with two phases can be generated with a variety of morphologies that affect the properties of the blend. Interfacial adhesion is one factor among several that determine the morphology of the blend. Compositions with polyester-grafted starch as compatibiliser may be used in the manufacture of biodegradable films for bags.

6.2.2 Film Blowing and Casting

The two main processes used commercially for making film from thermoplastics are blowing and casting. Most blown film is used for food and trash bags. Blown film is extruded as a tube and the tube is filled with air to expand the tube to the desired size [75]. The tubular film is cooled, flattened, and extruded again over an isolated bubble of air. Typical film thicknesses are 0.007-0.125 mm. Blown film processing requires a high melt viscosity resin so the melt can be pulled from a die in an upward direction [75].

The process for making cast film involves drawing a molten web of resin from a die onto a roll for controlled cooling. The cast film process is used to make a film with gloss and sparkle. The melt temperature in the cast film process is higher than in the blown film process, a higher melt temperature imparts better optical properties [75].

Many of the biodegradable polymers described in this chapter are suitable for film blowing and casting; although modifications are often necessary and productivities may not be as high as conventional thermoplastics. For example, starch-based Mater-Bi® films can be produced by film blowing and casting equipment traditionally used for LDPE with minor or no modifications. Film production productivity is reported to be 80-90% of LDPE [87]. The main difference from traditional PE film production is the lower welding temperatures, therefore small to medium sized production lines with good cooling capacity are the best for processing starch-based films [88].

PLA films with thicknesses of 8-510 µm have been obtained from commercial film casting equipment [89]. PLA can be difficult to process into a film due to instability at elevated processing temperatures. According to a recent Cargill patent, melt stable PLA suitable for processing into films can be made by controlling the polymer composition as well as adding stabilising or catalyst-deactivating agents [90]. The polymer MW plays a role in its processability. Also, polymer morphology is very important. Semi-crystalline PLA is suitable for processing into films with desirable barrier properties. The desired range of compositions for semi-crystalline PLA is less than 15 wt% meso-lactide and the remaining weight percent being L-lactide [90].

Crystallisation of a thermoplastic must occur quickly, i.e., in a few seconds, for efficient film processing. Cargill has patented four methods to increase the rate of PLA crystallisation [88]:

- Adding a plasticising agent such as dioctyl adipate

- Adding a nucleating agent such as talc

- Orientation by drawing during film casting or blowing or after it is cast or blown

- Heat setting, which involves holding constrained oriented film at temperatures above T_g.

Until recently, the only route to high performance PVOH film has been the expensive solution casting method. As previously mentioned, EPG has patented an extrusion process together with PVOH formulation technology to produce thermoplastic PVOH pellets which can be converted into film and sheet products [67]. According to Smith, dual extrusion is also possible using this technology allowing films to be produced combining layers of PVOH film with different water solubility characteristics [91].

6.2.3 Moulding

Thermoplastics can be moulded into articles by injection moulding or blow moulding. In injection moulding high pressure is used to inject molten thermoplastic into a mould where it solidifies. Blow moulding is the most common process for making hollow articles such as bottles [75]. In this type of moulding, a molten tube of resin is extruded, a mould is closed around the tube, and air is fed into the tube to expand it into the mould.

Most of the biodegradable polymers discussed in this chapter can be used for making moulded articles. One historical example is the processing of PHBV into injection moulded articles. It was found that the degree of crystallinity is a result of the processing history during the injection moulding process [92]. In what is known as the fountain flow effect, hot melt flows into a cold mould and quickly forms a frozen layer on the surface of the mould while material in the centre of the sample does not cool as quickly. According to Parikh and co-workers, this difference in cooling rate and orientation causes a difference in the crystallisation between the material close to the surface and material closer to the core. The degree of crystallinity of injection moulded PHBV affects both the properties of the article as well as its biodegradability [92]. This result is also true for many other biodegradable thermoplastics.

PLA is a biodegradable polymer that may not be well suited for injection moulding. Lunt reports that the rate of crystallisation is too slow to allow cycle times typical of those for commodity thermoplastics such as PS [36]. Stress induced crystallisation that can enhance PLA crystallisation is better suited to processes such as fibre spinning or biaxial orientation of film.

6.2.4 Fibre Spinning

The most commonly used commercial processes for making fibres are melt spinning, dry spinning and wet spinning. Melt spinning is the most economical, but can only be applied to polymers that are stable at temperatures sufficiently above their melting point to be extruded in the molten state without degradation. The properties of crystalline polymers can be improved when made into fibre form by the process of orientation or drawing. The result is the increased strength, stiffness, and dimensional stability associated with synthetic fibres.

Over the past decade the properties of PLA fibres have been studied intensively and these fibres are now commercially available from Cargill Dow in the United States and Kanebo Gohsen in Japan. PLA fibre properties compare well with both PET and rayon fibres as indicated in **Table 6.3**. Conditions that the polymer are subjected to during the spinning process do impact on fibre properties such as tensile strength and elongation [93]. Fambri and co-workers have found that polymer degradation takes place during the melt spinning process even when using dry polymer with less than 0.005% water content [94]. Fibres produced by dry spinning undergo very slight degradation. Studies by Schmack and co-workers indicate that PLA can be spun both in a high speed spinning process with a take-up velocity of up to 5000 m/min and in a spin drawing process up to a draw ratio of 6 [95].

6.3 Industrial Applications

One of the main obstacles to widespread use of biodegradable polymers has been the high cost of these polymers. For this reason, industrial applications tend to be specialist applications with unique environmental considerations. Loose-fill packaging and compost bags are the two major end uses constituting nearly 90% of demand. Several other applications offer strong market potential for the future, primarily in Europe.

6.3.1 Loose-Fill Packaging

Environmental concerns have driven cushioning material manufacturers to develop cleaner, more environmentally sound alternatives to traditional foamed PS packaging. Biodegradable packaging has intrigued packagers because it is degradable upon contact with water and can be easily disposed by composting or rinsing it down a drain. Since its introduction in 1990, demand for starch-based loose-fill packaging has grown to capture up to 25% of the 4.2 million m^3 per year US loose-fill market. Starch-based loose-fill packaging is also used in Europe and Japan, but to a lesser degree than in the United States.

The biodegradable polymer of choice for loose-fill packaging is starch-based due to its low price relative to other commercially available biodegradable polymers. Warner-Lambert was the pioneer of biodegradable loose-fill foam packaging with its starch-based Novon® N2002 product. The company closed down its business in late 1993 and terminated production in early 1994. With the departure of Warner-Lambert as a supplier, other companies in the US, Europe and Japan have acquired manufacturing and distribution rights for Novon® N2002 from Warner-Lambert. Novamont and National Starch have emerged as the two primary suppliers of starch-based loosefill.

6.3.2 Compost Bags

Compostable bags for organic waste has been one of the most natural applications for biodegradable polymers. In Europe this demand is driven by the European Council ban on landfilling or incinerating waste with more than 10% organic content. Germany bans from landfill anything with more than 5% organic content. As a result of these regulations, composting infrastructure is becoming highly developed in Germany and other parts of Europe such as Austria, Belgium, The Netherlands and Italy. Disposal of solid wastes through composting is about 50% lower than landfilling in Europe [96].

With the growth of composting has come the demand for compostable bags for collecting organic kitchen and garden waste. Compost bags have been marketed in Germany since 1995. They are primarily sold in supermarkets, but in some cases are also sold by local authorities and composting plants. The main suppliers of compost bags in Germany are Natura Verpackungs GmbH, Novamont Deutschland GmbH, Wentus Kunststoff GmbH, Wolff Walsrode AG and BASF AG. Prices range from 0.09 to 0.30 euros per bag. Recently, a compostability logo based on DIN 54900 [61] certification was introduced in Germany.

In 1994 separate collection of organic waste was made mandatory in The Netherlands. Most compost facilities are reluctant to receive organic waste in bags. Velca Trading sells compost bags on the Dutch market for 0.30 euros per 15 litre bag.

As a response to National Decree 22/97 in Italy, more than 1,500 Italian municipalities have begun source separation programs for food waste. Biodegradable bags and PE bags are being used to collect food waste. However, those composting facilities which accept PE bags have a higher tipping fee than those accepting biodegradable bags. As a result, more than 95% of the municipalities in Milan and Northern Italy have adopted biodegradable bags.

Demand for biodegradable polymers in compost bag applications has not grown to the same extent in the United States as it has in Europe. The only significant need for compost bags in the United States is to collect garden waste. Garden waste is collected by four different methods: in paper bags, in PE bags, in biodegradable bags, or mechanically in bulk. The advantage of biodegradable bags in this application is that they can be simply disposed of, along with the waste, in a composting facility. Nonbiodegradable plastic bags have to be emptied and then disposed of separately. Paper bags tend to absorb moisture from the waste and then lose their strength as a result. Also, paper bags tend to take up more shipping volume than plastic bags. In spite of some of these performance flaws, the majority of compost bags used in the US are paper due to price advantages.

Although compost bags are a target application for biodegradable polymer producers in Japan, compost infrastructure is still lacking. Composting has not yet achieved significant use as a system for treating solid waste, but it is being explored as an alternative to incineration. The Ministry of International Trade and Industry (MITI) conducted a model composting enterprise in Hiroshima Prefecture in 1994-1995. This model system included evaluating seven kinds of biodegradable polymers. Several small local governments have adopted the use of composting and begun using compost bags made of succinate polyesters.

6.3.3 Other Applications

A variety of other biodegradable polymer applications are under development, but have not reached significant market size in any region of the world yet. Many of these are especially promising in Europe due to the developing composting infrastructure as well as influences such as the German Packaging Ordinance and the European Union Packaging Waste Directive. Companies in Europe that currently use plastics for single use packaging are studying biodegradable polymers as the possible solution. If biodegradable polymers become widely accepted in composting systems and the price of these polymers fall, there is a large potential demand in Europe for biodegradable polymers in applications such as food packaging and disposable dishes and cutlery.

German dairy product manufacturer Danone launched a biodegradable yogurt package, known as the Eco Cup, in 1998. The base material was polylactic acid and the cups initially cost about three times as much as PP or PS cups. In 1999 Danone withdrew the Eco Cup from the market due to confusion about which waste bin consumers should use for disposal of the package.

One market which may provide potential for biodegradable polymers includes institutions such as theme parks or special events that must manage their own solid wastes. This concept was showcased at the Sydney Olympic Games in 2000 where 40 million food service items made of starch-based polymers were collected after use and composted [97]. Disposable cutlery and dishes used at the Sydney Olympic Games were supplied by Biocorp Inc., Novamont's North American distributor.

Farmers use plastic agricultural film in their fields for a variety of reasons, such as to heat up the soil, eliminate the growth of weeds or cover soil fumigated with methyl bromide. The film allows for earlier and better crops. One drawback is that the film must eventually be removed from the field and disposed of in some manner. The dominant material used for agricultural films is LDPE. Biodegradable polymers offer an advantage over LDPE because they can be left in the field to degrade. If biodegradable polymer prices become more competitive with LDPE, this could become a large potential application. Other potential agricultural applications include twine and nonwoven sheets for weed control.

6.4 Market Evolution

The evolution of the biodegradable polymers market is a relatively recent phenomena when compared to the decades long history of conventional plastics. Beginning in the 1970s, industry attempted to develop products that could be used in single-use, throwaway applications that would degrade after disposal. However, the efficacy of such products was questionable. Well-defined testing protocols for verifying degradability claims and environmental fate were lacking. As a result, the Federal Trade Commission, a group of state attorneys-general, state legislatures, and the US Congress became concerned about the various degradability and environmental claims made in relation to waste management [98]. The resulting publicity was a serious set back for market development, especially in the United States.

Environmental concerns associated with conventional, non-degradable polymers continued through the 1980s. Of primary concern is the solid waste problem associated with decreasing availability of landfills around the world. Other concerns include the benefits of sustainable or renewable raw material sources rather than petrochemical sources and the issue of global warming caused by increasing amounts of carbon dioxide in the atmosphere. In response to these environmental concerns, commercial introduction of second generation biodegradable polymers occurred at about the same time in the United States as in Europe - around 1990. These improved biodegradable polymers would not have been developed without the efforts of several leading companies. Some of these companies have since dropped their biodegradable polymer businesses when the realisation of the long development time set in.

One of the first pioneer companies to develop a biodegradable polymer that was completely biodegradable and also have good properties was ICI in the United Kingdom. The combined efforts of ICI's Agricultural Division, which was experienced in developing single-cell proteins and running large-scale fermentation processes, and its Plastics Division, experienced in polymer processing and evaluation, led to the commercialisation of BIOPOL™ polymers in 1981. Monsanto purchased the BIOPOL™ business from Zeneca Bio Products (formerly ICI) in April of 1996. Monsanto manufactured PHBV at its Knowsley, UK, fermentation facility until 1999. At that time Monsanto elected to leave the biodegradable polymers business after failing to make BIOPOL™ costs more competitive with petroleum-based polymers. Monsanto indicated that it cost roughly $8.8 to make a kilogram of PHBV through fermentation and that its goal of using bio-engineered crops to lower production costs was at least 5-7 years away [99]. Monsanto is no longer actively researching PHBV - Metabolix licensed its technology in 2001.

Warner-Lambert, a large pharmaceutical company in the United States, played an important role in the early 1990s in developing starch-based polymer technology and promoting the benefits of biodegradable polymers. Warner-Lambert scientists in Switzerland discovered starch-based polymers while they were researching injection-mouldable materials that could be substituted for gelatin in pharmaceutical capsules. NOVON® biodegradable polymers were introduced commercially in 1990 and a large 45,000 ton per year manufacturing facility was built in Rockford, Illinois, USA. In spite of its large investment, Warner-Lambert announced that it was suspending operations of its Novon Products Group in 1993 [100]. The company took a $70 million before tax charge on its $100 million investment in the biodegradable polymers business.

After the departure of Warner-Lambert, Novamont in Italy emerged as the dominant starch-based polymer company. Novamont started its research activity on starch-based polymers in 1989 as part of the Montedison group in Italy [101]. Mater-Bi® was commercialised with the startup of a 4,000 ton per year plant in Terni, Italy in 1990. The capacity was doubled in 1997 and then again in 2001 [97]. Novamont's leading position in starch-based polymers was further consolidated by the 1997 acquisition of the patent portfolio and related worldwide licenses formerly belonging to Warner-Lambert [101]. Bastioli reports that the global market for starch-based polymers had grown to 12,000 tons per year by the late 1990s [102].

The most dramatic development in the market evolution of biodegradable polymers in the 1990s came with the creation of Cargill Dow Polymers by Dow Chemical and Cargill, a large U.S. agribusiness company. The joint venture's product is NatureWorks™ PLA. Cargill began production of PLA on a semi-commercial scale at a 4,500 tons per plant in Savage, Minnesota, USA, in 1994. At the end of 1997 Dow and Cargill formed Cargill Dow Polymers to develop and market PLA on a large scale. They are investing $300 million in the venture and most of those funds are being used to build a 140,000 tons per year PLA plant in Blair, Nebraska, USA. With the startup of Cargill Dow Polymers' large production facility, the biodegradable polymers market could begin evolving from a higher priced niche type market to more commodity-like. Cargill Dow Polymers anticipates that the price of PLA will fall to $1-2 per kilogram after its new plant starts up [42].

In contrast to PHBV, starch-based polymers and PLA, which are primarily derived from renewable agricultural feedstocks; during the later 1990s a number of large chemical companies have introduced biodegradable polymers derived from petrochemical feedstocks. These include DuPont's Biomax®, Eastman's Eastar Bio copolyester, Bayer's BAK (production ceased in 2001) and BASF's Ecoflex®. These polymers offer improved performance properties and at the same time also offer somewhat lower costs because they are derived from commodity feedstocks such as adipic acid and dimethyl terephthalate. The success of these biodegradable polymers depends to a large extent on the importance of sustainability and renewable feedstocks in the market.

In 1998, total demand for biodegradable polymers in the United States, Western Europe and Japan reached 18 thousand metric tons valued at over $70 million according to SRI's Chemical Economics Handbook (CEH) [103]. **Table 6.10** summarises the 1998 supply/demand situation for biodegradable polymers in the major producing and consuming regions of the world.

Historically in the 1990s, the United States was the dominant market for biodegradable polymers, accounting for about one-half of world consumption; Western Europe accounted for about 40% and Japan accounted for less than 10%. This is primarily due to the demand for biodegradable loose-fill in the United States. However, a large proportion (over 60%) of the world's 1998 production capacity was located in Western Europe in anticipation of market growth there. In 1998, there was little production in Japan. In 1998, the starch-based family of biodegradable polymers was dominant; however, new plant investment will bring PLA to a much more prominent position in the market by 2001 or 2002.

Forecasting future demand for any new material in the early stages of development is always challenging, and the biodegradable polymers market is no exception. SRI has projected total consumption of biodegradable polymers in the three major regions will increase to about 70 thousand metric tons in 2003, representing an average annual growth rate of over 35% over the five-year period from 1998 to 2003 [103]. This growth projection assumes that approximately 140 thousand metric tons per year of new production capacity is brought on stream prior to 2003, allowing producers to achieve dramatic price reductions. Regulation and legislation will play a large role in market growth, especially in Europe. Bastioli indicates that the potential European market could be as much as 500,000 tons per year [102]. However, if there is no significant change in legislation there, she indicates that European sales volumes will probably not exceed 40,000 tons per year in the next few years.

Table 6.10 Supply/demand for biodegradable polymers by major region- 1998 (thousands of metric tons)				
	United States	Western Europe	Japan	Total
Annual Capacity	11	29	6	46
Production	10	8	1.5	19.5
Consumption	9	7	2	18
Source: CEH estimates [103]				

References

1. R. Narayan in *Proceedings of the International Composting Research Symposium*, Columbus, OH, USA, 1992.

2. D.L. Kaplan, J.M. Mayer, D. Ball, J. McCassie, A.L. Allen and P. Stenhouse in *Biodegradable Polymers and Packaging*, Eds., C. Ching, D.L. Kaplan and E.L. Thomas, Technomic, Lancaster, PA, USA, 1993, 1.

3. R.J. Muller, *DECHEMA Monographs*, 1996, **133**, 211.

4. ASTM D3826-98, *Standard Practice for Determining Degradation End Point in Degradable Polyethylene and Polypropylene Using a Tensile Test*, 2002.

5. M. Agarway, K.W. Koelling and J.J. Chalmers, *Biotechnology Progress*, 1998, **14**, 517.

6. J.E. Potts in *Kirk-Othmer Encyclopedia of Chemical Technology*, Ed., J. Kroschwitz, Wiley-Interscience, New York, 1984, Suppl. Vol., 626.

7. G. Swift in *Kirk-Othmer Encyclopedia of Chemical Technology*, Ed., J. Kroschwitz, Wiley-Interscience, New York, 1996, 19, 968.

8. A.L. Andrady in *Physical Properties of Polymers Handbook*, Ed., J.E. Mark, AIP Press, Woodbury, New York, 1996, 625.

9. C.L. McCormick in *Encyclopedia of Polymer Science and Technology*, Ed., J.I. Kroschwitz, John Wiley & Sons, New York, 1989, **17**, 745.

10. J.N. BeMiller in *Kirk-Othmer Encyclopedia of Chemical Technology*, Ed., J.I. Kroschwitz, Wiley-Interscience, New York, 1992, 4, 934.

11. B. Dobler and R.F.T. Stepto, inventors; Warner-Lambert, assignee; GB 2,214,516A, 1989.

12. C. Bastioli, *Polymer Degradation and Stability*, 1998, **59**, 263.

13. K.W. Suh and co-workers in *Encyclopedia of Polymer Science and Engineering*, Ed., J.I. Kroschwitz, Wiley & Sons, New York, 1985, 3, 40.

14. C. Bastioli, V. Bellotti, G. Del Tredici, A. Montino and R. Ponti, inventors; Novamont, assignee; EP 696,611A2, 1996.

15. H.G. Franke and D.R. Bittner, inventors; Free-Flow Packaging, assignee; US 5,512,090, 1996.

16. M. Stauderer and H. Zieger, inventors; K&S Bio-Pack, assignee; DE 4,429,269, 1996.

17. Y. Akamatu and M. Tomori, inventors; Nippon Synthetic Chemical Industries, assignee; US5,308,879, 1994.

18. J.M. Liddel in *Chemical Industry: Friend to the Environment?*, Ed., J.A.G. Drake, Special Publication No.103, Royal Society of Chemistry, Cambridge, UK, 1992, p.10.

19. *Modern Plastics*, 1981, **58**, 7, 90.

20. P.A. Holmes, *Physics Technology*, 1985, **16**, 32.

21. J.N. Baptist, inventor; W.R. Grace, assignee; US 3,036,959, 1962.

22. J.N. Baptist, inventor; W.R. Grace, assignee; US 3,044,942, 1962.

23. J.N. Baptist and F.X. Werber, inventors; W.R. Grace, assignee; US 3,107,172, 1963.

24. L. Hughes and K.R. Richardson, inventors; ICI, assignee; US 4,433,053, 1984.

25. P.A. Holmes, S.H. Collins and L.F. Wright, inventors; ICI, assignee; US 4,477,654, 1984.

26. K.R. Richardson, inventor; ICI, assignee; EP 114,086B1, 1988.

27. G.M. Bohlmann and R.G. Bray, *Biodegradable Polymers*, Process Economics Program Report No.115C, SRI Chemical and Health Business Services, Menlo Park, CA, USA, 1998.

28. Y. Poirier, D.E. Dennis, K. Klomparens and C. Somerville, *Science*, 1992, **256**, 520.

29. I. Noda, inventor; The Procter & Gamble Co., assignee; EP0846184B1, 2002.

30. M. Roberts, *Chemical Week*, 1994, **154**, 23, 23.

31. *Monsanto Annual Report*, Monsanto Company, St. Louis, MO, 1996, 12.

32. C.H. Holten, *Lactic Acid Properties and Chemistry of Lactic Acid and Derivatives*, Verlag Chemie, Vienna, Germany, 1971.

33. S.J. Huang in *Encyclopedia of Polymer Science and Engineering*, Eds., H.F. Mark, N.M. Bikales, C.G. Overberger, G. Menges and J. I. Kroschwitz, Wiley & Sons, New York, NY, USA, 1985, 2, 220.

34. X. Zhang, U.P. Wyss, D. Pichora and M.F.A. Goosen, *Journal of Macromolecular Science, Pure and Applied Chemistry A*, 1993, **30**, 12, 933.

35. C.G. Pitt in *Biodegradable Polymers and Plastics*, Eds., M. Vert, J. Feijen, A. Albertsson, G. Scott and E. Chiellini, Royal Society of Chemistry, Cambridge, UK, 1992, p.7.

36. J. Lunt, *Polymer Degradation and Stability*, 1998, **59**, 145.

37. P.R. Gruber and co-workers, inventors; Cargill, assignee; WO 95/9879, 1995.

38. P.R. Gruber, E.S. Hall, J.J. Kolstad, M.L. Iwen, R.D. Benson and R.L. Borchardt, inventors; Cargill, assignee; US 5,357,035, 1994.

39. I.D. Fridman, J. Kwok, R.J. Downey and S.P. Nemphos, inventors; Camelot Technologies, assignee; US 5,357,034, 1994.

40. W.G. O'Brien, L.A. Cariello and T.F. Wells, inventors; Ecological Chemical Products, assignee; US 5,521,278, 1996.

41. K. Enomoto, M. Ajioka and A. Yamaguchi, inventors; Mitsui Toatsu Chemicals, assignee; US 5,310,865, 1994.

42. *Plastics Information Europe*, 2000, **24**, 4, 6.

43. G.M. Bohlmann, *Chemicals from Renewable Resources*, Process Economics Program Report 236, SRI Consulting, Menlo Park, CA, USA, 2001.

44. Cargill Dow, www.cdpoly.com, 2000.

45. R.D. Lundberg and co-workers, inventors; Union Carbide, assignee; Canadian 900,092, 1972.

46. Solvay, www.solvay.com, 2001.

47. *Technical Brochure F-607-45, PO-8114*, Union Carbide, Danbury, CT, USA, 1990.

48. T. Fujimaki, *Polymer Degradation and Stability*, 1998, **59**, 209.

49. E. Takiyama, T. Fujimaki, S. Seki, T. Hokari and Y. Hatano, inventors; Showa Highpolymer, assignee; US 5,310,782, 1994.

50. R.J. Müller, I. Kleeberg and W-D. Deckwer, *Journal of Biotechnology*, 2001, **86**, 87.

51. U. Witt, R.J. Müller and W-D. Deckwer, *Macromolecular Chemistry and Physics*, 1996, **197**, 1525.

52. U. Witt, R.J. Müller and W-D. Deckwer, *Journal of Environmental Polymer Degradation*, 1997, **5**, 2, 81.

53. V. Warzelhan, G. Pipper, U. Seeliger, P. Bauer, U. Pagga and M. Yamamoto, inventors; BASF AG, assignee; WO 96/25446, 1996.

54. Warzelhan, G. Pipper, U. Seeliger, P. Bauer, D.B. Beimborn and M. Yamamoto, BASF, assignee; WO 96/25448, 1996.

55. C.M. Buchanan, R.M. Gardner and A.W. White, inventors; Eastman Chemical, assignee; US 5,446,079, 1995.

56. F.G. Gallagher, inventor; DuPont, assignee; WO 95/14740, 1995.

57. F.G. Gallagher, C.J. Hamilton, S.M. Hansen, H. Shin and R.F. Tietz, inventor; DuPont, assignee; US 5,171,309, 1992.

58. F.G. Gallagher, C.J. Hamilton, S.M. Hansen, H. Shin and R.F. Tietz, inventor; DuPont, assignee; US 5,171,308, 1992.

59. R.F. Tietz, inventor; DuPont E.I. de Nemours, assignee; US 5,097,005, 1992.

60. E. Grigat, R. Koch and R. Timmermann, *Polymer Degradation and Stability*, 1998, **59**, 223.

61. DIN 54900, *Testing of the Compostability of Plastics*, 1998.

62. R. Timmermann, R. Dujardin and R. Koch, inventors; Bayer, assignee; EP 641,817A2, 1995.

63. J.M. Mayer and D.L. Kaplan, *Trends in Polymer Science*, 1994, **2**, 7, 227.

64. S. Matsumura, S. Maeda, J. Takahashi and Yoshikawa, *Kobunshi Robunshu*, 1988, **45**, 4, 317.

65. N. Hodgkinson and M. Taylor, *Materials World*, 2000, **8**, 4, 24.

66. *British Plastics and Rubber*, 1999, **11**, 13.

67. R.L. Jack, inventor; British Technology Group, assignee; GB 2,291,831B, 1997.

68. G.J.L. Griffin, *Proceedings of the 166th meeting of the American Chemical Society*, Division of Organic Coatings and Plastics Chemistry, Chicago, IL, USA, 1973, p.159.

69. G.J.L. Griffin, *Advances in Chemistry Series*, 1974, **134**, 159.

70. G.J.L. Griffin, *Pure & Applied Chemistry*, 1980, **52**, 399.

71. C.L. Swanson, R.P. Westhoff, W.M. Doane and F.H. Otey, *Polymer Preprints*, 1987, **28**, 2, 105.

72. G.F. Fanta and F.H. Otey, inventors; US Department of Agriculture, assignee; US 4839450, 1989.

73. K. Portnoy, *Chemical Week*, 1987, **140**, 20, 36.

74. M. Knights, *Plastics Technology*, 2000, **46**, 6, 66.

75. P.N. Richardson in *Encyclopedia of Polymer Science and Engineering*, Ed., J.L. Kroschwitz, Wiley & Sons, New York, 1988, **11**, 262.

76. G.A. Kruder in *Encyclopedia of Polymer Science and Technology*, Ed., J.L. Kroschwitz, John Wiley & Sons, New York, 1986, **6**, 571.

77. M. Trznadel, *International Polymer Science and Technology*, 1995, **22**, T/58.

78. C. Bastioli, V. Bellotti, G. Del Tredici, R. Lombi, A. Montino and R. Ponti, inventors; Novamont, assignee; WO92/19680, 1992.

79. C. Bastioli in *Degradable Polymers*, Ed., G. Scott and D. Gilead, Chapman and Hall, London, UK, 1995, 112.

80. C. Ko, *Proceedings of Antec '98*, Atlanta, GA, USA, 1998, Volume 3, 3557.

81. U. Funke, W. Bergthaller and M.G. Lindhauer, *Polymer Degradation and Stability*, 1998, **59**, 293.

82. W. Weidmann, *Kunststoffe Plast Europe*, 1994, **84**, 8, 18.

83. C. Bastioli, V. Bellotti, A. Montino, G.D. Tredici, R. Lombi and R. Ponti, inventors; Novamont, assignee; US 5,412,005, 1995.

84. M. Thobor, inventor; Metraplast, assignee; WO 96/6886.

85. T. Shitaohzono, A. Muramatsu and J. Hino, inventors; Nihon Shokuhin Kako, assignee, EP 704,495A2, 1996.

86. R. Narayan, P. Dubois and M. Krishman, inventors; Board of Trustees Operating Michigan State University, assignee; US 5,540,929, 1996.

87. Anon, *Italian Technology Plast*, 2000, **1**, 232.

88. E. Schroeter, *Kunststoffe Plast Europe*, 1998, **6**, 892.

89. R.E. King and A. Gupta, *Proceedings of the Polymers, Laminations and Coatings Conference*, Toronto, Ontario, Canada, 1997, Book 2, 443.

90. P.R. Gruber, J.J. Kolstad, C.M. Ryan, E.S. Hall and R.S. Eichen Conn, inventors; Cargill, assignee; US 6,121,410, 2000.

91. P. Smith, *Panorama*, 1999, **23**, 1, 22.

92. M. Parikh, R.A. Gross and S.P. McCarthy, *Journal of Injection Moulding Technology*, 1998, **2**, 1, 30.

93. K. Yamanaka, *Chemical Fibers International*, 1999, **49**, 6, 501.

94. L. Fambri, A. Pegoretti, R. Fenner, S.D. Incardona and C. Migliaresi, *Polymer*, 1997, **38**, 1, 79.

95. G. Schmack, B. Taendler, R. Vogel, R. Beyreuther, S. Jacobsen and H.G. Fritz, *Journal of Applied Polymer Science*, 1999, **73**, 14, 2785.

96. *Macplas International*, 1999, **10**, 42.

97. R.D. Leaversuch and M. DeFosse, *Modern Plastics International*, 2000, **30**, 11, 28.

98. R. Narayan, *Proceedings of ACS National Meeting*, Chicago, IL, USA, 1993.

99. P. Van Arnum, *Chemical Market Reporter*, 1999, **255**, 7, 16.

100. C. Tollefson, *Chemical Marketing Reporter*, 1993, **224**, 20, 9.

101. Anon, *Italian Technology Plast*, 2000, **1**, 182.

102. C. Bastioli, *Macromolecular Symposia*, 1998, **135**, 193.

103. G.M. Bohlmann and Y. Yoshida, *Biodegradable Polymers*, CEH Marketing Research Report, SRI Chemical and Health Business Services, Menlo Park, CA, USA, 2000.

7 Polyhydroxyalkanoates

Kumar Sudesh and Yoshiharu Doi

Polyhydroxyalkanoates (PHA) is the term given to a family of polyesters produced by microorganisms. The most well known among them is the thermoplastic poly[*R*-3-hydroxybutyrate] (P[3HB]). In this chapter, an attempt has been made to summarise the present state of research and development of these interesting microbial polyesters. Several types of PHA homopolymers and copolymers with useful physical properties have been identified. It is now possible to tailor-make these PHA using suitable carbon sources in both wild type and/or recombinant microorganisms. The basic principles underlying the biosynthesis of various PHA will be emphasised. This will include the major biochemical pathways involved in the conversion of various carbon sources into suitable monomers that are polymerised by the key enzyme of PHA biosynthesis, PHA synthase. In addition, the development of various potential methods for the large-scale production of PHA will be compared. The most attractive property of PHA is its biodegradability, which is a crucial factor in today's polymer technology. Accordingly, this chapter will also include a brief overview on the biodegradation of PHA as well as the mechanisms involved. Finally, the many new potential applications of PHA especially in the medical field will be discussed.

7.1 Introduction

In conditions of excess nutrients, many microorganisms usually assimilate and store them for future consumption [1]. Various storage materials have been identified in microorganisms, which include glycogen, sulfur, polyamino acids, polyphosphate, and lipids [2]. PHA are lipoidic material [3] accumulated by a wide variety of microorganisms in the presence of an abundant carbon source. The assimilated carbon sources are biochemically processed into hydroxyalkanoate units, polymerised and stored in the form of water insoluble inclusions in the cell cytoplasm. The ability to carry out this polymerisation process is dependent on the presence of a key enzyme known as PHA synthase. The product of this enzyme is a high molecular weight (MW) optically active crystalline polyester. The latter is intriguingly maintained in an amorphous state *in vivo* [4]. Upon isolation however, this microbial polyester is a crystalline thermoplastic with properties comparable to that of polypropylene [5, 6].

219

The research and development concerning PHA can be traced back to the beginning of the 20th century. An historical overview is available elsewhere [7]. Tremendous progress has been made in the past four decades, mainly motivated by the environmentally friendly properties of PHA. Unlike the present commodity plastics, PHA are produced from renewable resources. Petrochemical-based plastics currently in wide use are being regarded as a major threat of pollution. Plastics have found widespread application in our daily life because they are chemically inert and durable. Over the years however, these properties gave rise to the accumulation of plastic materials in our environment. Now, these non-biodegradable polymers contribute to the pollution of the environment and therefore some attempts at recycling have been made. Nevertheless, a considerable amount ends up on beaches, in the oceans or clog landfill sites. Attempts to dispose of them by other means, i.e., incineration, produce different kinds of equally unacceptable pollution. These problems teach us that it is essential for mankind to develop and use materials that are compatible with our natural ecosystem. This has been the primary motivating factor in the research and development of PHA as a potential substitute for petrochemical-based plastics. PHA are biocompatible as well as biodegradable, and its degradation product, 3-hydroxyalkanoate is a normal mammalian metabolite [8].

Much work concerning PHA is in progress in many developed countries such as USA, Germany and Japan where waste disposal is becoming an increasingly serious problem. Cost factors will be critical in determining whether in the long term, PHA can enter into widespread use in fields presently dominated by conventional commodity plastics. Research on this microbial polyester has been and still is a great challenge to the scientists in the fields of biotechnology and polymer chemistry. The final goal is to be able to produce in a cost-effective manner various kinds of PHA from renewable carbon sources.

7.2 The Various Types of PHA

Microorganisms in nature are capable of synthesising various types of PHA depending on the types of carbon sources available and the biochemical pathways that are operating in the cell. Ever since monomers other than *R*-3-hydroxybutyrate (3HB) were identified in environmental samples [9, 10], much effort has been directed to identify all the various types of monomers that can possibly be incorporated by PHA synthase. It is now possible to synthesise various PHA homopolymers and copolymers that have a certain monomer composition.

7.2.1 Poly[R-3-hydroxybutyrate] (P[3HB])

P[3HB] (**Figure 7.1**) is the first type of PHA to be identified [11-13]. Today, it is known that P[3HB] is the most common PHA found in nature. Based on the MW of the biosynthesised

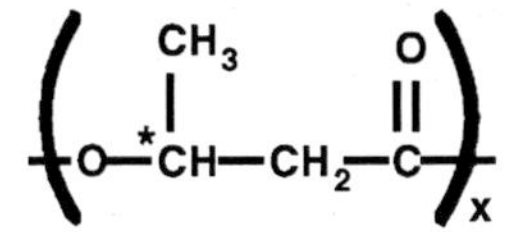

R-3HB

X = 120-200: low molecular weight P[3HB]

X = 1,000-20,000: high molecular weight P[3HB]

X ~ 100,000: ultrahigh molecular weight P[3HB]

Figure 7.1 Chemical structure of poly [R-3-hydroxy butyrate] (P[3HB])

P[3HB], they can be divided into three distinct groups, i.e., low MW P[3HB] [14-16], high MW P[3HB] [1, 8, 12], and ultra-high molecular weight (UHMW) P[3HB] [17, 18].

The low MW P[3HB] which is also known as complexed P[3HB] (cP[3HB]) is an ubiquitous cell constituent that exists in Eubacteria, archaebacteria, and eukaryotes [15, 16, 19, 20]. Recent studies have also revealed the presence of cP[3HB] in humans [21]. This cP[3HB] consists of about 120-200 3HB units and have a MW of about 12,000 Da [22]. Depending on the strength of their association with macromolecules, chloroform-soluble and chloroform-insoluble cP[3HB] have been identified [23]. The former forms a weakly bound (non-covalent) complex with polyphosphate salts while the latter is usually strongly bound (covalent) complex with proteins. These complexes are thought to function as ion (Ca^{2+}) transport channels across cell membranes and also may facilitate the uptake of extracellular deoxyribonucleic acid (DNA) material [23-25].

In contrast to the low MW cP[3HB], high MW P[3HB] is synthesised and accumulated in the form of water-insoluble inclusion bodies in microbial cell cytoplasm. They serve as carbon and energy storage compounds for the microorganisms. The MW of this storage P[3HB] are in the range of 200,000 to 3,000,000 Da and the precise value depends on the microorganism and its growth conditions [26]. In the 1960s and 1970s, much attention was directed to the high MW P[3HB] because of its thermoplastic property.

Recently, the production of UHMW P[3HB] (MW > 3,000,000) has been achieved by using a recombinant *Escherichia coli* cultivated under specific fermentation conditions [17]. Unlike the high MW P[3HB] that is characterised by stiffness and brittleness, the UHMW P[3HB] seems to show improved characteristics [18]. In addition, it was also found that films prepared from this UHMW P[3HB] were completely degraded at 25 °C in a natural freshwater river within three weeks [27].

High MW P[3HB] (MW = 200,000-3,000,000) was the first type of PHA to be identified, and because of its widespread occurrence, much work has been done to determine its physical properties and explore its potential applications. It is well established that P[3HB] samples obtained from various biological sources were all characterised by exceptional stereochemical regularity. They are linear polyesters and their chiral centres possess only the *R* absolute configuration [D(–) in traditional nomenclature]. The biosynthesised P[3HB] is therefore perfectly isotactic and upon extraction from the microorganisms, have a crystallinity of about 55-80% with a melting point at around 180 °C [28-30]. The P[3HB] molecules in the crystalline regions has the conformational structure corresponding to a 2_1 left-handed helix (two monomers being present in every one helical twist) [28, 31-33].

Despite having similar physical properties to polypropylene [34], the P[3HB] homopolymer produced by microorganisms is rather brittle and thermally unstable [35]. The brittleness is due to the formation of large crystalline domains in the form of spherulites. The formation of large spherulites is a special property of this biologically synthesised P[3HB] probably because of its exceptional purity. This makes the microbial P[3HB] an ideal system for the study of spherulites [36] but is definitely a major drawback to the commercial use of this homopolymer [30]. The brittleness can however be reduced to a certain extent by using suitable processing conditions, enabling the production of ductile films [37].

Interestingly, the recently found UHMW P[3HB] seems to possess better characteristics [27]. Mechanical properties of the stretched and annealed UHMW P[3HB] films remained unchanged for six months at room temperature [38]. Further work is in progress to determine the applicability of this UHMW P[3HB]. At the moment, the P[3HB] homopolymer is yet to acquire any significant economic importance.

7.2.2 Poly[3-hydroxybutyrate-co-3-hydroxyvalerate] (P[3HB-co-3HV])

Significant improvement in the properties of P[3HB] was achieved by the incorporation of a second monomer into the 3HB sequence. Initially it was thought that microorganisms can only synthesise the P[3HB] homopolymer. Hence, the widespread use of the term 'PHB' in early publications to refer to this class of microbial storage material [2]. Today, it is known that although P[3HB] is the most common, it is but one type in a huge family of microbial PHA [39]. 3-Hydroxyvalerate (3HV) was first identified as a member of the PHA family in the 1970s [10]. This eventually led to the development by Imperial Chemical Industries (ICI) of a biosynthesis process that is capable of producing a random copolymer of P[3HB-*co*-3HV] (**Figure 7.2**) containing various amounts (0-30 mol%) of 3HV units [6, 30]. The microorganism that was selected for this fermentation process is a nonpathogenic *Ralstonia eutropha* strain (formerly known as *Alcaligenes eutrophus*) that grows on glucose. Propionic acid was added to the culture medium as a precursor carbon source that gives rise to the incorporation of 3HV units [6]. This, for the first time enabled the production of PHA with properties that can be altered by controlling the content of the second monomer [6, 40, 41]. The development of P[3HB-*co*-3HV] copolymers also led to the discovery of a unique cocrystallisation behaviour known as isodimorphism [40, 42, 43]. No other forms/types of microbial PHA show this behaviour. The P[3HB-*co*-3HV] copolymers were then marketed under the trade name BIOPOL [44].

The easiest way to control the content of 3HV units in P[3HB-*co*-3HV] copolymer is by changing the concentration of the carbon source that contributes to the formation of 3HV units. Doi and co-workers found that by using a combination of butyric and pentanoic acids, *R. eutropha* (NCIB 11599) can be made to produce P[3HB-*co*-3HV] copolymers containing a wide range (0-85 mol%) of 3HV units (**Table 7.1**). The P[3HB-*co*-3HV]

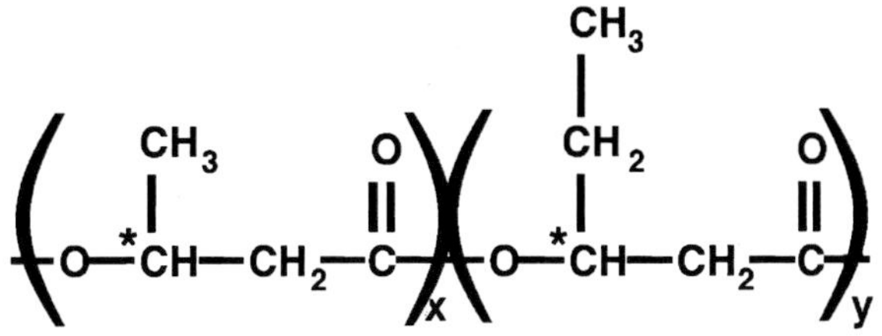

Figure 7.2 Chemical structure of poly[*R*-3-hydroxybutyrate-*co*-*R*-3-hydroxyvalerate] (P[3HB-*co*-3HV])

Table 7.1. Production of P[3HB-*co*-3HV] copolymers using *R. eutropha*[a]						
Carbon source (g/l)		Cell dry weight (g/l)	PHA content[b] (wt%)	PHA composition[c] (mol%)		Ref.
Butyric acid	Pentanoic acid			3HB	3HV	
20	0	7	48	100	0	[45, 46]
15	5	8	55	85	15	
8	12	6	37	70	30	
6	14	6	48	55	45	
2	18	6	43	40	60	
0	20	7	46	15	85	
[a]*R. eutropha cells were grown in a rich medium for 24 h and then transferred to a nitrogen-free medium containing the above carbon sources* [b]*PHA content in cells incubated for 48 h in nitrogen-free medium* [c]*Composition determined by* [1]*H NMR*						

copolymers were also shown to have a statistically random distribution of 3HB and 3HV units [47]. Recently the homopolymer of poly [*R*-3-hydroxyvalerate] (P[3HV]) has also been produced biologically using wild type microorganisms such as *Rhodococcus sp.* [48], and *Chromobacterium violaceum* [49], and also by using recombinant *R. eutropha* PHB–4 [50]. Single crystals of these biologically synthesised P[3HV] were found to be more perfect than those of synthetic P[3HV] although they both have a square shape as opposed to the characteristic lath shape of P[3HB] crystals [51].

7.2.3 Poly[3-hydroxybutyrate-co-4-hydroxybutyrate] (P[3HB-co-4HB])

Another type of PHA copolymer that shows useful physical properties is poly[3-hydroxybutyrate-co-4-hydroxybutyrate] (P[3HB-co-4HB]) (**Figure 7.3**). Like 3HB, 4HB is a normal mammalian metabolite. 4HB has been found in extracts of brain tissue of rat, pigeon and man [52]. Synthetic 4HB in the form of sodium salt was first made available in the early 1960s [53]. Approximately two decades later, Doi and co-workers reported the assimilation of this compound by *R. eutropha* to produce a random copolymer of P[3HB-co-4HB] [54, 55]. Subsequent studies resulted in the production of P(3HB-co-4HB) having a wide range of 4HB contents (**Table 7.2**) [56-60]. Other carbon sources such as, 4-chlorobutyric acid [54], γ-butyrolactone, 1,4-butanediol, 1,6 hexanediol, 1,8 octanediol, 1,10-decanediol, and 1,12-dodecanediol also resulted in the incorporation of 4HB units [61].

By increasing the content of 4HB in P[3HB-co-4HB], the physical property of the copolymer (based on solvent-cast films) changes from one that is characterised by high crystallinity to one that is a strong elastomer [57, 59]. In addition, an increase in the 4HB content up to about 70 mol% was accompanied by an increased rate of enzymic degradation [59, 62]. Further increase in the 4HB content however decreased the rate of enzymic degradation of P[3HB-co-4HB]. In general, there is an increase in the rate of enzymic degradation of PHA films following a decrease in its crystallinity [63]. For P[3HB-co-4HB] films, crystallinity decreases with an increase in the 4HB content up to about 70 mol%. Further increase in the 4HB content contributed to an increase in the copolymer film crystallinity [60]. This

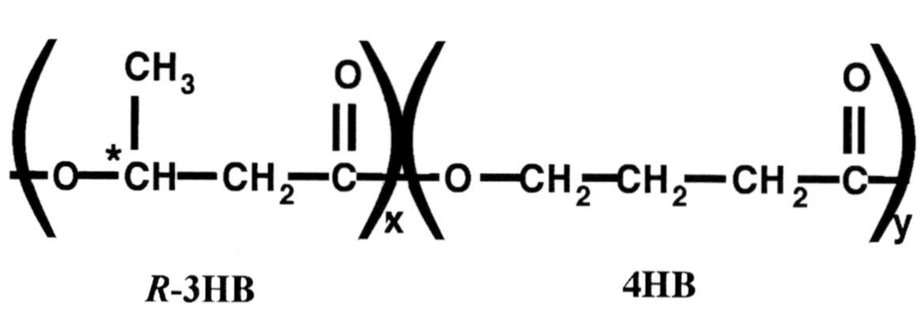

Figure 7.3 Chemical structure of poly(*R*-3-hydroxybutyrate-co-4-hydroxybutyrate]
(P[3HB-co-4HB])

Table 7.2. Production of P[3HB-*co*-4HB] copolymers containing various 4HB contents

Microorganism	Carbon source (g/l)		Cell dry weight (g/l)	PHA content[b] (wt%)	PHA composition[c] (mol%)		Ref.
	3HBA	4HBA			3HB	4HB	
R. eutropha[a]	20	0	10	51	100	0	[45, 58]
	12	6.4	8	52	90	10	
	8	9.6	7	43	82	18	
	0	20	5	16	67	33	[59, 60]
C. acidovorans[a]	2	8	3	27	44	56	
	1.5	8.5	3	26	27	73	
	0.5	9.5	3	23	17	83	
	0	10	3	17	0	100	

[a]*R. eutropha and C. acidovorans cells were grown in a rich medium for 24 h and then transferred to a nitrogen-free medium containing the above carbon sources*
[b]*PHA content in cells incubated for 48 h in nitrogen-free medium*
[c]*Composition determined by* [1]*H NMR*
3HBA, 3-hydroxybutyric acid; 4HBA, 4-hydroxybutyric acid

shows that the accelerated enzymic degradation of P[3HB-*co*-4HB] films may be due to a decrease in its crystallinity.

Similar properties are also shown by the incorporation of 3-hydroxypropionate (3HP) into the P[3HB] sequence [64]. Like 4HB, 3HP do not possess chirality. Since the discovery of 3HP as a member of the PHA family [65], much work has been done to produce P[3HB-*co*-3HP] containing various amounts (0-88 mol%) of 3HP units [66, 67]. The investigation of solid-state structure [68] and biodegradability [67] of these copolymers with various 3HP units showed much similarity to the copolymers of P[3HB-*co*-4HB] [69, 70]. However, in contrast to P[4HB] homopolymer, the homopolymer of P[3HP] was hardly eroded in river water [69] but could be degraded by the P[3HB] depolymerase (EC 3.1.1.75) (see Section 7.5) purified from *Alcaligenes faecalis* [71].

7.2.4 Other PHA Copolymers with Interesting Physical Properties

Figure 7.4 shows the chemical structure of PHA copolymers composed of 3HB units and C_6-C_{14} numbered *R*-3-hydroxyalkanoate units. The latter group of monomers have been classified as the medium-chain length (MCL) *R*-3-hydroxyalkanoates and the PHA made of these monomers are termed PHA$_{MCL}$. In contrast to PHA$_{SCL}$ made of short-chain length hydroxyalkanoates (C_4-C_5), PHA$_{MCL}$ are thermoplastic elastomers with melting points of about 45-60 °C and glass transition temperatures near −40 °C [72]. The biosynthesis of PHA copolymers containing both the SCL and the MCL units are relatively rare. This is because of the substrate specificity of the polymerising enzyme, PHA synthase [73]. Most of the PHA synthases can efficiently polymerise only either SCL or MCL units. Recently, an increased number of PHA copolymers containing both the SCL and the MCL units are being documented [50, 74-79]. These copolymers that are produced by PHA synthases having a broad range of substrate specificity show attractive physical properties (**Table 7.3**). Matsusaki and co-workers [78, 79] have shown that it is possible to produce these P(3HB-*co*-3HA$_{MCL}$) copolymers containing a wide range of 3HA$_{MCL}$ units.

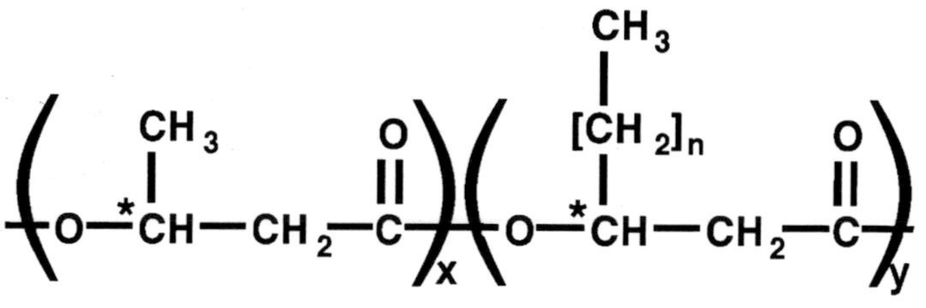

95 mol% 3HB   **5 mol% 3HA$_{MCL}$(n= 2, 4, 6, 8, 10)**

Figure 7.4 Chemical structure of PHA copolymer containing short-chain length (SCL) monomer (3HB) as well as medium-chain length (MCL) monomers (3PHA$_{MCL}$) (P[3HB-*co*-3PHA$_{MCL}$)

Table 7.3. Characteristics of PHA copolymers containing both the SCL and the MCL *R*-3-hydroxyalkanoates in comparison to other polymers				
Polymer samples	T$_g$ (°C)	T$_m$ (°C)	Tensile strength (MPa)	Elongation to break (%)
P[3HB]	4	180	43	5
P[3HB][a]	4	185	62	58
P[3HB-*co*-20% 3HV]	−1	145	20	50
P[3HB-*co*-16% 4HB]	−7	150	26	444
P[3HB-*co*-10% 3HHx]	−1	127	21	400
P[3HB-*co*-6% 3HD]	−8	130	17	680
Polypropylene	0	176	38	400
LDPE[b]	−30	110	10	620
[a]*Ultra-high molecular weight P(3HB)[38]* [b]*Low-density polyethylene* *3HB, 3-hydroxybutyrate; 4HB, 4-hydroxybutyrate; 3HHx, 3-hydroxyhexanoate; 3HD, 3-hydroxydecanoate*				

It must be noted that some Pseudomonads produce a blend of PHA$_{SCL}$ and PHA$_{MCL}$ [80]. Such blends cannot be distinguished from a copolymer solely by gas chromatographic analysis of the dried cells. However, the polymers once extracted from the cells can be separated by using suitable solvents. Thermal analyses of the extracted polymer are also frequently used to distinguish between blends and copolymers. In the case of blends, it has been shown that the PHA$_{SCL}$ and PHA$_{MCL}$ are stored in separate inclusions in the bacterial cell cytoplasm [81]. When cells containing these blends are subjected to freeze-fracture electron microscopy, the PHA$_{SCL}$ granules often deform plastically resulting in needle-type protruding structures [82-84]. On the other hand, PHA$_{MCL}$ granules show distinct mushroom-type deformation structures [85]. Based on these morphological differences co-existing granules of PHA$_{SCL}$ and PHA$_{MCL}$ can be distinguished quite accurately. However, the temperature at which the freeze-fracture process is carried out can greatly affect the deformation morphology, whereby, at −160 °C most P[3HB] granules show mushroom-type deformation structures although they show needle-type deformations at −110 °C [84]. Besides that, granules of P[4HB] show both mushroom-type and needle-type deformations when fractured at either −110 °C or −160 °C [84, 85]. **Figure 7.5** shows the morphology of P[3HB-*co*-3HV] granules in *Comamonas acidovorans* cells. As mentioned earlier, both the needle-type and mushroom-type deformation structures can be observed. It has been shown that mushroom-like deformations do not always represent the PHA$_{MCL}$ granule.

7.2.5 Uncommon PHA Constituents

Besides the common PHA constituents mentioned earlier, various other monomer constituents have been identified [39]. In total, about 150 different hydroxyalkanoates are known to be the members of the ever-growing PHA family [87]. Many of the uncommon monomers are incorporated only when related precursor substrates are supplied as carbon sources to the microorganisms. An interesting addition to the PHA family is the identification of a new class of sulfur-containing PHA with thioester linkages [88]. The poly[3-hydroxybutyrate-*co*-3-mercaptopropionate] (P[3HB-*co*-3MP]) was produced by *R. eutropha* when 3-mercaptopropionic acid or 3,3′-thiodipropionic acid was provided as carbon source in addition to fructose or gluconic acid under nitrogen-limited culture conditions. What is striking about this finding is the fact that the polymerising enzyme, PHA synthase, has a very versatile catalytic centre.

The PHA synthases of the Pseudomonads are probably the most versatile with broad substrate specificity. In *Pseudomonas oleovorans* for example, it is possible to synthesise PHA consisting of saturated, unsaturated, halogenated, branched and aromatic 3-hydroxyalkanoates with 6-14 carbon atoms [89-92].

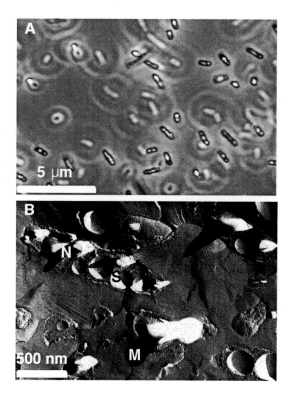

Figure 7.5 Morphologies of bacterial cells containing PHA granules in the cytoplasm. A: Phase contrast light microscopy picture of *Comamonas acidovorans* cells containing 38 wt.% of the dry cell weight P[3HB-*co*-71% 3HV]. B: Freeze-fracture electron micrograph of the same sample. The fracture process was carried out at –160 °C. N - needle-type; M - mushroom type; S - crater like holes in the cell cytoplasm resulting from granules that have been completely scooped out

7.3 Mechanisms of PHA Biosynthesis

Unlike other microbial storage materials like glycogen or polyphosphate that have been studied in detail for their physiological importance, only the early studies focused on the physiology of PHA biosynthesis [1, 93, 94]. The objectives of most recent studies have been to produce efficiently various kinds of PHA from simple and renewable carbon sources. With this goal in mind, much effort has been directed to understand the enzymes, metabolic pathways, and conditions that generate substrates for the PHA synthase. In recent years, recombinant DNA technologies are increasingly used to further understand complex regulatory mechanisms that affect PHA biosynthesis [95].

7.3.1 Conditions that Promote the Biosynthesis and Accumulation of PHA in Microorganisms

Early studies have revealed that the rate of PHA accumulation can be increased by increasing the ratio of carbon source to nitrogen source [96]. Eventually it became evident that PHA accumulation usually occurs when cell growth is impaired due to depletion of an essential nutrient such as sulfate, ammonium, phosphate, potassium, iron, magnesium, or oxygen [1, 97-99]. Suzuki and co-workers [100] studied 51 methylotrophs for their ability to produce P[3HB] from methanol. Similar nutrient limitations was found to stimulate the formation of P[3HB]. However, a kinetic study of the production of P[3HB] by a fed-batch culture of *Protomonas extorquens* showed that a nitrogen source was necessary even in the P[3HB] production phase [101]. Feeding with a small quantity of ammonia resulted in a more rapid increase of intracellular P[3HB] than was the case without ammonia feeding. Excessive feeding of ammonia, however, caused not only degradation of accumulated P[3HB] but also reduction of microbial P[3HB] synthetic activity.

PHA accumulation can also take place during active cell growth, but this ability is limited to only a few microorganisms such as *Alcaligenes latus* that can accumulate P[3HB] up to 80% of the dry cell weight without limitation of any nutrient [102, 103]. This characteristic may be due to a low activity of the β-ketothiolase, which catalyses the cleavage of acetoacetyl-CoA [104]. Besides *A. latus*, *Paracoccus denitrificans* also shows growth-associated PHA accumulation depending on the type of carbon sources available to the bacterium. Kim and co-workers [105], tested linear primary C_1-C_9 alcohols and linear C_2-C_{10} monocarboxylic acids and found that growth-associated synthesis of PHA could be obtained only with the carbon sources with an odd number of carbon, except for methanol.

The advantage of using a bacterium that shows growth-associated PHA accumulation for large-scale production is a shorter fermentation time. In addition, it also avoids the extra operations associated with the two-step fermentation process for PHA-accumulation under nutrient-limited conditions. By using *A. latus* btF-96, Chemie Linz was able to produce more than 1000 kg of P[3HB] in a week in a 15 m^3 fermenter [103]. ICI on the other hand chose *R. eutropha* as the production organism although this bacterium accumulates PHA under non-growth conditions. *R. eutropha* was chosen over *Azotobacter* and *Methylobacterium* because of higher polymer content, good molecular mass and also because of easier PHA recovery [44].

7.3.2 Carbon Sources for the Production of PHA

An attractive feature of the microbial PHA is the ability to produce them using renewable carbon sources. The plastic materials widely in use today are synthesised from fossil fuels such as petroleum and natural gas. PHA on the other hand can be produced using renewable carbon sources such as sugars and plant oils, which is an indirect way of utilising the atmospheric CO_2 as the carbon source. Various waste materials are also being considered as potential carbon sources for PHA production. Among them are whey [106, 107], molasses [108-110], and starch [111, 112]. The carbon source available to a microorganism is one of the factors (others being the PHA synthase substrate specificity and the types of biochemical pathways available) that determine the type of PHA produced. For industrial scale production, the carbon source significantly contributes to the final cost. This makes the carbon source one of the most important component in the production of PHA and is therefore a prime target for potential cost reduction.

Figure 7.6 shows three general systems for the production of biodegradable polymers using CO_2 as the starting material. The 3-step process involves the utilisation of plant sugars derived from photosynthetically fixed CO_2 as carbon sources in the fermentation of organic acids, alcohols and amino acids. These substances are then used as building blocks for the chemical synthesis of polymers. Example of polymers produced using this three-step process includes polylactic acid, polybutylene succinate, and polyaspartic acid.

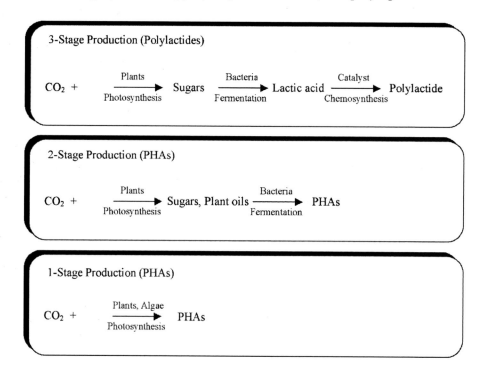

Figure 7.6 Systems for the production of biodegradable polymers from CO_2 as the starting material

On the other hand, the two-step process involves the direct conversion of plant sugars and plant oils into polymers by microorganisms. At present, the biosynthesis of PHA is largely carried out through this two-step process. Compared to the three-step process of polymer production, the two-step process can be more cost effective provided that excellent producers of PHA are identified and the fermentation process is highly optimised. It has been calculated that 2.5 kg of glucose must be used for each kilogram of polymer produced [113]. Recent studies in the Author's laboratory have shown that plant oils may be a better carbon source whereby a kilogram of oil can give rise to a kilogram of polymer. A recombinant strain of *R. eutropha* PHB–4 (a PHA-negative mutant), harbouring the PHA synthase gene from *Aeromonas caviae*, could produce a random copolymer of 3HB and 3-hydroxyhexanoate (3HHx) from plant oils such as olive oil, palm oil and corn oil. The P[3HB-*co*-3HHx] content was approximately 80% of the dry cell weight and the 3HHx mole fraction was 4-5 mol% regardless of the structure of the triglycerides fed [114]. The results demonstrate that inexpensive renewable plant oils are excellent carbon sources for the efficient production of PHA.

Ideally, producing environmentally friendly polymers directly in plants would be the most energy efficient process (one-step process) (**Figure 7.6**), provided that suitable technologies are available for the extraction and downstream purification processes of the polymers from plant materials. At present however, plant derivatives such as sugars and oils are the most popular carbon sources for the production of PHA by microbial fermentation.

7.3.3 Biochemical Pathways Involved in the Metabolism of PHA

In order to tap the full potential of microbial systems for PHA production, it is necessary that the existing metabolic pathways in a particular microorganism are modified. This is to ensure that the major portions of the supplied carbon sources are channelled towards PHA biosynthesis. Recent knowledge of the complete genetic makeup of several microorganisms [115-117] is facilitating the engineering of novel metabolic pathways. New pathways can be constructed by introducing relevant genes into suitable microorganisms. Likewise, unnecessary pathways can be shutdown by inactivating the enzyme(s) involved in a certain reaction. Such manipulations have to be carried out judiciously to achieve maximum PHA production in the shortest possible time using cheap and readily available carbon sources, without compromising the cell growth. Another important factor that is often overlooked in the experimental stage is the stability of the genetically modified strains over many generations. Recombinant strains that do not have this characteristic will not be attractive as an industrial strain for large-scale production of PHA.

Figure 7.7 shows the common metabolic pathways that are frequently encountered in the biosynthesis of PHA in various microorganisms. Along with the type of carbon source and the specificity of the PHA synthase, the metabolic pathways play a crucial role in determining

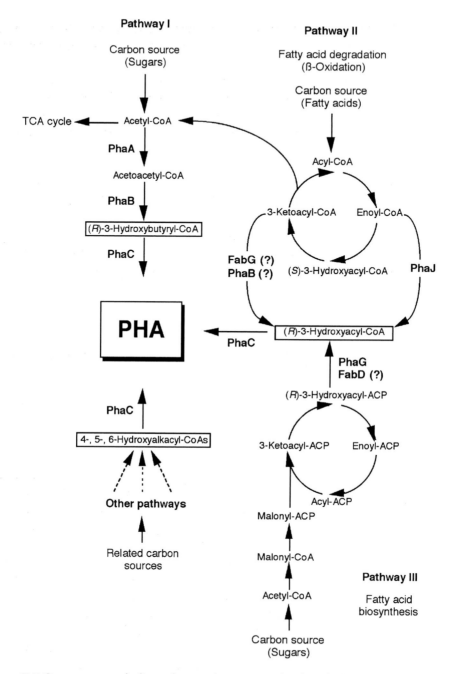

Figure 7.7 Common metabolic pathways that are involved in the biosynthesis of PHA in microorganisms. PhaA: β-ketothiolase; PhaB: NAOH-dependent acetoacetyl-CoA reductase; PhaC: PHA synthase; PhaG: 3-hydroxyacyl-ACP: CoA transferase; PhaJ: (*R*)-enoyl-CoA hydratase; FabC: malonyl-CoA: ACP transcylase; FabG: 3-Ketoacyl-CoA reductase

the type of PHA that can be produced by a particular microorganism. Most of the P[3HB] producing microorganisms possess Pathway I through which acetyl-CoA is converted into (R)-3-hydroxybutyryl-CoA and subsequently polymerised by the PHA synthase. Recently it has been shown that a similar pathway also operates in the cyanobacterium *Synechocystis sp.* PCC6803 [118]. In some microorganisms, (S)-isomers of 3-hydroxybutyryl-CoA are generated instead of the (R)-isomers. Since the PHA synthase is active only towards the (R)-isomers, additional reaction steps catalysed by enoyl-CoA hydratases are present in microorganisms such as *Rhodospirillum rubrum* to convert the (S)-isomers into the (R)-isomers [119].

Fatty acid β-oxidation pathway (Pathway II) is known to generate substrates that can be polymerised by the PHA synthases of pseudomonads. Pseudomonads that belong to the ribosomal nucleic acid (rRNA)-homology group I can synthesise PHA_{MCL} from various alkanes, alkanols, and alkanoates. The monomer composition of the PHA_{MCL} produced is often related to the type of carbon sources. Most of the pseudomonads belonging to this group, except *P. oleovorans*, can also derive (R)-3-hydroxyacyl-CoA substrates for PHA biosynthesis from unrelated carbon sources such as carbohydrates. Huijberts and co-workers [120] presented evidence showing that the PHA synthase responsible for PHA synthesis from fatty acids are also involved in PHA synthesis from glucose. It was then presumed that there are at least two distinct substrate supply routes for PHA synthesis in *Pseudomonas putida*, i.e., via the intermediates of fatty acid biosynthesis (Pathway III) [121] and via the intermediates of β-oxidation (Pathway II).

Although it was known that the intermediates of the β-oxidation cycle are channelled towards PHA biosynthesis, only recently the precursor sources were identified. In *A. caviae*, the β-oxidation intermediate, *trans*-2-enoyl-CoA is converted to (R)-3-hydroxyacyl-CoA via (R)-specific hydration catalysed by a (R)-specific enoyl-CoA hydratase [122, 123]. Subsequently, Tsuge and co-workers [124] reported the identification of similar enoyl-CoA hydratases in *P. aeruginosa*. In the latter case, two different enoyl-CoA hydratases with different substrate specificities channelled both SCL and MCL enoyl-CoA towards PHA biosynthesis. In recombinant *E. coli* it was further shown that 3-ketoacyl-CoA intermediates in the β-oxidation cycle can also be channelled towards PHA biosynthesis by a NADPH-dependent 3-ketoacyl-acyl carrier protein (ACP) reductase [125]. A similar pathway was also identified in *P. aeruginosa* [126]. In addition, it was also reported that the acetoacetyl-CoA reductase (PhaB) of *R. eutropha* can also carry out the conversion of 3-ketoacyl-CoA intermediates in Pathway II to the corresponding (R)-3-hydroxyacyl-CoA in *E. coli* [127]. The results clearly indicate that several channelling pathways are available to supply substrates from β-oxidation cycle to the PHA synthase. This explains why it was not possible to obtain mutants that completely lack PHA accumulation ability, unless the mutation occurred in the PHA synthase gene [128].

Among the various metabolic pathways that are involved in PHA biosynthesis, the fatty acid *de novo* biosynthesis pathway (Pathway III) is of particular interest because of its ability to supply various types of hydroxyalkanoate monomers from simple carbon sources such as gluconate, fructose, acetate, glycerol and lactate. It can be envisaged that the potential future production of PHA_{MCL} by using photosynthetic organisms will benefit through the exploitation of such pathways. This is because acetyl-CoA is the starting material (**Figure 7.6** Pathway III) that is used to generate hydroxyalkanoate monomers for PHA_{MCL} biosynthesis, and acetyl-CoA is a universal metabolite present in all living organisms. However it must be noted that the intermediates of fatty acid *de novo* biosynthesis pathway are in the form of (*R*)-3-hydroxyacyl-ACP, which is not recognised by the PHA synthase.

Studies of PHA_{MCL} biosynthesis in *P. putida* from glucose as the sole carbon source has identified an enzyme that is capable of converting 3-hydroxydecanoyl-ACP to 3-hydroxydecanoyl-CoA. The enzyme was referred to as a 3-hydroxyacyl-ACP:CoA transferase (PhaG) [129]. Since then, similar enzymes have been identified in several other pseudomonads [130-132]. *P. oleovorans* does not have the ability to synthesise PHA_{MCL} from gluconate but shows this ability upon the introduction of the PhaG gene of *P. putida*. The genes for PhaG and PHA synthase from *P. aeruginosa* were expressed together in a non-PHA producing pseudomonad, *P. frugi*. This resulted in the ability to produce PHA_{MCL} by *P. frugi* from gluconate as the sole carbon source. Besides the PhaG protein, overexpressions of transacylating enzymes such as malonyl-CoA-ACP transacylase (FabD) in *E. coli*, also seem to generate monomers for P[3HB] biosynthesis [133].

Besides the three main pathways mentioned above, there are several other metabolic pathways that can be manipulated to produce substrates for PHA biosynthesis. In recombinant *E. coli*, it has been shown that 4-hydroxybutyryl-CoA can be derived from the intermediates of tricarboxylic acid (TCA) cycle [134]. By providing external precursor substrates such as 4-hydroxybutyric acid, 1, 4-butanediol, and γ-butyrolactone to certain wild type [56, 57, 59, 60] and recombinant microorganisms [86, 135], 4HB monomers can be incorporated more efficiently.

7.3.4 The Key Enzyme of PHA Biosynthesis, PHA Synthase

Without doubt, PHA synthase is the key enzyme in the biosynthesis of PHA. Unfortunately, the mechanism of this important enzyme is not yet fully understood. Based on genetic analysis, the primary structures of PHA synthases from a large number of microrganisms are available [73]. The PHA synthases have been classified into three groups based on their primary structures and the types of PHA that they produce. The PHA synthases of *R. eutropha* and *P. oleovorans* represents the groups I and II, respectively, while that of *Chromatium vinosum* represents the group III. The latter differs from the two former groups by the fact that group III synthases consist of two different subunits (PhaC and

PhaE) while the members of groups I and II only have one subunit (PhaC). As for the types of PHA produced, PHA synthases of groups I and III are efficient in the synthesis of PHA_{SCL}, while those of group II are superior in the synthesis of PHA_{MCL}. A few exceptions to the above classification are the PHA synthases of *A. caviae* [122], *Thiocapsa pfennigii* [136], and *Pseudomonas* sp. 61-3 [137]. These PHA synthases are capable of producing PHA copolymers containing both the SCL- and the MCL-PHA. These exceptional PHA synthases are of great interest because they can be used to biosynthesise PHA copolymers containing novel compositions that show promising physical properties [78, 79].

In order to elucidate the mechanism of PHA synthase, various site-specific [138, 139] and random mutagenesis [140] studies have been carried out. Results show that many of the amino acid substitutions that affected PHA accumulation occurred in the conserved regions within an 'α/β hydrolase fold', while changes in the first 100 amino acid sequence at the N-terminal region did not show any significant effect. The α/β hydrolase fold is the characteristic of a superfamily of proteins that includes lipases [141].

Initially it was thought that the PHA synthase catalytic mechanism might resemble that of fatty acid synthases [142, 143]. The present knowledge dictates that lipase-based catalytic mechanism is perhaps more suitable as a model of PHA synthase [144].

7.4 Genetically Modified Systems and Other Methods for the Production of PHA

E. coli offers a well-defined physiological environment for the production of recombinant proteins and other bioproducts because the physiology, biochemistry and genetics of this bacterium have been studied in great detail [145, 146]. Likewise, the production of PHA using photosynthetic organisms is attractive because atmospheric CO_2 can directly be converted into plastic material [147]. In the past decade, both these recombinant systems have been explored by various research groups with the hope of reducing the cost of PHA production.

7.4.1 Recombinant Escherichia coli

In the case of recombinant *E. coli* various strategies are available to achieve high cell-density cultures [148]. It is important to select the most suitable *E. coli* strain for a particular purpose. Detailed studies by Lee and co-workers [149] revealed that the recombinant *E. coli* strains, XL1-Blue (pSYL105) and B (pSYL105) were the best candidates for P[3HB] production. Although strain JM109 (pSYL105) produced the highest P[3HB] content, the cell mass was low. In any case, high gene dosage is necessary to produce high concentrations of P[3HB] in recombinant *E. coli* [150, 151]. Recently, the PHA biosynthetic genes of *A. latus* were cloned and used to produce P[3HB] in *E. coli* [152]. It was found that the genes of *A. latus* resulted in better P[3HB] production in *E. coli* compared to the PHA biosynthetic genes of *R. eutropha*. It must be noted that, recombinant *E. coli* can produce P[3HB] during active growth in nutrient-rich conditions [150, 151] just like *A. latus* [104].

Ever since the first successful expression of the *R. eutropha* PHA biosynthetic genes [153-155], various other heterologous genes have been introduced into *E. coli* resulting in the ability to produce both PHA homopolymers and copolymers. Among the PHA$_{SCL}$ other than P[3HB] that have been produced in recombinant *E. coli* are, P(4HB) homopolymer and P(3HB-*co*-4HB) copolymer [134, 135, 156] and poly(4-hydroxyvalerate) P[4HV] homopolymer and copolymers [156]. Besides that, the production of PHA$_{MCL}$ have also been demonstrated [157-161]. In addition, Fukui and co-workers [162] demonstrated the production of P(3HB-*co*-3HHx) copolymer in *E. coli* by the co-expression of *A. caviae* PHA synthase gene and (*R*)-enoyl-CoA hydratase gene. The *P. aeruginosa* (*R*)-enoyl-CoA hydratase (PhaJ1) (EC 4.2.1.17) [124] can also be used to supply 3HHx monomers for the production of P(3HB-*co*-3HHx) copolymer in *E. coli* [163].

In recombinant microorganisms, plasmid stability is of crucial importance for continued PHA production. Some attention has been directed to this problem in recombinant *E. coli* [151, 164]. Another advantage of using *E. coli* as the production host is the relative ease at which the accumulated PHA can be extracted from the cells [165].

7.4.2 Transgenic Plants

Ever since the first successful expression of the PHA biosynthetic enzymes in plants [166], which resulted in the accumulation of small amounts (<0.1% of the dry cell weight) of P[3HB], much knowledge about the potential of this system has been obtained. Subsequent studies showed that, when PHA accumulation is targeted to the plastids, P[3HB] content of up to 14% of the dry cell weight (DCW) was accumulated by one of the transgenic plant [167]. Later, by the expression of a threonine deaminase protein in addition to enzymes of PHA biosynthesis, it was demonstrated that plants can be directed to produce a copolymer of P[3HB-*co*-3HV] [168]. This achievement is of considerable interest because of the poor physical properties of P[3HB] homopolymer. Later, it was shown that the targeting of a PHA synthase from *P. aeruginosa* into the peroxisomes of *Arabidopsis thaliana*, results in the accumulation of PHA$_{MCL}$ [169]. These achievements show that it is indeed possible to produce various kinds of PHA homopolymers and copolymers in transgenic plants.

Transgenic plants appear to be the most cost-effective PHA producer because in principle the PHA are produced from carbon dioxide, water and sunlight. If transgenic plants were to become the major approach for producing PHA cost-effectively, competition between the use of fertile agricultural land for food and for plastics production would be inevitable. While it may be argued that PHA production in trangenic plants can be directed to the parts of food crops that are normally unused such as the leaf for example, the efficiency of this process is questionable. Furthermore the extraction of PHA accumulated in plant materials will not be as easy as extracting PHA from microrganisms. In such a scenario, photosynthetic microorganisms such as cyanobacteria may be another option for PHA production [170-172].

7.4.3 In vitro Production of PHA

The ability to synthesise various high MW polymers in an aqueous environment at room temperature is an interesting idea. This *in vitro* system has been shown to be possible using the PHA synthases purified from various sources [138, 173-175]. At present, it is possible to produce homopolymers and copolymers containing 3HB, 3HV, 4HV, and 3-hydroxydecanoate (3HD) [87]. Multiple-enzyme systems have been developed that can utilise cheaper substrates as well as recycle expensive cofactors such as coenzyme A. Nevertheless the *in vitro* systems are still expensive to use to produce PHA for applications as commodity plastics. However, considering the fact that these systems are free from contaminants such as endotoxins [176], the PHA produced in this manner might be more suitable for applications in the medical field where the cost factor is usually overshadowed by material quality and performance.

7.5 Biodegradation of PHA

One of the principal reasons for the continued research on PHA as environmentally friendly plastic is because they are biodegradable in landfill, compost, and aquatic systems [177-180]. Various investigations have been directed to the study of environmental factors that influence biodegradation, the enzymology of the process, and the importance of polymer composition. Water content and temperature have been found to be important along with the microbial activity in any given environment. In aquatic ecosystems, even under extreme conditions (such as seasonal changes of the oxygen concentration from anoxic to oxic, low temperatures, high hydrostatic pressure, no sunlight) plastic articles made from PHA were degraded [181].

PHA-degrading enzymes (extracellular depolymerase) are excreted by a number of bacteria and fungi in the environment (soil [182-185, freshwater [186], sludge [187], seawater [188, 189], hot-springs [190], compost [178], air [191]). Electron microscopy analysis of PHA films revealed that degradation occurs at the surface by enzymic hydrolysis. The degradation is therefore a function of the surface area available for microbial colonisation.

Upon degradation, oligomers and monomers of PHA are produced, which are then assimilated by the microorganisms as nutrients.

Besides the environmental factors, the microstructure and properties of the PHA materials themselves can significantly affect the degradation rates [192-194]. This includes factors such as composition, crystallinity, additives, and surface area.

Several kinds of extracellular depolymerases have been purified and characterised from various microorganisms [187, 189, 191, 195-197]. All the depolymerases are comprised of an N-terminal catalytic domain, a C-terminal substrate binding domain, and a linker region connecting the two domains. Similar catalytic and binding domains have also been identified in other depolymerising enzymes that hydrolyse water-insoluble polysaccharides such as cellulose [198], xylan [198, 199], and chitin [200]. The catalytic domain contains a lipase box pentapeptide [Gly-X_1-Ser-X_2-Gly] as the active site, which is common for serine hydrolase [201]. Further detailed aspects on the structure and mechanisms of PHA depolymerase (EC 3.1.1.76) can be found elsewhere [202].

7.6 Applications of PHA

For practical reasons, petroleum-based plastic materials have to a great extent replaced natural materials such as wood, silk, cotton etc., during the first six decades of the 20th century. This has resulted in the rapid accumulation of non-degradable materials in our environment. In recent years, natural polymers are gaining much attention as the preferred materials because of environmental concerns. The biodegradability and sustainability of natural polymers are viewed as important characteristics that should be possessed by 21st century materials. In the past decade, environmental concerns have also given rise to the concept of Life Cycle Assessment (LCA), which is a tool used to gauge environmental burdens associated with a product, a process or an activity by identifying and quantifying energy, material used and wastes released to the environment.

Microbial PHA first received widespread attention during the petroleum crisis of the 1970s as a potential substitute for petrochemical-based plastics. Besides being a thermoplastic with properties comparable to that of polyethylene, PHA are also completely biodegradable. The ability to produce PHA from renewable carbon sources also ensures a sustainable 'green chemistry' process. Much work has been directed to the production of various types of PHA for applications as commodity plastics. The identification of monomers other than 3HB and the ease at which PHA copolymers can be designed and synthesised, have resulted in the development of materials having interesting physical and thermal properties [45, 203]. By using specialised carbon sources, Fuller, Lenz and their co-workers [89, 204-208] demonstrated the biosynthesis of PHA in *P. oleovorans* containing various functional groups in the side chain. Recent studies have used *P. putida* to produce similar PHA [209, 210]. The ability to further chemically modify the functional groups in these PHA broadens their scope of application as biodegradable polymers as well as bioabsorbable materials for biomedical purposes. An extensive review on the application of PHA$_{MCL}$ produced by the Pseudomonads is available elsewhere [211].

Biomedical applications of PHA were realised as early as in the mid-1960s [212]. The main drawback that is being faced by PHA for applications as biodegradable commodity plastics is the cost of production, which is about 5-10 times the cost of petrochemical-based plastics currently in use. On the other hand, the application of PHA in the medical field might not be hampered by the production costs because material performance rather than costs receive top priority. Lately, the potential applications of PHA in the medical field are being investigated in greater detail. This is evident from the establishment of an American company, Tepha, Inc., that is actively pursuing the development of tissue engineered products using PHA (http://tepha.com).

For applications such as tissue engineering scaffolds, the suitable material must possess properties such as biocompatibility, support cell growth, guide and organise the cells, allow tissue ingrowth and should finally degrade to non-toxic products [213]. Unlike the PHA extracted for applications as biodegradable plastic materials, biomedical applications require pharmaceutical grade purity. The widely used methods of extraction and purification of PHA are not designed to obtain materials of pharmaceutical grade and may contain bacterial endotoxins and other undesired substances. Such contaminants have been identified and can be removed feasibly using suitable extraction techniques [176]. Other methods of purifying microbial PHA for biomedical applications include, oxidisation using hydrogen peroxide and treatment with supercritical fluids such as supercritical carbon dioxide [213]. Highly pure PHA materials have been obtained by a combination of these methods. The resulting PHA have been widely used in the development of cardiovascular products such as pericardial patches [214], patch material in the pulmonary circulation [215, 216], and as vascular grafts [217]. Recently, impressive results have been obtained in the development of cell-seeded tissue engineered heart valves using PHA such as poly[3-hydroxyoctanoate] and P[4HB] [218-222].

PHA are also a potential material for applications in controlled drug release systems [223, 224]. The biocompatibility and biodegradability properties of PHA make them attractive as materials for drug delivery. A wide variety of monomers can be incorporated into PHA, resulting in various physical properties that range from highly crystalline materials to strong elastomers. By judiciously controlling the monomer composition of PHA, the degradation rate can be indirectly controlled. The enzymic degradation of PHA is usually catalysed by the bacterial PHA depolymerase. However, the occurrence of this enzyme in higher organisms has not been reported. It is most likely that the latter do not have this enzyme. Some PHA like P(4HB) can also be degraded by bacterial lipases [225]. A recent study reported the detection of lipase activities in the rat gastrointestine near the PHA implant, suggesting the involvement of lipases in the metabolism of PHA *in vivo* [226].

The fact that P(3HB-*co*-4HB) and P(4HB) are also polymers with potential therapeutic applications have been pointed out in a recent review [7]. The 4HB units are pharmacologically active compounds, which have been used in the treatment of alcohol withdrawal syndrome [227, 228] and narcolepsy [229]. Other potential applications include the treatment of patients with chronic schizophrenia, catatonic schizophrenia, atypical psychoses, chronic brain syndrome, neurosis, drug addiction and withdrawal, Parkinson's disease and other neuropharmacological illnesses, hypertension, ischemia, circulatory collapse, radiation exposure, cancer, and myocardial infarction [230].

Recently, PHA are also gaining much attention as a source of enantiomerically pure compounds although this possibility was realised almost a decade ago [231, 232]. Due to the specificity of the PHA synthase, biosynthesised PHA only contain the (*R*)-isomers of 3-hydroxyacids. Close to 150 different monomer constituents of PHA have been identified to date and therefore various enantiomerically pure compounds can theoretically be obtained by depolymerising the polymers. The depolymerisation can be performed either by chemical or biological methods. Chemical depolymerisation of P[3HB] to generate methyl esters of (*R*)-3HB have been performed by acidic alcoholysis using sulfuric acid [232] and hydrochloric acid [233], whereby the latter method proved to be more effective.

On the other hand, the biological method of depolymerisation uses either the intracellular [234] or the extracellular depolymerase [202]. Recently, it has been shown that the intracellular PHA depolymerisation mechanisms can be exploited *in vivo* to generate (*R*)-monomers from PHA_{SCL} and PHA_{MCL} accumulated by several bacteria such as *A. latus*, *R. eutropha*, *P. oleovorans* and *P. aeruginosa* [235].

7.7 Conclusions and Outlook

Lately, Life Cycle Assessment is becoming an essential tool to evaluate the performance of a product or process and their impact on the ecosystem. The 21st century will therefore be a time to seriously adapt the 'cradle to grave' concept whereby the complete life cycle of a product is carefully designed to be compatible with its environment. The vast amount of knowledge obtained from several decades of enthusiastic studies puts us in a better position to assess the full potential of these interesting microbial PHA. Initially, interests on PHA were mainly focussed on their potential as a biodegradable thermoplastic material. While this is still the main driving force behind research on PHA, various new specialised area of applications have emerged. These include the application of PHA in the medical field where the quality and performance of materials outweigh their production costs.

At present, PHA can be produced from renewable carbon sources using wild type and recombinant microorganisms, transgenic plants, and *in vitro* processes. By optimising all these systems, it will be possible to produce PHA in various countries for many different applications. No one option is likely to predominate because each production system will have its strong points. The characteristics (such as MW) of the PHA produced by these systems will be different. It may not be possible to efficiently produce PHA having special functional groups in transgenic plants. The purity of PHA produced by *in vitro* systems will be medically important and might also be more appealing to consumers.

It is now possible to produce substantial amounts of various PHA, which will enable them to be tested for different applications. At the fundamental level, the elucidation of the PHA synthase mechanism will probably enable us to have more control over the design and synthesis of novel PHA in the future.

Acknowledgements

The authors acknowledge all collaborators and members (both past and present) of the Polymer Chemistry Laboratory of RIKEN Institute for contributions to the research and development of this interesting microbial polyester.

References

1. E.A. Dawes and P.J. Senior, *Advances in Microbial Physiology*, 1973, **10**, 135.

2. J.M. Shively, *Annual Reviews of Microbiology*, 1974, **28**, 167.

3. K.L. Burdon, *Journal of Bacteriology*, 1946, **52**, 665.

4. G.N. Barnard and J.K. Sanders, *Journal of Biological Chemistry*, 1989, **264**, 6, 3286.

5. D. Brown in *Biotechnology, the Science and the Business*, Eds., V. Moses and R.E. Cape, Harwood Academic Publishers, Chur, Switzerland, 1991, 341.

6. P.A. Holmes, *Physics in Technology*, 1985, **16**, 1, 32.

7. K. Sudesh, H. Abe and Y. Doi, *Progress in Polymer Science*, 2000, **25**, 10, 1503.

8. E.A. Dawes, *Bioscience Reports*, 1988, **8**, 537.

9. R.H. Findlay and D.C. White, *Applied and Environmental Microbiology*, 1983, **45**, 71.

10. L.L. Wallen and W.K. Rohwedder, *Environmental Science and Technology*, 1974, **8**, 576.

11. M. Lemoigne, *Annales de l'Institute Pasteur*, 1925, **39**, 144.

12. M. Lemoigne, *Bulletin de la Societe de Chimie Biologique*, 1926, **8**, 770.

13. M. Lemoigne, *Annales de l'Institute Pasteur*, 1927, **41**, 148.

14. R.N. Reusch, *Canadian Journal of Microbiology*, 1995, **41**, (Supplement 1), 50.

15. R.N. Reusch, T.W. Hiske and H.L. Sadoff, *Journal of Bacteriology*, 1986, **168**, 2, 553.

16. R.N. Reusch and H.L. Sadoff, *Journal of Bacteriology*, 1983, **156**, 2, 778.

17. S. Kusaka, H. Abe, S.Y. Lee and Y. Doi, *Applied Microbiology and Biotechnology*, 1997, **47**, 2, 140.

18. S. Kusaka, T. Iwata and Y. Doi, *Journal of Macromolecular Science: Pure and Applied Chemistry*, **A35**, 2, 319.

19. R.N. Reusch, *FEMS Microbiological Reviews*, 1992, **9**, 2-4, 119.

20. R.N. Reusch, A.W. Sparrow and J. Gardiner, *Biochimica et Biophysica Acta*, 1992, **1123**, 1, 33.

21. R.N. Reusch, R. Huang and D. Kosk-Kosicka, *FEMS Letters*, 1997, **412**, 3, 592.

22. D. Seebach, A. Brunner, H.M. Bürger, J. Schneider and R.N. Reusch, *European Journal of Biochemistry*, 1994, **224**, 317.

23. R. Huang and R.N. Reusch, *Journal of Biological Chemistry*, 1996, **271**, 36, 22196.

24. R. Huang and R.N. Reusch, *Journal of Bacteriology*, 1995, **177**, 2, 486.

25. R.N. Reusch, *Biochemistry (Moscow)*, 2000, **65**, 3, 280.

26. D. Byrom in *Plastics from Microbes: Microbial Synthesis of Polymers and Polymer Precursors*, Ed., D.P. Mobley, Hanser, Munich, Germany, 1994, 5.

27. S. Kusaka, T. Iwata and Y. Doi, *International Journal of Biological Macromolecules*, 1999, **25**, 1-3, 87.

28. J. Cornibert and R.H. Marchessault, *Journal of Molecular Biology,* 1972, **71**, 735.

29. J. Cornibert, R.H. Marchessault, H. Benoit and G. Weill, *Macromolecules*, 1970, **3**, 6, 741.

30. P.A. Holmes in *Developments in Crystalline Polymers—2*, Ed., D.C. Bassett, Elsevier, London, UK, 1988, p.1.

31. S. Brückner, S.V. Meille, L. Malpezzi, A. Cesaro, L. Navarini and R. Tombolini, *Macromolecules*, 1988, **21**, 4, 967.

32. K. Okamura and R.H. Marchessault in *Conformation of Biopolymers*, Ed., G.M. Ramachandran, Academic Press, New York, NY, USA, 1967, **2**, p.709.

33. M. Yokouchi, Y. Chatani, H. Tadokoro, K. Teranishi and H. Tani, *Polymer*, 1973, **14**, 6, 267.

34. R. Alper, D.G. Lundgren, R.H. Marchessault and W.A. Cote, *Biopolymers*, 1963, **1**, 6, 545.

35. N.C. Billingham, T.J. Henman and P.A. Holmes in *Developments in Polymer Degradation-7*, Ed., N. Grassie, Elsevier, London, UK, 1987, p.81.

36. J.K. Hobbs, T.J. McMaster, M.J. Miles and P.J. Barham, *Polymer*, 1998, **39**, 12, 2437.

37. P.J. Barham, P. Barker and S.J. Organ, *FEMS Microbiological Letters*, 1992, **103**, 2-4, 289.

38. T. Iwata, S. Kusaka and Y. Doi in *Polymers from Renewable Resources: Biopolyesters and Biocatalysis*, Eds., C. Scholz and R.A. Gross, ACS Symposium Series No.764, American Chemical Society, Washington, DC, USA, 2000, **764**, p.67.

39. A. Steinbüchel and H.E. Valentin, *FEMS Microbiological Letters,* 1995, **128**, 3, 219.

40. T.L. Bluhm, G.K. Hamer, R.H. Marchessault, C.A. Fyfe and R. P. Veregin, *Macromolecules*, 1986, **19**, 11, 2871.

41. A.J. Owen, *Colloid and Polymer Science*, 1985, **263**, 10, 799.

42. N. Kamiya, M. Sakurai, Y. Inoue, R. Chûjô and Y. Doi, *Macromolecules*, 1991, **24**, 9, 2178.

43. H. Mitomo, P. J. Barham and A. Keller, *Polymer Journal*, 1987, **19**, 11, 1241.

44. D. Byrom, *FEMS Microbiological Letters*, 1992, **103**, 2-4, 247.

45. Y. Doi, *Microbial Polyesters*, VCH, New York, NY, USA, 1990.

46. Y. Doi, A. Tamaki, M. Kunioka and K. Soga, *Applied Microbiology and Biotechnology*, 1988, **28**, 330.

47. Y. Doi, M. Kunioka, Y. Nakamura and K. Soga, *Macromolecules*, 1986, **19**, 11, 2860.

48. G.W. Haywood, A.J. Anderson, D.R. Williams, E.A. Dawes and D.F. Ewing, *International Journal of Biological Macromolecules*, 1991, **13**, 2, 83.

49. A. Steinbüchel, E.M. Debzi, R.H. Marchessault and A. Timm, *Applied Microbiology and Biotechnology*, 1993, **39**, 443.

50. T. Fukui, T. Kichise, Y. Yoshida and Y. Doi, *Biotechnology Letters*, 1997, **19**, 11, 1093.

51. R.H. Marchessault, E.M. Debzi, J-F. Revol and A. Steinbüchel, *Canadian Journal of Microbiology*, 1995, **41**, (Supplement 1), 297.

52. S.P. Bessman and W.N. Fishbein, *Nature*, 1963, **200**, 1207.

53. H. Laborit, *International Journal of Neuropharmacology*, 1964, **3**, 4, 433.

54. Y. Doi, M. Kunioka, Y. Nakamura and K. Soga, *Macromolecules*, 1988, **21**, 9, 2722.

55. M. Kunioka, Y. Nakamura and Y. Doi, *Polymer Communications*, 1988, **29**, 6, 174.

56. Y. Doi, A. Segaura and M. Kunioka, *Polymer Communications*, 1989, **30**, 6, 169.

57. Y. Doi, A. Segawa and M. Kunioka, *International Journal of Biological Macromolecules*, 1990, **12**, 2, 106.

58. M. Kunioka, Y. Kawaguchi and Y. Doi, *Applied Microbiology and Biotechnology*, 1989, **30**, 569.

59. Y. Saito and Y. Doi, *International Journal of Biological Macromolecules*, 1994, **16**, 2, 99.

60. Y. Saito, S. Nakamura, M. Hiramitsu and Y. Doi, *Polymer International*, 1996, **39**, 3, 169.

61. Y. Doi, A. Segawa, S. Nakamura and M. Kunioka in *Novel Biodegradable Microbial Polymers*, Ed., E. A. Dawes, Kluwer Academic, Dordrecht, The Netherlands, 1990.

62. S. Nakamura, Y. Doi and M. Scandola, *Macromolecules*, 1992, **25**, 17, 4237.

63. Y. Kumagai, Y. Kanesawa and Y. Doi, *Makromoleculare Chemie*, 1992, **193**, 1, 53.

64. Y. Doi, H. Abe, E. Shimamura, M. Hiramitsu and S. Nakamura in *Polymeric Materials Encyclopedia*, Ed., J. C. Salamone, CRC Press, Boca Raton, FL, USA, 1996, **6**, p.4304.

65. S. Nakamura, M. Kunioka and Y. Doi, *Macromolecular Reports*, 1991, **A28**, 15.

66. M. Hiramitsu and Y. Doi, *Polymer*, 1993, **34**, 22, 4782.

67. E. Shimamura, M. Scandola and Y. Doi, *Macromolecules*, 1994, **27**, 16, 4429.

68. M. Ichikawa, K. Nakamura, N. Yoshie, N. Asakawa, Y. Inoue and Y. Doi, *Macromolecular Chemistry and Physics*, 1996, **197**, 8, 2467.

69. Y. Doi and H. Abe, *Macromolecular Symposia*, 1997, **118**, 725.

70. Y. Doi, K. Kasuya, H. Abe, N. Koyama, S. Ishiwatari, K. Takagi and Y. Yoshida, *Polymer Degradation and Stability*, 1996, **51**, 3, 281.

71. K-I. Kasuya, Y. Inoue and Y. Doi, *International Journal of Biological Macromolecules*, 1996, **19**, 1, 35.

72. R.H. Marchessault, C.J. Monasterios, F.G. Morin and P.R. Sundararajan, *International Journal of Biological Macromolecules*, 1990, **12**, 2, 158.

73. B.H.A. Rehm and A. Steinbüchel, *International Journal of Biological Macromolecules*, 1999, **25**, 1, 3.

74. Y. Doi, S. Kitamura and H. Abe, *Macromolecules*, 1995, **28**, 14, 4822.

75. G. Kobayashi, T. Shiotani, Y. Shima and Y. Doi in *Biodegradable Plastics and Polymers*, Eds., Y. Doi and K. Fukuda, Elsevier, Amsterdam, The Netherlands, 1994, p.410.

76. E.Y. Lee, D. Jendrossek, A. Schirmer, C.Y. Choi and A. Steinbüchel, *Applied Microbiology and Biotechnology*, 1995, **42**, 6, 901.

77. M. Liebergesell, F. Mayer and A. Steinbüchel, *Applied Microbiology and Biotechnology*, 1993, **40**, 292.

78. H. Matsusaki, H. Abe and Y. Doi, *Biomacromolecules*, 2000, **1**, 1, 17.

79. H. Matsusaki, H. Abe, K. Taguchi, T. Fukui and Y. Doi, *Applied Microbiology and Biotechnology*, 2000, **53**, 4, 401.

80. A. Steinbüchel and S. Wiese, *Applied Microbiology and Biotechnology*, 1992, **37**, 691.

81. H. Preusting, J. Kingma, G. Huisman, A. Steinbüchel and B. Witholt, *Journal of Environmental Polymer Degradation*, 1993, **1**, 1, 11.

82. U.B. Sleytr and A.W. Robards, *Journal of Microscopy*, 1977, **110**, 1.

83. U.B. Sleytr and A.W. Robards, *Journal of Microscopy*, 1982, **126**, 101.

84. K. Sudesh, T. Fukui, T. Iwata and Y. Doi, *Canadian Journal of Microbiology*, 2000, **46**, 4, 304.

85. M.J. De Smet, G. Eggink, B. Witholt, J. Kingma and H. Wynberg, *Journal of Bacteriology*, 1983, **154**, 2, 870.

86. K. Sudesh, T. Fukui, K. Taguchi, T. Iwata and Y. Doi, *International Journal of Biological Macromolecules*, 1999, **25**, 1-3, 79.

87. A. Steinbüchel, *Macromolecular Bioscience*, 2001, **1**, 1, 1.

88. T. Lütke-Eversloh, K. Bergander, H. Luftmann and A. Steinbüchel, *Microbiology*, 2001, **147**, 1, 11.

89. H. Brandl, R.A. Gross, R.W. Lenz and R.C. Fuller, *Applied and Environmental Microbiology*, 1988, **54**, 1977.

90. G.W. Haywood, A.J. Anderson and E.A. Dawes, *FEMS Microbiological Letters*, 1989, **57**, 1, 1.

91. G.W. Huisman, O. de Leeuw, G. Eggink and B. Witholt, *Applied and Environmental Microbiology*, 1989, **55**, 8, 1949.

92. R.G. Lageveen, G.W. Huisman, H. Preusting, P. Ketelaar, G. Eggink and B. Witholt, *Applied and Environmental Microbiology*, 1988, **54**, 2924.

93. M. Doudoroff and R.Y. Stanier, *Nature*, 1959, **183**, 1440.

94. R.A. Slepecky and J.H. Law, *Journal of Bacteriology*, 1961, **82**, 37.

95. B. Kessler and B. Witholt, *Journal of Biotechnology*, 2001, **86**, 2, 97.

96. R.M. Macrae and J.R. Wilkinson, *Journal of General Microbiology*, 1958, **19**, 210.

97. R. Repaske and A.C. Repaske, *Applied and Environmental Microbiology*, 1976, **32**, 585.

98. A. Steinbüchel and H.G. Schlegel, *Applied Microbiology and Biotechnology*, 1989, **31**, 168.

99. A.C. Ward, B.I. Rowley and E.A. Dawes, *Journal of General Microbioliogy*, 1977, **102**, 61.

100. T. Suzuki, T. Yamane and S. Shimizu, *Applied Microbiology and Biotechnology*, 1986, **23**, 322.

101. T. Suzuki, H. Deguchi, T. Yamane, S. Shimizu and K. Gekko, *Applied Microbiology and Biotechnology*, 1988, **27**, 487.

102. G. Braunegg and B. Bogensberger, *Acta Biotechnologica*, 1985, **5**, 4, 339.

103. O. Hrabak, *FEMS Microbiological Letters*, 1992, **103**, 2-4, 251.

104. B. Maekawa, N. Koyama and Y. Doi, *Biotechnology Letters*, 1993, **15**, 7, 691.

105. B.K. Kim, S.C. Yoon, J.D. Nam and R.W. Lenz, *Journal of Microbiology and Biotechnology*, 1997, **7**, 6, 391.

106. W.S. Ahn, S.J. Park and S.Y. Lee, *Applied and Environmental Microbiology*, 2000, **66**, 8, 3624.

107. H.H. Wong and S.Y. Lee, *Applied Microbiology and Biotechnology*, 1998, **50**, 1, 30.

108. W.J. Page, *FEMS Microbiological Letters*, 1992, **103**, 2-4, 149.

109. W.J. Page, N. Bhanthumnavin, J. Manchak and M. Rumen, *Applied Microbiology and Biotechnology*, 1997, **48**, 1, 88.

110. H. Zhang, V. Obias, K. Gonyer and D. Dennis, *Applied and Environmental Microbiology*, 1994, **60**, 4, 1198.

111. M.A. Hassan, Y. Shirai, A. Kubota, M.I.A. Karim, K. Nakanishi and K. Hashimoto, *Journal of Fermentation Bioengineering*, 1998, **86**, 1, 57.

112. J. Yu, *Journal of Biotechnology*, 2001, **86**, 2, 105.

113. H.M. Müller and D. Seebach, *Die Angewandte Chemie*, 1993, **32**, 4, 477.

114. T. Fukui and Y. Doi, *Applied Microbiology and Biotechnology*, 1998, **49**, 3, 333.

115. F.R. Blattner, G. Plunkett, C.A. Bloch, N.T. Perna, V. Burland, M. Riley, J. Collado-Vides, J.D. Glasner, C.K. Rode, G.F. Mayhew, J. Gregor, N.W. Davis, H.A. Kirkpatrick, M.A. Goeden, D.J. Rose, B. Mau and Y. Shao, *Science*, 1997, **277**, 5331, 1453.

116. T. Kaneko, S. Sato, H. Kotani, A. Tanaka, E. Asamizu, Y. Nakamura, N. Miyajima, M. Hirosawa, M. Sugiura, S. Sasamoto, T. Kimura, T. Hosouchi, A. Matsuno, A. Muraki, N. Nakazaki, K. Naruo, S. Okumura, S. Shimpo, C. Takeuchi, T. Wada, A. Watanabe, M. Yamada, M. Yasuda and S. Tabata, *DNA Research*, 1996, **3**, 3, 109.

117. C.K. Stover, X. Q. Pham, A.L. Erwin, S.D. Mizoguchi, P. Warrener, M.J. Hickey, F.S. Brinkman, W.O. Hufnagle, D.J. Kowalik, M. Lagrou, R.L. Garber, L. Goltry, E. Tolentino, S. Westbrock-Wadman, Y. Yuan, L.L. Brody, S.N. Coulter, K.R. Folger, A. Kas, K. Larbig, R. Lim, K. Smith, D. Spencer, G.K. Wong, Z. Wu and I.T. Paulsen, *Nature*, 2000, **406**, 6799, 959.

118. G. Taroncher-Oldenburg, K. Nishina and G. Stephanopoulos, *Applied and Environmental Microbiology*, 2000, **66**, 10, 4440.

119. G.J. Moskowitz and J.M. Merrick, *Biochemistry*, 1969, 8, 7, 2748.

120. G.N. Huijberts, G. Eggink, P. de Waard, G.W. Huisman and B. Witholt, *Applied and Environmental Microbiology*, 1992, **58**, 536.

121. G.N. Huijberts, T.C. de Rijk, P. de Waard and G. Eggink, *Journal of Bacteriology*, 1994, **176**, 6, 1661.

122. T. Fukui and Y. Doi, *Journal of Bacteriology,* 1997, **179**, 15, 4821.

123. T. Fukui, N. Shiomi and Y. Doi, *Journal of Bacteriology,* 1998, **180**, 3, 667.

124. T. Tsuge, T. Fukui, H. Matsusaki, S. Taguchi, G. Kobayashi, A. Ishizaki and Y. Doi, *FEMS Microbiological Letters,* 2000, **184**, 2, 193.

125. K. Taguchi, Y. Aoyagi, H. Matsusaki, T. Fukui and Y. Doi, *FEMS Microbiological Letters,* 1999, **176**, 1, 183.

126. Q. Ren, N. Sierro, B. Witholt and B. Kessler, *Journal of Bacteriology,* 2000, **182**, 10, 2978.

127. Q. Ren, N. Sierro, M. Kellerhals, B. Kessler and B. Witholt, *Applied and Environmental Microbiology,* 2000, **66**, 4, 1311.

128. Q. Ren, B. Kessler, F. van der Leij and B. Witholt, *Applied Microbiology and Biotechnology,* 1998, **49**, 6, 743.

129. B. H. A. Rehm, N. Krüger and A. Steinbüchel, *Journal of Biological Chemistry,* 1998, **273**, 37, 24044.

130. N. Hoffmann, A. Steinbüchel and B.H.A. Rehm, *FEMS Microbiological Letters,* 2000, **184**, 2, 253.

131. N. Hoffmann, A. Steinbüchel and B.H.A. Rehm, *Applied Microbiology and Biotechnology,* 2000, **54**, 5, 665.

132. K. Matsumoto, H. Matsusaki, S. Taguchi, M. Seki and Y. Doi, *Biomacromolecules,* 2001, **2**, 1, 142.

133. K. Taguchi, Y. Aoyagi, H. Matsusaki, T. Fukui and Y. Doi, *Biotechnology Letters,* 1999, **21**, 7, 579.

134. H.E. Valentin and D. Dennis, *Journal of Biotechnology,* 1997, **58**, 1, 33.

135. S. Hein, B. Söhling, G. Gottschalk and A. Steinbüchel, *FEMS Microbiological Letters,* 1997, **153**, 2, 411.

136. M. Liebergesell, S. Rahalkar and A. Steinbüchel, *Applied Microbiology and Biotechnology,* 2000, **54**, 2, 186.

137. H. Matsusaki, S. Manji, K. Taguchi, M. Kato, T. Fukui and Y. Doi, *Journal of Bacteriology,* 1998, **180**, 24, 6459.

138. T.U. Gerngross, K.D. Snell, O.P. Peoples, A.J. Sinskey, E. Csuhai, S. Masamune and J. Stubbe, *Biochemistry*, 1994, **33**, 31, 9311.

139. A. Hoppensack, B.H. Rehm and A. Steinbüchel, *Journal of Bacteriology*, 1999, **181**, 5, 1429.

140. S. Taguchi, A. Maehara, K. Takase, M. Nakahara, H. Nakamura and Y. Doi, *FEMS Microbiological Letters*, 2001, **198**, 1, 65.

141. J.D. Schrag and M. Cygler, *Lipases, Part A: Biochemistry, Methods in Enzymology*, 1997, **284**, 85.

142. R. Griebel and J.M. Merrick, *Journal of Bacteriology*, 1971, **108**, 782.

143. R. Griebel, Z. Smith and J.M. Merrick, *Biochemistry*, 1968, 7, 10, 3676.

144. Y. Jia, T.J. Kappock, T. Frick, A.J. Sinskey and J. Stubbe, *Biochemistry*, 2000, **39**, 14, 3927.

145. J. Hodgson, *Nature BioTechnology*, 1993, **11**, 8, 887.

146. G. Sandmann, M. Albrecht, G. Schnurr, O. Knorzer and P. Boger, *Trends in Biotechnology*, 1999, **17**, 6, 233.

147. Y. Poirier, *Current Opinion in Biotechnology*, 1999, **10**, 2, 181.

148. S.Y. Lee, *Trends in Biotechnology*, 1996, **14**, 98.

149. S.Y. Lee, K.M. Lee, H.N. Chan and A. Steinbüchel, *Biotechnology and Bioengineering*, 1994, **44**, 11, 1337.

150. S.Y. Lee, H.N. Chang and Y.K. Chang, *Annals of the New York Academy of Sciences*, 1994, **721**, 43.

151. S.Y. Lee, K.S. Yim, H.N. Chang and Y.K. Chang, *Journal of Biotechnology*, 1994, **32**, 2, 203.

152. J-I. Choi, S.Y. Lee and K. Han, *Applied and Environmental Microbiology*, 1998, **64**, 12, 4897.

153. O.P. Peoples and A.J. Sinskey, *Journal of Biological Chemistry*, 1989, **264**, 26, 15298.

154. P. Schubert, A. Steinbüchel and H.G. Schlegel, *Journal of Bacteriology*, 1988, **170**, 12, 5837.

155. S.C. Slater, W.H. Voige and D.E. Dennis, *Journal of Bacteriology*, 1988, **170**, 10, 4431.

156. S-J. Liu and A. Steinbüchel, *Applied and Environmental Microbiology*, 2000, **66**, 2, 739.

157. S. Klinke, Q. Ren, B. Witholt and B. Kessler, *Applied and Environmental Microbiology*, 1999, **65**, 2, 540.

158. S. Langenbach, B.H.A. Rehm and A. Steinbüchel, *FEMS Microbiological Letters*, 1997, **150**, 2, 303.

159. Q. Qi, B.H.A. Rehm and A. Steinbüchel, *FEMS Microbiological Letters*, 1997, **157**, 1, 155.

160. Q. Qi, A. Steinbüchel and B.H.A. Rehm, *FEMS Microbiological Letters*, 1998, **167**, 1, 89.

161. B.H.A. Rehm and A. Steinbüchel, *Applied Microbiology and Biotechnology*, 2001, **55**, 2, 205.

162. T. Fukui, S. Yokomizu, S. Kobayashi and Y. Doi, *FEMS Microbiological Letters*, 1999, **170**, 1, 69.

163. S.J. Park, W.S. Ahn, P.R. Green and S.Y. Lee, *Biomacromolecules*, 2001, **2**, 1, 248.

164. M.A. Prieto, M.B. Kellerhals, G.B. Bozzato, D. Radnovic, B. Witholt and B. Kessler, *Applied and Environmental Microbiology*, 1999, **65**, 8, 3265.

165. J-I. Choi and S.Y. Lee, *Biotechnology and Bioengineering*, 1999, **62**, 5, 546.

166. Y. Poirier, D.E. Dennis, K. Klomparens and C. Somerville, *Science*, 1992, **256**, 520.

167. C. Nawrath, Y. Poirier and C. Somerville, *Proceedings of the National Academy of Sciences of the USA*, 1994, **91**, 26, 12760.

168. H.E. Valentin, D. L. Broyles, L.A. Casagrande, S.M. Colburn, W.L. Creely, P.A. DeLaquil, H.M. Felton, K.A. Gonzalez, K.L. Houmiel, K. Lutke, D.A. Mahadeo, T.A. Mitsky, S.R. Padgette, S.E. Reiser, S. Slater, D.M. Stark, R.T. Stock, D.A. Stone, N.B. Taylor, G.M. Thorne, M. Tran and K.J. Gruys, *International Journal of Biological Macromolecules*, 1999, **25**, 1-3, 303.

169. V. Mittendorf, E.J. Robertson, R.M. Leech, N. Kruger, A. Steinbuchel and Y. Poirier, *Proceedings of the National Academy of Sciences of the USA*, 1998, **95**, 23, 13397.

170. Y. Asada, M. Miyake, J. Miyake, R. Kurane and Y. Tokiwa, *International Journal of Biological Macromolecules,* 1999, **25**, 1-3, 37.

171. M. Miyake and Y. Asada, *Nippon Nogeikagaku Kaishi,* 1998, **72**, 4, 528.

172. M. Miyake, M. Erata and Y. Asada, *Journal of Fermentation and Bioengineering,* 1996, **82**, 5, 512.

173. R. Jossek, R. Reichelt and A. Steinbüchel, *Applied Microbiology and Biotechnology,* 1998, **49**, 3, 258.

174. R. Jossek and A. Steinbüchel, *FEMS Microbiological Letters,* 1998, **168**, 2, 319.

175. Q. Qi, A. Steinbüchel and B.H.A. Rehm, *Applied Microbiology and Biotechnology,* 2000, **54**, 1, 37.

176. S.Y. Lee, J-I. Choi, K. Han and J.Y. Song, *Applied and Environmental Microbiology,* 1999, **65**, 6, 2762.

177. Y. Doi, Y. Kanesawa, N. Tanahashi and Y. Kumagai, *Polymer Degradation and Stability,* 1992, **36**, 2, 173.

178. J. Mergaert, C. Anderson, A. Wouters and J. Swings, *Journal of Environmental Polymer Degradation,* 1994, **2**, 3, 177.

179. J. Mergaert, A. Webb, C. Anderson, A. Wouters and J. Swings, *Applied and Environmental Microbiology,* 1993, **59**, 3233.

180. J. Mergaert, A. Wouters, C. Anderson and J. Swings, *Canadian Journal of Microbiology,* 1995, **41**, (Supplement 1), 154.

181. H. Brandl and P. Püchner in *Novel Biodegradable Microbial Polymers,* Ed., E.A. Dawes, Kluwer Academic, Dordrecht, The Netherlands, 1990, p.421.

182. C.L. Bructo and S.S. Wong, *Archives of Biochemistry and Biophysics,* 1991, **290**, 497.

183. F.P. Delafield, M.M. Doudroff, N.J. Palleroni, C.J. Lusty and R. Contopoulos, *Journal of Bacteriology,* 1965, **90**, 1455.

184. M. Matavulj and H.P. Molitoris, *FEMS Microbiological Letters,* 1992, **103**, 2-4, 323.

185. D.W. McLellan and P.J. Halling, *FEMS Microbiological Letters,* 1988, **52**, 3, 215.

186. K. Mukai, K. Yamada and Y. Doi, *Polymer Degradation and Stability,* 1994, **43,** 3, 319.

187. T. Tanio, T. Fukui, Y. Shirakura, T. Saito, K. Tomita, T. Kaiho and S. Masamune, *European Journal of Biochemistry,* 1982, **124,** 71.

188. K. Kita, K. Ishimaru, M. Teraoka, H. Yanase and N. Kato, *Applied and Environmental Microbiology,* 1995, **61,** 1727.

189. K. Mukai, K. Yamada and Y. Doi, *Polymer Degradation and Stability,* 1993, **41,** 1, 85.

190. M. Takeda, J-I. Koizumi, K. Yabe and K. Adachi, *Journal of Fermentation Bioengineering,* 1998, **85,** 4, 375.

191. K. Yamada, K. Mukai and Y. Doi, *International Journal of Biological Macromolecules,* 1993, **15,** 4, 215.

192. H. Abe and Y. Doi, *Macromolecules,* 1996, **29,** 27, 8683.

193. H. Abe and Y. Doi, *International Journal of Biological Macromolecules,* 1999, **25,** 1-3, 185.

194. H. Abe, Y. Doi, H. Aoki and T. Akehata, *Macromolecules,* 1998, **31,** 6, 1791.

195. D. Jendrossek, I. Knoke, R.B. Habibian, A. Steinbüchel and H.G. Schlegel, *Journal of Environmental Polymer Degradation,* 1993, **1,** 1, 53.

196. Y. Shirakura, T. Fukui, T. Saito, Y. Okamoto, T. Narikawa, K. Koide, K. Tomita, T. Takemasa and S. Masamune, *Biochimica et Biophysica Acta,* 1986, **880,** 46.

197. M. Uefuji, K-I. Kasuya and Y. Doi, *Polymer Degradation and Stability,* 1997, **58,** 3, 275.

198. N.R. Gilkes, B. Henrissat, D.G. Kilburn, J.R. Miller and A.J. Warren, *Microbiological Reviews,* 1991, **55,** 2, 303.

199. L.E. Kellet, D.M. Poole, L.M. Ferreira, A.J. Durrant, G.P. Hazlewood and H.J. Gilbert, *The Biochemical Journal,* 1990, **272,** 369.

200. T. Watanabe, K. Suzuki, W. Oyanagi, K. Ohnishi and H. Tanaka, *Journal of Biological Chemistry,* 1990, **265,** 15659.

201. M. Nojiri and T. Saito, *Journal of Bacteriology,* 1997, **179,** 6965.

202. D. Jendrossek, A. Schirmer and H.G. Schlegel, *Applied Microbiology and Biotechnology*, 1996, **46**, 5-6, 51.

203. K. Sudesh and Y. Doi, *Polymers for Advanced Technology*, 2000, **11**, 8-12, 865.

204. K. Fritzsche, R.W. Lenz and R.C. Fuller, *International Journal of Biological Macromolecules*, 1990, **12**, 2, 92.

205. K. Fritzsche, R.W. Lenz and R.C. Fuller, *International Journal of Biological Macromolecules*, 1990, **12**, 2, 85.

206. R.A. Gross, C. DeMello, R.W. Lenz, H. Brandl and R.C. Fuller, *Macromolecules*, 1989, **22**, 3, 1106.

207. Y.B. Kim, R.W. Lenz and R.C. Fuller, *Macromolecules*, 1992, **25**, 7, 1852.

208. J.M. Curley, B. Hazer, R.W. Lenz and R.C. Fuller, *Macromolecules*, 1996, **29**, 5, 1762.

209. G.A. Abraham, A. Gallardo, J.S. Roman, E.R. Olivera, R. Jodra, B. García, B. Miñambres, J.L. García and J.M. Luengo, *Biomacromolecules*, 2001, **2**, 2, 562.

210. D.Y. Kim, Y.B. Kim and Y.H. Rhee, *International Journal of Biological Macromolecules*, 2000, **28**, 1, 23.

211. G.A.M. van der Walle, G.J.M. de Koning, R.A. Weusthuis and G. Eggink, in *Biopolyesters, Advances in Biochemical Engineering/Biotechnology*, Eds., W. Babel and A. Steinbüchel, Springer Verlag, Berlin, Germany, 2001, Volume 71, p.263.

212. J.N. Baptist and J.B. Ziegler, inventors; WR Grace and Co.; assignee; US3,225,766, 1965.

213. S.F. Williams, D.P. Martin, D.M. Horowitz and O.P. Peoples, *International Journal of Biological Macromolecules*, 1999, **25**, 1-3, 111.

214. O. Duvernoy, T. Malm, J. Ramström and S. Bowald, *The Journal of Thoracic and Cardiovascular Surgery*, 1995, **43**, 271.

215. U.A. Stock, M. Nagashima, P.N. Khalil, G.D. Nollert, T. Herden, J.S. Sperling, A. Moran, J. Lien, D.P. Martin, F.J. Schön, J.P. Vacanti and J.E. Mayer, Jr., *The Journal of Thoracic and Cardiovascular Surgery*, 2000, **119**, 732.

216. U.A. Stock, T. Sakamoto, S. Hatsuoka, D.P. Martin, M. Nagashima, A.M. Moran, M.A. Moses, P.N. Khalil, F.J. Schön, J.P. Vacanti and J.E. Mayer, Jr., *The Journal of Thoracic and Cardiovascular Surgery*, 2000, **120**, 6, 1158.

217. S.P. Hoerstrup, G. Zund, R. Sodian, A.M. Schnell, J. Grunenfelder and M.I. Turina, *European Journal of Cardiothoracic Surgery*, 2001, **20**, 1, 164.

218. S.P. Hoerstrup, R. Sodian, S. Daebritz, J. Wang, E.A. Bacha, D.P. Martin, A.M. Moran, K.J. Guleserian, J.S. Sperling, S. Kaushal, J.P. Vacanti, F.J. Schön and J.E. Mayer, Jr., *Circulation*, 2000, **102**, (Supplement 3), III-44.

219. R. Sodian, S.P. Hoerstrup, J.S. Sperling, S. Daebritz, D. P. Martin, A.M. Moran, B.S. Kim, F.J. Schön, J.P. Vacanti and J.E. Mayer, Jr., *Circulation*, 2000, **102**, (Supplement 3), III-22.

220. R. Sodian, S.P. Hoerstrup, J.S. Sperling, S.H. Daebritz, D.P. Martin, F.J. Schön, J.P. Vacanti and J.E. Mayer, Jr., *The Annals of Thoracic Surgery*, 2000, **70**, 1, 140.

221. R. Sodian, S.P. Hoerstrup, J.S. Sperling, D.P. Martin, S. Daebritz, J.E.J. Mayer, Jr. and J.P. Vacanti, *ASAIO Journal*, 2000, **46**, 1, 107.

222. R. Sodian, J.S. Sperling, D.P. Martin, A. Egozy, U. Stock, J.E. Mayer, Jr., and J.P. Vacanti, *Tissue Engineering*, 2000, **6**, 2, 183.

223. G.A.R. Nobes, R.H. Marchessault and D. Maysinger, *Drug Delivery*, 1998, **5**, 167.

224. C. Scholz in *Polymers from Renewable Resources: Biopolyesters and Biocatalysis*, Ed., C. Scholz and R.A. Gross, ACS Symposium Series No.764, American Chemical Society, Washington, DC, USA, 2000, 328.

225. K. Mukai, Y. Doi, Y. Sema and K. Tomita, *Biotechnology Letters,* 1993, **15**, 6, 601.

226. M. Löbler, M. Sass, P. Michel, U.T. Hopt, C. Kunze and K. P. Schmitz, *Journal of Materials Science: Materials in Medicine*, 1999, **10**, 12, 797.

227. G. Addolorato, G. Balducci, E. Capristo, M.L. Attilia, F. Taggi, G. Gasbarrini and M. Ceccanti, *Alcoholism: Clinical and Experimental Research*, 1999, **23**, 10, 1596.

228. G. Addolorato, G.F. Stefanini, G. Casella, L. Marsigli, F. Caputo and G. Gasbarrini, *Alcologia - European Journal of Alcohol Studies*, 1995, **7**, 233.

229. A.M. Mamelak, M.B. Scharf and M. Woods, *Sleep*, 1986, **9**, 1, 285.

230. S.F. Williams and D.P. Martin, inventors; Tepha, Inc., assignee; WO 2001019361, 2001.

231. J. Crosby in *Chirality in Industry*, Eds., A.N. Collins, G.N. Sheldrake and J. Crosby, Wiley, Chichester, UK, 1992, p.1.

232. D. Seebach, A.K. Beck, R. Breitschuh and K. Job, *Organic Synthesis*, 1992, **71**, 39.

233. Y. Lee, S.H. Park, I.T. Lim, K. Han and S.Y. Lee, *Enzyme and Microbial Technology*, 2000, **27**, 1-2, 33.

234. J.M. Merrick and M. Doudoroff, *Journal of Bacteriology*, 1964, **88**, 60.

235. S.Y. Lee, Y. Lee and F. Wang, *Biotechnology and Bioengineering*, 1999, **65**, 3, 363.

8 Starch-Based Technology

Catia Bastioli

8.1 Introduction

In nature, the availability of starch is just second to cellulose. Starch represents a link with the energy of the sun, which is partially captured during photosynthesis. It serves as a food reserve for plants and provides a mechanism by which non-photosynthesising organisms, such as man, can utilise the energy supplied by the sun.

The most important industrial sources of starch are corn, wheat, potato, tapioca and rice. Today, starch is inexpensive and is available annually from such crops, in excess of current market needs in the United States and Europe [1].

Corn production, has risen over time, as higher yields followed improvements in technology and in production practices.

US corn production in 2003 passed the level of 10 billion bushels (560 billion kg) and from 1970 to 2000 the bushels per acre increased from 80 to 140 (4480 to 7840 kg) (National Agricultural Statistics Service, USDA). In Europe (15 member states) corn starch production in 2001 was 8.4 million tons.

Approximately 75% of US domestic corn use is allocated to livestock feed. Food, seed, and industrial uses of corn comprise 25% of domestic utilisation. The market for food made from corn is mature, and food uses of corn are expected to expand at the rate of population growth. Besides starch, corn is also processed by wet millers into high-fructose corn syrup (HFCS), glucose, dextrose, corn oil, beverage alcohol and fuel ethanol.

In the last decade there was a significant decrease in the price of corn and potato starch (in Europe and the USA); whereas in Europe during the period 1990-2001, the price of wheat starch remained almost unchanged.

The low price and the availability of starch associated with its very favourable environmental profile in the last 15 years aroused a renewed interest in starch-based polymers as an attractive alternative to polymers based on petrochemicals.

Starch is totally biodegradable in a wide variety of environments and permits the development of totally degradable products for specific market demands. Degradation or incineration of starch products recycles atmospheric carbon dioxide trapped by starch-producing plants and does not increase potential global warming.

The most relevant achievements in this sector are related to thermoplastic starch polymers resulting from the processing of native starch by chemical, thermal and mechanical means and to its complexation with other co-polymers: the resulting materials show properties ranging from the flexibility of polyethylene (PE) to the rigidity of polystyrene, and can be soluble or insoluble in water as well as insensitive to humidity. Such properties explain the leading position of starch-based materials in the bio-based polymers market.

This chapter reviews the main results obtained in the fields of starch-filled plastics and thermoplastic starch with particular attention to the concept of gelatinisation, destructurisation, extrusion cooking, and the complexation of amylose by means of polymeric complexing agents with the formation of specific supra-molecular structures. The behaviours of products now in the market are considered in terms of processability, physical-chemical and physical-mechanical properties and biodegradation rates.

8.2 Starch Polymer

Starch consists of two major components: amylose, a mostly linear α-D(1-4)-glucan and amylopectin, an α-D-(1-4) glucan which has α-D(1-6) linkages at the branch point (See **Figure 8.1**).

The linear amylose molecules of starch have a molecular weight of 0.2-2 million, while the branched amylopectin molecules have molecular weights as high as 100-400 million [2-3].

Starch is unique among carbohydrates because it occurs naturally as discrete granules. This is because the short branched amylopectin chains are able to form helical structures which crystallise. Starch granules exhibit hydrophilic properties and strong inter-molecular association via hydrogen bonding due to the hydroxyl groups on the granule surface. The melting point of native starch is higher than the thermal decomposition temperature: hence the poor thermal stability of native starch and the need for conversion to starch-based materials with a much improved property profile.

In nature, starch is found as crystalline beads of about 15-100 μm in diameter, in three crystalline modifications designated A (cereal), B (tuber), and C (smooth pea and various beans), all characterised by double helices: almost perfect left-handed, six-fold structures, as elucidated by X-ray diffraction experiments [2, 4-6].

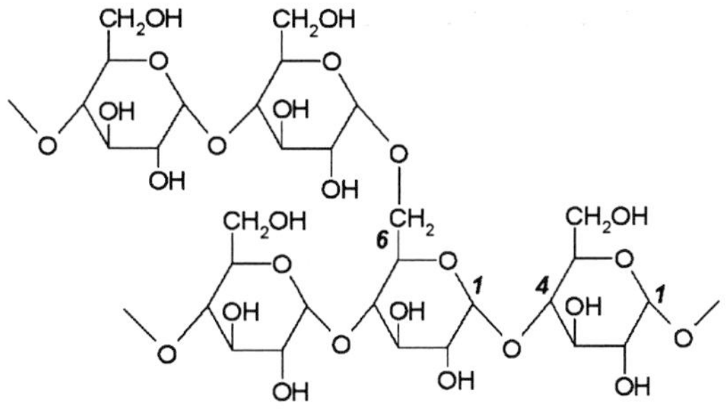

Figure 8.1 Molecular structure of amylopectin

Crystalline starch beads in plastics can be used as fillers or can be transformed into thermoplastic starch which can be processed alone or in combination with specific synthetic polymers. To make starch thermoplastic, its crystalline structure has to be destroyed by pressure, heat, mechanical work or plasticisers such as water, glycerine or other polyols. Three main families of starch polymers can be used: pure starch, modified starch and partially fermented starch polymers.

The production of starch polymers begins with the extraction of starch. Taking as an example corn: starch is extracted from the kernel by wet milling. The kernel is first softened by steeping it in a dilute acid solution, then ground coarsely to split the kernel and remove the oil-containing germ. Finer milling separates the fibre from the endosperm which is then centrifuged to separate the less dense protein from the more dense starch. The starch slurry is then washed in a centrifuge, dewatered and dried. Either prior or subsequent to the drying step, the starch may be processed in a number of ways to improve its properties.

The addition of chemicals leading to alteration of the structure of starch is generally described as 'chemical modification'. Modified starch is starch which has been treated with chemicals so that some hydroxyl groups have been replaced by for example ester or ether groups. High starch content plastics are highly hydrophilic and readily disintegrate when in contact with water. Very low levels of chemical modification can significantly improve hydrophilicity, as well as change other rheological, physical, and chemical properties of starch. Crosslinking, in which two hydroxyl groups on neighbouring starch molecules are linked chemically is also a form of chemical modification. Crosslinking inhibits granule swelling on gelatinisation and gives increased stability to acid, heat treatment, and shear forces. Chemically modified starch may be used directly in pelletised or otherwise dried form for conversion to a final product.

In the past, the study of starch esters and ethers [7-14] was abandoned due to the inadequate properties of these materials in comparison with cellulose derivatives for most applications.

More recently, starch graft copolymers [15-19], starch plastic composites [20-21], and starch itself [22-25], have been proposed as plastic materials.

Starch can also be modified by fermentation as used in the Rodenburg process. In this case [26] the raw material is a potato waste slurry originating from the food industry. This slurry mainly consists of starch, the rest being proteins, fats and oils, inorganic components and cellulose. The slurry is held in storage silos for about two weeks to allow for stabilisation and partial fermentation. The most important fermentation process that occurs is the conversion of a small fraction of starch to lactic acid by means of the lactic acid bacteria that are naturally present in the feedstock. The product is subsequently dried to a final water content of 10% and then extruded.

8.3 Starch-filled Plastics

Starch can be used as a natural filler in traditional plastics [20, 27-37] and particularly in polyolefins. When blended with starch beads, PE films [38] biodeteriorate on exposure to a soil environment. The microbial consumption of the starch component, in fact, leads to increased porosity, void formation, and the loss of integrity of the plastic matrix. Generally [36, 39-42], starch is added at fairly low concentrations (6-15%); the overall disintegration of these materials is achieved by the use of transition metal compounds, soluble in the thermoplastic matrix, as pro-oxidant additives which catalyse the photo- and thermo-oxidative process [43-48].

Starch-filled PE containing pro-oxidants are commonly used in agricultural mulch film, in bags and in six-pack yoke packaging. Commercial products based on this technology were sold first by Ecostar and Archer Daniels Midland Companies [49, 50]. In the St Lawrence Starch [51, 52] technology, bought by Ecostar, regular corn starch was treated with a silane coupling agent to make it more compatible with hydrophobic polymers, and then dried to less than 1% of water content. It was then mixed with the other additives such as an unsaturated fat or fatty acid auto-oxidant to form a masterbatch which is added to a commodity polymer. The polymer can then be processed by convenient methods, including film blowing, injection moulding and blow moulding.

The temperature has to be kept below 230 °C to prevent decomposition of the starch, and exposure of the masterbatch to air had to be minimised to avoid water absorption. Direct addition of starch and autoxidant without the masterbatch step can also be used: as this requires some specific equipment, it is only practical for large volumes [46]. It is claimed that under appropriate conditions, the disintegration time of a buried carrier bag, containing an Ecostar additive of up to 6% starch, will be reduced from hundreds of years to three to six years [42]. However there is no evidence of a compliance of such materials with the norms of biodegradability and compostability already in place at the international level. Moreover, the destabilisation of PE induced by the pro-oxidants may significantly affect its in-use performances as a function of time.

Within the field of starch-filled materials other systems were studied, some of which a completely biodegradable such as starch/poly(ε-caprolactone) (PCL) [53], others which are partially biodegradable, such as starch/polyvinylchloride/PCL and its derivatives [54] or starch/modified polyesters [55]. In all these cases starch granules are used to increase the surface area available for attack by microorganisms.

8.4 Thermoplastic Starch

Starch can be gelatinised by extrusion cooking technology [56-70]. As described by Conway in 1971, extrusion cooking and forming is characterised by sufficient work and heat being applied to a cereal-based product to cook or gelatinise completely all the ingredients. In general the main components of high pressure cooking extruders are feeders, compression screws, barrels, dies, and heating systems [56]. The effects of processing conditions on the gelatinisation of starch and on the texture of the extruded product have been studied by several researchers [57-74]. Gelatinised materials with different starch viscosity, water solubility and water absorption have been prepared by altering the moisture content of the raw product and the temperature or the pressure in the extruder. It was demonstrated that an extrusion-cooked starch can be solubilised without any formation of maltodextrins, and that the extent of solubilisation depends on extrusion temperature, moisture content of the starch before extrusion and the amylase:amylopectin ratio. Mercier [73] determined the properties of different types of starch and considered the influence of the following parameters: moisture content between 10.5 and 28%, barrel temperature between 65 and 250 °C, residence time between 20 seconds and 2 minutes, in a twin-screw extruder. Corn starch, after extrusion cooking, gave a solubility lower than 35% [73], while potato starch solubility was up to 80%.

Starch gelatinisation is a difficult term to clearly define and it was used in the past to describe loss of crystallinity of starch granules, notwithstanding the process conditions applied [2], namely, extrusion cooking, spray drying or heating of diluted starch slurries. The work carried out by Donovan in 1979 [75] and by Colonna and Mercier in 1985 [76] gave, however, a clear explanation of two different conditions for the loss of crystallinity of starch. Colonna reported that all starches exhibit a pure gelatinisation phenomenon, which is the disorganisation of the semicrystalline structure of the starch granules during heating in the presence of a water volume fraction > 0.9. For normal genotypes, gelatinisation occurs in two stages. The first step, at around 60-70 °C, corresponds mainly to swelling of the granules, with limited leaching. Loss of birefringence, demonstrating that macromolecules are no longer oriented, occurs prior to any appreciable increase in viscosity. By contrast, differential scanning calorimetry (DSC) permits the determination of the gelatinisation temperature more easily and precisely than microscopy and, additionally, the energy input needed to disorganise the crystalline structure of the granules. The second step, above 90 °C, implies the complete disappearance of granular integrity by excessive swelling and solubilisation. Nevertheless this last transition is not detectable by DSC. Only at this stage can the swollen granules be destroyed by shear.

As observed by Donovan [75] and Colonna [76], at low water volume fractions, loss of crystallinity occurred by two (pea and high amylose maize) or three (standard maize) crystalline melting steps, according to the Flory equation:

$$1/T_m - 1/T_m{}^0 = R/\Delta Hu \cdot Vu/V_1[V_1 - X_1 V_1{}^2]$$

where: R is the gas constant,

ΔHu the fusion enthalpy per repeating unit (anhydroglucose), Vu/V_1 the ratio of the molar volume of the repeating unit to the molar volume of the diluent (water),

T_m (K) the melting point of the crystalline polymer plus diluent, T_{m0} (K) the true melting point of undiluted polymer crystallites, V_1 the volume fraction of the diluent and

X_1 the Flory-Huggins interaction parameter.

At high water volume fractions, melting of crystallites and swelling are co-operative processes.

On the contrary, according to Colonna, during extrusion cooking and mainly under the conditions described by Mercier (water volume fraction < 0.28) [73] starch undergoes a real melting process.

In the patent literature the term 'destructurised starch' [77-95] refers to a form of thermoplastic starch described as molecularly dispersed in water [96]. Destructurisation of starch is defined as melting and disordering of the molecular structure of the starch granules and as a molecular dispersion [96, 79]. The molecular structure of the starch granules is molten and consequently the granular structure disappears. This is achieved by heating the starch above the glass transition temperature (T_g) and the melting temperature (T_m) of its components until they undergo endothermic transitions. In the melt stage both the crystalline and the granular structure of the starch are destroyed and the starch - water system forms a single phase in which no structure is discernible microscopically. The disappearance of the molecular structure of the starch granule may be determined using conventional light microscopy techniques [97].

If starch is heated above the T_g and T_m in the presence of plasticisers the endothermic transition can be replaced by an exothermic transition. Destructurised starch, in simple terms, is a form of thermoplastic starch suitable for applications in the sector of plastics, with minimised defects tied to the granular structure of native starch [26, 98-103].

A recent patent claims to obtain a thermoplastic material, more resistant to water, starting from native starch and plasticisers based on the formation of nanocomposites by using a clay having a layered structure [91].

Thermoplastic starch alone can be processed as a traditional plastic [73, 96, 104]; its sensitivity to humidity, however, makes it unsuitable for most applications [105]. Starch can be also made thermoplastic at water contents lower than 10% by weight, in the presence of high boiling point plasticisers [23, 25], to avoid expansion phenomena at the die.

Another term which can be found in the literature is 'thermoplastically processable starch' (TPPS), defined as a thermoplastic starch that is substantially water free. TPS is a modified native starch which is obtained without water, since instead of water, use is made of a plasticiser or additive. The starch is thermoplastically processed together with the additive and the thermal transition taking place here is exothermic [106-111].

Starch can be destructured in combination with different synthetic polymers to satisfy a broad spectrum of needs for the market. In this case it is possible to reach starch contents higher than 40%. Otey has studied ethylene-acrylic acid copolymer (EAA)/ thermoplastic starch composites since 1977 [112-122] and has demonstrated that the addition of ammonium hydroxide to EAA makes it compatible with starch. Urea, in these formulations, enhances the film tear propagation resistance and reduces ageing phenomena due to segmental motions in amorphous starch [123, 124]. The films obtained with a plasticised starch level of about 50% showed good tensile properties (**Table 8.1**) [117]. The sensitivity to environmental changes and in particular the susceptibility to tear propagation precluded their use in most packaging applications [123]; moreover EAA is not biodegradable at all.

In 1989, studies on EAA-thermoplastic starch films, containing 40% by weight of EAA (acrylic acid content 20% by weight), processed at water contents lower than 2%, led to improved processability and film properties with elongation at break up to 200% [98]. Using microscopic analysis it was possible to observe at least three different phases: one consisting of destructured starch, one consisting of the synthetic polymer alone, and a third one described as 'interpenetrated', characterised by a strong interaction between the two components. As a confirmation, phase changes observed by DSC and nuclear magnetic resonance (NMR) [118, 122, 125-129], for starch-EAA-PE films showed at least four phases. DSC endotherms and extraction of free starch with hot water demonstrated the existence of a starch phase. DSC showed melting of an EAA phase and a low-density polyethylene (LDPE) phase but did not indicate the presence of EAA in amorphous regions of the PE. NMR, X-ray diffraction and extraction indicated the presence of an insoluble

Table 8.1 Influence of starch/EAA ratio and of partial replacement of EAA with PE or PVOH on the tensile strength and elongation of starch/EAA films [117, 123]					
Starch (phr)	EAA (phr)	PE (phr)	PVOH (phr)	Elongation (%)	UTS (MPa)
10	90	-	-	260	23.9
30	70	-	-	150	22.2
40	60	-	-	92	26.7
40	40	20	-	66	23.9
40	25	25	-	85	21.7
40	20	40	-	34	20.1
40	55	-	5	97	32.0
40	40	-	20	59	39.7
UTS: *ultimate tensile strength* PVOH: *polyvinyl alcohol*					

starch-EAA complex [129]. It was demonstrated that a fraction of starch interacts [127, 128] with EAA when EAA is salified by ammonium hydroxide or other salts during extrusion cooking, providing not only partial miscibility between the two polymers but also the formation of molecular complexes.

Rheological studies were performed on a product consisting of 60% of starch and natural additives and 40% of EAA copolymer, containing 20% by mole of acrylic acid [130]. A strong non-Newtonian behaviour was shown by the viscosity curves at high shear rates. At intermediate shear rates the material seemed to approach a Newtonian plateau, while at low shear rates a viscosity upturn was observed, suggesting the presence of yield stress. Breaking-stretching data for the same material are also reported in the literature, together with those of LDPE [130].

Starch/vinyl alcohol copolymer systems [99, 131-137], can generate a wide variety of morphologies and properties, depending on the processing conditions, the starch type and the copolymer composition. Different microstructures were observed, from droplet-like to layered, as a function of the different hydrophilicity of the synthetic copolymer. Furthermore, for this type of composite, materials containing starch with an amylase: amylopectin ratio > 20:80 *w/w* do not dissolve even with stirring in boiling water. Under these conditions a microdispersion, consisting of microsphere aggregates is produced, whose individual particle diameter is under 1 µm (**Figure 8.2**). A droplet-like structure is also confirmed by transmission electron microscopy (TEM) analysis of film slices [132]. The droplet size is comparable with that of the microdispersion obtained by boiling.

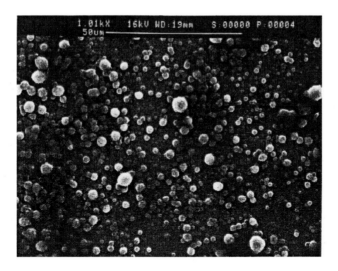

Figure 8.2 Droplet-like structure of thermoplastic corn starch/EVOH blend in film from, after disaggregation in boiling water [134]

For these products, high levels of melt elasticity are monitored by exit pressure data, whereas its recoverable fraction is almost negligible (low die swell) [134, 136]. The morphology of materials in film form, containing starch with an amylase:amylopectin ratio lower than 20:80 *w/w*, gradually looses the droplet-like form, generating layered structures (**Figure 8.3**). In this case no microspheres are produced by boiling and the starch component becomes partially soluble. Fourier transform infrared (FTIR) second derivative spectra of materials with droplet-like structure, in the range of starch ring vibrations between 960 and 920 cm^{-1}, gives an absorption peak at about 947 cm^{-1} (**Figure 8.4**). This peak, observed also when starch is complexed with butanol, is attributed by Cael and co-workers [4] to ring vibrations, which result when amylose assumes a conformation known as the V-form (a left-handed single helix).

Therefore, the absorption at 947 cm^{-1} does not correspond to crystalline or gelatinised amylose, but to a complexed one (V-type complex), as in the presence of low molecular weight molecules such as butanol and fatty acids [4, 134]. Starch-based materials with an amylose content close to zero, even in the presence of vinyl alcohol copolymers, do not show any peak at 947 cm^{-1}, demonstrating that vinyl alcohol copolymers, as well as butanol, leave the amylopectin conformation unchanged.

On the other hand, the V-complex formed by starch, having an amylase:amylopectin ratio higher than 20% by weight, with ethylene-vinyl alcohol (EVOH) copolymers makes even amylopectin insoluble in boiling water. The experimental evidence was accounted

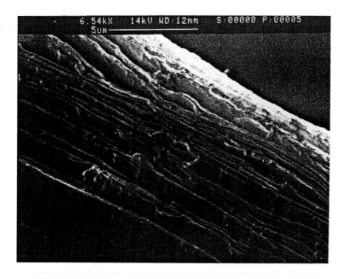

Figure 8.3 Layered structure of thermoplastic waxy maize/EVOH film after three days of soil burial test [134]

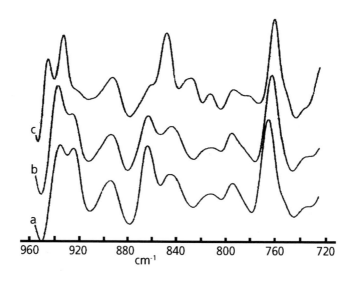

Figure 8.4 FTIR second derivative spectrum of corn starch. (a) crystalline, (b) gelatinised; (c) blended with EVOH [134]

for by a model considering large individual amylopectin molecules interconnected at several points per molecule as a result of hydrogen bonds and entanglements by chains of amylose/EVOH copolymer V-complexes [134]. The biodegradation rate of starch in these materials is inversely proportional to the content of amylose/vinyl alcohol complex. Furthermore FTIR second derivative spectra show the 947 cm^{-1} peak increasing with biodegradation, which means a delayed microbial attack of complexed amylose relative to amylopectin [134]. In addition, water permeability of starch/EVOH films is a function of the V-type complex and can range from about 820 to 334 gr30 μm/m^2/24 h (according to Lyssy method) [136].

A general study of shear flow characteristics was performed on a material containing about 60% of starch and natural additives and 40% of EVOH copolymer 40/60 mol/mol [137]. A strong pseudoplastic behaviour at high shear stresses as well as yield stress at lower ones was detected. The non-linear Bingham fluid model [138] described its viscous behaviour well over a wide range of shear rates. High levels of melt elasticity were detected from steady shearing tests, whereas it's recoverable fraction was almost negligible, at least for a reasonable time scale. The peculiar viscous and elastic behaviour has been explained on the basis of the droplet-like morphology generated by the ability of starch to form V-complexes in the presence of EVOH. Notwithstanding the peculiar rheological behaviour shown by starch/EVOH systems, traditional processing techniques such as film blowing can be easily applied.

The products based on starch/EVOH show mechanical properties good enough to meet the needs of specific industrial applications [139]. Their mouldability is comparable with that of traditional plastics such as polystyrene (PS) and acrylonitrile-butadiene-styrene copolymer (ABS). Nevertheless, they continue to be highly sensitive to low humidities, especially when in film form, with evident embrittlement.

In terms of biodegradation, ten months of aerobic biological treatment performed by a high sensitivity respirometric test, provoked the degradation of more than 90% *w/w* of a product consisting of 60% of maize starch and natural additives and by 40% of EVOH copolymer at 40% mol/mol of ethylene. Furthermore it has also been demonstrated that the synthetic component was degraded to about 80% *w/w*, notwithstanding interrupting the test when carbon dioxide evolution was still relevant [132, 133]. A material with the same composition, containing an EVOH copolymer, characterised by a lower ethylene content (29% instead of 40% mol/mol) and, therefore, by a reduced ability to generate interpenetrated structures showed, in the Sturm test, an initial biodegradation rate significantly higher [132]. The Semi-continuous Activated Sludge (SCAS) test and biodegradation in lake water of a product consisting of 70% maize starch and natural additives and 30% EVOH supports the hypothesis of a substantially different biodegradation mechanism for the two components [133]:

- the natural component, even if significantly shielded by the interpenetrated structure, appeared to be initially hydrolysed by extracellular enzymes;

- the synthetic component appeared to be biodegraded through surface adsorption of microorganisms, assisted by the increase of available surface area during the hydrolysis of the natural component.

Other limited evidence for the disappearance of EVOH copolymers have been produced by Römesser [140] and Kaplan and co-workers [141]. The presence of starch improves the biodegradation rate of these synthetic polymers - a fundamental role is also played by size and distribution of ethylene blocks. The degradation rate is too slow to consider these materials as compostable [133].

Specific types of plasticisers were selected in order to avoid migration phenomena and physical ageing [139]. The possibility of speeding up the biodegradation process was considered by modifying the EVOH copolymer by introducing a carbon monoxide group making it more sensitive to photodegradation [142]. The transparency of the material was also improved by adding additives such as boric acid, borax and other saline compounds [143]. Surface treatment by wax lamination or co-extrusion was also considered [144]. With this kind of material it is possible to obtain finished parts by film blowing, injection moulding, blow moulding, thermoforming, etc. It is also possible to make foamed parts [145], particularly by an expansion process based on injection moulding technology. The technology consists of a breathable mould connected with a vacuum pump, applied to an ordinary injection moulding mandrel [146]. Cushioning characteristics of these materials are close to expanded polystyrene (EPS-55); moreover, the foam density is 0.040 g/ml

Starch can also be destructured in the presence of more hydrophobic polymers such as aliphatic polyesters [147]. It is known that aliphatic polyesters with low melting points are difficult to process by conventional techniques for thermoplastic materials, such as film blowing and blow moulding. With reference, particularly to PCL and its copolymers, films produced using it are tacky as extruded, and rigid, and have a low melt strength at temperatures over 130 °C. Moreover, due to the slow crystallisation rate of such polymers, the crystallisation process proceeds for a long time after production of the finished articles giving an undesirable change of properties with time.

Novamont's Mater-Bi starch-based technology implies processing conditions able to almost completely destroy the crystallinity of amylose and amylopectine, in the presence of macromolecules which are able to form a complex with amylose such as specific polyesters. They can be of natural or synthetic origin, and are biodegradable. The complex formed by amylose with the complexing agent is generally crystalline and it is characterised by a single helix of amylose formed around the complexing agent. Unlike, amylose, amylopectine does not interact with the complexing agent and remains in its amorphous state. The specification of the starch, i.e., the ratio between amylose and amylopectin, the nature of the additives, the processing conditions and the nature of the complexing agents allows engineering of various supramolecular structures with very different properties.

In **Figure 8.5** and **Figure 8.6** it is reported the molecular model proposed for the droplet-like and layered structure mentioned previously, which can be produced as a result of Mater-Bi technology is shown. The 'droplet like' structure (**Figure 8.4**) has a core of an almost amorphous amylopectine molecule surrounded by complexed amylose molecules which render amylopectine unsoluble [148-150].

The layered structure is consists of by sub-micron layers of amylopectin molecules intercalated by layers of complexing agent, such layers being compatibilised by complexed amylose (**Figure 8.6**). The two structures and the many others derived from them explain the wide range of mechanical, physical-chemical, and rheological properties and the different biodegradation rates of Mater-Bi products.

Blending of starch with aliphatic polyesters improves their processability and biodegradability. Particularly suitable polyesters are PCL and its copolymers, or polymers of higher melting point formed by the reaction of 1,4 butandiol with succinic acid or

Figure 8.5 Mater-Bi technology: Droplet-like structure

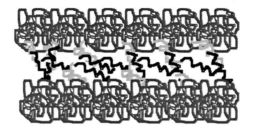

Figure 8.6 Mater-Bi technology: Layered structure

with sebacic acid, azelaic acid or poly(lactic acid), poly hydroxyalkanoates and aliphatic-aromatic polyesters.

The compatibilisation between starch and aliphatic polyesters can be promoted either by the processing conditions and/or by the presence of compatibilisers between starch and aliphatic polyesters. Examples of preferred compatibilisers are amylose/EVOH V-type complexes and starch-grafted polyesters as well as chain extenders such as diisocyanates, epoxides, and layered silicate organoclay [134, 135]. These types of materials are characterised by excellent compostability, mechanical properties and reduced sensitivity to water.

Thermoplastic starch can also be blended with polyolefins [151]. In this case about 50% of thermoplastically processable starch is mixed with 40% of PE and 10% of ethyl acrylate - maleic anhydride copolymer. During this mixing process an esterification reaction takes place between the maleic anhydride groups in the copolymer and the free hydroxyl groups in starch. Other studies have been performed on polyamide/high amylose [79, 152, 153] and acrylic copolymers/high amylose starch systems [79, 153, 154]. The problem of partial biodegradability and a too high sensitivity to humidity persists.

In the last five years Asian countries, and specifically China and Korea, have performed impressively in the sector of blends of thermoplastic starch with polyolefins, in terms of intellectual property and products range offered to the market. The non compliance of these products with the international norms of biodegradability and compostability, however, did not permit a significant market growth in western countries where low environmental impact products have more market potential.

Starch/cellulose derivative systems are also reported in other publications [143, 147, 155, 156], particularly, cellulose acetate and butyrate/starch blends in presence of glycerine and epoxidised soybean oil [155].

The combination of starch with a water soluble polymer such as PVOH and/or polyalkylene glycols has been widely considered since 1970 [157]. Recently, the system, thermoplastic starch/PVOH has been studied mainly for producing starch-based loose fillers as a substitute for expanded PS [158-164]. As an example, Lacourse and Altieri developed a technology based on hydroxy propylated high amylose starch containing small amounts of PVOH for improving foam resiliency and density [158-162]. In this case loose fill was produced directly by a twin-screw extruder. Recently more advanced processes and alloys have been developed which have resulted in foams with lower foam densities (8-6 kg/m^3) and better performance [165-167]. Other applications of modified starch/PVOH can be in the sector of sheet extrusion/thermoforming.

8.5 Starch-Based Materials on the Market

In 2003 the market of destructurised and complexed starch-based bioplastics accounted for about 30000 tons/year, 75% of which was for packaging applications and included soluble foams for industrial packaging and films for bags and sacks. The market share of these products accounted for about 70% of the global market of bioplastics [174].

Leading producers with well established products in the market are Novamont, National Starch (main Novamont partner and licensee in the sector of loose-fills and of foamed sheets), and Biotec. Following the recent start-up of its third line dedicated to the production of Mater-Bi film grades in Terni, Novamont's internal production capacity is of 20000 tons/year. The total capacity, including the network of licensees in the sector of loose fills, is of about 35000 tons/year. The technology for the production of starch-based loose fills is licensed together with National Starch and Chemical Co.

The wide patent portfolio of Novamont covers the technologies of complexed starch developed by Novamont and of destructurised starch developed by Warner-Lambert and acquired by Novamont in 1997 after the exit of Warner-Lambert from the market in 1993. Moreover, in August 2001, Novamont acquired the film technology of Biotec which included an exclusive license of the Biotec's patents on TPPS in the sector of film [174].

Biotec, the German company which acquired in 1994 the patents of Fluntera, was acquired by Essem Kashoggi Industries (EKI) in 1998. Biotec, after the sale of the film business to Novamont in 2000 offers materials for food serviceware (cutlery, plates, cups) and for pharmaceutical applications under the Bioplast trademark. Its production capacity is of 2000 tons/year.

BIOP Biopolymer Technologies recently entered the market with a starch-based material containing an additive consisting of a vinyl-alcohol/vinyl-acetate copolymer [175]. The material is sold under the Biopar trademark. The real production capacity is unknown.

Plantic is an Australian research company which is starting to offer a starch/PVOH product for thermoforming [176].

Other companies like Japan Corn Starch and Nihon Shokuhin Kako are involved in R&D as well as pilot demonstration projects. Their production capacity is unknown and there is no information on how their materials differ from the other starch-based products offered in the market. The trademarks of their materials are Cornpole and Placorn, respectively.

In recent years, companies such as Earthshell, Apack, Avebe and Potatopak dedicated significant efforts to the development of food containers through 'baking technology'.

Market tests are in place in USA and Europe to check their performances [177].

In the Netherlands, Rodenburg built up a plant for the transformation of potato wastes generated by the fried potatoes industry. The waste is fermented and the resulting granulate is used for the injection moulding of slow release devices. The claimed capacity is of 40000 tons/year. Rodenburg material is sold under the Solanyl trademark.

The price of starch-based bioplastics ranges from 1.25 to 4 Euro/kg, with the possibility of competing even with traditional materials in some limited areas [177].

The properties available for starch-based bioplastics in certain applications and the commitment of the companies today dealing with this family of bioplastics give more confidence in the future possibilities of this market sector. Bioplastics from renewable origin, either biodegradable or not biodegradable, still constitute a niche market which requires great efforts in the areas of material and application development. The technical and economic breakthroughs achieved in the last three years, however, open new possibilities for such products in the mass markets. Novamont is the market leader and boasts a diversified portfolio of industrial tailor-made materials for a wide range of applications [174].

Under the Mater-Bi trademark today Novamont produces a wide range of materials, divided into five families, by processing technology: film, extrusion/thermoforming, injection moulding, foaming, tyre technology. Mater-Bi products are mainly used in specific applications where biodegradability is required. Examples include: composting bags and sacks, fast food tableware (cups, cutlery, plates, straws, etc.), packaging (soluble and unsoluble foams for industrial packaging), film wrapping, laminated paper, food containers, agricultural film products (mulch film, nursery pots, plant labels, slow release devices), hygiene (nappy backsheet and topsheet, primary packaging, cotton swabs).

Other applications are also growing outside the sector of biodegradability, driven by the different technical performance of some Mater-Bi products *versus* traditional materials, as in the case of breathable films with silky feel for nappies, chewable items for pets or biofillers for low rolling resistance tyres. The tyre Biotred GT3 , launched by Goodyear in 2001 in Europe, and the Goodyear tyres launched in 2003 in Japan under the 'Hybrid' trademark offer an example of the high tech performances reached by starch-based materials [178].

Mater-Bi starch-based materials are characterised by the following properties:

- significant reduction of environmental impact, particularly with respect to carbon dioxide emissions and energy consumption, in comparison with traditional materials in specific uses [179, 180].

- in use performances similar to traditional plastics.

- processability similar or improved in comparison with traditional plastic materials [174].

- soft, silky feel.

- wide range of permeability to water vapour (from 250 to 1400 g 30 $\mu m/m^2/24$ h. (The method used in this case is Lyssy).

- wide range of mechanical properties from soft and tough materials to rigid ones, with no significant ageing after one year of storage (see **Tables 8.2** and **8.3**) [177].

- antistatic behaviour.

- colourability with food contact approved pigments.

- compostability in a wide range of composting conditions: from home composting and static windrows (heaps) to rotary fermenting reactors.

They are biodegradable and compostable according to the present European standards and are certified by AIB-Vincotte in Belgium, by DIN CERTCO in Germany and by IIP in Italy, according to EN13432 [181], and UNI 10785 standards [182], respectively.

After the acquisition of Enpac in 1998 and the subsequent agreement with Novamont, National Starch is licensing two technologies for the production of loose-fills: one from hydroxypropylated high amylose starch and a second from almost unmodified starch. The loose-fills' densities range from 6 to 10 kg/m^3. The main licensees are Unisource, American Excelsior, Storopack and Flow Pack in USA.

Table 8.2 Some physical properties of Mater-Bi grades for film, in comparison with traditional plastics				
TEST	PROCEDURE	UNIT	MATER-BI	LDPE
MFI	ASTM D1338 [168]	g/10 min	2-8*	0.1-22†
Strength at break	ASTM D882 [169]	MPa	24-30	8-10
Elongation at break	ASTM D882	%	200-1000	150-600
Young's Modulus	ASTM D882	MPa	100-400	100-200
Tear strength	ASTM D1938 [170]			
• Primer		N mm^{-1}	30-90	60
• Propagation		N mm^{-1}	30-90	60
*: 150 °C, 5 kg †: 190 °C, 2.16 kg				

	Table 8.3 Some physical properties of Mater-Bi grades for injection molding, in comparison with traditional plastics.					
TEST	PROCEDURE	UNIT	MATER-BI*	PP†	PS‡
MFI	ASTM D1238 [171]	g/10 min	20-10	0.3-40	1.2-25
Strength at break	ASTM D638 [172]	MPa	20-30	23	30-60
Elongation at break	ASTM D638	%	20-500	400-900	1-4.5
Young's modulus	ASTM D638	MPa	200-2000	1400-1800	3000-3500
IZOD (notched impact)	ASTM D256 [173]	kJ/m^2	1-80	3-10	2-3
*: 170 °C, 5 kg †: 230 °C, 2.16 kg ‡: 200 °C, 5 kg					

8.6 Conclusions

Starch-based bioplastics constitute a new generation of materials able to significantly reduce the environmental impact in terms of energy consumption and greenhouse effect in specific applications, to perform as traditional plastics when in use, and to completely biodegrade within a composting cycle through the action of living organisms when engineered to be biodegradable. They offer a possible alternative to traditional materials when recycling is unpractical or not economical or when environmental impact has to be minimised.

After more than 15 years of research and development starch-based materials start to fulfil specific in-use performances in different application sectors. They are able to offer original solutions both from the technical and the environmental point of view.

Today some of the bioplastics available in the market are used in specific applications where biodegradability is required such as the sectors of composting (bags and sacks), fast food tableware (cups, cutlery, plates, straws, etc.), packaging (soluble foams for industrial packaging, film wrapping, laminated paper, food containers), agriculture (mulch film, nursery pots, plant labels), hygiene (nappy back sheet, cotton swabs), slow release of active molecules in the agricultural and pharmaceutical sectors. Moreover new sectors are growing outside biodegradability, driven by improved technical performances *versus* traditional materials, as in the case of biofillers for tyres and chewable items for pets.

The world market for biodegradable plastics is still small, but it has grown significantly in the last few years reaching about 43,000 tons/year in the year 2003; products totally or partially from renewable resources represent nearly 85-90% of this market [183].

References

1. *Corn Annual Report,* Corn Refiners Association, Inc., Washington, DC, USA, 1991, p.20.

2. F.H. Otey and W.M. Doane in *Starch Chemistry and Technology,* Eds., R.L. Whistler, J.N. BeMiller and E.F. Paschall, Academic Press, Orlando, FL, USA, 1984, p.154 and p.667.

3. F.H. Otey and W.M. Doane in *Starch Chemistry and Technology,* Eds., R.L. Whistler, J.N. BeMiller and E.F. Paschall, Academic Press, Orlando, FL, USA, 1984, p.397.

4. J.J. Cael, J.L. Koenig and J. Blackwell, *Biopolymers,* 1975, **14**, 1885.

5. B. Casu and M. Reggiani, *Die Starke,* 1966, **7**, 218.

6. H.F. Zobel in *Starch Chemistry and Technology,* Eds., Eds., R.L. Whistler, J.N. BeMiller and E.F. Paschall, Academic Press, Orlando, FL, USA, 1984, p.287.

7. J. Seiberlich, *Journal of Modern Plastics,* 1941, **18**, 7, 64.

8. E.B. Bagley, G.F. Fanta, W.M. Doane, L.A. Gugliemelli and C.R. Russell, inventors; USDA, assignee; US 4,026,849, 1977.

9. B.S. Barach, inventor; USDA, assignee; US 4,382,813, 1983.

10. D. Trinmell and B.S. Shasha, inventors; USDA, assignee; US 4,439,488, 1984.

11. B.S. Shasha, W.M. Doane and C.R. Russell, inventors; USDA, assignee; US 4,344,857, 1982.

12. B.S. Shasha and T.P. Abbott, inventors; USDA, assignee; US 43,48,492, 1982.

13. G.F. Fanta, W.M. Doane and E.I. Stout, inventors; USDA, assignee; US 4,483,950, 1984.

14. J.C. Rankin, inventor; USDA, assignee; US 4,330,443, 1982.

15. E.B. Bagley, G.F. Fanta, R.C. Burr, W.M. Doane and C.R. Russell, *Polymer Engineering and Science,* 1977, **17**, 5, 311.

16. D.R Patil and G.F. Fanta, *Journal of Applied Polymer Science,* 1993, **47**, 10, 1765.

17. D. Trimnell, G.F. Fanta, and J.H. Salch, *Journal of Applied Polymer Science*, 1996, **60**, 3, 285.

18. R.A. De Graaf and L.P.B.M. Janssen, *Polymer Engineering and Science*, 2000, **40**, 9, 2086.

19. J.L. Willett and V.L. Finkenstadt, *Polymer Engineering and Science*, 2003, **43**, 10, 1666.

20. G.J.L Griffin, inventor; Coloroll Limited, assignee; US 4,016,117, 1977.

21. F.H. Otey, R.P. Westhoff and C.R. Russell in *Proceedings of a Technical Symposium on Nonwoven Product Technology*, Miami Beach, FL, USA, 1975, p.40.

22. R.F.T. Stepto and I. Tomka, *Chimia*, 1987, **41**, 3, 76.

23. I. Tomka, inventor; I. Tomka, assignee; WO 9005161A1, 1990.

24. N.L. Lacourse and P.A. Altien, inventors; National Starch and Chemical Investment Holding Company, assignee; EP 0376201B1, 1996.

25. C. Bastioli, R. Lombi, G.F. Del Tredici and I. Guanella, inventors; Novamont, assignee; EP 0 400 531B1, 1997.

26. R. Jongboom in *Proceedings of the* EPN *Biodegradable Plastics 2002 Conference*, Frankfurt,Germany, 2002, Session 5.

27. G.J.L. Griffin, inventor; Epron Industries Limited, assignee; WO 9104286, 1991.

28. G.J.L. Griffin, inventor; Coloroll Limited, assignee; US 4,016,117, 1977.

29. G.J.L. Griffin, inventor; Coloroll Limited, assignee; US 4,021,388, 1977.

30. No inventor; Coloroll Limited, assignee; GB 1,485,833, 1977.

31. G.J.L. Griffin, inventor; Coloroll Limited, assignee; US 4,125,495, 1978.

32. G.J.L. Griffin, inventor; Coloroll Limited, assignee; US 4,324,709, 1982.

33. G.J.L. Griffin, inventor; Epron Industries Limited, assignee; assignee; EP363,383B1, 1995.

34. G.J.L. Griffin, inventor; Epron Industries Limited, assignee; assignee; WO 9201741A2, 1992.

35. G.J.L. Griffin, *Progress in Biotechnology,* 1985, **1**, 201.

36. R. Warren, *Bridges,* 1989, **5**, 23.

37. G.J.L. Griffin, *Pure and Applied Chemistry,* 1980, **52**, 2, 399.

38. G.J.L. Griffin, *Biodegradable Fillers in Thermoplastics,* ACS Advances in Chemistry Series, No.134, ACS, Washington, DC, USA, 1974, p.159.

39. J.D. Evans and S.K. Sikdar, *Chemtech,* 1990, **20**, 38.

40. R. Johnson in *Proceedings of SPI Symposium on Degradable Plastics,* Washington, DC, USA, 1987, p.6.

41. G.J.L. Griffin , *Starch as a Filler in HDPE, Iranian Journal of Polymer Science & Technology,* 1992, **1**, 1, 45.

42. W.J. Maddever and G.M. Chapamn in *Proceedings of Symposium on Degradable Plastics,* Washington, DC, USA, 1987, p.41.

43. No inventor; G. Scott, assignee; GB 1,356,107, 1974.

44. D. Gilead and G. Scott, inventors; No assignee; US 4,519,161, 1985.

45. A.J. Sipinen, J.T. Jaeger, D.R. Rutherford and E.C. Edblom, inventors; Minnesota Mining and Manufacturing Company, assignee; US 5,216,043, 1993.

46. P.J. Hocking *Journal of Macromolecular Science C,* 1992, **32** ,1, 35.

47. A.L. Andrady, J.E. Pegram and S.J. Nakatsuka, *Environmental Polymer Degradation,* 1993, **1**, 1, 31.

48. A.L. Andrady, J.E. Pegram and Y.J. Tropsha, *Environmental Polymer Degradation,* 1993, **1**, 3, 171.

49. R. Wilder, *Modern Plastics International,* 1989, **19**, 9, 74.

50. R. Narayan and W. Lafayette, *Kunstoffe-German Plastics,* 1989, **79**, 10, 92.

51. R.H. Thomas, *Polymer News,* 1989, **14**, 9, 271.

52. G. Chapman, *Kontakt and Studium,* 1994, **425**, 74.

53. Chuo Chemical Co., assignee; JP 146953, 1992.

54. J-M. Aime, G. Mention and A. Thouzeau, inventors; Charbonnages de France, assignee; US 4,873,270, 1989.

55. F.G., Gallagher, H. Shin, and R.F. Tietz, inventors; EI DuPont de Nemours, assignee; US 5,219,646, 1993.

56. H.F. Conway, *Food Product Development,* 1971, **5**, 3, 14.

57. B.Y. Chiang and A.J. Johnson, *Cereal Chemistry,* 1977, **54**, 3, 429.

58. H.F. Conway, E.B. Lancaster and G.N. Bookwalter, *Food Engineering,* 1968, **40**, 11, 102.

59. C. Mercier, *Feedstuffs,* 1971, **4**, 33.

60. B.T. Lawton, G.A. Henderson and E.J. Derlatka, *Canadian Journal of Chemical Engineering,* 1972, **50**, 168.

61. R.A. Anderson, H.F. Conway, V.F. Pfeifer and E.L Griffin, *Cereal Science Today,* 1969, **14**, 1, 4.

62. R.L. Derby, B.S. Miller, B.F. Miller and H.B. Trimbo, *Cereal Chemistry,* 1975, **52**, 5, 702.

63. D.V. Harmann and J.M. Harper, *Transactions of the ASAE,* 1973, **7**, 6, 1175.

64. J. F. De La Gueriviere, *Bulletin Des Anciens Eleves de L'Ecole Francaise de Meumerie,* 1976, **276**, 305.

65. D.J. Stevens and G.A.H. Elton, *Die Starke* 1971, **23**, 1, 8.

66. R.A. Anderson, H.F. Conway and A.J. Peplinski, *Die Starke,* 1970, **22**, 4, 130.

67. H.H. Mottern, J.J. Spadaro and A.S. Gallo, *Food Technology,* 1969, **23**, 169.

68. R.M. Shetty, D.R. Lineback and P.A. Seib, *Cereal Chemistry,* 1974, **51**, 364.

69. B.S. Miller, R.I. Derby and H.B. Trimbo, *Cereal Chemistry,* 1973, **50**, 271.

70. B.Y. Chiang and A.J. Johnson, *Cereal Chemistry,* 1977, **54**, 3, 436.

71. R.A. Anderson, H.F. Conway, V.F. Pfeifer and E.L. Griffin, *Cereal Science Today,* 1969, **14**, 372.

72. C.J. Sterling, *Journal of Texture Studies,* 1978, **9**, 225.

73. C. Mercier and P. Feillet, *Cereal Chemistry,* 1975, **52**, 3, 283.

74. C. Mercier, *Die Starke*, 1977, **2**, 48.

75. J.W. Donovan, *Biopolymers,* 1979, **18**, 263.

76. P. Colonna and C. Mercier, *Phytochemistry*, 1985, **24**, 8, 1667.

77. J-P. Sacchetto, R. F. T.Stepto and H. Zeller, inventors; Warner-Lambert Company, assignee; US 4,900,361, 1990.

78. R.F.T. Stepto and B. Dobler, inventors; Warner-Lambert Company, assignee; EP 0,326,517B1, 1994.

79. G. Lay, J. Rehm, R.F.T. Stepto, and M. Thoma, inventors; Warner-Lambert Company, assignee; EP 327,505B1, 1997.

80. J-P. Sacchetto, M. Egil, R.F.T. Stepto and H. Zeller, inventors; Warner-Lambert Company, assignee; EP 0,391,853A3, 1992.

81. J-P. Sacchetto, D.J Lentz and J. Silbiger, inventors; Warner-Lambert Company, assignee; EP 0,404,723B1, 1994.

82. J. Silbiger, J-P. Sacchetto and D.J Lentz, inventors; Warner-Lambert Company, assignee; EP 0,404,728A3, 1990.

83. J-P. Sacchetto and J. Rehm, inventors; Warner-Lambert Company, assignee; EP 0,407,350A3, 1991.

84. J. Silbiger, J-P. Sacchetto and D.J Lentz, inventors; Warner-Lambert Company, assignee; EP 0,408,501A3, 1991.

85. J-P. Sacchetto, J. Silbiger, and D.J Lentz, inventors; Warner-Lambert Company, assignee; EP 0,408,502A3, 1991.

86. J. Silbiger, D.J Lentz and J-P. Sacchetto, inventors; Warner-Lambert Company, assignee; EP 0 408 503 B, 1994.

87. J-P. Sacchetto, J. Silbiger, and D.J Lentz, inventors; Warner-Lambert Company, assignee; EP 0,409,781B1, 1994.

88. D.J Lentz, J-P. Sacchetto and J. Silbiger, inventors; Warner-Lambert Company, assignee; EP 0,409,782B1, 1994.

89. J-P. Sacchetto, J. Silbiger and D.J Lentz, inventors; Warner-Lambert Company, assignee; EP 0,409,783B1, 1994.

90. D.J Lentz, J-P. Sacchetto and J. Silbiger, inventors; Warner-Lambert Company, assignee; EP 0,409,788B1, 1994.

91. H.R. Fischer and S. Fischer, inventors; TNO, assignee; EP 1,134,258, 2001.

92. M. Egli and E.T. Cole, inventors, Warner-Lambert Company, assignee; EP 0,500,885B1, 1997.

93. G.L. Loomis, M.J. Izbicki and A. Flammino, inventors, Novon International Inc., assignee; EP 0,679,172A3, 1995.

94. C. Bastioli, R. Lombi, M. Nicolini, M. Tosin and F. Degli Innocenti, inventors; Novamont SpA, assignee; EP 1,109,858B1, 2003.

95. C. Bastioli, V. Bellotti, G. Del Tredici, I. Guanella and R. Lombi, inventors; Novamont SpA, assignee; EP 0,965,615B1, 2002.

96. F. Wittwer and I. Tomka, inventors; Warner-Lambert Company, assignee; EP 0,118,240A2, 1984

97. H.B. Wigman, W.W. Leathen and M.J. Brackenbeyer, *Food Technology,* 1956, **10,** 179.

98. C. Bastioli, V. Bellotti, L. Del Giudice, G.F. Del Tredici, R. Lombi and A. Rallis, inventors; Novamont SpA, assignee; US 5,262,458, 1993.

99. C. Bastioli, V. Bellotti, L. Del Giudice and R. Lombi, inventors; Novamont SpA, assignee; EP0,400,532B1, 1994.

100. C. Bastioli, V. Bellotti, L. Del Giudice and R. Lombi, inventors; Novamont SpA, assignee; EP 0437 561 B1, 1994.

101. C. Bastioli, V. Bellotti, G.F. Del Tredici, inventors; Novamont SpA, assignee; EP 0437 589 B1, 1996.

102. C. Bastioli, V. Bellotti, A. Montino, G.F. Del Tredici, and R. Lombi, inventors; Novamont SpA, assignee; EP 0,494,287B1, 1995.

103. F. Wittwer and I. Tomka, inventors; Warner-Lambert Company, assignee; US 4,673,438, 1987.

104. R.L. Mellies and I.A. Wolff, inventors, USDA, assignee; US 2,788,546, 1957.

105. C. Van den Berg, *Vapour Sorption Equilibria and other Water-Starch Interactions; a Physico-Chemical Approach,* Agricultural University Wageningen, The Netherlands, 1981, p.49. [PhD thesis]

106. I. Tomka, inventor; Bio-tec Biologische Naturverpackungen GmbH, assignee; EP 0,397,819B1. 1995.

107. I. Tomka, inventor; Bio-tec Biologische Naturverpackungen GmbH, assignee; EP 0,539,544B1, 1994.

108. I. Tomka, J. Meissner and R. Menard, inventors; Bio-tec Biologische Naturverpackungen GmbH, assignee; EP 0,537,657B1, 1997.

109. I. Tomka, inventor; Bio-tec Biologische Naturverpackungen GmbH, assignee; EP 0,542,155B1 (1998).

110. I. Tomka, inventor; Bio-tec Biologische Naturverpackungen GmbH, assignee; US 6,242102, 2001.

111. I. Tomka, inventor; Bio-tec Biologische Naturverpackungen GmbH, assignee; EP 0,596,437B1, 1997.

112. F.H. Otey and R.P. Westhoff, inventors; USDA, assignee; US 4,133,784 (1979),

113. F.H. Otey and R.P. Westhoff, inventors; USDA, assignee; EP 0,132,299B1, 1989.

114. F.H. Otey and R.P. Westhoff, inventors; USDA, assignee; US 4,454,268, 1984.

115. F.H. Otey and R.P. Westhoff, inventors; USDA, assignee; US 4,337,181, 1982.

116. F.H. Otey, R.P. Westhoff and W. M. Doane, *Industrial and Engineering Chemistry Research,* 1987, **26**, 8, 1659.

117. F.H. Otey, R.P. Westhoff and W.M. Doane, *Industrial and Engineering Chemistry Product Research and Development,* 1980, **19**, 4, 592.

118. R.P. Westhoff, F.H. Otey, and C.L. Mehltretter and C.R. Russell, *Industrial and Engineering Chemistry Product Research and Development,* 1974, **13**, 2, 123.

119. F.H. Otey, *Polymer-Plastics Technology and Engineering*, 1976, **7**, 2, 221.

120. F.H. Otey, R.P. Westhoff and C.R. Russell, *Industrial and Engineering Chemistry Product Research and Development*, 1976, **15**, 2, 139.

121. F.H. Otey, A.M. Mark, C.L. Mehltretter and C.R. Russell, *Industrrial and Engineering Chemistry Product Research and Development*, 1974, **13**, 1 90.

122. G.F. Fanta and F.H. Otey, inventors; USDA, assignee; US 4,839,450, 1989.

123. C.L. Swanson, R.L. Shogren, G.F. Fanta and S.H. Imam, *Journal of Environmental Polymer Degradation*, 1993, **1**, 2, 155.

124. R.L. Shogren, *Carbohydrate Polymers*, 1992, **19**, 2, 83.

125. G.F. Fanta, C.L. Swanson and W.M. Doane, *Polymer Preprints*, 1988, **29**, 2, 453.

126. G.F. Fanta, C.L. Swanson and W.M. Doane, *Journal of Applied Polymer Science*, 1990, **40**, 5-6, 811.

127. G.F. Fanta, C.L. Swanson and W.M. Doane, *Carbohydrate Polymers*, 1992, **17**, 1, 51.

128. G.F. Fanta and C.L. Swanson and R.L. Shogren, *Journal of Applied Polymer Science*, 1992, **44**, 11, 2037.

129. R.L. Shogren, A.R. Thompson, F.C. Felker, R.E. Harry-Okuru, S.H. Gordon, R.V. Greene and J.M. Gould, *Journal of Applied Polymer Science*, 1992, **44**, 11, 1971.

130. C. Bastioli, A. Rallis, F. Cangialosi, F.P. La Mantia, G. Titomanlio and S. Piccarolo in *Proceedings of the Polymer Processing Society - European Regional Meeting*, Palermo, Italy, 1991, p.15.

131. C. Bastioli and V. Bellotti, *Agrobiotecnologie e Nuove Produzioni Chimiche da Risorse Rinnovabili*, (*New Biodegradable Materials From Starch*), Accademia Nazionale dei Lincei, Rome, Italy, 1990, p.26. [In Italian]

132. C. Bastioli, V. Bellotti, L. Del Giudice and G. Gilli, *Journal of Environmental Polymer Degradation*, 1993, **1**, 3, 181.

133. C. Bastioli, V. Bellotti, L. Del Giudice and G. Gilli in *Biodegradable Polymers and Plastics*, Ed., M. Vert, J. Feijem, A. Albertsson, G. Scott and E. Chiellini, The Royal Society of Chemistry, Cambridge, UK, 1992, p.101.

134. C. Bastioli, V. Bellotti, M. Camia, L. Del Giudice and A. Rallis, *Biodegradable Plastics and Polymers,* Eds., Y. Doi and K. Fukuda, Elsevier, Amsterdam, The Netherlands, 1994, p.200.

134. P. Halley, S. McGlashan and J. Gralton, inventors; Food and Packaging Centre Management, assignee; WO 02083784, 2002.

136. C. Bastioli, V. Bellotti, G.F. Del Tredici, A. Montino, R. Ponti inventors; assignee; IT T092 A000199, 1992.

137. C. Bastioli, V. Bellotti and A. Rallis, *Rheologica Acta,* 1994, **33**, 4, 307.

138. W.H. Herschel and R. Bulkley, *Proceedings of the ASTM,* 1926, **26**, 621.

139. C. Bastioli, V. Bellotti and A.Montino, inventors; Novamont SpA, assignee; WO9214782, 1992.

140. J. Römesser, *Presentation at Plastics Waste Management Workshop*, American Chemical Society, Polymer Chemistry Division, New Orleans, LA, 1991.

141. D.L. Kaplan, J.M. Mayer and D. Ball in the *Proceedings of the Biodegradable Packaging Symposium,* Natick, MA, USA, 1992.

142. C. Bastioli, V. Bellotti, L. Del Giudice and R. Lombi, inventors; Novamont SpA, assignee; EP 0,437,561B1, 1994.

143. C. Bastioli, V. Bellotti, A. Montino, G. Del Tredici and R. Lombi, inventors; Novamont SpA, assignee; EP 0,494,287B1, 1995.

144. C. Bastioli, V. Bellotti, G.C. Romano and M. Tosin, inventors; Novamont SpA, assignee; EP 0,495,950B1, 1998.

145. C. Bastioli, V. Bellotti, G. Del Tredici, R. Lombi and A. Rallis, inventors; Novamont SpA, assignee; EP 0,436,689 B1, 1995.

146. Y. Yoshida, and T. Uemura in *Biodegradable Plastics and Polymers*, Eds., Y. Doi and K. Fukuda, Elsevier, Amsterdam, The Netherlands, 1994, p.443.

147. C. Bastioli, V. Bellotti, G.F. Del Tredici, R. Lombi, A. Montino and R. Ponti, inventors; Novamont SpA, assignee; WO 9219680, 1992.

148. C. Bastioli *Polymer Degradation and Stability,* 1998, **59**, 1-3, 263.

149. C. Bastioli in *Proceedings of Renewable Bioproducts, Industrial Outlets and Research for the 21st Century*, Wageningen, The Netherlands, 1997.

150. C. Bastioli in *Proceedings of Global Status for Biodegradable Polymers Production – 7thAnnual Meeting of the Bio/Environmentally Degradable Polymer Society*, Cambridge, MA, USA, 1998, p.50.

151. I. Tomka, inventor, Bio-tec Biologische Naturverpackungen GmbH, assignee; EP 0,539,544B1, 1994.

152. F.S. Buehler, V. Baron, E. Schmid, P. Meier and H.J. Schultze, inventors; EMS-INVENTA AG, assignee; EP 0,541,050A3, 1993.

153. E. Schmid, F.S. Buehler and H.J. Schultze, inventors; EMS-INVENTA AG, assignee; EP 0,522,358A3, 1993.

154. C. Bastioli, V. Bellotti and G. Del Tredici, inventors; Butterfly Srl, assignee; EP 0,437,589A1, 1991.

155. I. Tomka, inventor; Bio-tec Biologische Naturverpackungen GmbH, assignee; EP 0,542,155, 1994.

156. J. Schröter, inventor, Buck Werke GmbH, assignee; EP 0551 125A1, 1993.

157. C.S. Maxwell, *Tappi Journal*, 1970, **53**, 8, 1464.

158. N.L. Lacourse and R.A. Altieri, inventors; National Starch and Chemical Corporation, assignee; US 4,863,655, 1989.

159. N.L. Lacourse and R.A. Altieri, inventors; National Starch and Chemical Corporation, assignee; EP 0,375,831, 1990.

160. N.L. Lacourse and R.A. Altieri, inventors; National Starch and Chemical Corporation, assignee; EP 0,376,201A1, 1990.

161. N.L. Lacourse and R.A. Altieri, inventors; National Starch and Chemical Investment Holding Company, assignee; US 5,035,930, 1991.

162. N.L. Lacourse and R.A. Altieri, inventors; National Starch and Chemical Investment Holding Company, assignee; US 5,043,196, 1991.

163. P.E. Neumann and P.A. Seib, inventors, Kansas State University Research Foundation, assignee; US 5,185,382, 1993.

164. J.R. Anfinsen and R.R. Garrison, inventors; EI DuPont De Nemours & Company, assignee; WO 9208759, 1992.

165. C. Bastioli, V. Bellotti, G.F. Del Tredici and A. Rallis, inventors; Novamont SpA, assignee; EP 0,667,369A1, 1995.

166. C. Bastioli, V. Bellotti, G.F. Del Tredici, A. Montino and R. Ponti, inventors; Novamont SpA, assignee; EP 0,696,611A3, 1996.

167. C. Bastioli, V. Bellotti, G. Del Tredici and A. Rallis, inventors; Novamont SpA, assignee; EP 0,696,612A2, 1996.

168. ASTM D1338, *Practice for Working Life of Liquid or Paste Adhesives by Consistency and Bond Strength*, 1999.

169. ASTM D882, *Test Method for Tensile Properties of Thin Plastic Sheeting*, 2002.

170. ASTM D1938, *Test Method for Tear-Propagation Resistance (Trouser Tear) of Plastic Film and Thin Sheeting by a Single-Tear Method*, 2002.

171. ASTM D1238, *Test Method for Melt Flow Rates of Thermoplastics by Extrusion Plastometer*, 2004.

172. ASTM D638, *Test Method for Tensile Properties of Plastics*, 2003.

173. ASTM D256, *Test Methods for Determining the Izod Pendulum Impact Resistance of Plastics*, 2004.

174. C. Bastioli and S. Facco, *Proceedings of Biodegradable Plastics 2001 Conference*, Frankfurt, Germany, 2001, Session No.3.

175. W. Berger, L. Jeromin, U. Mierau and G. Opitz, inventors; BIOP Biopolymer GmbH, assignee; WO 9,925,756, 1999.

176. L. Yu, G.B.Y. Christie and S. Coombs, inventors; Food and Packaging Centre Management Limited, assignee; WO 0036006, 2000.

177. C. Bastioli in *Proceedings of Actin Conference Biopolymers: Packaging - A New Generation*, Birmingham, UK, 2001.

178. *Automotive News Europe*, 2001, October 22, 15.

179. R. Estermann, B. Schwarzwalder and N. Gysin, *Life Cycle Assessment of Mater-Bi Bags for the Collection of Compostable Waste*, Composto, Switzerland, 1998.

180. R. Estermann, B. Schwarzwalder and N. Gysin, *Life Cycle Assessment of Mater-Bi and EPS Loose-Fills*, Composto, Switzeland, 2000

181. EN 13432, *Packaging - Requirements for packaging recoverable through composting and biodegradation - Test scheme and evaluation criteria for the final acceptance of packaging*, 2000.

182. UNI 10785, *Compostability of Plastics – Requirements and Test Methods*, 1999. (in Italian)

183. *Actual Situation and Prospects of EU Industry using Renewable Rraw Materials*, Ed., J. Ehrenbergh, European Commission, DG Enterprise/E.1, Brussels, Belgium, 2002.

9 Poly(Lactic Acid) and Copolyesters

Samuel J. Huang

9.1 Introduction

Traditional applications of synthetic polymers are mostly based on their inertness to environmental degradations (hydrolysis, oxidation, biodegradation, and so on). The rapid increase in the volume of use of synthetic polymers has contributed to the solid waste management problems in recent years. Total management of polymer wastes requires complementary combinations of recycling, incineration for energy, and biodegradation [1]. Polymers prepared from renewable and sustainable resources can be easily designed, synthesised, and engineered by environmentally compatible routes and can be disposed after use by biodegradation (composting, etc.) [2-6]. Biodegradable polymers are necessary in the design, synthesis and applications of biomedical implants and drug release systems. Among those received increasing attention since 1970s are aliphatic polyesters such as microbial polyhydroxybutyrate (PHB), and its copolymers [7, 8] and polylactic acid (PLA), and its copolymers [9-12].

PHB is the energy storage material for certain bacteria and efforts to commercialise it and its copolymers as structural and package materials have not been successful due to their high costs and the difficulty in thermal processing. PLA from the polymerisation of lactic acid, a fermentation product of low cost polysaccharides, is a product which is produced from a combination of biotechnology and chemical technology. PLA and its copolymers are the subjects of this review. There have been many reviews published recently, especially related to the biomedical application areas, and thus only major current work is covered here.

9.2 Synthesis

9.2.1 Homopolymers

L-Lactic acid is metabolic intermediate and can be obtained at low cost from the fermentation of agriculture and food by products containing carbohydrates [13, 14].

Thermal dehydration polymerisation of L-lactic acid gives poly(L-lactic acid) (PLLA). This requires high energy and PLLA of low molecular weight (MW)(few thousands) is obtained [14]. Ring opening polymerisation of lactic acid dimer, lactide, with a suitable catalyst results in high molecular PLLA with useful properties. These can proceed through coordination, anionic, or cationic mechanisms. Among the effective catalysts/initiators are Lewis acids in form of metal salts of aluminium, tin, titanium and zinc, and rare earth metals [15-21]; alkali metal alkoxides and supermolecular complexes [19, 22-23]; and acids [24]. Coordination ring opening polymerisation is the most effective route for the bulk polymerisation of lactide. It is generally agreed by researchers that transition metal ions such as tin catalyse the polymerisation proceed via an insertion mechanism [19, 25]. At temperatures above 150 °C the transesterification between cyclic lactide and PLA proceeds through acyl cleavage and results in high degree of retention of the stereo-chemistry of the lactide monomers. Tin catalysts are easily available and effective. They can be used for large scale producers of PLA. Stannous (II) chlorides and stannous (II) 2-ethylhexanoate are approved for food additives and are thus more often used than the others. Glycols are often used as co-initiators to obtain polyester chain growth from both hydroxy terminals of the glycols. Multi-functional glycol co-initiators can be used to obtain star shaped and highly branched PLA [25]. Sufficient reaction time generally results in PLA with molecular weight dispersity of 1.5-2.0.

Better polymerisation of lactic acid with tin salts as catalysts can be carried out in multiple steps. Lactic acid is heated at 150 °C with tin catalyst to obtain oligomeric PLA (with a degree of polymerisation of 1-8). The oligomers are then heated at 180 °C under vacuum (1333 Pa) for 5 hours to give PLA of high MW (100,000). Finally the third step is carried out at solid state above the crystallisation temperature, T_c, (105 °C, 66 Pa, 0.5-2 hours) and annealing 150 °C for 10-30 hours. A PLA of MW up to 600,000 is obtained. Solution polymerisation in diphenyl ether results in a PLA of MW of 140,000.

A considerable amount of effort has been directed towards the research on catalysts for ring opening polymerisation of lactide. Alkoxides such as aluminium triisopropoxide are effective catalysts. The anionic polymerisation gives PLA of MW up to 100,000 with MW/M_n around 1.4. When this polymerisation is carried solvent dispersion systems microspheres of well defined size of PLA can be obtained. Direct condensation of lactic acid with high boiling point solvent and ring opening polymerisation of lactide were studied and both were found to be effective and PLA of MW of 300,000 was obtained. PLA from obtained by using different methods were compared and found to have different properties. Both had glass transition temperature (T_g) of around 58-59 °C but the direct process PLA had melting temperature, T_m, 163 °C and was relatively stable whereas the PLA prepared by ring opening had a higher T_m of 178 °C but was less stable. This was attributed to the presence of catalyst and impurities [22, 26-32].

9.2.2 Copolymers

High MW PLA from prepared from PLLA, are partially crystalline with a T_m of 175-180 °C, T_g of 60 °C and a crystallisation temperature (T_c) of 100-105 °C, and a decomposition temperature (T_d) of 185-190 °C. It is brittle and undergoes unzipping to lactide when thermally processed. Copolymerisation with D-lactic acid and other hydroxyacids to obtain polyesters with a lower T_m and thus better thermal processing characteristics has been the common approach to obtain useful PLA. Stereo copolymers of L-lactic acid and D-lactic acid have lower crystallinity and T_m than the homopolymer of L-lactic acid and the polyester properties vary with the optical purity with the 50/50 DL polylactic acid (PDLLA) having no crystallinity at all. Variation of the optical purity is the most commonly used means to produce PLA of different property.

Copolymers of lactic acid with glycolic acid were the first commercialised biodegradable polymers to be used as biomaterials and are used as sutures, wound dressings, and drug release systems since the 1970s [33, 34]. Copolymers of lactic acid with other aliphatic polyesters specially those with cyclic esters, ethers, and anhydrides have become the most studied biodegradable polymers. Ring opening polymerisation with other cyclic monomers is the best method. Thermal polymerisation with mixtures of monomers generally gives copolymers with random sequence with less crystallinity and lower T_m and T_g than PLLA. Sequential addition of monomers into the polymerisation, in some cases, results in block copolymers. Most of these aliphatic polyesters are compatible with each other at low MW but tend not to be compatible at high MW and thus complex morphology is observed for many block copolymers of PLA.

Poly(ε-caprolactone) (PCL) with T_g at −60 °C and T_m at 60 °C, are commercially available in large quantity and its biodegradation was studied in detail in terms of morphology and microbial variety [35-37]. It is more flexible and hydrophobic than PLA. It was reasoned that copolymers of PLA and PCL with the proper compositions and sequences could be prepared which would have better flexibility, hydrophilic/hydrophobic balance, and impact strength than homopolymers of PLA. The need for a biodegradable replacement for poly(dimethylsiloxane) as a sustainable drug release systems was behind research on PLA/PCL copolymers as biomaterials [38]. Block copolymers of PLA and PCL are easily obtained by using PCL-diols as co-initiator with stannous catalysts [39-41] in lactide ring opening polymerisation. The expected trends in T_m, Young's moduli, material strength and ultimate elongation were observed up to 50 wt% of PCL. Bulk polymerisation of mixtures of lactides and caprolactone with stannous 2-ethylhexanoate catalyst resulted in copolymers with thermal properties of phase-separated block structures [42]. Chain extensions can be used to expand the range of MW, composition and properties [43, 44]. Solution polymerisation with aluminium tris(isopropoxide) catalyst have been studied [45-47]. Anionic initiators, including lithium *t*-butoxide, were also studied. Results from different research groups do not agree, and this is likely to be due to the different extents of ester exchange reaction during the polymerisation [48, 49]. Using poly(ethylene glycol) (PEG), as co-initiator block copolymers of PLA/PEG have been prepared [50]. Copolymers of L-lactide and 1,5-dioxepan-2-one were prepared with a tin catalyst [51]. These tri-block copolymers behave like elastomers.

9.2.3 Functionalised Polymers

It is acknowledged by researchers that the practice of using metallic implants for bone fracture fixation has serious problems [51-53]. Most serious ones are osteoporosis due to stress shielding caused by the mismatch of the metallic properties with that of bones and necessitate second operations for the removal of the implants. To alleviate problems, use of biodegradable polyesters were explored [3, 4, 54-58]. Although it can be used as a suture poly(glycolic acid) undergoes hydrolysis too fast in various forms to be effective as implants. The presence of methyl side groups in PLA as the longer methylene unit in PCL slows down the rate of hydrolysis for PCL and PLA as compared with polyglycolic acid (PGA) and use of various copolymers of PLA with PGA and PCL have been explored as implant materials. These copolymers are generally partially crystalline. During hydrolysis and biodegradation the amorphous regions are degraded faster than the crystalline regions resulting in the formation of highly crystalline fragments and catastrophic loss of mechanical properties [59, 60]. It was reasoned that polyester networks will be less crystalline and also suffer less loss of mechanical properties during degradation [61]. Crosslinkable polyesters and copolyesters with unsaturated maleic acid, fumaric acid, and itaconic acid units were synthesised from reactions of corresponding unsaturated anhydrides for networks and composites formation [61-63]. Methacrylate terminated oligomeric polyesters can be obtained from polymerisations with co-initiators with a methacrylate group [64, 65]. These are starting materials for the graft copolymer of PLA [66]. Hydroxy groups containing terminals are generally present in PLA polymerised with glycols as co-initiator. Those with hydroxy side groups were obtained from co-polymerisation of lactide with tartaric acid [59] and cyclic carbonate with ketal groups which upon hydrolysis yields hydroxy groups [67]. PLA with hydroxy terminals have been converted into degradable polyurethanes [68-70]. PLA with amino, carboxylic, and chloro terminals were prepared from the PLA with hydroxy terminals [71, 73].

9.3 Structure, Properties, Degradation, and Applications

9.3.1 Physical Properties

As mentioned in the previous section properties of PLA are greatly dependent on the optical purity (**Table 9.1**). PLA with 100% L-unit, PLA 100, is partially crystalline (45-70%) with a T_m of around 180-184 °C [73, 74]. The degree of crystallinity and T_m of PLA decrease with decreasing optical purity. PLA of less than 87.5% optical purity are amorphous. PLA of high optical purity has similar T_m to that of two other polymers with methyl side groups, microbial PHB, and isotactic polypropylene (iPP). All three polymers are helical in the crystalline form. The T_g of high MW PLA with different optical purity is within 55-61.5 °C range, which is higher than that of PHB and iPP. PLA is strong but brittle. Although PLA is soluble in chlorinated organic solvents and can be solution processed thermal processing of PLA with 96% or less optical purity (injection moulding or extrusion) are preferred. Properties of PLA are compared with those of common thermal plastics in **Table 9.2**.

Table 9.1 Optical purity of PLA and properties				
% L form of PLA	T_g, °C, DSC	T_m, °C, DSC	ΔHf, J/g	Density, g/cm³
100	60	184		
98	61.5	176.2	56.4	1.2577
92.2	60.3	158.5	35.8	1.2601
87.5	58	ND	ND	
80	57.5	ND	ND	1.2614
45	49.2	ND	ND	1.2651
ND: amorphous Data from [9, 25 and 73]				

Table 9.2 Comparison of PLA (96% optical purity) with thermoplastics				
	Tensile modulus (MPa)	Notch Izod impact (J/m)	Flexural modulus (MPa)	Elongation at break (%)
PLA	3834	24.6	3689	4
Polystyrene	3400	27.8	3303	2
iPP	1400	80.1	1503	400
High density polyethylene	1000	128.16	800	600
Data from [74]				

Star-shaped PLA have lower crystallinity than linear PLA with the same optical purity [39]. The T_c of PLA with various structures are around 115-125 °C. Stereo-complexation have been observed for L-and D-PLA [39, 75-78]. The complex has a T_m at 220 °C.

High MW PLA (100,000 and up) can be processed into fibres, non-woven, and articles with rigidity and strength, which are potentially useful at commodity scale if the initial high costs can be reduced as the volume increases [10-12]. A considerable amount of effort has been directed toward packaging films of PLA with mixed results. The addition of suitable plasticiser to lower the T_g of PLA is necessary for obtaining flexible films. Low MW PLA and lactide are known to act as plasticisers for high MW PLA [9, 10]. Various biodegradable monomeric and oligomeric aliphatic esters have been studied as plasticisers for PLA. Addition of citric acid esters of MW 200-600 lowers the T_g and T_c of PLA with the increase of crystallinity with no definite trend observed [79]. Blends of PLA and PCL have been studied in detail [80]. Low MW PCL (MW of 530) is compatible with PLA and is an effective plasticiser for PLA. PCL of higher MW than 2,000 is partially compatible with PLA and tri-phase morphology (crystalline PLA, crystalline PCL and amorphous) is observed. The presence of PCL in blends increases the ductility of PLA. Thermal processing of the PLA/PCL blends results in ester exchange, resulting in block copolymers of PLA/PCL. Oligomeric poly(ethylene succinate) (PHS) of MW 1,300 is compatible with PLA up to 20% and is an effective plasticiser for PLA [81]. Blends of PLA and poly(ethylene/butylene succinate) have been utilised as films. They are immiscible blends [82] with some increase of the ductility of PLA. Low MW PEG and poly(propylene glycol) can act as plasticisers for PLA [83]. However, the presence hydrophilic polyethers increases the hydrolysis rate of PLA.

9.3.2 Chemical Properties

The most important degradation of PLA is hydrolysis. Under dry conditions pure PLA 100 can last more than 10 years [4, 9, 57, 85-92]. The rate of hydrolysis varies with many factors. The changes of properties of PLA during hydrolysis have been studied [95] and are shown in **Table 9.3**. In thin film rapid changes due to hydrolysis were observed in 35 days and the changes levelled off. Increase in crystallinity can be attributed to in the increase of mobility of oligomers formed which can crystallise themselves or induce the crystallisation of larger size PLA. The hydrolysis of PLA with smaller surface/volume ratios is much m slower and complicated. PLA/GA copolymers are hydrolysed much faster than PLA and have become the main biodegradable polymeric materials for biomedical applications such as sutures, implants, tissue engineering and drug release when fast rates of hydrolysis are desirable whereas poly(lactide-*co*-caprolactone), PLA/CL, are more suitable for slower hydrolysis than PLA. The hydrolysis of PLA, PLA/CL, and poly(lactide-*co*-glycolide), PLA/GA, like that of many hydrophobic aliphatic polyesters, is rather complex. The hydrolysis of the amorphous

Table 9.3 Effect of hydrolysis, pH 7.4 at 37 °C, on PLA properties						
Days	Wt loss (%)	M_n (Da)	M_w (Da)	T_g (°C)	T_m (°C)	ΔHf (J/g)
0		65,000	80,000	64	155.8	0
7	1	14,000	35,000	56.1	154.7	8
14	4	2,000	4,000	50	149.7	14
21	14	1,100	2,200	48.7	146.3	45
28	27	1,000	2,000	51.9	142.8	47
35	28	1,000	2,000	51.9	143.4	45
Data from [93] Low D PLA from Cargill press film (0.5 x 4 x 10 mm)						

regions are much faster than the crystalline regions. The crystallinity of copolymers decreases rapidly with increasing amount of the second component in the copolymers. Typically little crystallinity is observed for copolymers with less than 80% PLA and the rate of hydrolysis increases accordingly. The hydrolysis of PLA and its hydrophobic copolymers is subjected to auto-catalysis by the acid groups attached to oligomers formed during the hydrolysis [9, 92]. The internal part of a device had been observed to undergo hydrolysis faster than the outer part and resulting in a hollow partially degraded device which weight loss is relatively small with little volume change. These are good characteristics for implants. The rate of hydrolysis also varies with the hydrophilicity/hydrophobicity of the second component. PLA/GA are hydrolysed much faster than PLA which in tern is hydrolysed faster than the more hydrophobic PLA/CL. Sutures of various ages have been produced from PLA copolymers of glycolic acid, caprolactone, trimethylene carbonate and dioxanone [94]. PLA provides the crystallinity and strength, PGA the fast rate of hydrolysis and the others the flexibility. An increasing order of rate of hydrolysis was observed: PLLA<poly (D, L-lactide) [PDLA]<PLA/GA network [61].

Biodegradations of PLA have been a subject of interest and so far proteinase K (EC 3.4.21.64) is the only reported enzyme that will degrade PLA amorphous regions of low MW [95]. Microbial degradation studies of PLA have been inclusive [96]. Although most microorganisms studied can utilise lactic acid and its dimer, microbial degradation of oligomers and polymers of PLA have not yet been observed at appreciable rates. A microbial degradation study on PLA/CL only showed the degradation of the PCL segments [93]. Compost, field and environmental degradations of PLA are primarily due to hydrolysis [97].

Thermal degradation of PLA can proceed via different mechanisms. Hydroxy-terminated PLA might undergo 'back-biting' transesterification resulting in 'unzipping' of the PLA to lactide. A common method of forming lactide is the thermal decomposition of oligomeric PLA. Inter- and intra-molecular transesterifications, both facilitated by the presence of polymerisation catalysts, is commonly observed. Finally, fragmentation of PLA, might also happen. Stabilisation by the addition of suitable stabilisers is an area of ongoing research as biocompatible additives for polymers are not commonly available.

9.3.3 Applications

Hydrogels have received increasing interest for biomedical and consumer products application [98]. PLA and PEG hydrophilic/hydrophobic block copolymers are especially promising for soluble hydrophilic/hydrophobic system that becomes an insoluble microsphere when injected into the body as drug release systems [99]. The hydrolysis and biodegradation of these copolymers are subjects of ongoing research.

As is generally true for new polymers, costs for PLA and copolymers are relatively high for large volume applications. However, they are from renewable resources and environmentally compatible. All factors considered they are polymers for the future. Mixing with low cost biopolymers such as starch to lower the cost and increase biodegradation rates, was successfully done for PCL and cellulose esters [100], has had only mixed results as PLA and copolymers are not hydrolytically stable enough at high temperatures when the mixing has to be carried out. Reactive coupling of PLA with starch unfortunately adds to the cost [101].

9.4 Conclusions

PLA and copolymers can be derived from renewable resources and in many cases be environmentally and biodegradable degradable and they are important in biomedical applications. Extensive processing research is still needed for linear PLA to become a large volume polymer. However, PLA can be synthesised and recycled from used PLA into methacrylate functionalised oligomers [102] by thermal ester exchange with caprolactone ethyl methacrylate [102]. Methacrylate terminated PLA can be then be copolymerised with itaconic anhydride [68]. These can be used as high added value materials in specialty applications such as adhesives, coatings, blends, and composites. They will become very useful in the near future.

References

1. S.J. Huang, *Journal of Macromolecular Science - Pure and Applied Chemistry*, 1995, **A32**, 4, 593.

2. M. Heyde, *Polymer Degradation and Stability,* 1998, **59**, 1-3, 3.

3. S.J. Huang in *Encyclopedia of Polymer Science and Engineering,* Ed., J.I. Kroschwitz and H.F. Mark, John Wiley & Sons, New York, NY, USA, 1985, Volume 2, p.220-243.

4. S.J. Huang in *The Encyclopedia of Advanced Materials,* Eds., D. Bloor, R.J. Brook, M.C. Flemings and S. Mahajan, Pergamon, Oxford, UK, 1994, p.338-249.

5. *Degradable Polymers, Recycling, and Plastics Waste Management,* Eds., A-C. Albertsson and S.J. Huang, Marcel Dekker, Inc., New York NY, USA, 1995.

6. *Polymers from Renewable Resources: Biopolyesters and Biocatalysis,* Eds., C. Scholz and R. Gross, ACS Symposium Series No.764, American Chemical Society, Washington, DC, USA, 2000.

7. Y. Doi, *Microbial Polyesters,* VCH, New York, NY, USA, 1990.

8. A. Steinbuchel, B. Fuchtenbusch, V. Gorenflo, S. Hein, R. Jossek, S. Langenbach and B.H.A. Rehm, *Polymer Degradation and Stability,* 1998, **59**, 1-3, 177.

9. M. Vert, G. Schwarch and J. Coudane, *Journal of Macromolecular Science - Pure and Applied Chemistry,* 1995, **A32**, 4, 787.

10. R.G. Sinclair, *Journal of Macromolecular Science - Pure and Applied Chemistry,* 1996, **A33**, 5, 585.

11. J.L. Runt, *Polymer Degradation and Stability,* 1998, **59**, 1-3, 145.

12. M. Ajioka, H. Suizu, C. Higuchi and T. Kashima, *Polymer Degradation and Stability,* 1998, **59**, 1-3, 137.

13. J.M. Brady, D.E. Cutwright, R.A. Miller, G. Battistone, *Journal of Biomedical Materials Research,* 1973, 7, 155.

14. S. Asakura, Y. Katayama, *Journal of the Chemical Society of Japan,* 1964, **67**, 956.

15. W.H. Carothers and J.W. Hill, *Journal of the American Chemical Society,* 1932, **54**, 1559.

16. P. Degee, P. Dubois, R. Jerome, S. Jacobsen and H-G. Fritz, *Macromolecular Symposia,* 1999, **144**, 289.

17. H.R. Kricheldorf, I. Kreiser-Saunders and C. Boettcher, *Polymer*, 1995, **36**, 6, 1253.

18. J. Dahlmann and G. Rafler, *Acta Polymerica*, 1993, **44**, 2, 103.

19. A. Duda and S. Penczek, *Macromolecules*, 1990, **23**, 6, 1636.

20. F.E. Kohn, J.W.A. van den Berg, G. van de Ridder and J. Feijen, *Journal of Applied Polymer Science*, 1984, **29**, 12, 4265.

21. L. Trofimoff, T. Aida and S. Inoue, *Chemistry Letters*, *Chemical Society of Japan*, 1987, **16**, 991.

22. S.I. Moon, G.W. Lee, M. Miyamota and Y. Kimura, *Journal of Polymer Science - Polymer Chemistry Edition*, 2000, **38**, 9, 1673.

23. N. Spassky, V. Simic, L.G. Hubert-Pfalzgraf and M.S. Montaudo, *Macromolecular Symposia*, 1999, **144**, 257.

24. Z. Jedlinski, P. Kurcok and R.W. Lenz, *Journal of Macromolecular Science - Pure and Applied Chemistry*, 1995, **A32**, 4, 797.

25. H.R. Kricheldorf and R. Dunsing, *Makromolekulare Chemie*, 1986, **187**, 7, 1611.

26. Y. Kimura in *Proceedings of Biopolymers - Advances in Medical and Material Science Applications*, Cambridge, MA, USA, 2001.

27. E. Lillie and R.C. Schulz, *Die Makromolekulare Chemie*, 1975, **176**, 6, 1901.

28. S.H. Kim and Y.H. Kim, *Macromolecular Symposia*, 1999, **144**, 277.

29. H.R. Kricheldorf, I. Kreiser-Saunders, D-O. Damrau, *Macromolecular Symposia*, 1999, **144**, 269.

30. S. Slomkowski, S. Sosnowski and M. Gadzinowski, *Polymer Degradation and Stability*, 1998, **59**, 1-3, 153.

31. M. Ajioka, K. Enomoto, K. Suzuki and A. Yamaguchi, *Journal of Environmental Polymer Degradation*, 1995, **3**, 4, 225.

32. J.R. Dorgan, J. Palade, J.S. Williams, D. Knauss, J. Wegner and S. Dec in *Proceedings of Poly Millennial 2000*, Waikoloa, HI, USA, Abstract No. 275.

33. M. Vert, *Angewandte Makromolekulare Chemie*, 1989, **166/167**, 155.

34. K.M. Benabdillah, M.N. Boustta, J. Coudane and M. Vert in *Polymers from Renewable Resources: Biopolyesters and Biocatalysis*, Eds., C. Scholz and R. Gross, ACS Symposium Series No.764, American Chemical Society, Washington, DC, USA, 2000, p.200-220.

35. W.J. Cook, J.A. Cameron, J.P. Bell and S.J. Huang, *Journal of Polymer Science - Polymer Letters Edition*, 1981, **19**, 4, 159.

36. C.V. Benedict, W.J. Cook, P. Jarrett, J.A. Cameron, S.J. Huang and J.P. Bell, *Journal of Applied Polymer Science*, 1983, **28**, 1, 327.

37. C.V. Benedict, J.A. Cameron and S.J. Huang, *Journal of Applied Polymer Science*, 1983, **28**, 1, 335.

38. C.G. Pitt, R. Jeffcoat, R.A. Zweidinger and A. Schindler, *Journal of Biomedical Materials Research*, 1979, **13**, 3, 497.

39. M. Spinu, C. Jackson, M.Y. Keating and K.H. Gardner, *Journal of Macromolecular Science - Pure and Applied Chemistry*, 1996, **A33**, 10, 1497.

40. M.R. Lostocco and S.J. Huang, *Polymer Modification*, Ed., G. Swift, Plenum Press, New York, NY, USA, 1997, p.45-47.

41. D.W. Grijpma, R.D.A. Van Hofslot, H. Super, A.J. Nijenhuis and A.J. Pennings, *Polymer Engineering and Science*, 1994, **34**, 22, 1674.

42. G. Perego and T. Vercellio, *Die Makromolekulare Chemie*, 1993, **194**, 9, 2463.

43. M. Spinu, inventor; EI DuPont de Nemours and Company, assignee; US Patent 5,202,413, 1993.

44. W.M. Stevels, A. Bernard, P. Van de Witte, P.J. Dijkstra and J. Feijen, *Journal of Applied Polymer Science*, 1996, **62**, 8, 1295.

45. J-M. Vion, R. Jerome, P. Teyssie, M. Aubin and R.E. Prud'homme, *Macromolecules*, 1986, **19**, 7, 1828.

46. J. Kasperczyk and M. Bero, *Die Makromolekulare Chemie*, 1991, **192**, 1777.

47. C.X. Song and X.D. Feng, *Macromolecules*, 1984, **17**, 12, 2764.

48. W.M. Stevels, M.J.K. Ankone, P.J. Dijkstra and J. Fiejen, *Makromolecular Chemistry and Physics*, 1995, **196**, 4, 1153.

49. M. Bero, G. Adamus, J. Kasperczyk and H. Janeczek, *Polymer Bulletin*, 1993, **31**, 1, 9.

50. Z. Jedlinski, P. Kurcok, W. Walach, H. Janeczek and I. Rodecka, *Die Makromolekulare Chemie*, **194**, 6, 1681.

51. K. Stridsberg and A-C. Albertsson, *Journal of Polymer Science Part A: Polymer Chemistry Edition*, 2000, **38**, 10, 1774.

52. S.A. Brown and M.B. Mayor, *Journal of Biomedical Research*, 1978, **12**, 1, 67.

53. S.L-Y. Woo, W.H. Akeson, B. Levenetz, R.D. Coutts, J.V. Mattews and D. Amiel, *Journal of Biomedical Materials Research*, 1974, **8**, 5, 321.

54. D.F. Williams, *Journal of Materials Science*, 1982, **17**, 5, 1233.

55. C.C. Chu and M. Louie, *Journal of Applied Polymer Science*, 1985, **30**, 8, 3133.

56. K.R. Huffman and D.J. Casey, *Journal of Polymer Science Part A: Polymer Chemistry Edition*, 1985, **23**, 7, 1939.

57. D.K. Gilding in *Biocompatibility of Clinical Implant Materials*, Volume 2, Ed., D.F. Williams, CRC Press, Boca Raton, FL, USA, 1982.

58. R.M. Rice, A.F. Hegyeli, S.J. Gourley, C.W.R. Wade, J.G. Dillon, H. Jaffe and R.K. Kulkarni, *Journal of Biomedical Materials Research*, 1978, **12**, 1, 43.

59. S.J. Huang and P.G. Edelman in *Degradable Polymers, Principles and Application*, Eds., G. Scott and D. Gilead, Chapman & Hall, London, UK, 1995, p.18-28.

60. S. Li and M. Vert, *Degradable Polymers, Principles and Application*, Eds., G. Scott and D. Gilead, Chapman & Hall, London, UK, 1995, p.43-87.

61. Y-K. Han, P.G. Edelman and S.J. Huang, *Journal of Macromolecular Science - Pure and Applied Chemistry*, 1988, A25, 5-7, 847.

62. F.O. Eschbach and S.J. Huang, *Polymer Preprints*, 1993, **34**, 1, 848.

63. M. Ramos and S.J. Huang, *Polymeric Materials Science and Engineering*, 2001, **84**, 432.

64. S.J. Huang and J.M. Onyari, *Journal of Macromolecular Science - Pure and Applied Chemistry*, 1996, **A33**, 5, 571.

65. J.A. Wallach and S.J. Huang, *Polymeric Materials Science and Engineering*, 1999, **80**, 596.

66. J.A. Wallach and S.J. Huang, *Biomacromolecules*, 2000, **1**, 2, 174.

67. X. Chen, Y. Shen and R. Gross in *Polymers from Renewable Resources, Biopolyesters and Biocayalysis*, Eds., C. Scholz and R.A. Gross, ACS Symposium Series No.764, ACS, Washington, DC, USA, 2000, p.129-134.

68. P.G. Edelman, *Synthesis and Characterisation of Three Reactive Polymer Systems, Poly(amide enamines), Diacetylenencontaining Polymers, Biodegradable Polyester Resins*, University of Connecticut, Connecticut, CT, USA, 1986, pp. 109-111. [PhD Thesis]

69. S. Owen, M. Masaoka, R. Kawamura and N. Sakota in *Degradable Polymers, Recycling, and Plastics Waste Management*, Eds., A-C. Albertsson and S.J. Huang, Marcel Dekker, Inc., New York, NY, USA, 1995, p.251-258.

70. M. Harkonen, K. Hiltunen, M. Malin and J.V. Seppala in *Degradable Polymers, Recycling, and Plastics Waste Management*, Eds., A-C. Albertsson and S.J. Huang, Marcel Dekker, Inc., New York, NY, USA, 1995, p.265-270.

71. S.H. Kim, S.H. Lee, Y-K. Han and Y.H. Kim in *Proceedings of Poly Millennial 2000*, Waikoloa, HI, USA, 2000, Abstract No.296.

72. R.F. Storey, B.D. Mullen and K.M. Melchert, *Journal of Macromolecular Science - Pure and Applied Chemistry*, 2001, **A38**, 9, 897.

73. K. Jamshidi, S-H. Hyon and Y. Ikada, *Polymer*, 1998, **29**, 12, 2229.

74. J.R. Dorgan, H. Lehermeier and M. Mang, *Journal of Polymers and the Environment*, 2000, **8**, 1, 1.

75. D. Bendix, *Polymer Degradation and Stability*, 1998, **59**, 1-3, 129.

76. H. Tsuji, F. Horii, S.H. Hyon and Y. Ikada, *Macromolecules*, 1991, **24**, 10, 2719.

77. J.R. Murdoch and G.L. Loomis, inventors; EI DuPont de Nemours and Company, assignee; US Patent 4,719,246, 1988.

78. G.L. Loomis, J.R. Murdoch and K.H. Gardner, *Polymer Preprints*, 1990, **31**, 2, 55.

79. R.A. Kumar, C.L. Yue, R.A. Gross and S.P. McCarthy in *Proceedings of the 5th Annual Meeting of the BioEnvironmental Degradable Polymer Society*, 1996, p.16.

80. M.R. Lostocco, A. Borzacchiello and S.J. Huang, *Macromolecular Symposia*, 1998, **130**, 151.

81. M.R. Lostocco and S.J. Huang, *Journal of Macromolecular Science - Pure and Applied Chemistry*, 1997, **A34**, 11, 2165.

82. X. Liu, M. Dever, N. Fair and R.S. Benson, *Journal of Environmental Polymer Degradation*, 1997, **5**, 4, 225.

83. S.P. McCarthy and X. Song in *Proceedings of Biopolymers – Advances in Medical and Material Science Applications*, Cambridge, MA, USA, Knowledge Foundation, 2001.

84. M. Vert, *Degradable Materials: Perspectives, Issues and Opportunities*, Eds., S.A. Barenberg, J.L. Brash, R. Narayan and A.E. Redpath, CRC Press, Boca Raton, FL, USA, 1990, p.11.

85. D.K. Gilding and A.M. Reed, *Polymer*, 1979, **20**, 12, 1459.

86. S.J. Huang, L-H. Ho, M.T. Huang, M.F. Koening and J.A. Cameron in *Biodegradable Plastics and Polymers*, Eds., Y. Doi and K. Fukuda, Elsevier Science, Amsterdam, The Netherlands, 1994, p.3-7.

87. C.G. Pitt, F.I. Chasalow, Y.M. Hibionada, D.M. Klimas and A. Schindler, *Journal of Applied Polymer Science*, 1981, **26**, 11, 3779.

88. S.M. Li, H. Garreau and M. Vert, *Journal of Materials Science: Materials in Medicine*, 1990, **1**, 3, 123.

89. J. Kohn and R. Langer, *Biomaterials Science: An Introduction to Materials in Medicine*, Eds., B.D. Ratner, A.S. Hoffman, F.J. Schoen and J.E. Lemons, Academic Press, San Diego, CA, USA, 1996, p.64.

90. R. Langer and J.P. Vacanti, *Science*, 1993, **260**, 920.

91. J. Heller, R.V. Sparer and G.M. Zentner in *Biodegradable Polymers as Drug Delivery Systems*, Eds., M. Chasin and R. Langer, Marcel Dekker, New York, NY, USA, 1990, p.121-162.

92. X. Zhang, U.P. Wyss, D. Pichora and M.F.A. Goosen, *Journal of Bioactive and Biocompatible Polymers*, 1994, **9**, 1, 80.

93. M.R. Lostocco, C.A. Murphy, J.A. Cameron and S.J. Huang, *Polymer Degradation and Stability*, 1998, **59**, 1-3, 303.

94. M.S. Roby in *Proceedings of Biopolymers – Advances in Medical and Material Science Applications*, Cambridge, MA, USA, 2001.

95. M.S. Reeve, S.P. McCarthy, M.J. Downey and R.A. Gross, *Macromolecules*, 1994, **27**, 3, 825.

96. A. Torres, S.M. Li, S. Roussos and M. Vert, *Journal of Environmental Polymer Degradation*, 1996, **4**, 4, 213.

97. K-L.G. Ho, A.L. Pometto, A. Gadea-Rivas, J.A. Briceno and A. Rojas, *Journal of Environmental Polymer Degradation*, 1999, **7**, 4, 173.

98. *Journal of Macromolecular Science - Pure and Applied Chemistry*, 1999, **36**, 7-8.

99. B. Jeong, Y.K. Choi, Y.H. Bae, G. Zentner and S.W. Kim, *Journal of Controlled Release*, 1999, **62**, 1-2, 109.

100. C. Bastioli in *Degradable Polymers, Recycling, and Plastics Waste Management*, Eds., A-C. Albertsson and S.J. Huang, Marcel Dekker, Inc., New York NY, USA, 1995, p112-137.

101. Y. Kimura in *Proceedings of Biopolymers – Advances in Medical and Material Science Applications*, Cambridge, MA, USA, 2001.

102. J.A. Wallach and S.J. Huang in *Polymers from Renewable Resources: Biopolyesters and Biocatalysis*, Eds., C. Scholz and R. Gross, ACS Symposium Series No.764, American Chemical Society, Washington, DC, USA, 2000, p.281-292.

10 Aliphatic-Aromatic Polyesters

Rolf-Joachim Müller

10.1 Introduction

Biodegradable polymeric materials are dominated by polymers which contain hetero-atoms in their main chains. Chemical bonds such as ether-bonds, amide-bonds or ester-bonds are susceptible for hydrolytic attack and can lead to a primary reduction in molar mass of the polymers which is necessary to generate low molecular weight and water soluble intermediates able to penetrate biological membranes. The cleavage of the polymer chains can be catalysed by enzymes but also may take place without the action of the biological catalysts. The latter mechanism is predominantly observed when such polymers are degraded in human or animal bodies (medical applications) and also in this case the expression 'biodegradation' usually is used.

In the history of biodegradable plastics, polyesters played a dominant part just from the beginning of the development. One of the first products developed as biodegradable plastics from the beginning of the 1970s was based on a polyester belonging to the group of poly(hydroxyalkanoates) and is called polyhydroxybutyrate (PHB). PHB and a number of other aliphatic polyesters and copolyesters are produced and intracellularly accumulated by a number of microorganisms [1]. A copolyester of PHB and poly(hydroxy valerate) was commercially produced by fermentation and was available on the market under the trade name BIOPOL (Metabolix) for many years. However, the production of this polyester was recently stopped, probably due to the high price level of the product and also some problems in material properties.

Beside the natural polyesters a number of synthetic aliphatic polyesters have also been shown to be biologically degradable [2-5]. From the commercial point of view the most important synthetic biodegradable aliphatic polyester until now is poly(ε-caprolactone) (PCL), which is available under the trade name TONE (Union Carbide Corporation, Danbury, CT, USA). PCL is predominately used as component in polyester/starch blends [6], (e.g., Mater-Bi Z-grade, Novamont, Novara, Italy). Various aliphatic copolyesters based on succinate, adipate, ethylene glycol and 1,4-butanediol are produced by Showa Highpolymer in Japan (Bionolle Showa High Polymer Co., Ltd., Tokyo, Japan). Using lactic acid, which is produced on a large scale by fermentation, polylactic acid (PLA) can be synthesised by different means. Companies like Cargill/Dow (Cargill Dow Polymers LLC, Midland, MI, USA) or Mitsui Chemicals (Mitsui Chemicals Inc., Tokyo, Japan) are announcing a widespread application of these aliphatic polyesters, based on natural monomers.

However, most of the aliphatic polyesters presently commercially used for biodegradable materials exhibit serious disadvantages. Beside the relative high price level, material properties are often limited and exclude these materials from many applications. The very low melting point of PCL (about 60 °C) is an example.

For conventional technical application aromatic polyesters such as polyethylene terephthalate (PET) or polybutylene terephthalate (PBT) are widely used, (e.g., PET-bottles for beverages), but these polyesters are commonly regarded as biologically inert and, thus, not directly applicable for biodegradable plastics. Trying to combine both the excellent material properties of aromatic polyesters and the potential biodegradability of aliphatic polyesters, a number of aliphatic-aromatic copolyesters have been developed during the last 10 years and the first products are now going to be commercialised on a scale of a few thousand tons per year. Relative low product prices and the very good material properties for application and processing promise a successful future for this group of biodegradable plastics. However, due to the complex structure of these types of copolyester, which include structure elements that may exhibit a very slow degradability, intensive investigations on the biodegradability and degradation mechanisms are necessary and already under way.

10.2 Development of Biodegradable Aliphatic-Aromatic Copolyesters

Since aromatic polyesters turned out to be quite resistant to hydrolytic degradation under physiological conditions a number of attempts were made to implant structures open to biological attack in such polyesters. This was predominately done by introducing aliphatic acid components in the aromatic polyester chains [7].

Table 10.1 gives an overview of different aliphatic-aromatic copolyesters synthesised as degradable materials during the last few years. Part of the work reported in the literature dealt with hydrolytic degradation mechanisms which do not involve enzymic catalysis (chemical hydrolysis). This kind of degradation is often present in medical applications of polyesters, e.g., as implants in living tissues. Enzymic catalysed hydrolysis, in contrast, is usually connected to microbial degradation in the environment.

The first papers published on the biological degradation of aliphatic-aromatic copolyesters came to the conclusion that only at a quite high fraction of aliphatic monomers did the

Table 10.1 History of development of biodegradable aliphatic-aromatic copolyesters

Aromatic polymer	Aliphatic component	Mode of degradation	Refs.
PET (1979/1981)	Oxyethylene diols	Hydrolysis in buffer at 37 °C	[15, 16]
PET PBT PEIP (1981)	ε-Caprolactone	Hydrolysis with lipase from *Rhizopus arrhizus* in buffer at 37 °C	[4, 8, 11]
PBT (1989)	Glycolic acid	Hydrolysis in water at 60 °C	[17]
PBT (1990)	Oxalic acid	Hydrolysis in water at 33 °C and 50 °C	[18]
PET (1992)	Adipic acid	Hydrolysis in water at 25-80 °C	[19]
PET (1993)	L-Lactic acid, oxyethylene diols	Hydrolysis in buffer at 60 °C	[20]
PET (1994)	ε-Caprolactone	Hydrolysis with lipase from *Pseudomonas sp.*, in buffer at 37 °C; soil burial; composting	[9, 10]
PET (1994)	Adipic acid	Hydrolysis in water at 25-90 °C	[21]
PPT (1994/95)	Adipic acid, sebacic acid	Degradation in a synthetic liquid medium by microorganisms	[5, 12]
PET PPT PBT (1995)	Adipic acid, sebacic acid	Degradation in a synthetic liquid medium by microorganisms, soil burial, composting	[13]
PET (1995)	Oxyethylene diols, oxybutylene diol	10% NaOH at 70 °C	[22]
PET (1996)	Succinic acid	No data on degradation given	[23]
PBT (1997)	Adipic acid	Composting and agar plate test with pre-screened microorganisms	[14]
PET (1997)	Adipic acid, sebacic acid, ethylene glycol	Hydrolysis with lipase from *Rhizopus arrhizus* in phosphate buffer at 37 °C	[24]
PPP (1998)	Fumaric acid	Hydrolysis with lipase from *Chromobacterium viscosum* in potassium phosphate buffer at 40 °C	[25]
PET (1999)	Succinic acid, sebacic acid, 1,12 dodecane di-carboxylic acid	Hydrolysis with lipase from *Rhizopus arrhizus* in phosphate buffer at 37 °C	[26]
PBT	Succinic acid	Composting	[27]
PBT (2001)	Succinic acid, 1,4-cyclohexane dimethanol	Hydrolysis in buffer at pH 4, pH 7 and pH 10; composting	[28]
PBT (2001)	Succinic acid	Hydrolysis with lipase from *Rhizopus arrhizus* at 37 °C	[29]
PHT (2001)	ε-Caprolactone	No degradation experiments	[30]
PEIP: poly(ethylene isophthalate)		*PPP: poly(1,2-propanediyl phthalate)*	
PPT: poly(propylene terephthalate)		*PHT: poly(hexamethylene terephthalate)*	

copolymers exhibit a significant degradability [4, 8-11]. However, these authors only investigated quite short degradation times (degradation with lipases (EC 3.1.1.3) for a few days, composting for about 15 days) and thus, the relative slow degradation processes of the copolyesters were not detectable under such non-optimised test conditions.

In 1994, Witt and co-workers [5] first reported a microbial degradation of a block-copolyester [poly(trimethylene decanoate)-block-(trimethylene terephthalate)] with 50 mol% of terephthalic acid in the acid component. In a mineral medium inoculated with sewage sludge, Witt and co-workers observed a weight loss in polyester films of about 9% within four weeks. In 1995 the same authors published data about the degradation of random aliphatic-aromatic copolyesters from terephthalic acid, 1,3-propanediol and adipic acid or sebacic acid (30 mol% of terephthalic acid content) in a soil burial experiment [12]. The melting points of these statistical copolyesters were above 100 °C and, thus, promised better properties for applications than, for example, PCL. Generally, a decreasing degradation rate was observed when the fraction of aromatic component increased and this behaviour was correlated with the melting point of the materials, a correlation which was already demonstrated by Tokiwa and co-workers for different aliphatic polyesters [8]. In another paper the biodegradation of statistical copolyesters composed of ethylene glycol/adipic acid/terephthalic acid and 1,3-propanediol/adipic acid/terephthalic acid and 1,4-butanediol/adipic acid/terephthalic acid, was examined in a composting simulation test [13]. While the fraction of terephthalic acid in the copolymers predominantly determined the degradation behaviour, the kind of dihydroxylic monomer was shown to be of minor importance for the degradability of the material.

Random copolyesters composed of 1,4-butanediol, terephthalic acid and adipic acid (BTA-copolyesters) turned out to be the most promising materials for technical applications, not only for their degradation behaviour and their material properties, but also from the availability of the monomers and an estimated price level, as well. BTA-copolyesters were examined in more detail in 1997 for their biodegradation behaviour and preliminary material properties [14]. It was demonstrated, that in a range between 40 mol% to about 50 mol% of terephthalic acid (referred to the acid components) materials can be achieved which combine sufficient biodegradability with promising technical properties. Compared to a low density polyethylene material, a BTA-copolyester with about 40 mol% terephthalic acid exhibited a comparable mechanical strength but a significant higher flexibility (elongation at break). Also a chain extension of the polyester chains with diisocyanates up to a molecular weight of MW = 230,000 g/mol was possible without reducing the biodegradability. In the range of copolyester composition which is of interest for materials providing useful technical properties, a correlation of the rate of biodegradation (in terms of an erosion rate) was established for composting conditions. This correlation allows the rough estimation of the period of time which is necessary for the complete deterioration of items of BTA-copolyesters with different composition. The correlation is shown in **Figure 10.1.**

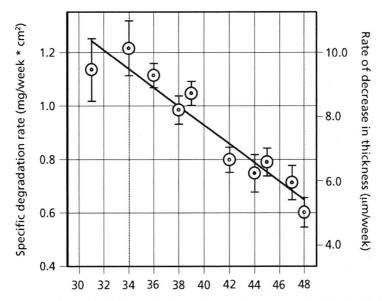

Figure 10.1 Dependence of the degradation rate of statistical polyesters of 1,4-butanediol, terephthalic acid and adipic acid from the content of aromatic dicarboxylic acid; degradation of polyester-films on mineral-agar inoculated with a mixed population from compost at 60 °C [14]. Degradation is given either as weight loss per film surface area (left y-axis; in mg/week cm^2) or as rate of surface erosion (right y-axis; in µm/week), calculated from weight loss data and material density

10.3 Degradability and Degradation Mechanism

10.3.1 General Mechanism/Definition

The term biodegradation of plastics is often used for totally different mechanisms of degradation phenomena and it is essential for a rational discussion about biodegradable plastics to differentiate properly between these different mechanisms.

10.3.1.1 Biocorrosion

When plastics come in contact with microorganisms it can cause changes (negative) in the properties of plastics. Typical material parameters which are altered by the microbial action are mechanical properties, (e.g., tensile strength, flexibility), or simply the colour. Biocorrosion is usually an unwanted process which is to be avoided, e.g., by the addition of biocides to the material. The mechanism of biocorrosion is often a selective degradation of one component of the plastic, e.g., a plasticiser, and is accompanied in many cases by other, non-biological mechanisms such as hydrolysis or oxidation.

Biocorrosion can lead to a complete deterioration of the material to (invisible) particles.

10.3.1.2 In Vivo Degradation

Polymers for medical applications which are degraded in living tissues or in the environment of the living body are also called biodegradable, because the degradation takes place in a biological environment However, very often degradation mechanisms here are solely abiotic, (e.g., non-enzymically catalysed hydrolysis). Other expressions used in this context are bio-resorbable or bio-compatible.

10.3.1.3 Biodegradation in the Environment

For environmental applications such as biodegradable packaging, biowaste bags or mulching films in agriculture, the degradation of biodegradable plastics is predominantly caused by microorganisms.

Usually plastic materials are not water soluble and even if they are soluble to some extent, the polymer chains have a high molar mass and thus they cannot be transported directly through the cell membranes into microbial cells to be biochemically converted there (**Figure 10.2**).

For that reason microorganisms excrete enzymes into the environment, which can attach to the polymer surface and cleave the polymer chains, as long as the degradation products become short enough to be water soluble (this biological system has been developed by the evolution to use natural polymers or other poorly bio-available substrates for microbes). Then these intermediates can diffuse into the surrounding environment of the plastics, be incorporated into the microbial cells and metabolised there to form biochemical end products such as water and carbon dioxide (and many others).

For the general understanding of the term biodegradable in connection with plastics it has been widely agreed, that the microbial induced attack of a polymer, (e.g., determined

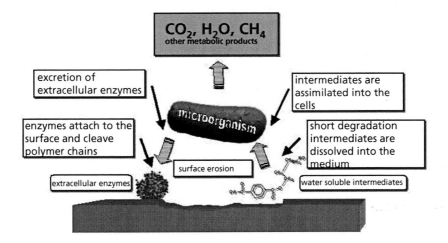

Figure 10.2 General scheme of microbial polymer degradation

as deterioration or weight loss), is not a sufficient characteristic but ideally a complete transformation of the entire plastics components into naturally occurring materials is necessary. This kind of view is also reflected for example in some definitions for biodegradable plastics, e.g., DIN V 54900 [31].

However, although microbes are always included in the degradation process, parts of the entire degradation mechanism can also be of abiotic nature. One example is the degradation of the aliphatic polyester PLA, which is hydrolysed under natural conditions by a non-enzymically catalysed process into short oligomers and monomers. These chemically produced intermediates then are metabolised by microorganisms to form products that become part of natural cycles.

The requirement, that all components of a plastic-composition must be degraded is of essential relevance for copolymers and especially for aliphatic-aromatic copolyesters. Here in one polymer chain structures are combined, which differ significantly in their degradation behaviour when the monomers are located in the corresponding homopolymers (aliphatic polyesters often are easily biodegraded while aromatic polyesters such as PET are quite biologically resistant). In this particular case it has to be ensured that no domains within the copolyester chains are poorly biodegradable and can accumulate in the nature.

Generally it is not possible to state that a plastic is biodegradable under all circumstances. The biological environment as well as the time frame of a claimed biodegradability has to be specified, since the environmental conditions in a composting process are different from those present in soil and also the time frame of a composting process (usually some months) is much shorter than the time which can be accepted for the degradation of a polymer in soil, e.g., in the range of years for agricultural applications.

The basic general requirements for biodegradable plastics and also biodegradable aliphatic-aromatic copolyesters can be summarised as follows:

- The material must keep its functionality during customer's usage.

- The material must disintegrate in the specific environment in a time frame defined by the process or the kind of usage.

- The entire material must be transformed into natural occurring metabolic products. Degradation rate and input rate into the specific environments should not cause any accumulation of material components or degradation intermediates.

- At any step of the degradation no toxic effect should be caused to the environment.

In conclusion it can be stated, that the biodegradation of plastics means the complete transformation into natural products, but the rate of degradation can only be specified according to the specific application of the biodegradable plastic. Thus, the specific biodegradability of a material always has to be defined and evaluated, taking into account the use of the end-products and the environment where degradation finally takes place.

This chapter is mainly focuses on the environmental degradation of aliphatic-aromatic copolyesters.

10.3.2 Degradation of Pure Aromatic Polyesters

Polyesters which solely contain aromatic acid components such as PET or PBT are commodity materials for many technical applications and are commonly regarded as quite resistant to any hydrolytic degradation. Only by applying very drastic chemical treatments, (e.g., sulfuric acid at 150 °C), which are a long way away from any physiological conditions, hydrolysis of such polymers can be achieved at reasonable rates, which can be used for recycling purposes [32]. Furthermore, chemical hydrolysis was applied in some cases for the analysis of aromatic polyesters [33].

From the point of view of the durability of aromatic polyesters investigations were performed to predict the life time of products in different environments. From a kinetic model, based on accelerated degradation experiments, some authors tried to characterise the long term behaviour of PET under ambient conditions [34, 35]. As one result the life time of PET in nature was predicted in a range of 16 up to 48 years [36-38]. In human and animal tissues the degradation of PET fibres was evaluated partially using *in vitro* experiments with ^{14}C-labelled PET [39, 40]. From this work the life time of the fibres in tissues was estimated to 30 years.

In both cases hydrolytic degradation of PET was performed without any enzymic catalysis. Up to now there are no reliable indications available in the literature that microbes and enzymes can attack aromatic polyesters such as PET, PBT or poly(ethylene naphthalate) [2, 41, 42].

From all the information in the literature it can be concluded that conventional aromatic polyesters used up to now for technical purposes are not subject to a biologically induced degradation at a reasonable degradation rate. This excludes such polyesters from applications in biological waste treatment processes (composting) or which use biodegradability as a new material property for novel polymer applications, (e.g., controlled release of active substances in agriculture).

However, recently a laid open specification of a German patent was published (DE 199 35 156 A1, 2000 [43]) where the inventors claim a procedure to disintegrate aromatic polyesters by means of special microbial strains (*Trichosporum* and *Arthrobacter*). The deterioration of the test specimen took place over some weeks. Because of the limited information given in this application concerning the polymeric test material and the behaviour of the samples without contact to the microbes (blank tests) it is not clear if the claimed effects are really related to an enzymic action of the microorganisms involved.

Some authors synthesised special aromatic polyesters, using long polyethylene glycols as dihydroxylic components for polycondensation and found hints to a certain biological susceptibility of these model materials to a microbial attack [44, 45].

10.3.3 Degradation of Aliphatic-Aromatic Copolyesters

10.3.3.1 Polymer Related Parameters Determining Biodegradation

While a number of aliphatic components which alter the biodegradation behaviour of aromatic polyesters have been tested, the aromatic component predominantly used was terephthalic acid. Also the materials which are commercially available on the market contain this aromatic dicarboxylic acid.

The degradation behaviour of aliphatic-aromatic copolyesters generally depends on the composition of the monomers as well as on the structure of the polymer chains at a given composition.

When introducing terephthalic acid units into an aliphatic polymer, at first an increase in the degradability can be observed at low levels of the aromatic monomer. Increasing the fraction of terephthalic acid, the degradation rate decreases and above a level in the range of 60-70 mol% terephthalic acid (with regard to the dicarboxylic acid components) no significant biological attack can be observed anymore [12]. An example of this behaviour is shown in **Figure 10.3**.

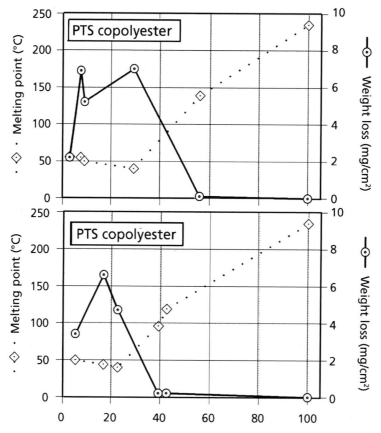

Figure 10.3 Weight losses and melting points of statistical copolyesters of 1,3-propanediol, terephthalic acid, sebacic acid and 1,3-propanediol, terephthalic acid, adipic acid as a function of the molar fraction of terephthalic acid in the copolyester. Degradation of 100 µm films at room temperature in an aerated mineral salt medium inoculated with an eluate from soil for 8 weeks [12]

This phenomenon is attributed to the melting point of the material. For many aliphatic polyesters a correlation of the degradability with the melting point was observed [2]. Marten [46] interpreted this with a decrease in the mobility of the polyester chains at lower temperatures; here the polymer chains are highly fixed in the polymer crystals and cannot adjust easily into the active sites of the extracellular enzymes. A random insertion of some aromatic monomers in aliphatic polymer chains disturbs the formation of crystals. The amount of crystals (crystallinity) is reduced and the melting point is lowered due to the less regular crystal structures. Both effects cause an increase in biodegradability. In contrast at higher contents in aromatic dicarboxylic acid the formation of crystals rich in terephthalic acid then leads to increasing melting points and decreasing degradability. In the range of 30 mol% to 60 mol% of terephthalic acid, which is of particular interest since such materials provide useful material properties, the degradation rate drops linearly with the content of the aromatic acid [14, 47] (see **Figure 10.2**).

While the amount of terephthalic acid strongly influences the degradation behaviour, the kind of aliphatic monomer is obviously of minor importance. For copolyesters of terephthalic acid and adipic acid differing in the dialcohol component (1,2-ethanediol, 1,3-propanediol and 1,4-butanediol) similar erosion rates were observed in soil and in compost [13] (**Figure 10.4**).

At the elevated temperatures under composting conditions a significant decrease in the molecular weight of the residual material was observed. This indicates, that parallel to the enzymic action, which takes place solely at the surface of the material, also a non-biological, pure chemical process of hydrolysis is involved at these temperatures in the degradation process. Water penetrates into the polymer matrix, hydrolyses the ester bonds and thus, lowers the molecular weight of the entire material. Also for the variation of the aliphatic dicarboxylic acid component in copolyesters with terephthalic acid a number of biodegradable materials are reported [46]. In this work Marten found, that the major parameter controlling the biodegradation rate of different aliphatic polyesters and aliphatic-aromatic copolyesters is the temperature difference between the melting point of the materials and the temperature at which the degradation takes place. This temperature difference is discussed as a measure of the mobility of the polymer chains, which is of great importance, since the chains have to fit into the active sites of the enzymes to by cleaved. In aliphatic-aromatic copolyesters, e.g., in PBT, the melting behaviour is mainly determined by the length of aromatic sequences in the polymer chains, which depends both on the composition and the structure [46, 48]. Besides the fixation of the polymer chains in the crystalline domains, the flexibility of the chain itself also influences the degradation behaviour to some extent. Copolyesters with long aliphatic dicarboxylic acids exhibit a somewhat higher degradation rate than those with shorter ones [14], however, this effect usually is masked by the much higher influence of the melting point. In the same work it was shown that the poor biodegradability of aromatic polyesters is not predominantly

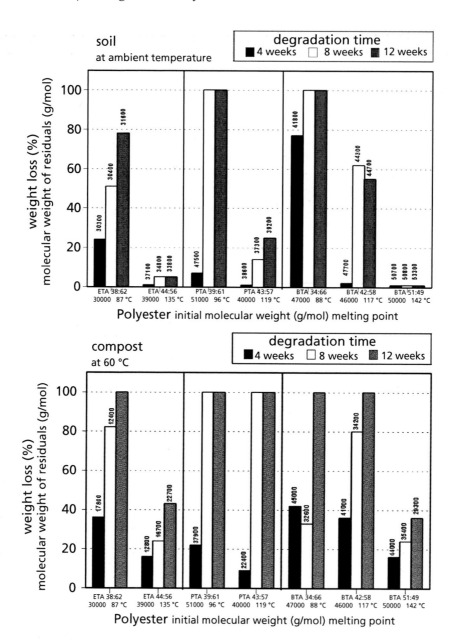

Figure 10.4 Weight losses of films of different aliphatic-aromatic copolyesters in soil at ambient temperature and mature compost at 60 °C; film thickness 100 μm [13]; components: E = > 1,2-ethanediol. P = > 1,3-propanediol, B: 1,4-butanediol, A: adipic acid, T: terepthalic acid; both numbers at the end of the identification reflect the ratio of aromatic/aliphatic acid component in mol%, (e.g., ETA38:62 copolyester from 1,2-ethanediol, adipic acid and terephthalic acid with 38 mol% terephthalic acid in the acid component)

caused by a steric hindrance of the enzymic attack to the ester-bonds caused by the vicinity of the quite bulky aromatic rings to the ester group. Marten demonstrated that a di-ester of 1,4-butanediol with benzoic acid, which represents a sequence in a PBT homopolyester, was attacked by a lipase from *Pseudomonas sp*. With a degradation experiment using a strictly alternating copolyester of terephthalic acid, adipic acid and 1,4-butanediol it was shown that even a sequence of only one aliphatic acid is sufficient to enable the enzyme to cleave the ester bonds within the polymer chain, while a block-copolyester of the same overall composition was not enzymically degraded although it could be suspected that the long aliphatic sequences would facilitate the degradation. In the alternating copolyester, which has a melting point of 85 °C the polymer chains are less fixed in the crystals than in the block-copolymers exhibiting melting points between 200 °C and 224 °C.

While the reduced enzymic susceptibility of the aromatic ester bonds is caused by the interaction of the long chains in the polymer, the chain length itself has no direct influence on the biodegradability above a minimum molar mass. Tests with copolyesters from terephthalic acid, adipic acid and 1,4-butanediol which were chain extended with hexamethylene-diisocyanide resulted not in a decrease in the biological degradation rate (in a compost simulation test) although the molar mass of the pre-polyesters of about 48,000 g/mol was increased to 232,000 g/mol by the chain extension [14].

10.3.3.2 Degradation Under Composting Conditions

Composting, the biological treatment of biowaste under controlled technical conditions, has been discussed as the major environment where biodegradable plastics and thus, biodegradable aliphatic-aromatic copolyesters will be degraded, e.g., as waste from biodegradable packaging or as biodegradable biowaste bags. For this reason a number of publications used test systems to evaluate the biodegradation of plastics, which reflect conditions similar to a composting process. Also some major standards concerning biodegradable materials are focussed on the degradation of plastics under composting conditions [31, 49, 50].

Jun and co-workers [10] studied the degradation of a copolyester of PET and PCL and came to the conclusion, that at a fraction of more than 50% (w/w) of aromatic component, no degradation took place. However, the incubation time was only 15 days and thus, too short to monitor slow degradation processes. However, Witt and co-workers [13] observed for a BTA-copolyester film (composed of 1,4-butanediol, adipic acid and terephthalic acid) with 51 mol% of terephthalic acid, a significant weight loss after a three month incubation in a compost at 60 °C (see Figure 10.4). With these experiments, the biodegradation of polyesters with a relative high content of aromatic components could be demonstrated for the first time. In 1998 BASF AG (Ludwigshafen, Germany) presented a respirometric measurement of the copolyester Ecoflex (approximately 45 mol% terephthalic acid in the acid component) that this material was more than 90% metabolised in compost within three months [51].

Kleeberg and co-workers succeeded in 1998 in isolating and identifying a number of thermophilic microorganisms from compost, which are able to depolymerise BTA-copolyesters [52]. Out of 61 isolates, 30 strains were able to attack a BTA40:60 (40 mol% terephthalic acid, 60 mol% adipic acid) copolyester at a rate sufficient for the detection method used (clear-zone method and weight loss of films on agar plates). It turned out, that under thermophilic conditions actinomycetes play a dominant role in copolyester degradation. Out of the 30 degrading isolates 25 belonged to the group of actinomycetes and only 5 to bacteria. Fungi were not found to be relevant in copolyester degradation in compost since most fungi only grow at temperatures less than 50 °C.

Two of the most active strains were identified and belong both to the genus *Thermobifida* and are consistent with the *Thermobifida fusca* taxon.

In a screening experiment with 1328 actinomycete-strains from the German Collection of Microorganisms and Cell Cultures (Braunschweig, Germany) 34 strains could be identified to attack a BTA40:60 copolyester [53]. The degrading strains were both mesophilic and thermophilic organisms, however, the thermophilic actinomycetes exhibited the highest degradation rates.

From the strain *Thermobifida fusca* DSM 43793 the copolyester depolymerising extracellular enzyme could be isolated and characterised. The enzyme exhibits a homology of 65% with a triacylglycerol-lipase from *Streptomyces sp (strain M11)* [54], has a molar mass of approximately 27 kDa and an optimal temperature for hydrolysing BTA – copolyesters of about 60 °C. The identification of the enzyme to have a lipase-like structure is in accordance with the observation, that many lipases are able to attack polyesters [2, 46, 55] and are probably also in nature predominantly responsible for the microbial induced depolymerisation of synthetic polyesters.

With this thermophilic actinomycete strain it was possible to investigate the degradation behaviour of BTA copolyesters very accurately over a time scale of a few weeks [56] and with the enzyme the hydrolysis of BTA copolyesters could be measured within less than 24 hours [53].

10.3.3.3 Degradation in Soil

Since the application of biodegradable plastics to soil, for example, as mulching films in agriculture, becomes more and more important, the characterisation of the degradation behaviour of the copolyesters in soil is currently of great interest.

Generally, compared to composting, degradation in soil is slower and less predictable. Reasons for this are the lower temperature in soil and the variability in environmental conditions, (e.g., humidity, temperature), and soil composition.

Little data about the degradation of aliphatic-aromatic copolyesters in soil has been published up to now. Witt and co-workers [13] could show that a number of copolyesters (acid components: adipic acid and terephthalic acid; alcohol components: 1,2-ethanediol, 1,3-propanediol and 1,4-butanediol) degrade in compost and in soil as well (see **Figure 10.4**). The weight losses of copolyester films (100 μm thickness) in soil (gardening soil, 60% humidity, ambient temperature, incubation times 1, 2 or 3 months) were significantly lower than in compost at 60 °C. While for a BTA-copolyester with about 40 mol% terephthalic acid in the acid component a weight loss of approximately 50% after 3 months was observed, the same material degraded completely in compost within 3 months. Increasing the aromatic monomer to 50 mol%, no weight loss of films could be detected in soil anymore within the 3-month period.

At 60 °C incubation in compost, the material of the test specimen exhibited a significant loss in molecular weight, indicating a contribution of non-enzymically catalysed hydrolysis (abiotic) which takes place not only at the surface but in the entire material. This abiotic hydrolysis was not observed in the soil burial experiments within the period of time for testing.

Using similar conditions in another study [57] the weight loss of 55 μm films of a BTA-based copolyester containing approximately 55 mol% terephthalic acid was tested in the same kind of soil and was compared with the degradation of poly(ε-caprolactone) (PCL Tone787, Union Carbide, films of 75 μm thickness) (**Figure 10.5**) in soil and in compost at 60 °C.

For the BTA-copolyester and PCL, the weight losses are much smaller in soil than in compost. However, for the copolyester with 55 mol% terephthalic acid, a degradation of the film of about 30% could be measured after an incubation of 10 weeks.

10.3.3.4 Degradation in Aqueous Environment

Degradation results obtained in a compost environment at elevated temperatures or in soil differ significantly from those determined in a liquid system. In liquid media the degradation rate is usually much slower for the copolyesters. Besides the lower temperature, which is in most cases at ambient temperature, a different microbial population may be responsible for the differences. In degradation tests in liquid environment, the inoculum to provide microbial activity is often taken from sewage sludge. Although in some cases eluates from soil or compost are used to inoculate the media, a liquid environment is not optimal to

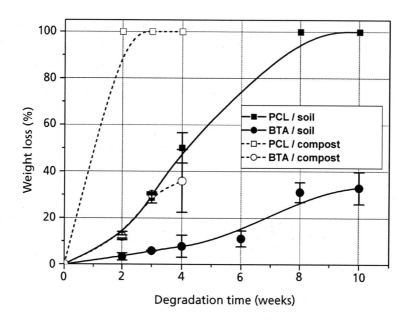

Figure 10.5 Degradation of a BTA-based copolyester (approximately 55 mol% terephthalic acid) and PCL in soil at ambient temperature and in compost at 60 °C. Weight loss of films (BTA: 55 μm thickness; PCL: 75 μm thickness)

grow a number of mycelium-forming microorganisms (fungi, actinomycetes), which have been shown to be important for the degradation of copolyesters in compost [53].

Van der Zee [58] reported only a very slow degradation in a modified Sturm Test (measurement of carbon dioxide which is produced by the microorganisms during the degradation of plastic) for copolyesters from terephthalic acid, adipic acid and 1,4-butanediol with levels of aromatic dicarboxylic acid of more than 30 mol% (with regard to the acid components). In a study concerning the applicability of different test methods the degradation of an aliphatic-aromatic copolyester was found to be strongly dependent on the specific conditions (especially the kind and pretreatment of the inoculum) of the aquatic test [59]. After 50 days of incubation a conversion of the polymer to carbon dioxide in a range from less than 10% up to more than 90% was observed depending on the inoculum.

10.3.3.5 Degradation Under Anaerobic Conditions

While for the degradation of plastics in the presence of oxygen a large number of investigations have been published, very few data exist for anaerobic biodegradation.

Some aliphatic polyesters such as poly(hydroxyalkanoates) or PCL also turned out to be biodegradable under anaerobic conditions [60]. However, aliphatic-aromatic copolyesters of the BTA-type seem to be quite stable in the absence of oxygen. For incubation of BTA-copolyesters with 40 mol% and 45 mol% of terephthalic acid, respectively, in anaerobic sewage sludge (at 37 °C) and also in an anaerobic high-solids sludge from an anaerobic biowaste treatment plant (at 50 °C), no significant biodegradation of BTA films could be observed within a test period of three months [61]. The small weight losses of less than 5% were obviously caused by abiotic effects such as migration of low molecular compounds or non-enzymic catalysed hydrolysis.

In the work mentioned previously, a number of anaerobic individual microbial strains degrading poly(hydroxyalkanoates), PCL, and a polyester from 1,3-propanediol and adipic acid (SP3:6) were isolated and identified. BTA-copolyesters were attacked by the individual strains only if the content of terephthalic acid did not exceed 20 mol%. Here it can be supposed that these organisms predominately attack the quite long aliphatic domains in these copolyesters.

At this point there is no definite proof that anaerobic microorganisms exist which can depolymerise BTA-copolyesters at compositions which are interesting for technical applications. The results reported up to now indicate, that a biological treatment of BTA-waste in anaerobic digestion plants will be difficult, especially since the residence time of the biowaste in anaerobic reactors is only in the range of a few weeks. However, in most anaerobic digestion plants an aerobic maturation step of the anaerobic sludge is included in the process; here it may also be possible that aliphatic-aromatic copolyesters degrade to such an extent that the final compost quality is not affected negatively.

10.3.3.6 Fate of Aromatic Sequences and Risk Assessment

A major point of criticism for aliphatic aromatic copolyesters is the final degradability of the aromatic sequences in the polymers. In such statistical copolyesters there exist domains in the polymer chains, where several aromatic dicarboxylic acids are linked with the alcohol component, without being interrupted by an aliphatic dicarboxylic acid. The distribution of sequence lengths depends on the ratio of aliphatic and aromatic dicarboxylic acids and can be calculated for an ideal random copolymerisation by:

$$W_{ar(n)} = \{([M_{ar}] / ([M_{ar}] + [M_{al}]))^{n-1}\} \ \{ [M_{al}] / ([M_{ar}] + [M_{al}])\} \qquad (10.1)$$

where:

$W_{ar(n)}$: Fraction of the aromatic dicarboxylic acid (in mol%) located in sequences of the length n

[M_{al}]: Fraction of the aliphatic dicarboxylic acid monomer in the polymer (in mol%)

[M_{ar}]: Fraction of the aromatic dicarboxylic acid monomer in the polymer (in mol%)

n: length of a sequence

As an example the distribution of the sequence lengths are listed for a BTA45/55 copolyester (45% terephthalic acid in the acid component) in **Table 10.2**.

Table 10.2 Sequence length distribution of aromatic domains in a statistical aromatic copolyester with 55 mol% terephthalic acid in the acid monomers

Length of aromatic sequence (number of repeating units)	Fraction of terephthalic acid monomers in the sequences of length n (mol% of terephthalic acid monomers)
1	0.550
2	0.248
3	0.111
4	0.050
5	0.023
6	0.010
7	0.005
8	0.002
9	0.001

If the biodegradation is monitored by weight loss or disintegration or even when the carbon conversion is determined in a respirometric test, it is problematic to decide if the ester sequences of pure aromatic acid are also subject to a biological attack. First attempts to evaluate the biodegradation behaviour of oligomeric aromatic intermediates from the copolyester degradation used especially synthesised aromatic model oligomers [62-64] for degradation experiments (**Figure 10.6**). It turned out that in a liquid mineral salt medium (inoculated with a mixed microbial population), in soil and in mature compost at 60 °C, oligomers with one or two repeating units (1: BTB and 2: BTBTB) were selectively removed from the synthetic oligomer mixture (average molar masses in the range from MW: 680 g/mol through MW: 2600 g/mol) while the amount of longer aromatic oligomers remained almost unchanged in these experiments.

The sharp change in the degradation behaviour correlates with the water solubility of the oligomers. Only mono- and di-esters of terephthalic acid and 1,4-butanediol (and 1,2-ethanediol and 1,3-propanediol) were water soluble to some extent (the oligomers were OH-terminated) and thus, could be directly transported into the microbial cells to

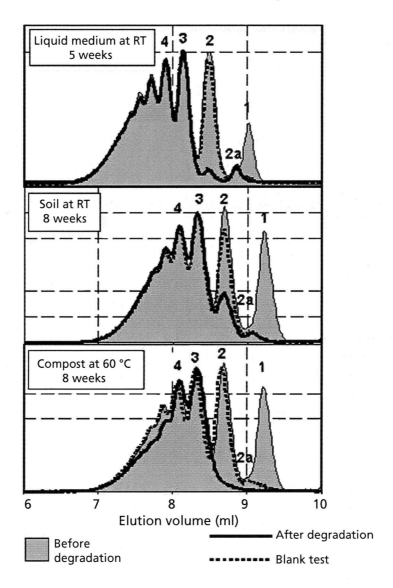

Figure 10.6 Gel permeation chromatography (GPC) chromatograms of a synthetic oligomer mixture synthesised from 1,4-butanediol and terephthalic acid before and after incubation in different microbial active environments (synthetic mineral medium at room temperature inoculated with an compost extract; soil at room temperature and compost at 60 °C. Gray area: GPC profile of the oligomer mixture before incubation; dotted line: GPC profile of the oligomer mixture after incubation in sterile water at conditions comparable to the degradation experiment (blank test); solid line GPC profile of the oligomer mixture after incubation in the different microbial environments [63]

be metabolised there. The microbial transformation could be confirmed by following the degradation of the soluble oligomers in a liquid medium by carbon dioxide measurements.

However, despite the results with the oligomeric model substances it was shown, that under temperature conditions similar to a composting process, longer aromatic oligomers, which were generated as intermediates from the biodegradation of the copolyesters, can be totally degraded by microorganisms. In a degradation experiment with a BTA 40:60 copolyester in a liquid medium and on agar plates which were inoculated with a pre-screened mixed microbial population from compost, the formation of various oligomeric intermediates could be proved by GPC measurements [64, 65] (**Figure 10.7**).

However, the quantitative analysis revealed that the different oligomers, (e.g., with two or three repeating units), were present in lower concentrations than could be calculated from equation (10.1). In **Figure 10.7** aromatic oligomers formed during degradation of the BTA polymer were identified by comparison with a synthetic BT-oligomer mixture. The concentrations of the BTBTB - oligomer (dimer) and the BTBTBTB - oligomer (trimer) were lower by a factor of 2.2 and 10.2, respectively, as it could be expected from the aromatic sequence length distribution [Equation (10.1)] without considering any degradation.

In a recent study [56] the formation and the degradation of aromatic intermediates could be monitored by GC-MS and GPC analysis. The commercial copolyester Ecoflex® was degraded in a synthetic liquid medium with a thermophilic microbial strain (see Section 3.3.2). These microorganisms are able to depolymerise the polymer very effectively, but cannot metabolise the intermediates. Under these conditions it was possible to generate high concentrations of degradation intermediates in the medium, allowing an accurate analysis.

After three weeks of incubation no residual polymer nor non-soluble intermediates could be detected. From the lower detection limit of the GPC method used, a 99.9% depolymerisation of the material was estimated. From equation (10.1) it can be calculated that in Ecoflex, a material based on a BTA45:55 copolyester, about 20 mol% of the terephthalic acid forms sequences of more than 2 aromatic repeating units, corresponding to about 10% (w/w) of material consisting of aromatic sequences which earlier were shown to be water insoluble and hardly biodegradable. Thus, it could be concluded that also these long aromatic intermediates were subject to degradation. In the medium of the degradation experiment the monomers (1,4-butanediol, terephthalate and adipate) could be detected, besides some short aliphatic and aromatic mono-, di- and tri-esters. However, adding a mixed population to the monomeric and oligomeric intermediates, a complete metabolisation could be observed within 14 days.

The different degradation behaviour of the synthetic aromatic model - oligomers and the aromatic oligomeric intermediates which are really formed during the polyester

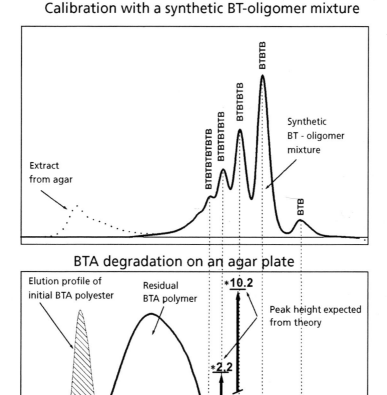

Calibration with a synthetic BT-oligomer mixture

Figure 10.7 GPC chromatogram of a BTA 39:61 copolyester after degradation for 11 weeks at 55 °C on a mineral medium agar plate, inoculated with a pre-screened microbial population. The upper diagram represents the calibration with a synthetic oligomer mixture of known composition. The molar mass distribution of the degraded sample (solid line) changed to lower masses during degradation compared with the initial molar mass distribution (hatched area). Two aromatic oligomers formed during degradation could by identified (BTBTB and BTBTBTB). Their theoretically expected concentrations according to equation (10.1) are marked with arrows and the factor of the theoretical concentration divided by the measured concentration is given above the arrows [65]

depolymerisation - may have various reasons. The lack in biological accessibility seems to be determined predominantly by intramolecular interactions. The aromatic intermediates formed during polymer degradation are embedded in a different environment than aromatic oligomers in the model substances. Furthermore, the model esters were mostly OH-terminated, while during polyester hydrolysis probably also COOH-endings of the oligomers will occur. Both effects can cause the final hydrolysis and final metabolisation of the aromatic copolyesters observed in the experiments with the individual strain.

In the same work it was also tried to estimate the environmental effect of a BTA-copolyester when treated in a composting plant and the resulting compost is used for agriculture. Based on toxicological tests with *Photobacterium phosphoreum* and *Daphnia magna* a risk assessment was calculated. For both test organisms no toxic effect of the intermediates produced during the degradation were detected. Quite conservative assumptions were made that the compost is loaded with 1% (w/w) polymer and that the entire polymer is depolymerised and transferred into oligomers and monomers, but these intermediates are not metabolised and remain in the compost material (actually it has been shown that the intermediates are rapidly metabolised by a mixed culture of microorganisms). After a weight reduction of the biowaste during composting of 50% (and thus an increase in concentration of the intermediates by a factor of two) 30 tons of compost per hectare are applied to the fields within three years (maximum value recommended by the German biowaste directive) and are supposed be ploughed 30 cm deep into the soil. The concentration of the degradation intermediates was then calculated to be 130 ppm. Toxic effects could be excluded up to a concentration of approximately 1400 ppm in the toxicity tests and thus, it can be expected that no toxic effect will result from the application of compost from the copolyester treatment.

The structure and morphology of BTA - copolyesters, including the crystal structure, were recently investigated by Kuwabara and co-workers [66] and Herrera and co-workers [67].

10.4 Commercial Products and Characteristic Material Data

In 2001 four materials based on aliphatic-aromatic copolyesters were advertised from companies as biodegradable plastics (some characteristic data are compiled in **Table 10.3**).

Trade name	Ecoflex F BX 7011	Eastar Bio GP	Biomax	EnPol G8000 [c]
Producer	BASF AG, Germany	Eastman Chemical Company, USA	DuPont Polyester Resins & Intermediates, USA	IRE Chemicals Ltd., Korea
Chemical basis	Modified copolyesters from 1,4-butanediol, adipic acid, terephthalic acid	Modified copolyesters from 1,4-butanediol, adipic acid, terephthalic acid	Copolyester based on PET with aliphatic dicarboxylic acids	Copolyester from terephthalic acid, adipic acid, succinic acid, 1,4-butanediol and/or 1,2-ethanediol
Density	1.25-1.27 g/cm^3 (ISO 1183 [68])	1.22 g/cm^3 (ASTM D1505 [69])	1.35 g/cm^3	1.25 g/cm^3 (ASTM D792 [70])
T_m	110-115 °C (DSC)	108 °C (ASTM D3418 [71])	200 °C	95-100 °C (ISO 11357-3 [72])
T_g		–33 °C (ASTM D3418 [71])		
MFI	3-6 cm^3/10 min (190 °C/2.16 kg) (ISO 1133 [73])	28 g/10 min (190 °C/ 2.16 kg) ASTM D1238 [74]		3-6 g/10 min (190 °C, 2.16 kg) (ASTM D1238 [74])
Tensile strength [e]	36/32 N/mm^2 [a] (ISO 527 [75-77])	22/20 MPa [b] (ASTM D882 [78])	15-50 MPa	>440/>320 kg/cm^2 (ASTM D638 [79])
Elongation at break [e]	580/820% [a] (ISO 527 [75-77])	700/730% [b] (ASTM D882 [78])	40-500%	>300/>700 % (ASTM D638 [79])
Modulus [e]		107/106 MPa [b] (tangent) (ASTM D882 [78])	60-2100 MPa	
Shore D hardness	32 (ISO 868 [80])			
Vicat VST A/50	80 °C (ISO 306 [81])			

Table 10.3 Compilation of some characteristic material data of commercial biodegradable aliphatic-aromatic copolyesters

Table 10.3 Continued...			
Trade name	Ecoflex F BX 7011	Eastar Bio GP	Biomax	EnPol G8000 [c]
Transparency	82% [a] (ASTM D1003 [82])	89% (ASTM D1003 [82])		
Oxygen permeation	1600 cm^3/(m^2 d bar) [a] (DIN 53380 [83])	80 cm^3 mm/(d atm) (ASTM D3985 [84])		
Water vapour permeation	140 g/(m^2 d) [a] (DIN 53122 [85])	280 g/(m^2 d) (ASTM E96D [86])		

(a) 50 µm film; elongation vertical to extrusion direction/elongation parallel to extrusion direction
(b) 37 µm film
(c) different grades available
(d) processing direction/vertical to processing direction
DSC: Differential scanning calorimetry
MFI: Melt flow index

10.4.1 Ecoflex

Producer/patents: BASF AG, Germany [87, 88]

Ecoflex is based on a copolyester from terephthalic acid, adipic acid and 1,4-butanediol. The content of terephthalic acid in the polymer is approximately 42-45 mol% (with regard to the dicarboxylic monomers). Modifications of the basic copolyester lead to a flexible material which is especially suitable for film applications. A down gauging to 10 µm films can be achieved [89] according to the producer.

The following (potential) applications for Ecoflex are announced by BASF AG:

- Biowaste bags

- Films for horticulture

- Films for agriculture

- Films for hygiene products

- Films for household applications

The biodegradation of Ecoflex was tested under composting conditions. After 100 days in a compost environment more than 90% of the carbon in the polymer was converted to carbon dioxide [90]. In a detailed investigation metabolisation of more than 99% could be proved for Ecoflex using a thermophilic actinomycete strain [56]. From these tests it can also be concluded that aromatic oligomers are subject to biodegradation under conditions present in a composting process. Ecotoxicological tests with *Photobacterium phosphorum* and *Daphnia magna* revealed no toxic effects of degradation intermediates. A risk assessment resulted in the statement, that no toxic effects can be expected from composting the copolyester (see Section 10.3.3.6).

Ecoflex meets the requirements of DIN V 54000 [31] as compostable material and is certified by DIN CERTCO (Ecoflex: max 120 μm films, Reg. No. 7W0011; Ecoflex CL: max 15% cellulose, max. 120 μm films Reg. No. 7W0020; Ecoflex TK: max. 45% talc, max. 120 μm films, Reg. No. 7W0019).

10.4.2 Eastar Bio

Producer/patents: Eastman Chemical Company, USA [91]

Like Ecoflex, the Eastman product is based on a copolyester composed of terephthalic acid, adipic acid and 1,4-butanediol, but due to some special modification the material properties are different.

The following (potential) applications are announced by Eastman:

- Disposable products

- Fast-food cups and containers

- Food packaging

- Food-contact applications

- Bio waste bags

- Agricultural film

- Mulch film

- Lawn and garden bags

- Seed mats

- Slow release agent

- Diapers

- Sanitary Napkins

- Body bags

Degradation of Eastar Bio was tested under composting conditions using ^{14}C labelled material [92]. After 210 days of composting (according to ASTM D5338 [93]), about 80% of the polymer-carbon was released as carbon dioxide. Taking into account the carbon of the biomass, a degradation in terms of metabolisation of more than 90% could be demonstrated in another study [94].

Eastar Bio meets the requirements of DIN V 54900 [31] as compostable material and is certified by DIN CERTCO (Eastar Bio 14766, max 127 µm films, Reg. No. 7W0022). The behaviour of carbon-black-filled mulching films of Eastar BIO 14766 under outdoor weathering conditions were investigated by Tocchetto and coworkers [95].

10.4.3 Biomax

Producer/patents: DuPont Polyester Resins & Intermediates, USA [96-99]

Biomax is, according to the producer, a standard PET with addition of special monomers to allow degradation to take place. Comparable to PLA, the degradation mechanism is described as an initial attack of water to the special monomers which are sensitive to hydrolysis. Oligomers formed by this first abiotic degradation step can be transported into microbial cells and there be metabolised (Biomax is claimed to be (hydro/ biodegradable).

However, although it seems that Biomax sufficiently disintegrates under composting conditions [100], the final metabolisation of the material in a reasonable time scale is still under discussion. The producer itself admits that Biomax in the current formulation (June 2000) does not degrade fast enough to meet the accepted standards and thus, will improve the material with regard to its biodegradability [101].

The following applications are mentioned by the producer:

- disposable cutlery

- paper coating

- thermoformable cups and trays

- films

The producers stress the superior barrier properties of the polyester compared to other biodegradable materials.

10.4.4 EnPol

Producer: IRe Chemicals, Korea [102, 103]

Based on a group of aliphatic copolyesters composed of adipic acid, succinic acid, 1,2-ethanediol and/or 1,4-butanediol [104] IRe Chemicals in Korea produces an aliphatic-aromatic copolyester where a part of the aliphatic dicarboxylic acids is substituted by terephthalic acid (G8000 grades). The producer states that EnPol polymers meet the specifications of the FDA for food contact and the USP specifications for medical device application.

The biodegradation of EnPol polymers was tested in a controlled laboratory composting test (according to ISO 14855 [105]). Within 45 days a carbon dioxide evolution of more than 90% of the carbon present in the copolyester was detected [106]

The following applications are mentioned by the producer:

- agricultural films

- shrinkable films

- plastic bags

- air-cushion films

10.4.5 Characteristic Material Data

From the data sheets provided by the producers some characteristic material parameters are compiled in **Table 10.3.**

References

1. Y. Doi, *Microbial Polyesters*, VCH Publishers, New York, NY, USA, 1990.

2. Y. Tokiwa and T. Suzuki, *Nature*, 1977, **270**, 76.

3. Y. Tokiwa, T. Ando, T. Suzuki and T. Takeda, *Polymer Materials Engineering and Science*, 1990, **62**, 988.

4. T. Tokiwa, T. Ando, T. Suzuki and T. Takeda in *Agricultural and Synthetic Polymers, Biodegradability and Utilisation*, Eds., J.E. Glass and G. Swift, ACS Symposium Series No.433, ACS, Washington, DC, USA, 1990, 136.

5. U. Witt, R-J. Müller, J. Augusta, H. Widdecke and W-D. Deckwer, *Macromolecular Chemistry and Physics*, 1994, **195**, 2, 793.

6. C. Bastioli, *Polymer Degradation and Stability*, 1998, **59**, 1-3, 263.

7. D. Kint and S. Munoz-Guerra, *Polymer International*, 1999, **48**, 5, 346.

8. Y. Tokiwa and T. Suzuki, *Journal of Applied Polymer Science*, 1981, **26**, 2, 441.

9. H.S. Jun, B.O. Kim, Y.C. Kim, H.N. Chang and S.I. Woo, *Journal of Environmental Polymer Degradation*, 1994, **2**, 1, 9.

10. H.S. Jun, B.O. Kim, Y.C. Kim, H.N. Chang and S.I. Woo, *Studies in Polymer Science*, 1994, **12**, 498.

11. Y. Tokiwa, T. Ando, T. Suzuki and T. Takeda, *Polymer Materials Engineering and Science*, 1990, **62**, 988.

12. U.Witt, R-J. Müller and W-D. Deckwer, *Journal of Macromolecular Science A*, 1995, **32**, 4, 851.

13. U. Witt, R-J. Müller and W-D. Deckwer, *Journal of Environmental Polymer Degradation*, 1995, **3**, 4, 215.

14. U. Witt, R-J. Müller and W-D. Deckwer, *Journal of Environmental Polymer Degradation*, 1997, **5**, 2, 81.

15. D.K. Gilding and A.M. Reed, *Polymer*, 1979, **20**, 12, 454.

16. A.M. Reed and D.K. Gilding, *Polymer*, 1981, **22**, 4, 499.

17. P. Sharma and B. Gordon, *Polymer Preprints*, 1989, **30**, 2, 197.

18. B. Gordon, P.P. Sharma and S.L. Hansen, *Polymer Preprints*, 1990, **31**, 1, 507.

19. S. Heidary and B. Gordon, *Polymer Materials Engineering and Science*, 1992, **67**, 190.

20. A. Niekraszewiecz, *Polymeri (Warswa)*, 1993, **38**, 8-9, 399.

21. S. Heidary and B. Gordon, *Journal of Environmental Polymer Degradation*, 1994, **2**, 1, 19.

22. T. Kiyotsukuri, T. Masuda, N. Tsutsumi, W. Sakai and M. Nagata, *Polymer*, 1995, **36**, 13, 2629.

23. D-K. Kim, Y.S. Shin, Y-T. Yoo and J-R. Huh, *Polymer (Korea)*, 1996, **20**, 3, 431.

24. M. Nagata, T. Kiyotsukuri, S. Minami, N. Tsutsumi and W. Sakai, *European Polymer Journal*, 1997, **33**, 10-12, 1701.

25. N. Valiente, T. Lalot, M. Brigodiot and E. Maréchal, *Polymer Degradation and Stability*, 1998, **61**, 3, 409.

26. M. Nagata, T. Machida, W. Sakai and N. Tsutsumi, *Journal of Polymer Science, Part A: Polymer Chemistry*, 1999, **37**, 13, 2005.

27. Y.J. Kim and O.O. Park, *Journal of Environmental Polymer Degradation*, 1999, **7**, 1, 53.

28. H.C. Ki and O.O Park, *Polymer*, 2001, **42**, 5, 1849.

29. M. Nagata, H. Goto, W. Sakai, N. Tsutsumi and H. Yamane, *Sen'i Gakkaishi*, 2001, **57**, 6, 178. (in Japanese)

30. C.C. Lefèvre, D. Villers, M.H.J. Koch and C. David, *Polymer*, 2001, **42**, 21, 8769.

31. DIN V 54900, *Testing of the Compostability of Plastics*, 1998.

32. T. Yoshioka, T. Sato and A. Okuwaki, *Journal of Applied Polymer Science*, 1994, **52**, 9, 1353.

33. M.E. Cagiao, F.J.B. Calleja, C. Vanderdonckt and H.G. Zachmann, *Polymer*, 1993, **34**, 10, 2024.

34. S.L. Greene and S.C. Nicastro, *Polymer Preprints*, 1992, **33**, 2, 298.

35. N.S. Allen, M. Edge, M. Mohammadian and K. Jones, *Polymer Degradation and Stability*, 1993, **41**, 2, 191.

36. M. Edge, M. Hayes, M. Mohammadian, N.S. Allen, T.S. Jewitt, K. Brems and K. Jones, *Polymer Degradation and Stability*, 1991, **32**, 2, 131.

37. N.S. Allen, M. Edge, M. Mohammadian and K. Jones, *Polymer Degradation and Stability*, 1994, **43**, 2, 229.

38. M. Mohammadian, N.S. Allen, M. Edge and K. Jones, *Textile Research Journal*, 1991, **61**,11, 690.

39. T.E. Rudakova, G.E. Zaikov, O.S. Voronkova, T.T. Daurova and S.M. Degtyareva, *Journal of Polymer Science - Polymer Symposia*, 1979, **66**, 277.

40. D.F. Williams, R. Smith, and R. Oliver in *Biological and Biomechanical Performance of Biomaterials*, Eds., P. Christel, A. Meunier and A.J.C. Lee, Advances in Biomaterials, Volume 6, Elsevier, Oxford, UK, 1986, p.239.

41. C. Lefevre, C. Mathieu, A. Tidjani, I. Dupret, C. Vander Wauven, W. De Winter and C. David, *Polymer Degradation and Stability*, 1999, **64**, 1, 9.

42. E. Chiellini, A. Corti, A. Giovannini, P. Narducci, A.M. Paparella and R. Solaro, *Journal of Environmental Polymer Degradation*, 1996, **4**, 1, 37.

43. K. Oda and Y. Kimura, inventors; Kyoto Institute of Technology, assignee; DE19935156A1, 2000.

44. S.J. Huang and C.A. Byrne, *Journal of Applied Polymer Science*, 1980, **25**, 9, 1951.

45. F. Kawai, *Journal of Environmental Polymer Degradation*, 1996, **4**, 1, 21.

46. E. Marten, Dissertation, 2000, Technical University Braunschweig, Germany, http://www.opus.tu-bs.de/opus/volltexte/2000/136.

47. R-J. Müller, U. Witt, E. Rantze and W-D. Deckwer, *Polymer Degradation and Stability*, 1998, **59**, 1-3, 203.

48. H-W. Hässlin, M. Dröscher and G. Wegner, *Makromolekulare Chemie*,1980, **181**, 2, 301.

49. ASTM D6002, *Guide for Assessing the Compostability of Environmentally Degradable Plastics*, 1996.

50. EN 13432, *Packaging - Requirements for Packaging Recoverable Through Composting and Biodegradation - Test Scheme and Evaluation Criteria for the Final Acceptance of Packaging*, 2000.

51. *Ecoflex – a Biodegradable Plastic from BASF*, Press release for K98, BASF AG, Düsseldorf, Germany, 17.03.98.

52. I. Kleeberg, C. Hetz, R.M. Kroppenstedt, R-J. Müller and W-D. Deckwer, *Applied Environmental Microbiology*, 1998, **64**, 5, 1731.

53. I. Kleeberg, Dissertation, 1999, Technical University Braunschweig, Germany, Internet: http://opus.tu-bs.de/opus/volltexte/2000/90

54. C. Perez, K. Juarez, E. Garcia-Castells, G. Soberon and L. Servin-Gonzales, *Gene*, 1993, **123**, 1, 109.

55. T. Walter, J. Augusta, R-J. Müller, H. Widdecke and J. Klein, *Enzyme and Microbial Technology*, 1995, **17**, 3, 218.

56. U. Witt, T. Einig, M. Yamamoto, I. Kleeberg, W-D. Deckwer and R-J. Müller, *Chemosphere*, 2001, **44**, 2, 289.

57. S. Basei, *Influence of Starch on the Aerobic and Anaerobic Biodegradation of an Aliphatic and an Aliphatic-Aromatic Polyester*, , University of Venice, Italy, 1999. [Graduation thesis]

58. M. van de Zee, *Structure-Biodegradability Relatiosnhips of Polymeric Materials*, University of Twente, The Netherlands, 1997. [Dissertation]

59. U. Pagga, A. Schäfer, R-J. Müller and M. Pantke, *Chemosphere*, 2001, **42**, 3, 319.

60. D-M. Abou-Zeid, R-J. Müller, W-D. Deckwer, *Journal of Biotechnology*, 2001, 86, 2, 113.

61. D-M. Abou-Zeid, Dissertation, 2000, Technical University Braunschweig, Germany, Internet: http://opus.tu-bs.de/opus/volltexte/2001.

62. E. Rantze, I. Kleeberg, U. Witt, R-J. Müller and W-D. Deckwer, *Macromolekulare Symposia*, 1998, **130**, 319.

63. U. Witt, R-J. Müller and W-D. Deckwer, *Journal of Environmental Polymer Degradation*, 1996, **4**, 1, 9.

64. U. Witt, Synthese, *Charakterisierung und Beurteilung der Biologischen Abbaubarkeit von anwendungsorientierten Biologische Abbaubaren Aliphatisch-Aromatischen Copolestern*, Technical University Braunschweig, Germany, 1996. [Dissertation]

65. R-J. Müller, U. Witt and W-D. Deckwer, *Fett/Lipid*, 1997, 99, **2**, 40.

66. K. Kuwabara, Z. Gan, T. Nakamura, H. Abe and Y. Doi, *Biomacromolecules*, 2002, **3**, 2, 390.

67. R. Herrera, L. Franco, A. Rodríguez-Galan and J. Puiggali, *Journal of Polymer Science, Part A: Polymer Chemistry*, 2002, **40**, 23, 4141.

68. ISO 1183, *Plastics - Methods for Determining the Density and Relative Density of Non-Cellular Plastics*, 1987.

69. ASTM D1505-98e1, *Standard Test Method for Density of Plastics by the Density-Gradient Technique*, 1998.

70. ASTM D792, *Standard Test Methods for Density and Specific Gravity (Relative Density) of Plastics by Displacement*, 2000.

71. ASTM D3418, *Standard Test Method for Transition Temperatures of Polymers By Differential Scanning Calorimetry*, 1999.

72. ISO 11357-3, *Plastics - Differential Scanning Calorimetry (DSC) - Determination of Temperature and Enthalpy of Melting and Crystallization*, 1999.

73. ISO 1133, *Plastics - Determination of the Melt Mass-Flow Rate (MFR) and the Melt Volume-Flow Rate (MVR) of Thermoplastics*, 1997.

74. ASTM D1238-01e1, *Standard Test Method for Melt Flow Rates of Thermoplastics by Extrusion Plastometer*, 2001.

75. ISO 527-1, *Plastics - Determination of Tensile Properties - General Principles*, 1994.

76. ISO 527-2, *Plastics - Determination of Tensile Properties - Test Conditions for Moulding and Extrusion Plastics*, 1994.

77. ISO 527-3, *Plastics - Determination of Tensile Properties - Part 3: Test Conditions for Films and Sheets*, 2001.

78. ASTM D882, *Standard Test Method for Tensile Properties of Thin Plastic Sheeting*, 2002

79. ASTM D638-02a, *Standard Test Method for Tensile Properties of Plastics*, 2002.

80. ISO 868, *Plastics and Ebonite - Determination of Indentation Hardness by Means of a Durometer (Shore Hardness)*, 1985.

81. DIN EN ISO 306, *Thermoplastic Materials - Determination of Vicat Softening Temperature (VST)*, 1997.

82. ASTM D1003, *Standard Test Method for Haze and Luminous Transmittance of Transparent Plastics*, 2000.

83. DIN 53380-1, *Determining the Gas Transmission Rate of Plastic Film by the Volumetric Method*, 2001.

84. ASTM D3985, *Standard Test Method for Oxygen Gas Transmission Rate Through Plastic Film and Sheeting Using a Coulometric Sensor*, 2002.

85. DIN 53122, *Determination of Water Vapour Transmission Rate of Plastic Film, Rubber Sheeting, Paper, Board and other Sheet Materials by Gravimetry*, 2001.

86. ASTM E96-00e1, *Standard Test Methods for Water Vapor Transmission of Materials*, 2000.

87. V. Warzelhan, G. Pipper, U. Seeliger, P. Bauer, U. Pagga and M. Yamamoto, inventors; BASF AG, assignee; WO 9625446, 1996

88. V. Warzelhan, G. Pipper, U. Seeliger, P. Bauer, D.B. Beimborn and M. Yamamoto, inventors; BASF AG, assignee; WO 9625448, 1996.

89. Preliminary Product Information: *Ecoflex F BX 7011 – The Biodegradable Polyester from BASF for Compostable Film*, KSS/BP-F 206, , BASF AG, Ludwigshafen, Germany, 2.11.2000.

90. U. Witt, M. Yamamoto, U. Seeliger, R-J. Müller and V. Warzelhan, *Angewandte Chemie International Edition*, 1999, 38, 10, 1438.

91. M. Ohta, S. Obuchi and Y. Yoshida, inventors; Mitsui Toatsu Chemicals, Inc., assignee; US 5,444,143, 1995.

92. J. Berting, *Proceedings of the 4th Conference on Biogically Degradable Materials*, 1997, Festung Marienberg, Würzburg, Germany, Presentation No.H.

93. ASTM D5338, *Test Method for Determining Aerobic Biodegradation of Plastic Materials under Controlled Composting Conditions*, 1998.

94. J. Sami, *Proceedings of the 4th Conference on Biogically Degradable Materials*, 1999, Festung Marienberg, Würzburg, Germany, Presentation No.D.

95. R.S. Tocchetto, R.S. Benson and M. Dever, *Journal of Polymers and the Environment*, 2001, 9, 2, 57.

96. F.G. Gallagher, D.S. Gray, C.J. Hamilton, R.F. Tietz and F.T. Wallenberger, inventors; E.I. DuPont De Nemours and Company, assignee, WO 9514740, 1995.

97. F.G. Gallagher, C.J. Hamilton, S.M. Hansen, H. Shin, Hyunkook and R.F. Tietz, inventors; E.I. DuPont de Nemours and Company, assignee; US 5,171,309, 1992.

98. F.G. Gallagher, C.J. Hamilton, S.M. Hansen, H. Shin, Hyunkook and R.F. Tietz, inventors; E.I. DuPont de Nemours and Company, assignee; US 5,171,308, 1992.

99. R.F. Tietz, inventor; E.I. DuPont de Nemours and Company, assignee; US 5,097,005, 1992.

100. *Product Information on Biomax*, DuPont Polyester Resins & Intermediates, USA, March 2001.

101. K. Atwood, *Proceedings of the Biodegradable Plastics 2000 Conference*, 2000, Frankfurt, Germany.

102. H.S. Chung, J.W. Lee, D.H. Kim, J.N. Jun, S.W. Lee, inventors; Ire Chemical Ltd., assignee; EP 1106640A2, 2001

103. H.S. Chung, J.W. Lee, D.H. Kim, J.N. Jun, S.W. Lee, inventors; Ire Chemical Ltd., assignee; EP 1108737A2, 2001.

104. H.S. Chung, J.W. Lee, D.H. Kim, J.N. Jun, S.W. Lee, inventors; Ire Chemical Ltd., assignee; WO 0011063, 2000.

105. ISO 14855, *Determination of the Ultimate Aerobic Biodegradability and Disintegration of Plastic Materials under Controlled Composting Conditions - Method by Analysis of Evolved Carbon Dioxide*, 1999.

106. *Product information on EnPol*, IRe Chemical Ltd., Korea, 01.08.2001.

11 Material Formed from Proteins

Stéphane Guilbert and Bernard Cuq

11.1 Introduction

Over the last two decades, there has been a renewal of interest in the development of recyclable, biodegradable and/or edible materials formed with raw materials of agricultural origin. These materials are often referred to as 'agromaterials' or 'biopackagings' (when used to make trays or packaging films). Natural biodegradable thermoplastic materials are commonly called 'bioplastics'. The 'agromaterial' concept generally involves the use of renewable raw materials that can be recycled after utilisation [1-3]. The main uses of agromaterials and bioplastics are reviewed in **Table 11.1**.

Proteins are natural polymers that have long been used empirically to produce edible packaging and materials (i.e., soybean lipoprotein sheets in Asia, collagen envelopes). At the beginning of the 20th century, proteins were considered interesting raw materials for making plastics to eventually replace cellulose. Formaldehyde crosslinking of milk casein (i.e., galalith) is a process that was invented as early as 1897 to make moulded objects such as buttons [4-5]. The first patents were taken out in the 1920s on the use of zein to formulate different materials (coatings, resins, textile fibres). At that time, formaldehyde was widely used in blends with soybean proteins and slaughterhouse blood to make automotive parts, especially distributor caps [6]. In addition, gelatin was used to produce films for foods, drug capsules and photography. Protein materials were then used for many applications (**Table 11.2**). In the 1960s, synthetic plastics posed a serious threat to proteins for many of these applications. The abandonment of protein materials (except for gelatin) lasted for the next 30 years.

Since the 1980s, the number of academic research programmes and industrial research and development projects on protein-based bioplastics have increased exponentially, as a result of the present interest in using some field crops for renewable and biodegradable materials for non-food applications, and also in order to explore the unique specific properties of proteins. The complexity of proteins and the diversity of their different fractions can be tapped to develop materials with original functional features that differ markedly from those of standard synthetic plastic materials. Apart from the previously mentioned proteins,

Table 11.1 Main applications of agromaterials and bioplastics (adapted from Guilbert [2])

Plastics to be composted or recycled

Food packaging (dried foods, short life cycle food, egg boxes, fresh or minimally processed fruits and vegetables, dairy products, organically grown products, etc.)
Paper or cardboard (windows from paper envelopes or for cardboard packaging, coating for paper or cardboard)
Hygienic disposable (nappies, sanitary napkins, sticks for cotton swabs, razors, toothbrushes, etc.)
Miscellaneous short life goods (pens, toys, gadgets, keys holders, etc.)
Dishes and cutlery (trays, spoons, cups, etc.)
Loose-fill packaging (shock absorbers)
Waste and carrier bags (compost bags)
Blister packaging

Plastics used in natural environment (no recovery)

Biodegradable/soluble/controlled release materials for agriculture and fisheries (mulching plastic, films for banana culture, twine, flower pots, materials for controlled release fertilisers or agrochemical, high water retention materials for planting, soluble sachets, biodegradable containers for fertilisers or agrochemical, fishing lines and nets)
Civil engineering, car industry and construction materials (heat insulators, noise insulators, car interior door casings, retaining walls or bags for mountain areas or sea, protective sheets and nets for tree planting)
Disposable leisure goods (golf tees, goods for marine or mountain sports)

Specialty plastics

Edible films and coatings (barrier internal layers, surface coatings, 'active' superficial layers, soluble sachets for instant dry food and beverages or food additives)
Matrix for controlled release systems (slow release of fertilisers, agrochemical, pharmaceuticals, food additives)
O_2 *barrier, selective* CO_2/O_2 *barrier*, aroma barrier (simple or multi-layer packaging)
Medical goods (bone fixation, suture threads, films, non-woven tissues, etc.)
Super-absorbent materials (material for plant planting in desert, nappies, etc.)
Adhesives (glue)

many other proteins (wheat and maize gluten, cottonseed flour, whey proteins, myofibrillar proteins, etc.), can be used as raw materials to produce films, moulded materials, and various hollow items. Materials formed from proteins are biodegradable and even edible when food-grade additives are used. In addition, they are generally biocompatible except for some traits associated with specific proteins, (e.g., gliadins in wheat gluten are allergenic), their processing, and the presence of impurities or additives.

Table 11.2 Main applications of protein-materials until the 1960s (adapted from Di Gioia [7])
Applications
Films, coatings, facings and adhesives: Waterproofing of paper and cardboard bags for food packaging Paper glazing for magazine covers and sleeves for long-playing records Edible coatings for pharmaceutical tablets Edible coating for food products (protection against water absorption or lipid oxidation) Photographic supports (or papers) Pigment binder for printing inks Adhesives for pasting (wallpapers), for sticking (labels on bottles), for wood veneers Adhesives for cork and chipboard, (e.g., 'hardened wood' based on egg white)
Textile fibres: 'Vicara' (corn zein-based fibres) Casein textiles
Moulded plastic items: Buttons, door handles, belt buckles, driving belts in car engines Jewellery (necklaces, earrings)

11.2 Structure of Material Proteins

Until recently, the research work on structure, properties and applications of proteins were mainly considered within the scientific field of Food Science. To reach a better understanding of properties and to define the potential applications of material proteins, it is essential to compare their structural features with those of chemically synthesised organic polymers used to produce plastic materials. Novel research on non-food uses of agricultural resources, and especially on 'material proteins', has led to the application of Polymer Science concepts and tools to investigate the structure-function relationships of these macromolecular organisations. This involves:

i) Investigating the three-dimensional structure of proteins at different levels (atomic, molecular and supra-molecular arrangements).

ii) Studying structural variations according to temperature and the presence of functional additives.

iii) Simulating the macroscopic properties of macromolecular arrangements (mechanical, optical, thermal and electrical properties).

Proteins (except homopolymers or copolymers in which one or two monomers are repeated) are heteropolymers comprising more than 20 different amino acids, each with specific sequences and structures. The structure of the 20 natural amino acids, shown in **Figure 11.1**, highlights the high chemical variability conferred by the lateral groups. Amino acids are generally classified by groups that could interact via hydrogen bonds (non-ionised polar amino acids), ionic interactions (ionised polar amino acids), non-polar interactions (non-polar amino acids) or covalent bonds (disulfide or dityrosine bonds). Amino acids are also classified on the basis of their relative hydrophobicity (**Table 11.3**).

The molecular diversity means that proteins have considerable potential for the formation of various interactions and links that differ according to their position, nature and/or energy [9, 10]. This heterogeneous structure provides many reaction sites for potential crosslinking or chemical grafting - it even facilitates modification of the film-forming properties and end-product properties. The amino acid sequence formed by peptide bonds is called the primary structure.

The secondary structure concerns the spatial pattern of the peptide chain. This involves λ-helix structures and λ structures, i.e., a zigzag structure that is more stretched than the λ-helix. These stretched chains bind to form folded structures. The lateral group structure of some amino acids upsets these ordered patterns, giving some of these proteins a random coiled 'less ordered' structure. The tertiary structure corresponds to a three-dimensional polypeptide chain organisation containing organised and unorganised secondary structure zones. In a polar solvent medium, hydrophilic amino acids are distributed over the surface of the molecule, while non-polar amino acids tend to be located within the structure and give rise to hydrophobic interactions. The so-called quaternary structure is formed by

Table 11.3 Relative hydrophobicity (or polarity) of the different amino acids (adapted from Rothfus [8])					
Amino acid	Relative hydrophobocity	Polarity	Amino acid	Relative hydrophobocity	Polarity
Arg	+176	Polar	Tyr	-2	-
Lys	+110	Polar	Cys	-4	Non-Polar
Asp	+72	Polar	Gly	-16	Non
Gln	+69	Polar	Ala	-25	Non
Asn	+64	Polar	Met	-26	-
Glu	+62		Trp	-37	Non
His	+40		Leu	-53	Non
Ser	+26		Val	-54	Non
Thr	+18		Phe	-61	Non
Pro	+7		Ile	-73	Non-Polar

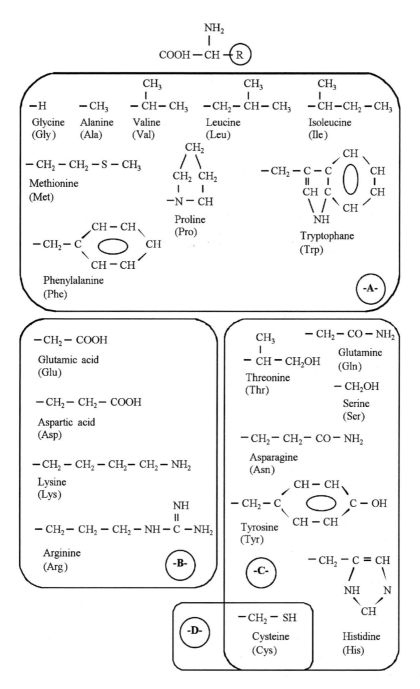

Figure 11.1 Biochemical structure of amino acids and of their different lateral chains. Classification: non-polar amino acids (-A-), ionised polar amino acids (-B-), non-ionised polar amino acids (-C-), amino acids able to form –SS– bond (-D-)

generally non-covalent associations of three-dimensional organised protein subunits that are sometimes identical.

Several different types of interactions help to stabilise the secondary, tertiary and quaternary structures of material proteins. Concerning low-energy interactions, van der Waals interactions (especially the London forces) have very little impact on protein arrangement structuring, as compared to synthetic polymers. As a comparison, **Table 11.4** gives the molar interaction energy of material proteins and synthetic organic polymers. Material proteins are mainly stabilised by ionic or hydrogen interactions. Lower overall values for proteins compared to synthetic polymers could be explained by their highly heterogeneous structure, thus reducing the frequency of ordered zones that promote such interactions. Hydrophobic interactions, which only take place in polar solvent solutions, should also be mentioned in connection with proteins. The energy of such interactions depends on the hydrophobic or hydrophilic amino acids and the type of solvent involved [12]. Material proteins often resemble elastomers due to the presence of disulfide bonds between cystein residues. These proteins are highly stabilised by hydrogen bonds, contrary to most elastomers that have very weak secondary bonds.

Material proteins could thus be defined as amorphous three-dimensional arrangements, stabilised by low-energy interactions, and that are eventually strengthened by covalent bonds (-SS- bonds). The thermomechanical behaviour of material proteins should be compared to the behaviour of thermoplastic compounds or thermoplastic elastomers. This means that they could be used to form materials by dissolution in a solvent film-forming solution, with subsequent spreading and drying, (i.e., the 'casting' process), or via thermal processes (extrusion, thermomoulding, injection, etc.). The functional properties of these material proteins depends on their structural heterogeneity, heat sensitivity and hydrophilic characteristics.

In organic polymers, macromolecules can form regular 'crystal network' type arrangements. These arrangements have a marked effect on the properties of polymers, especially their mechanical strength. For proteins, α-helix or β-sheet secondary structures are highly

Table 11.4 Comparison of molar interaction energy of material proteins and synthetic organic polymers (adapted from Oudet [11] and Phillips and co-workers [12]		
Interaction Type	Energy of interaction (kJ/mol)	
	Proteins	Synthetic polymers
Van der Waals	0.1 - 0.3	2 - 17
Hydrogen bonds	8.4 - 42	≈ 40
Ionic interactions	21 - 84	160 - 560

stabilised by hydrogen bonds and can resemble crystalline zones (**Figure 11.2**). Rothfus [8] demonstrated that the presence of β-sheets determines a cereal's potential use as an adhesive, coating or textile fibre. For example, X-ray studies revealed that the stretching of protein fibres can lead to the formation of crystalline structures, thus enhancing their mechanical resistance.

Protein molecular weights (MW) have a substantial effect on protein network structure. They also determine the presence of molecular entanglements, leading to the formation of physical nodes. As is the case for synthetic macromolecules, entanglements could occur beyond a critical molecular weight (M_c) of around 10^4 g/mol, and thus the material properties would be stable (**Figure 11.3**). **Table 11.5** compares the mean molecular weights of synthetic polymers and proteins commonly used in material production. A high mean molecular weight also hampers polymer flow during material formation, which can lead to defects in the end product. Synthetic polymers are characterised by heterogeneity of molecular weights, (i.e., polydispersity), that could be explained by the fact that chain growth is stopped randomly by the absence of residual monomers (polycondensation) or by termination reactions (polyaddition). On the other hand structure and molecular weight for proteins, are determined during synthesis (by using the genetic code) which means that

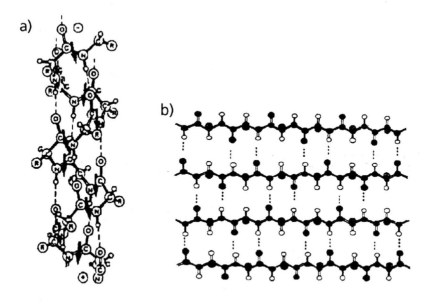

Figure 11.2 Structural organisation (α-helix and β-sheet of proteins. (a) Structure of α-helix (3.6 amino acid per turn, stabilised by H-bonds), (b) Structure of β-sheet (anti-parallel linear chains, stabilised by H-bonds) from Phillips and co-workers [12])

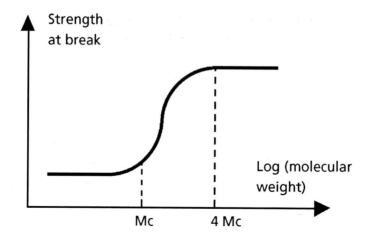

Figure 11.3 Influence of molecular weight on rupture strength for a olymeric material (adapted from Oudet [11])

Proteins	Nomenclature of main sub-units	Molecular Weight (kDa)	Ref
Corn gluten	α, β, γ, δ Zein A, B, C, D, E, F Glutelin	10 – 28 11 – 127	[13] [14]
Wheat gluten	α, γ, ω Gliadins Glutenins	30 – 80 200 – 2000	[15]
Soy proteins	Glycinin	363	[16]
Peanut proteins	Arachin	330	[16]
Cottonseed proteins	Albumins Globulins	10 – 25 113 – 180	[17]
Gelatin	–	3 – 200	[18]
Caseins	α_{s1}, α_{s2}, β, κ, γ Caseins	19 – 25	[15]
Myofibrillar proteins	Myosin Actin	16 – 200 42	[19]
Poly(methylmethacrylate)	–	100 – 200	[4]
Polyethylene	High density Low density	20 – 60 200 – 400	[4]

Table 11.5 Molecular weight and main sub-units of protein-materials used to form 'bioplastics' and comparison with some conventional synthetic plastics (adapted from Di Gioia [7])

for a given subunit there is very little variability. However for each protein family, there are generally several sub-units of different molecular weights, hence polydispersity. As for synthetic polymers, protein materials with high 'apparent' polydispersity are generally easier to process but have poorer mechanical properties.

11.3 Protein-Based Materials

Many plant and animal proteins have been considered as raw material for producing films and coatings, which are generally characterised by functional properties of great interest [3, 10, 10-27].

Corn zein has been the focus of considerable industrial interest, especially during the first half of the 20th century, in manufacturing films, lacquers, varnishes, adhesives, textile fibres and moulded plastic objects [28]. During World War 2, this protein was considered as a strategically important substance, being used as a substitute for shellac, which was in short supply [29]. Zein is one of the four proteins (along with milk casein, soy glycinin and peanut arachin) that have been used for fibre manufacturing. In the 1950s, this provided a major outlet for zein, with 1800 tonnes/year, sold to produce 'Vicara' fibres. A standard protein spinning process (developed especially for soy protein texturing) was used to produce zein fibres. The process involves an initial stage of protein solubilisation (in organic solvents or alkaline medium), spinning in an acidic coagulant bath, formaldehyde-induced hardening, and drying. Recently, soy protein and zein blends were studied to produce textile fibres with the aim of improving fibre properties and decreasing their manufacturing cost [30].

Zein is not water soluble because of its high non-polar amino acid content. Zein solutions are thus obtained by solubilisation in solvent (generally alcohol or a volatile organic solvent), or in alkaline medium, sometimes with a soap supplement. Zein water dispersions (zein 'latex') are now commercially available, (e.g., OptaGlaze from Opta Food Ingredients Inc., Bedford, MA, USA). Another technique involves direct hot press moulding of zein, after decreasing the glass transition temperature via the addition of plasticisers [31].

Many studies and patents have focused on the production of moulded or calendered and stamped zein-based plastics [32-34]. The manufacturing process generally involves hot mixing (40-90 °C) of zein with non-volatile plasticisers, water (up to 20% *w/w*) and other additives (crosslinking agents, pigments, extenders). The plasticised mass is hot press moulded (100-150 °C) for 1-2 minutes, or up to 15 minutes when formaldehyde is added (0.5-5% *w/w* as crosslinking agent. The mould is then cooled to around 90 °C and the piece is removed from the mould. When formaldehyde has been added, the piece can be left to harden for about 10 hours at atmospheric pressure and 60-90 °C. Zein has many amide functions that could form methylene bonds in presence of formaldehyde and at temperatures above 40 °C [35]. Free formaldehyde is thus inserted in the protein before moulding. Hardening is then achieved by heat treatment (> 40 °C) or acid catalysis [36].

Zein has been used to enhance the water resistance of starch-based plastics. It can be blended directly with starch (10-20% *w/w*) and the mixture plasticised and crosslinked by aldehydes or acid anhydrides [6]. A film-forming solution based on zein can also be spread on the surface of the starch objects [37]. In biodegradable packaging, it has also been shown that coating paper with zein gives heat-sealed products that are just as resistant as polyethylene-coated paper [38]. Beck and co-workers [39] recently conducted a study on film-coating of pharmaceutical drugs and demonstrated that zein-based coatings could be applied by a conventional extrusion coating technique. These coatings were also found to have mechanical and oxygen-barrier properties comparable or better than standard cellulose derivatives. Hot press moulding of plastics formed with corn gluten, (i.e., corn-starch industry co-product with around 60-70% protein content, 60% of which are zeins) was studied by Di Gioia and co-workers [40, 41] and Di Gioia and Guilbert [31]. The mechanical properties of corn gluten-based materials were found to be similar to those of polyvinyl chloride (PVC). Corn gluten materials are very inexpensive (about 0.5 €/kg).

The film-forming properties of corn zein have also been investigated in detail [42-47]. Zein films are water insoluble, relatively shiny and greaseproof, but they sometimes have unsuitable organoleptic properties (off-odour, chewiness, etc). Zein-based films and coatings to improve food shelf life, via their high barrier properties (to water vapour or especially to oxygen), or their retention or controlled-release properties of active additives on food surface. These films have been successfully used to protect dry fruits and various parts of frozen or intermediate moisture foods.

Many studies have focused on the film-forming properties of wheat gluten [44, 45, 48-55]. Wheat gluten films are generally formed by spreading and drying of protein hydro-alcohol solutions, in acidic or basic medium, usually in the presence of disruptive agents such as sulfite. Wheat gluten-based films have also been produced by skimming off the skin formed on the surface of heated protein solutions [56], and by wheat gluten extrusion, with or without the addition of disruptive agents [1, 57].

Wheat gluten-based films are water-resistant and their properties (thus their applications) are close to those of zein-based films. They have a more neutral taste and colour, but their use as an edible film or coating or as a packaging material in contact with a food product can be problematic for consumers with coeliac disease, (i.e., gluten intolerance). The film-forming properties of wheat gluten have been used particularly for encapsulating additives, enhancing the quality of cereal products and maintaining antioxidant and anti-microbial agents on the surface of food products [58]. The remarkable gas (O_2, CO_2) barrier properties of these materials, because of their exceptional selectivity, can be utilised to improve the shelf life of fresh or slightly processed vegetables [53, 59] (see Section 11.5).

The viscoelastic and flow properties of the plasticised malleable phases based on wheat gluten were investigated as a function of temperature, water content and time [57, 60, 61]. 'Plasticised gluten' resembles a structured viscoelastic solid with pseudo-plastic behaviour. The pseudo-plasticity index of plasticised gluten (m = 0.27-0.37) is comparable to that of plasticised starch (m = 0.32-0.37) and low-density polyethylene (LDPE; m = 0.4). The consistency (k = 18-47 kPa-s) is higher than that of LDPE (k = 9.7 kPa-s) but comparable to that of plasticised starch (k = 11-40.3 kPa-s). A study of rheological functions (G' and G'' moduli, complex viscosity) revealed that the time/temperature superposition can be applied. For given mixing conditions, the complex viscosity of plasticised gluten can be characterised by a power law function with temperature and plasticiser content as variables. Redl and co-workers [57, 60, 61] demonstrated that wheat gluten can form a homogeneous plasticised malleable phase under thermal [at temperatures above the glass transition (T_g)], mechanical (shear) and chemical (additives and degradation) treatments. Redl and co-workers [57, 61] carried out studies on the extrusion of gluten-based materials in a twin co-rotating screw extruder, with simulation of flow properties and extrusion conditions.

Due to the thermoplastic properties of wheat gluten and its high capability for chemical modifications, it is possible to adjust the extent of network crosslinking, the hydrophobicity of the network, (e.g., by using hydrophobic glutens obtained by lipophilisation treatment), and make it compatible with synthetic materials. This natural material could thus be developed for a wide range of non-food uses [1], e.g., window envelopes, paper coatings, biodegradable plastic films for agricultural applications, soluble bags for fertilisers, detergents or additives, moulded objects [1, 6, 62].

The film-forming properties of soy proteins are traditionally used in Asia to produce edible films. These traditional films, (i.e., 'yuba' in Japan) are obtained by skimming off the lipoprotein skin formed on the surface of heated soymilk [22, 63-66]. Proteins are the main components of these films, but significant quantities of polyosides (sucrose, raffinose and stachyose) and lipids (droplets trapped in the protein matrix) are also present. These films have good mechanical properties but are generally not very water-resistant. In addition, films have been formed from soy protein isolates dispersed in a hydro-alcohol solvent system [67-69] or by spreading and drying a thin layer of solution [43, 70-72]. Soy protein films are often proposed to improve the shelf life of many foods and for making soluble sachets. Biodegradable plastics have also been produced from soy protein isolates and concentrates by hot press-moulding techniques [6]. However, these materials are highly water sensitive unless a chemical crosslinking agent such as formaldehyde is used.

Peanut protein-based films and soluble sachets have been obtained by skimming off the skin on the surface of heated peanut milk, as described previously for soymilk [68, 69,

349

73]. Marquié and co-workers [74, 75] recently developed biodegradable and bioresorbable cottonseed protein-based films from a film-forming solution treated with different chemical crosslinking agents. A recent review on formulation and properties of cottonseed protein films and coating was proposed by Marquié and Guilbert [76].

Up until the 1960s, milk protein-based materials were used for making glossy record album covers, buttons and decorative items. Labels for some cheeses are still made with crosslinked casein. The film-forming properties of casein and whey proteins were investigated with the aim of developing edible films and coatings. Caseins dispersed in aqueous solutions can form transparent, flexible and neutral-flavoured films. Covalent bonds catalysed by peroxidases or transglutaminases have been proposed to improve the moisture resistance of casein materials and to immobilise active enzymes, (e.g., ß-galactosidase, α-mannosidase) [77, 78]. The film-forming properties of caseins have been utilised to improve the appearance of many foods, to make soluble sachets, quality labels for custom-cut cheeses, to maintain additives on the surface of intermediate moisture foods and to encapsulate polyunsaturated fats produced for livestock feed [43, 79-82]. The film-forming properties of whey proteins have been utilised to produce transparent, flexible, tasteless and odourless films [3, 83]. Mahmoud and Savello [84, 85] formed films by enzymic polymerisation of whey proteins using transglutaminases. Films have also been obtained by skimming off the skin formed on the surface of heated whey dispersions [68, 86]. Whey protein-based materials are stabilised by disulfide bonds and are therefore not water-soluble.

The film-forming properties of collagen are traditionally used in the meat industry for the extrusion of edible casings [87, 88]. Collagen-based materials are also used for medical applications [89]. Gelatin is conventionally used to produce transparent, flexible, and oxygen resistant and oxygen-proof films [43, 90, 91]. These films are formed after cooling and drying an aqueous gelatin solution. Film-forming applications of gelatin are common in the pharmaceutical industry, i.e., for producing pills and capsules (dry or soft). Gelatin is also a raw material for photographic films and micro-encapsulation of flavourings, vitamins and sweeteners [92]. In addition, studies were carried out to assess the use of gelatin films to protect frozen meats from oxidation [93], but the results showed a very limited protective effect unless an antioxidant was incorporated in the film.

Anker and co-workers [48] developed insoluble keratin-based films obtained by spreading and drying a thin layer of alkaline dispersions. The high cysteine content in keratin prompts the formation of many disulfide bonds that stabilise the protein network [69, 94]. However, consumer acceptance of keratin-based edible sachets for food products has been low [95].

The use of albumen proteins as a base for the encapsulation of hydrophobic organic compounds for cosmetic or food uses is the focus of several patents [96-98]. Application of albumen coatings can reduce raisin moisture loss in breakfast mixtures [99]. Albumen has also been used as an edible coating ingredient [100, 101]. Okamoto [69] reported the formation of films on the surface of heated albumen-based lipoprotein solutions, similar to the formation of soy films. The mechanical and water vapour properties of albumen-based films were studied by Gennadios and co-workers [101]. The materials are clearer and more transparent than wheat gluten-, soy protein- and corn zein-based materials. Albumen-based films could be used to produce soluble sachets to protect ingredients used in pharmaceutical, food and chemical industries.

Recent studies highlighted the film-forming properties of fish and meat myofibrillar proteins [102-109]. Films formed from an aqueous solution were found to be water-insoluble and completely transparent, with good mechanical and gas-barrier properties [53]. Their mechanical strength is close to that of polyethylene films. The thermoplastic features of myofibrillar proteins [110, 111] could also be tapped for industrial-scale production of these films using techniques commonly implemented to obtain synthetic thermoplastic polymers, (e.g., extrusion or thermoforming).

Table 11.6 gives a list of the main material proteins, along with the production techniques used for each of them. Other proteins have also been used for film-forming applications: rye, pea, barley, sorghum and rice proteins, silk fibroins, fish flesh proteins, and serum albumin [24, 69, 112, 113].

Table 11.6 Summary of the main proteins used as polymeric materials to form biopackagings (from [10])				
Proteins	Tested methods to obtain films			
	Film forming solution [a]	Collect the 'skin' [b]	Enzymic treatment [c]	Thermoplastic Extrusion
Corn zein	+			
Corn gluten	+	+		+
Wheat gluten	+	+		+
Soy proteins	+	+		
Peanut proteins				
Cottonseed protein				
Keratin	+			
Collagen	+	+	+	+
Gelatin	+	+	+	+
Caseins	+	+		
Whey proteins	+			
Egg albumin	+			
Myofibrillar proteins				
[a] Casting in thin layer and drying of a film-forming solution				
[b] Collect 'skin' formed after boiling protein solutions				
[c] Enzymic polymerisation				

11.4 Formation of Protein-Based Materials

Protein materials are obtained via the formation of a relatively organised, low hydrated and continuous macromolecular network. Interactions between proteins therefore have to be quite numerous and uniformly distributed. The probability of inter-protein links depends on the protein structure and denaturation conditions (solvent, pH and ionic strength, heat treatment, etc). High molecular weight proteins, (e.g., glutenins), and fibrous proteins, (e.g., collagen, glutenins) have attractive film-forming features [9]. Conversely, globular or pseudo-globular proteins, (e.g., gliadins, glycinin, caseins), have to be unfolded prior to network formation. At the present state of the knowledge, it would be unrealistic to try to predict the functional properties of material proteins on the basis of their primary structure [114]. Nevertheless, a good understanding of the main physicochemical characteristics of the raw materials is essential. The main protein raw material characteristics are summarised in **Table 11.7**.

Several steps are required to form a protein network:

i) rupture of low-energy intermolecular bonds stabilising systems in the native state,

ii) protein rearrangement, and

iii) the formation of a three-dimensional network stabilised by new interactions or bonds, after removal of the intermolecular bond scission agent. Two different technological strategies can be used to make protein-based materials: the 'wet process' or 'solvent process' involving a protein solution or dispersion, and the 'dry process' or 'thermoplastic process' using the thermoplastic properties of the proteins under low hydration conditions (**Figure 11.4**).

11.4.1 'Solvent Process'

The formation of materials by coacervation of a protein solution or dispersion, (i.e., the 'solvent process'), has been widely studied [20, 21, 120]. This process (which is fully controlled on a laboratory scale) often involves spreading a thin layer of protein solution, which is why this is often called a 'casting' or 'continuous flow' process. Solubility of proteins, as defined by Osborne [119], seems to be highly variable (**Table 11.7**). There are no specific solubilisation conditions for casting of protein-based solutions. It is generally useful to know the nature of the different intermolecular interactions before attempting to solubilise proteins [121]. For example, due to the presence of intermolecular disulfide bonds in keratin, disruptive agents have to be added to obtain homogeneous solutions [118]. The low water solubility of wheat gluten is also attributed to the low content of ionised polar amino acids (14%), to the many hydrophobic interactions between non-polar

Table 11.7 Main physico-chemical characteristics of the proteins used as polymeric materials to form biopackagings (from [10])

Proteins	Ref.	Amino acid ratios [a]			Main sub-units			
		A	B	C	Name	W_R	MW	S [b]
Corn zein	[15]	36	10	47	λ-Zein	80	21-25	IV
Wheat gluten	[15]	39	14	40	Gliadin	40	30-80	IV
					Glutenin	46	200-2000	III
Soy proteins	[16]	31	25	36	β-Conglycinin	35	185	II
					Glycinin	40	363	II
Peanut protein	[16]	30	27	32	Arachin	75	330	II
Cottonseed proteins	[17]	41	23	32	Albumin	30	10-25	I
					Globulin	60	113-180	II
Keratin	[115]	34	11	42	--	--	10	III
Collagen	[116]	13	13	40	Tropocollagen	--	300	III
Gelatin (A)	[18]	12	14	41	--	--	3-200	III
Caseins	[15]	31	20	44	α_{S1}, α_{S2}, β, κ, γ	--	19-25	[c]
Whey protein	[117,118]	30	26	40	β-lactoglobulin	60	18	I
					α-lactalbumin	20	14	I
Myofibril								
- sardine	[19]	27	31	35	Myosin	50	16-200	II
- beef meat	[15]	27	27	39	Actin	20	42	II

MW: the molecular weight (kD)
W_R *is the rate of sub-unit weight (%) in raw materials*
50% amidation rates of aspartic and glutamic acids are supposed for peanut proteins, keratin, collagen, and gelatin
[a] *Amino acid ratios (mol/100 mol): A: non-ionised polar (Asn, Cys, Gln, His, Ser, Thr, Tyr), B: ionised polar (Arg, Asp, Glu, Lys) and C: non-polar (Ala, Ile, Leu, Met, Phe, Pro, Trp, Val)*
[b] *S is the solubility of proteins according to Osborne [119]: I - in water, II - in diluted salt solutions, III - in diluted acidic or basic solutions, and IV - in ethanol (80%) solutions*
[c] *Miscellaneous associations*

amino acids (39.6%) and to the presence of disulfide bonds [29]. The water insolubility of zein is also linked with the high non-polar amino acid content (46.6%) [14, 122].

The properties and physicochemical characteristics of proteins in an aqueous solvent system depend on the pH conditions. Many protein-based materials are sensitive to pH variations, which could be linked with the relatively high proportion of ionised polar amino acids in protein raw materials (**Table 11.7**). Zein and keratin materials can, for

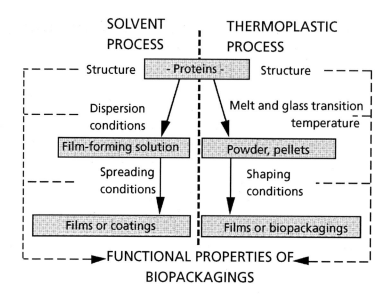

Figure 11.4 Schematic representation of the two technological processes used to form biopackagings based on proteins (adapted from Cuq and co-workers [10])

example, be produced within a broad pH range because these proteins (which have low ionised polar amino acid contents, 10% and 10.7%, respectively), are not very sensitive to pH variations [69]. Conversely, the high content of ionised polar amino acids in soy proteins (25.4%) limits film-forming applications to within a narrow pH range [123].

Coacervation of protein dispersions involves separation of the film-forming material in the solvent phase by precipitation or a phase change, through:

i) modifications of the solvent system (pH or polarity modification, addition of electrolytes),

ii) heat treatments, or

iii) solvent removal.

Films can be formed by solvent removal as a result of an increase in the polymer concentration in the solution, leading to molecule aggregation and formation of a three-dimensional network. Films that are obtained by skimming of the skin formed on the surface of heated protein milks, result from the polymerisation of heat-denatured proteins associated with solvent evaporation [22, 63]. Coacervation is said to be 'simple' when a single molecule is involved. This is the main process applied for producing protein-based materials. For 'complex' coacervation, at least two macromolecules of opposite charge are combined to obtain a blend of insoluble molecules. Associating proteins with chitosans could, for instance, be very appealing for material formation.

Solvent systems used for preparing film-forming solutions or dispersions are generally water- and/or ethanol-based, sometimes even acetone-based. Dispersion of molecules in solvent medium sometimes requires the addition of disruptive agents (mercaptoethanol, sodium sulfite, cysteine, sodium borohydride, N-ethylmaleimide), pH adjustment by the addition of acids (lactic, hydrochloric, acetic acids, etc.), or bases (ammonium, sodium, potassium, or triethylamine hydroxide, etc.), or controlling the ionic strength by adding electrolytes. The functional properties of materials formed by the solvent process depend on the production conditions: molecule concentration in the solution, pH, choice of additives, polarity of the solvent system, solution drying temperature and rate [49, 124, 125].

The film-forming solution can be directly applied on to food with a brush, by spraying, by coating using a falling film system, by immersion and then draining, by spinning or fluidisation, etc. In some cases, the food product is subjected to a second processing phase with a crosslinking solution that stabilises the film [126]. When a draining step is required, products are heated (to decrease the coating viscosity) using a vibrating grate, centrifugal drainage, or forced ventilation. The coating or film is then hardened by drying or cooling. Relatively quick film hardening is generally required for industrial reasons. It is still important to control the cooling temperature, or the drying conditions, so that the film does not harden too quickly, i.e., quick hardening can lead to an irregular coating that could tear or become wrinkled. Coating techniques generally require a high level of skill and experience. For direct coating, it is sometimes difficult to properly moisten the support, e.g., for protection of a food with a greasy surface (peanuts, etc.). In such cases, a surfactant can be applied to the support or incorporated in the film-forming system. Another solution is to pre-coat the product with a suitable material that will stick to each component.

Films can be preformed by the solvent process without the food support by spreading the solution on a smooth flat surface. This technique is applied by continuous feed on 'carpets' for industrial production of soluble films, especially with a wheat gluten base. Protein films are easier to unstick from some surfaces, depending on the surface material (metal, polyethylene, polycarbonate, Teflon, etc.) and on surface properties. Drying the film-forming solution on a drum dryer can also be used to make films. It is also essential to carefully prepare the support, (i.e., the food surface for coatings and the mould for films). Mould-release agents can be required when moulding a film on a support.

11.4.2 'Thermoplastic Process'

The thermoplasticity of material proteins has been utilised to produce materials by thermal or thermomechanical processes under low hydration conditions, as already used for starch- or polyolefin-based materials [111, 127]. According to the thermoplastic behaviour of synthetic polymers, the glass transition (T_g) of the proteins involves sudden variations in their physical properties (thermal, mechanical, dielectric properties, etc). The molecular response associated with the transition from the glassy to the rubbery state involves an overall increase in the free volume and in macromolecule mobility [128, 129]. As for synthetic polymers, the T_g of proteins is affected by the molecular weight, chain rigidity, size and polarity of the lateral groups, presence of intermolecular bonds or crystalline zones, and also by the plasticiser type and concentration [130, 131].

The T_g values of native proteins or materials developed from proteins are given in **Table 11.8**. Protein T_g values are obtained by differential scanning calorimetry (DSC) or dynamic mechanical thermal analysis (DMTA) [54]. They can be predicted [31] on the basis of the amino acid composition using the method described by Matveev [132].

The T_g of proteins is highly affected by moisture content (around 10 °C decrease for 1% added water) because of their hydrophilic nature, which varies between proteins. In practice, once proteins contain more than 15% water, (i.e., which generally occurs when they are in equilibrium with 85% relative humidity at ambient temperature), the T_g of protein material is close to the ambient temperature (**Figure 11.5**). This effect is even more obvious in the presence of plasticisers.

The effect of adding water (or another plasticiser) on the T_g can be described by the equation developed by Couchman and Karasz [146] when the molar fractions of components in the blend, the T_g values and heat capacity change at the T_g of the 'pure' components are known. Other empirical equations such as the equations of Gordon and Taylor [147] or Kwei [148] can also be applied. For example, **Figure 11.6** highlights the impact of relative humidity on the T_g and on the complex viscosity of a wheat gluten-based film. A 'critical' water content or relative humidity (**Figure 11.6**) thus indicates a sharp change in the mechanical and barrier properties (see **Section 11.5**) of the material that can be estimated from this diagram.

In order to be able to describe and predict the changes in physical-chemical properties of proteins during dry processing according to temperature and relative humidity, it is essential to construct the state diagram relative to the water (or plasticiser) content [1, 149]. **Figure 11.7** shows the different steps involved in the formation of protein-based materials using the dry process [11, 111, 127, 150]:

i) Plasticiser addition.

ii) Heating the plasticised material above its T_g.

iii) Mechanical energy input to form a homogeneous blend and to shape the product.

iv) Cooling to ambient temperature to retransform the rubbery product into a vitreous material with a more or less rigid structure.

Table 11.8 Glass transition temperature of protein materials in dry conditions (adapted from Di Gioia [7])			
Proteins	Analytical Methods	T_g (°C)	Ref.
Corn gluten	DSC/DMTA	174 - 182	[41]
Gluteline	DSC/DMTA	198 - 209	[41]
Corn zein	DSC/DMTA	164 - 168	[41]
Purified zein	DSC	165	[133]
Purified zein	DSC	165	[134]
Commercial zein	DSC	139	[133]
Commercial zein	DSC	167 [a]	[135]
Wheat gluten	DMTA	190	[136]
Wheat gluten	DMTA	180 [a]	[137]
Wheat gluten	DSC	180 [a]	[138]
Wheat gluten	DSC	160 [a]	[139]
Glutenin	DMTA	175	[140]
Alkylated glutenin	DSC	138	[141]
HMW glutenins [b]	DSC	139	[141]
Gliadin	DMTA	121	[142]
Gliadin	DSC	125	[143]
α- Gliadin	DSC	144	[141]
γ-Gliadin	DSC	124	[141]
ω-Gliadin	DSC	145	[141]
Caseins	DMTA/DSC	140 - 150 [a]	[144]
Sodium caseinate	DMTA/DSC	130 [a]	[144]
Myofibrillar proteins	DMTA	215 - 250	[111]
Gelatin	DSC	200 [a]	[145]
Collagen	DSC	180 - 210 [a]	[145]
[a] *Extrapolated values at 0% moisture content* [b] *High molecular weight glutenins* *DSC: dynamic scanning colorimetry* *DMTA: dynamic mechanical thermal analysis*			

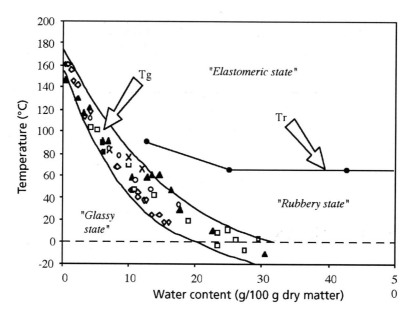

Figure 11.5 Effect of water content on the glass transition temperature (T$_g$) and on minimum thermosetting temperature (T$_r$) for wheat gluten proteins. From Hoseney and co-workers [138] (◇); Kalichevski and co-workers [139] (□); Nicholls and co-workers (○); Cherian and Chinachoti [129] (Δ); Pouplin and co-workers [162] (X). The thermosetting temperature (●) was determined for wheat glutenins [141]

Thus, protein-based bioplastics can be obtained by extrusion, calendering, extrusion blow moulding, injection and thermoforming processes. These 'thermoplastic processes' are derived from synthetic material production processes.

However, when compared to standard synthetic thermoplastic polymers, proteins have markedly different thermoplastic properties. The complex and specific molecular organisation of proteins could explain their specific behaviour during thermomechanical treatments in low hydrated conditions. Polydispersity, heterogeneous intermolecular interactions, common presence of physical nodes and entanglements in protein chains, and the formation of some intermolecular covalent bonds are generally considered. The specific behaviour of proteins is characterised by high elastic modulus values around the rubbery plateau, by the absence of mass flow region, by the large T$_g$ range, and by the apparent reversibility of T$_g$ [133, 141, 142, 151].

The T$_g$ of proteins is partly reversible depending on the density of covalent interactions (usually disulfide bonds) established as a result of heating treatments or variations in redox potential. Heat treatments associated with 'thermoplastic processing' of film-forming materials facilitate formation of covalent bonds [152, 153]. For wheat gluten,

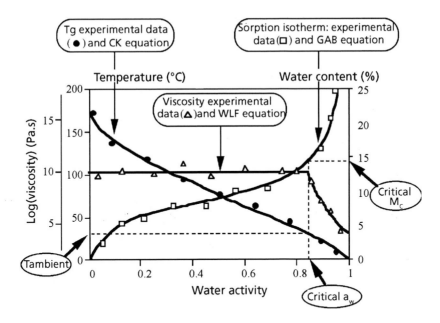

Figure 11.6 Schematic representation of relationships between water activity, water content, glass transition temperature and viscosity for wheat gluten based films. Calculated values were obtained by using the GAB equation [151], Couchman and Karasz equation [146] and Williams Landel and Ferry equation [128]. The critical water activity (a_w) and moisture content (M_C) are indicated when T_g is equal to the ambient temperature

the crosslinking activation energy is weaker under intense shear conditions. Hence, for example, the thermal crosslinking activation energy of wheat gluten is 70 kJ/°C during film formation by extrusion, whereas the activation energy is just 30 kJ/°C in a static mixer with a high shear rate [154]. In some cases, for example, to produce wheat gluten materials, extrusion is easier when disruptive agents (such as cystein or SO_2) are added, which break intermolecular disulfide bonds that stabilise native proteins. Covalent bonds will (re)form after extrusion and removal of the disruptive agents, or after the addition of crosslinking agents. In addition, Micard and co-workers [155] and Morel and co-workers [156] demonstrated that structural rearrangements could occur as protein materials age. The crosslinking rate of a wheat gluten network thus increases with storage and levels off after 72 hours. This crosslinking is due to the gradual oxidation of cysteine residues that are not yet involved in disulfide bonds.

In order to optimise the process parameters (temperature, plasticiser content, residence time, etc.), during transformation of material proteins, the specific characteristics of each protein should be determined (thermal, mechanical and chemical sensitivities, and high

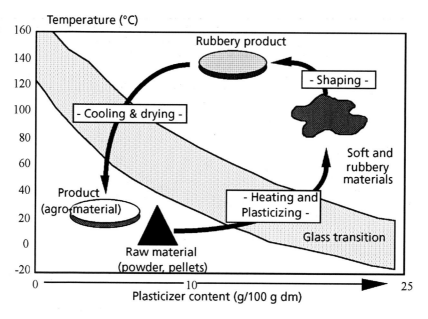

Figure 11.7 Schematic representation of the thermoplastic process applied to shape agro-packagings based on wheat gluten proteins in relation to the glass transition temperature (adapted from Cuq and co-workers [10])

viscosities of the rubbery phases above the T_g) for these new raw materials. However, the physicochemical factors involved in these processes are unclear because very little is currently known about protein modifications that take place when processing at high temperature under low hydration conditions [157]. This has mainly been done for wheat gluten-based materials [57, 60, 61].

Plasticisers are generally required for the formation of protein-based materials. These agents are small, relatively non-volatile molecules that can modify the three-dimensional structure of a polymer, and prompt a decrease in the attractive molecular bond energy, an increase in the intermolecular space and in chain mobility. Plasticisers modify the functional properties of protein-based materials, generally with a decrease in resistance, rigidity and barrier properties, and an increase in flexibility and maximal elongation of the materials [50, 86, 107, 158-161]. The main plasticisers used for protein materials and the observed effects are shown in **Table 11.9**. Adding water, polyhydroxyl compounds, or amphipolar agents, is called 'external' plasticisation. Plasticisers are generally used at concentrations ranging from 10 to 50% (weight base). Water is the most efficient plasticiser in weight-base terms. Polyols, (e.g., glycerol, sorbitol, polyethylene glycol (PEG)), mono-, di-, or oligo-saccharides, di- and tri-ethanolamine, and urea are the most common plasticisers for protein-based materials [31, 40]. For these polar compounds, the best plasticising effect on a molar base is often obtained for compounds with

Proteins	Plasticiser	Effect	Ref.
Table 11.9 Main plasticisers used with proteins materials and the observed effects (adapted from Di Gioia [7])			
Corn gluten	Glycerol, sorbitol, PEG 300, PEG 600, di- tri-ethanolamin, urea, octanoic acid, palmitic acid, dibutyl tartrate and phthalate, mono- di- tri-glycerids esters	C, D	[40, 41]
Wheat gluten Gliadin	- Water	D	[54]
	- Water, glycerol, sorbitol	D	[136]
	- Water, sucrose, glucose, fructose, caproic acid, hydrocaproic acid	D	[139,163]
	- Di-, tri-, tetra-ethylene glycol, glycerol, 1,3-propane diol, 1,4-butane diol, 1,5-pentane diol	B, C	[164]
Soy proteins	Water, glycerol, glycerol mono-ricinoleate, triethanolamine, urea, triethylenen glycol, PEG	D	[165]
Lupin, colza protein	Water	D	[166]
Myofibrillar proteins	Water, glycerol, sorbitol, sucrose	A, B, C, D	[107,110]
Caseins	Water, triethanolamine	A, B, C	[167]
Sodium caseinate	Water	D	[143]
Whey proteins	Water, glycerol	A, B	[83]
Elastins	Water, ethylene glycol, di-, tri-, tetra-ethylene glycol	D	[168]
A: decrease in elastic modulus B decrease in strength at break C: decrease in glass transition temperature D: decrease in shaping temperature			

a high number of hydrophilic groups [40, 162]. Amphipolar plasticisers such as octanoic and palmitic acids, dibutyl tartrate and phthalate, and mono-di- and triglyceride esters are also very efficient, at least for highly non-polar proteins like zein and wheat gluten. In such cases, the plasticising effect (characterised by a T_g drop) at a constant molar concentration seems to be proportional to the molecular weight and inversely proportional to the percentage of hydrophilic groups in the plasticiser [40]. Generally, the effect of the T_g drop can be modelled on the basis of the number of potential hydrogen bonds between the plasticiser and the protein, or according to the respective hydrophilic/lipophilic ratios [31]. The plasticiser migration rate in the protein matrix during the formation process (mixing, extrusion, etc.), is highly dependent on the physicochemical characteristics of the plasticiser. Polar substances therefore quickly interact with readily accessible polar amino acids, while amphipolar plasticisers interact more slowly with non-polar zones, which are often masked and not readily accessed. These kinetic aspects can be very important if non-polar or amphipolar plasticisers are used when water-soluble compounds are not recommended, e.g., to limit plasticiser loss and thus changes in the properties (especially mechanical) of the protein material that could come in contact with water or an aqueous product.

Chemical modifications are often aimed at enhancing water resistance and reducing the effects of relative humidity on protein material properties. However, no significant T_g modifications and especially no improvement in barrier and water resistance properties have been noted for materials formed with chemically lipophilised glutens, at different lipophilisation levels [151].

Crosslinking agents are often used to improve water resistance, cohesion, rigidity, mechanical strength and barrier properties of materials, but in general to the detriment of the product appearance [43, 74, 126, 158, 169-171]. Thus, the functional properties of casein-based materials are substantially improved when calcium is added [126, 172]. The most common covalent crosslinking agents are glutaraldehyde, glyceraldehyde, formaldehyde, gossypol, tannic acid and lactic acid. Standard crosslinking agents (formol, glyoxal and glutaraldehyde) and specially designed crosslinking agents (bifunctional monosaccharides of variable carbon chain length (n = 2, 4, and 6), i.e., N,N'-suberoyl glucosamine, N,N' hexamethylene glucuronamide or *bis* 1,1 [1,8 octyl] glucofuranosidurono-6,3-lactone types) have been used to crosslink wheat gluten materials. Enzymic crosslinking treatments involving transglutaminases or peroxydases were undertaken to stabilise protein materials [77].

Proteins crosslinked via heat treatments, crosslinking agents or radiation treatments (UV, gamma, etc.), form insoluble and infusible networks, characterised by elastomeric or thermosetting thermomechanical behaviour according to the covalent crosslinking density. Collagen-based materials obtained by extrusion can be chemically crosslinked, and casein often used to be crosslinked by formaldehyde to form 'galalith'. In practice, crosslinking treatments substantially modify the mechanical properties and solubility of protein materials but have very little effect on their water vapour barrier properties [155]. All treatments used with protein materials are shown in **Table 11.10**. The use of crosslinking agents, however, is unsuitable for edible films and coatings and even those designed for contact with food products.

11.5 Properties of Protein-Based Materials

The macroscopic properties of the protein-based, three-dimensional macromolecular networks partially depend on system-stabilising interactions. The water solubility of protein materials depends on the nature and density of intermolecular interactions. Materials are soluble in water when the energy of inter-protein bonds is lower than the energy of

Table 11.10 Main physico-, chemical and enzymic treatments applied to the protein materials and their effects on properties (from Cuq and co-workers [111]; Di Gioia [7]; Kolster and co-workers [173]; Micard and co-workers [155])

Treatments	Main effects
Physical treatments	
- Fractionation (ultra filtration, centrifugation) - Mechanical treatments (high pressures, shear), - Irradiation (UV, microwave) - Heating	- Changes in protein composition - Unfolding, changes in texturation properties, crosslinking, desulfuration, desamidation
Chemical treatments	
Chemical reactions - Grafting - Acids - Reducing agents - Crosslinking agents - Solvent Interactions with other components - Proteins - Glucides - Salts - Pigments - Plasticiser	- Hydrophobisation - Hydrolysis - Formation of S-sulfone derivatives - Covalent bonds - Variable effects as a function of solvent type - Aggregation - Maillard reaction - Conformation changes - Colour changes - Decrease in density of low energy interactions
Enzymic treatments	
- Transglutaminase, peroxidase	- Specific modifications of primary structure (covalent bonds, chemical gels)

interactions that could be established between water and polar groups not involved in the network. The presence of 'physical nodes', (i.e., chain entanglements), covalent intermolecular bonds and/or a high interaction density is sufficient to produce films that are completely or partially insoluble in water [63]. For example, the presence of intermolecular covalent bonds in wheat gluten- or keratin-based materials makes them insoluble.

The mechanical properties of protein-based materials can partly be related to the distribution and intensity of inter- and intra-molecular interactions that take place in primary and spatial structures. The cohesion of protein materials mainly depends on the distribution and intensity of intra- and inter-protein interactions, as well as interactions with other components. For example, in soy-based materials, hydrophobic interactions between soy proteins and lipids have a key role in network stability [174]. Cooperative phenomena are generally involved to achieve optimal thermodynamic stability within the system. Interaction effects depend on their occurrence probability and the energy involved. The mechanical properties of materials are relatively dependent on potential controlling interactions that stabilise the network. When covalent bonds stabilise the network or when binding energy is high, materials are basically very resistant and relatively elastic, (e.g., keratin films). Conversely, when low-energy inter-protein interactions are mainly involved, the resulting materials are highly ductile.

The mechanical properties of protein-based materials are substantially lower than those of standard synthetic materials, such as poly(vinylidine chloride) or polyester (**Table 11.11**). The mechanical properties of protein-based materials were measured and modeled as a function of film characteristics [60, 106, 107]. For 'stronger' materials, (e.g., based on wheat gluten, corn gluten and myofibrillar proteins), critical deformation (DC = 0.7 mm) and elastic modulus (K = 510 N/m) values are slightly lower than those of reference materials such as LDPE (DC = 2.3 mm; K = 135 N/m), cellulose (DC = 3.3 mm; K = 350 N/m), or even PVC films. Mechanical properties of corn gluten-based material are close to those of PVC.

In **Figure 11.8**, general mechanical properties of various wheat gluten based films (obtained by casting and by thermomoulding), are compared with properties of conventional plastics, synthetic biodegradable plastics, and biodegradable materials derived from agricultural

Table 11.11 Mechanical properties of various films based on proteins and comparison with synthetic films (adapted from Cuq and co-workers [10])

Films	Ref.	Tensile Strength (MPa)	Elongation (%)	X	T	RH
Myofibrillar proteins	[102]	17.1	22.7	34	25	57
Whey protein isolate	[86]	13.9	30.8	–	23	50
Soy proteins	[125]	1.9	35.6	88	25	50
Wheat gluten proteins	[125]	0.9	260	88	25	50
Corn zein proteins	[45]	0.4	–	81	26	50
Methylcellulose	[175]	56.1	18.5	–	25	50
Polyesters	[176]	178	85	–	–	–
PVDC	[176]	93.2	30	–	–	–
HDPE	[176]	25.9	300	–	–	–
LDPE	[176]	12.9	500	–	–	--

PVDC: poly(vinylidene chloride)　　　　*HDPE: high density polyethylene*
LDPE: low density polyethylene　　　　*X: film thickness (μm)*
RH: relative humidity (%)　　　　*T: temperature (°C)*

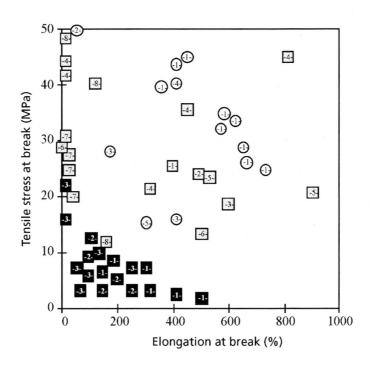

Figure 11.8 Mechanical properties of selected protein-based films compared with some biodegradable and non-biodegradable materials (adapted from Guilbert and co-workers [55]). All non referenced data are from Saechtling [4] and from commercial data sheets.

Synthetic materials (O): -1- thermoplastic polyurethane elastomer (Dow Chemical); -2- polyvinylchloride; -3- PVC plasticised with di-2-ethylhexylphthalate; -4- polypropylene; -5- low density polyethylene.

Synthetic Biodegradable materials (□): -1- BAK 1095: polyester amide (Bayer, G); -2- ECOFLEX: 1,4 butandiol adipinic-dicarbonic and terephthalate copolyester (BASF, G); -3- EASTAR 14766: poly(tetramethylene adipate-*co*-terephthalate) (Eastman, USA); -4- Bionolle 3000: polybutylene succinate/adipate (Showa, Japan).

Biodegradable materials from agricultural origin (□: -5- BIOTEC: starch/polyester (Biotec, G); -6- Materbi: starch/polycaprolactone (Novamont, Italy); -7- Biopol: polyhydroxybutyrate (Monsanto, Italy); -8- Lacea: polylactic acid (Mitsui, Japan).

Protein materials (■): -1- Cast gluten films [159]; -2- Moulded gluten (unpublished results); -3- Moulded soy protein isolate materials [168, 180]

products. This figure shows that mechanical properties of plastic materials can be classified in the following order: conventional synthetic (PVC, polyethylene (PE)) > biodegradable synthetic (BAK, Eastar) > biodegradable agricultural-based materials (wheat gluten-based materials). It is also interesting to note that protein-based materials have either high strength at break or high elongation at break but never both simultaneously.

Mechanical properties of protein-based films can be markedly improved by adding fibres, (i.e., composite materials). Mechanical properties are always highly dependent on the temperature and relative humidity of the protein material (**Figure 11.9**). This modification, (i.e., sharp increase in deformation at break and decrease in mechanical strength), occurs suddenly when the material crosses the T_g range [149].

The barrier properties of protein materials depend on the nature and density of the macro-molecular network, and more particularly on the proportion and distribution of non-polar amino acids relative to polar amino acids [9, 25]. The protein composition and structural organisation of the network enables some chemical groups to remain free, which means that they are sites of potential interactions with permeating molecules. Generally for protein-based materials, most free hydrophilic groups are able to interact with water vapour and to permit water transfer phenomena, to the detriment of hydrophobic gas transfers, (e.g., nitrogen, oxygen).

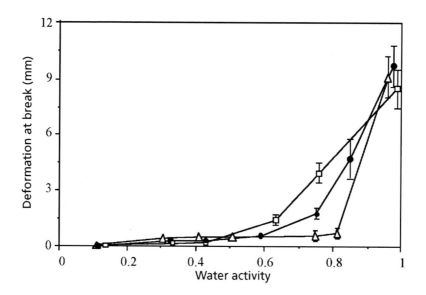

Figure 11.9 Influence of temperature (at 5°C (Δ), 25°C (●), and 50°C (□) and equilibrium relative humidity on the mechanical properties of myofibrillar protein-based films (from Cuq and co-workers[106])

Protein-based materials generally have high water vapour permeability (**Table 11.12**). Water vapour permeation through protein films is facilitated by the systematic presence of hydrophilic plasticisers, which promote water molecule adsorption. Protein-based materials have much higher water vapour permeability (around 5×10^{-12} mol/m/s/Pa) than synthetic materials (0.05×10^{-12} mol/m/s/Pa for LDPE). This feature could still be interesting for coatings on materials that need to 'breathe', (e.g., packaging of fresh products and films for agricultural or cosmetic applications). These properties can be significantly improved to resemble those of PE films by adding lipid compounds, (e.g., beeswax, paraffin), to the film formulation [51, 52, 171]. As already noted for the mechanical properties, water barrier properties are highly dependent on the temperature and relative humidity of the protein material and decrease suddenly when materials cross the T_g range [149] (**Figure 11.10**).

The gas barrier properties (O_2, CO_2 and ethylene) of protein-based materials are highly attractive since they are minimal under low relative humidity conditions. Oxygen permeability (around 1 amol/m/s/Pa) is comparable to the ethylene vinyl alcohol (EVOH) properties (0.2 amol/m/s/Pa) and much lower than the properties of LDPE (1000 amol/m/s/Pa) [53]

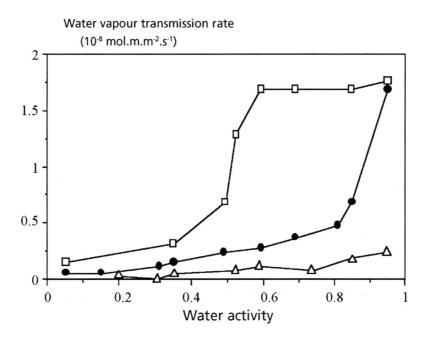

Figure 11.10 Influence of water activity and temperature on the water vapor barrier properties of wheat gluten-based films. At 5 °C (Δ), 20 °C (●) or 50 °C (□) (from Gontard and co-workers [50])

(**Table 11.13**). The O_2 permeability of protein films is about 10-fold higher that EVOH-based films, mainly due to the high plasticiser content of protein-based films.

While the barrier properties of synthetic materials remain quite stable at high relative humidity, the gas-barrier properties of material proteins (as for all properties of

Table 11.13 Oxygen permeability (10^{18} mol/m/s/Pa) and carbon dioxide permeability (10^{18} mol/m/s/Pa) of various films based on proteins and comparison with synthetic films and edible films (from Cuq and co-workers [10])					
Film	Ref.	P(O_2)	P(CO_2)	T	a_w
LDPE	[182]	1003	4220	23	0
HDPE	[182]	285	972	23	0
Polyester	[182]	12	38	23	0
EVOH	[183]	0.2	-	23	0
Methylcellulose	[178]	522	29900	30	0
Beeswax	[159]	480	-	25	0
Hydroxypropylcellulose	[178]	470	28900	30	0
Carnauba wax	[159]	81	-	25	0
Corn zein	[45]	35	216	38	0
Wheat gluten protein	[45]	3	-	38	0
Soy protein	[184]	2	-	23	0
Wheat gluten	[53]	1	7	25	0
Fish myofibrillar protein	[53]	1	9	25	0
Chitosan	[53]	0.6	-	25	0
HDPE	[185]	224	-	23	1
Cellophane	[186]	130	-	23	0.95
Polyester	[185]	12	-	23	1
EVOH	[187]	6	-	23	0.95
Pectin	[53]	1340	21300	25	0.96
Wheat Gluten	[53]	1290	36700	25	0.95
Starch	[177]	1085	-	25	1
Fish myofibrillar protein	[53]	873	11100	25	0.93
Chitosan	[53]	472	8010	25	0.93
HDPE: high density polyethylene *EVOH: ethylene vinyl alcohol* *T: temperature (°C)* *a_w: equilibrium water activity*					

hydrocolloid-based materials) are highly relative humidity- and temperature-dependent (**Figure 11.11**). The O_2 and CO_2 permeabilities are about 1000-fold higher for moist films than for films stored at 0% relative humidity. For proteins, this effect is much greater for 'hydrophilic' gases (CO_2) than for 'hydrophobic' gases (O_2). Changes in RH of temperature modify the CO_2/O_2 selectivity coefficient, which rises from three to more than 50 when the relative humidity rises from 0 to 100% and the temperature from 5 to 45°C, as compared to constant values of around 3-5 for standard synthetic films (**Figure 11.11**). Gas permeability differences in protein materials are partly due to gas solubility differences in the film matrix, and could be mainly explained by the high

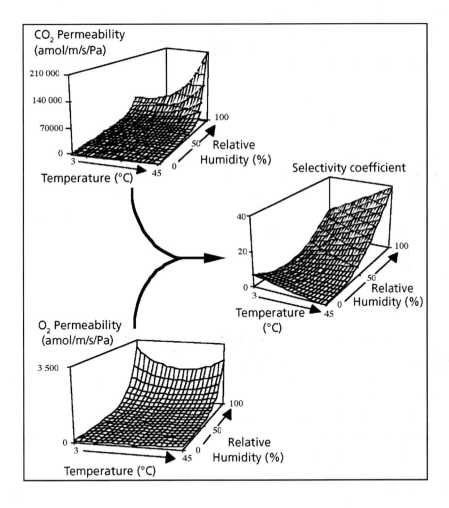

Figure 11.11 Changes in gas barrier properties of wheat gluten based films as a function of temperature and relative humidity (from Barron and co-workers [59])

affinity between CO_2, the polypeptide chain and many lateral amino acid groups [188]. A film with good O_2 barrier properties is interesting for the protection of oxidisable foods (rancidification, loss of oxidisable vitamins, etc.). However, some extent of permeability to O_2 and especially to CO_2 is required to decrease the metabolic activity of many fresh fruits and vegetables. The development of protein films with selective gas permeability features could thus be highly promising, especially for controlling respiratory exchange and improving the shelf-life of fresh or minimally processed fruits and vegetables [189]. Wheat gluten-based films were tested with the aim of creating atmospheric conditions suitable for preserving fresh vegetables. Measurements of changes in the gas composition of modified atmosphere packaged mushrooms under wheat gluten films confirmed the high selectivity of such materials, i.e., the CO_2 and O_2 composition ranged from 1-2% despite product respiration [59].

The aroma barrier properties of protein-based materials seem especially interesting for blocking non-polar compound permeation. However, it is hard to determine the relationship between the physicochemical properties of aroma compounds and their retention by protein films [190].

Solute retention properties (especially anti-microbial and anti-oxidant agents) were investigated and modelled for wheat gluten-based films [58] and the results indicated potential applications for the controlled release of functional agents. The anti-microbial efficacy of edible wheat gluten-based films containing anti-microbial agents has been very well documented [191-193]. The use of these films on high moisture model foods extended their shelf-life by more than 15 days at 4 °C and 30 °C. Many patents and publications recommend adding antioxidant agents to protein films and coatings, as already done in some commercial edible films. Guilbert [43] measured α-tocopherol retention in gelatin films applied to the surface of margarine blocks. No migration was noted after 50 days storage when the film was pretreated with a crosslinking agent (tannic acid), whereas α–tocopherol diffusivity was around 10 to 30 x 10^{-11} m²/s without the film.

Few studies have focused on the biodegradability and environment-friendly aspects of protein-based products that degrade naturally or in compost. The construction of protein networks can induce marked changes in the conformation and resistance to enzymic hydrolysis and chemical attacks of proteins [194]. However, Garcia-Rodenas and co-workers [195] showed that the susceptibility of casein and wheat gluten-based films to *in vitro* proteolysis did not significantly differ from that noted for native proteins.

11.6 Applications

Proteins could be used as raw material for bioplastics with a wide range of agricultural, agri-food, pharmaceutical and medical industry applications. The functional properties (especially optical, barrier and mechanical) of these protein-based materials are often specific and unique.

Plant proteins are generally inexpensive (0.5-1 €/kg for corn and wheat glutens, with 70 - 80% protein content, respectively), widely available and relatively easy to process. Animal proteins are more expensive (2-10 €/kg), but sometimes have no functional substitutes, (e.g., gelatin).

The casting process is generally adapted for coating seeds, drug pills, and foods, for making cosmetic masks or varnishes, and pharmaceutical capsules. Heat casting of protein-based materials by techniques usually applied for synthetic thermoplastic polymers (extrusion, injection, moulding, etc.), is more cost-effective. This process is often applied for making flexible films, (e.g., films for agricultural applications, packaging films, and cardboard coatings) or objects, (e.g., biodegradable materials), that are sometimes reinforced with fibres (composite bioplastics for construction, automobile parts, etc.).

The complexity of proteins and the broad range of protein fractions could be used to produce materials with unique functional properties that differ markedly from those of conventional plastic materials. Protein-based materials are biodegradable and even edible when food-grade additives are used. Moreover, they are often biocompatible, barring some protein-specific aspects (e.g., allergenic features of wheat gluten gliadins), processing aspects, and the presence of impurities or additives.

Protein materials are generally homogeneous, transparent, resistant and water insoluble. Their high moisture permeability is especially attractive for cheese, fruit and vegetable packaging, and for agricultural material and cosmetic applications. Protein-based materials have slightly lower mechanical properties than reference materials such as LDPE or plasticised PVC, but the addition of fibres (composite materials) can considerably improve them. The thermoplastic properties of proteins and their water resistance (for insoluble proteins) are especially interesting for natural resin uses to produce chipboard, medium and particleboard type materials.

Gas barrier properties (O_2, CO_2 and ethylene) of protein-based materials can be utilised in designing selective or active materials for modified atmosphere packaging of fresh products (such as fruits, vegetables, cheese). Solute retention properties (especially anti-microbial and antioxidant agents) are attractive for designing controlled-release systems for functional additives in food, (e.g., active coatings, encapsulation), agriculture, (e.g., coated seed), pharmacy (drug delivery) and cosmetic industries.

Multilayer 'protein/paper' and 'protein/biodegradable polyester' (polycaprolactone, polylactic acid (PLA), etc.), materials can be produced using some highly amphipolar proteins with a wide compatibility range. Composite agromaterials combining proteins with cottonseed, sisal, coconut and straw fibres were successfully tested (excellent compatibility) and have a considerable application potential.

Multilayer materials based on modified polyethylene and proteins can be obtained by thermomoulding processes. Thermosetting protein/resin composite materials can also be produced. Materials could thus be developed that combine the unique gas- vapour- and solute-permeability properties of protein films with the mechanical performances of conventional synthetic materials. Material protein properties can generally be modified in a wide range of ways, via raw material choices and combinations, the proper use of fractionating techniques and rheological modifying additives, and also by adjusting the product formation process variables.

References

1.   S. Guilbert and N. Gontard in *Foods and Packaging Materials:Chemical Interactions*, Ed., P. Ackermann, M. Jägerstad, and T. Ohlsson, The Royal Society of Chemistry, Cambridge, UK, 1995, p.159.

2.   S. Guilbert, *Bulletin of the Research Institute of Food Science, Kyoto University*, 1999, 56, 38.

3.   *Protein-based Films and Coatings*, Ed., A. Gennadios, CRC Press, Boca Raton, FL, USA, 2001.

4.   W. Woebcken, *Saechtling International Plastics Handbook for the Technologist, Engineer and User*, 3rd Edition, Hanser Publishers, Munich, Germany, 1995.

5.   I. Kruszewska, *Clean Production Action Report of Greenpeace International Toxic Campaign*, Greenpeace, Amsterdam, The Netherlands, 1997.

6.   J.L. Jane, S.T. Lim, I. Paetau, K. Spence and S. Wang in *Polymers from Agricultural Coproducts*, Ed., M.L. Fishman, R.B. Friedman, and S.J. Huang, ACS Symposium Series No.575, ACS, Washington, DC, USA, 1994, p.92.

7.   L. Di Gioia, *Obtention et Etude de Biomatériaux à Base de Protéines de Maïs*, ENSA Montpellier, France, 1998. [PhD Thesis]

8.   J.A. Rothfus, *Journal of Agricultural and Food Chemistry*, 1996, 44, 10, 3143.

9.   S. Guilbert and J. Graille in *Valorisations Non Alimentaires des Grandes Productions Agricoles*, Les colloques No.71, Ed., J. Gueguen, INRA Editions, Paris, Frances, 1994, p.195.

10. B. Cuq, N. Gontard and S. Guilbert, *Cereal Chemistry*, 1998, **75**, 1, 1.

11. C. Oudet, *Polymères - Structure et Propriétés*, Masson, Paris, France, 1994.

12. L.G. Phillips, D.M. Whitehead and J. Kinsella in *Structure-Function Properties of Food Proteins*, Eds., L.G. Phillips, D.A. Whitehead and J. Kinsella, Academic Press, Inc., San Diego, CA, USA, 1994.

13. P.R. Shewry and A.S. Tatham, *Biochemical Journal*, 1990, **267**, 1, 1.

14. J.S. Wall and J.W. Paulis in *Advances in Cereal Science and Technology*, Volume 2, Ed., Y. Pomeranz, VCH, New York, NY, USA, 1978, p.135.

15. F.W. Sosulski and G.I. Imafidon, *Journal of Agricultural and Food Chemistry*, 1990, **38**, 6, 1351.

16. C.E. Boldwell and D.T. Hopkins in *New Protein Foods*, Volume 2, Eds., A.M. Altschul and H.L. Wilcke, Academic Press, Inc., Orlando, 1985, p.152.

17. C. Defromont in *Proceedings of 'Colloque CIDT'*, Abidjan, Ivory Coast, 1985, p.112.

18. P.I. Rose in *Encyclopedia of Polymer Science and Engineering*, Volume 7, Eds., H.F. Mark, N.M. Bikales, C.G. Overberger and G. Menges, John Wiley and Sons, Inc., New York, NY, USA, 1987, p.488.

19. E. Orban, G.B. Quaglia, I. Casini, I. Caproni and E. Moneta, *Lebensmittel-Wissenschaft und -Technology*, 1992, **25**, 371.

20. J.J. Kester and O. Fennema, *Food Technology*, 1986, **40**, 47.

21. S. Guilbert and B. Biquet in *L'Emballage des Denrées Alimentaires de Grande Consommation*, Ed., G. Bureau and J-L. Multon, Technique et Documentation, Lavoisier, Paris, France, 1989, p.320.

22. A. Gennadios, T.H. McHugh, C.L. Weller and J.M. Krochta in *Edible Coatings and Films to Improve Food Quality*, Ed., J.M. Krochta, E.A. Baldwin, and M.O. Nisperos-Carriedo, Technomic Publishing Company, Lancaster, PA, 1994, p.201.

23. N. Gontard and S. Guilbert in *Food Packaging and Preservation*, Ed., M. Mathlouthi, Blackie Academic and Professional, Glasgow, UK, 1994, p.159.

24. J.A. Torres in *Protein Functionality in Food Systems*, Ed., N.S. Hettiarachchy and G.R. Ziegler, Marcel Dekker Inc., New York, NY, USA, 1994, p.467.

25. B. Cuq, N. Gontard and S. Guilbert in *Active Food Packagings*, Ed., M.L. Rooney, Blackie Academic and Professional, Glasgow, UK, 1995, p.111.

26. F. Debeaufort, J.A. Quezada-Gallo and A. Voilley, *Critical Reviews in Food Science and Nutrition*, 1998, **38**, 4, 299.

27. S. Guilbert and B. Cuq in *L'Emballage des Denrées Alimentaires de Grande Consommation*, Ed., J-L. Multon and G. Bureau, Technique et Documentation, Lavoisier, Paris, France, 1998, p.471.

28. D.M. Rathmann in *Mellon Institute Bibliographic Series*, Bulletin No.7, Mellon Institute, Pittsburgh, PA, USA, 1954, p.1891.

29. R.A. Reiners, J.S. Wall and G.E. Inglett in *Industrial Uses of Cereals*, Ed., Y. Pomeranz, Proceedings of the 58th Annual Meeting of the AACC, St. Paul, MN, USA, 1973, p.285.

30. M. Zhang, C.A. Reitmeier, E.G., Hammond and D.J. Myers, *Cereal Chemistry*, 1997, **74**, 5, 594.

31. L. Di Gioia and S. Guilbert, *Journal of Agricultural and Food Chemistry*, 1999, **47**, 3, 1254.

32. D.N. Martin, inventor; Corn Products Co., assignee; US 3,497,369, 1970.

33. N. Hasegawa, K. Suzuki, T. Ishii, M. Hayashi and G. Danno, inventors; Sanei Touka KK, assignee; JP 6,192,577, 1994.

34. L.F. Pelosi, inventor; EI Du Pont de Nemours & Company, assignee; WO 9610582, 1996.

35. J. Mossé, *Annales de Physiologie Végétale*, 1961, **3**, 105.

36. W.L. Morgan, inventor; American Maize Products Company, assignee; US 2,427,503, 1947.

37. K. Yamada, H. Takahashi and A. Noguchi, *International Journal of Food Science and Technology*, 1995, **30**, 599.

38. T. Trezza and P.J. Vergano, *Packaging Technology and Engineering*, 1995, **3**, 33.

39. M.I. Beck, I. Tomka and E. Waysek, *International Journal of Pharmaceutics*, 1996, **141**, 1-2, 137.

40. L. Di Gioia, B. Cuq and S. Guilbert, *Cereal Chemistry*, 1998, **75**, 4, 514.

41. L. Di Gioia, B. Cuq and S. Guilbert, *International Journal of Biological Macromolecules*, 1999, **24**, 4, 341.

42. H. Takenaka, H. Ito, H. Asano and H. Hattori, *Gifu Yakka Daigaku Kiyo*, 1967, **17**, 142.

43. S. Guilbert in *Food Preservation by Moisture Control*, Eds., C.C. Seow, T.T. Teng, and C.H. Quah, Elsevier Applied Science Publishers, London, UK, 1988, p.199.

44. A. Gennadios and C.L. Weller, *Food Technology*, 1990, **44**, 63.

45. T.P. Aydt, C.L. Weller and R.F. Testin, *Transactions of the ASAE*, 1991, **34**, 1, 207.

46. T.J. Herald, K.A. Hachmeister, S. Huang and J.R. Bowers, *Journal of Food Science*, 1996, **61**, 2, 415.

47. L. Di Gioia, B. Cuq and S. Guilbert, *Journal of Materials Research*, 2000, **15**, 1.

48. C.A. Anker, G.A. Foster, Jr., and M.A. Loader, inventors; General Mills Inc., assignee; US 3,653,925, 1972.

49. N. Gontard, S. Guilbert and J-L. Cuq, *Journal of Food Science*, 1992, **57**, 190.

50. N. Gontard, S. Guilbert and J-L. Cuq, *Journal of Food Science*, 1993, **58**, 1, 206.

51. N. Gontard, C. Duchez, J-L. Cuq and S. Guilbert, *International Journal of Food Science and Technology*, 1994, **29**, 39.

52. N. Gontard, S. Marchesseau, J-L. Cuq and S. Guilbert, *International Journal of Food Science and Technology*, 1995, **30**, 49.

53. N. Gontard, R. Thibault, B. Cuq and S. Guilbert, *Journal of Agricultural and Food Chemistry*, 1996, **44**, 4, 1064.

54. N. Gontard and S. Ring, *Journal of Agricultural and Food Chemistry*, 1996, **44**, 11, 3474.

55. S. Guilbert, N. Gontard, M.H. Morel, P. Chalier, V. Micard and A. Redl in *Protein-based Films and Coatings*, Ed., A. Gennadios, CRC Press, Boca Raton, FL, USA, 2002.

56. K. Watanabe and S. Okamoto, *Nippon Shokuhin Kogyo Gakkaishi*, 1973, **20**, 66.

57. A. Redl, M.H. Morel, J. Bonicel, B. Vergnes and S. Guilbert, *Cereal Chemistry*, 1999, **76**, 3, 361.

58. A. Redl, N. Gontard and S. Guilbert, *Journal of Food Science*, 1996, **61**, 1, 116.

59. C. Barron, S. Guilbert, N. Gontard and P. Varoquaux, *Journal of Food Science*, 2001, **67**, 251.

60. A. Redl, S. Guilbert and B. Vergnes, *Cahiers de Rhéologie*, 1997, **15**, 339.

61. A. Redl, M.H. Morel, J. Bonicel, S. Guilbert and B. Vergnes, *Rheological Acta*, 1999a, **38**, 4, 311.

62. J.A. Bietz and G.L. Lookhart, *Cereal Foods World*, 1996, **41**, 376.

63. D. Fukushima and J. Van Buren, *Cereal Chemistry*, 1970, **47**, 687.

64. L.C. Wu and R.P. Bates, *Journal of Food Science*, 1972, **37**, 36.

65. L.C. Wu and R.P. Bates, *Journal of Food Science*, 1972, **37**, 40.

66. R.P. Bates and L.C. Wu, *Journal of Food Science*, 1975, **40**, 425.

67. S.J. Circle, E.W. Meyer and R.W. Whitney, *Cereal Chemistry*, 1964, **41**, 3, 157.

68. L.C. Wu and R.P. Bates, *Journal of Food Science*, 1973, **38**, 783.

69. S. Okamoto, *Cereal Foods World*, 1978, **23**, 256.

70. H.O. Jaynes and W.N. Chou, *Food Product Development*, 1975, **9**, 86.

71. A. Gennadios, H.J. Park and C.L. Weller, *Transactions of the ASAE*, 1993, **36**, 1867.

72. Y.M. Stuchell and J.M. Krochta, *Journal of Food Science*, 1994, **59**, 6, 1332.

73. Y. Aboagye and D.W. Stanley, *Canadian Institute of Food Science and Technology Journal*, 1985, **18**, 12.

74. C. Marquié, C. Aymard, J-L. Cuq and S. Guilbert, *Journal of Agricultural and Food Chemistry*, 1995, **43**, 10, 2762.

75. C. Marquié, A-M. Tessier, C. Aymard and S. Guilbert, *Journal of Agricultural and Food Chemistry*, 1997, **45**, 3, 922.

76. C. Marquié and S. Guilbert in *Protein-Based Films and Coatings*, Ed., A. Gennadios, CRC Press, Boca Raton, FL, USA, 2002.

77. M. Motoki, H. Aso, K. Seguro and N. Nio, *Agricultural and Biological Chemistry*, 1987, **51**, 993.

78. M. Motoki, H. Aso, K. Seguro and N. Nio, *Agricultural and Biological Chemistry*, 1987, **51**, 997.

79. M.S. Cole, inventor; Archer Daniels Midland Co, assignee; US 3,479,191, 1969.

80. L.J. Tryhnew, K.W. Gunaratne and J.V. Spencer, *Journal of Milk and Food Technology*, 1973, **36**, 272.

81. J.R. Ashes, L.J. Cook and G.S. Sidhu, *Lipids*, 1984, **19**, 159.

82. R.J. Avena-Bustillos, J.M. Krochta, M.E. Salveit, R.J. Rojas-Villegas and J.A. Sauceda-Perez, *Journal of Food Engineering*, 1994, **21**, 2, 197.

83. G. Galietta, L. Di Gioia, S. Guilbert and B. Cuq, *Journal of Dairy Science*, 1998, **81**, 1, 3123.

84. R. Mahmoud and P.A. Savello, *Journal of Dairy Science*, 1992, **75**, 4, 942.

85. R. Mahmoud and P.A. Savello, *Journal of Dairy Science*, 1993, **76**, 1, 29.

86. T.H. McHugh and J.M. Krochta, *Journal of Agricultural and Food Chemistry*, 1994, **42**, 4, 841.

87. A. Courts in *The Science and Technology of Gelatin*, Eds., A.G. Ward and A. Courts, Academic Press, New York, NY, USA, 1977, p.395.

88. L.L. Hood, *Advances in Meat Research*, 1987, **4**, 109.

89. J.F. Cavallaro, P.D. Kemp and K.H. Kraus, *Biotechnology and Bioengineering*, 1994, **43**, 8, 781.

90. E. Bradbury and C. Martin, *Proceedings of the Royal Society*, 1952, A**214**, 183.

91. G.D. Hebert and O.E. Holloway, inventors; Nabisco, Inc., assignee; US 5,149,562, 1992.

92. L.L. Balassa and G.O. Fanger, *CRC Critical Reviews of Food Technology*, 1971, **2**, 245.

93. A.A. Klose, E.P. Mecchi and H.L. Hanson, *Food Technology*, 1952, **6**, 308.

94. A. Gennadios, C.L. Weller and R.F. Testin, *Cereal Chemistry*, 1993, **70**, 465.

95. R. Daniels, *Edible Coatings and Soluble Packaging*, Noyes Data Corporation, Park Ridge, NJ, USA, 1973.

96. S. Soloway, inventors; Fabergé, Inc., assignee; US 3,137,631, 1964.

97. J. Kosar and G.M. Atkins, Jr., inventors; Keuffel & Esser Co., assignee; US 3,406,119, 1968.

98. R.C. Baker, J.M. Darfler and D.V. Vadehra, *Poultry Science*, 1972, **51**, 1220.

99. H.R. Bolin, *Journal of Food Science*, 1976, **41**, 1316.

100. J.R. Durst, inventor; The Pillsbury Co., assignee; US 3,434,843, 1969.

101. A. Gennadios, C.L. Weller, M.A. Hanna and G.W. Froning, *Journal of Food Science*, 1996, **61**, 3, 585.

102. B. Cuq, C. Aymard, J-L. Cuq and S. Guilbert, *Journal of Food Science*, 1995, **60**, 6, 1369.

103. B. Cuq, N. Gontard, J-L. Cuq and S. Guilbert, *Lebensmittel-Wissenschaft und -Technology*, 1996, **29**, 4, 344.

104. B. Cuq, N. Gontard, J-L. Cuq and S. Guilbert, *Journal of Food Science*, 1996, **61**, 3, 580.

105. B. Cuq, N. Gontard, J-L. Cuq and S. Guilbert, *Journal of Agricultural and Food Chemistry*, 1996, **44**, 4, 1116.

106. B. Cuq, N. Gontard, C. Aymard and S. Guilbert, *Polymer Gels and Networks*, 1997, **5**, 1, 1.

107. B. Cuq, N. Gontard, J-L. Cuq and S. Guilbert, *Journal of Agricultural and Food Chemistry*, 1997, **45**, 3, 622.

108. B. Cuq, N. Gontard, J-L. Cuq and S. Guilbert, *Nahrung/Food*, 1998, **42**, 3-4, 260.

109. B. Cuq in *Protein-Based Films and Coatings*, Ed., A. Gennadios, CRC Press, Boca Raton, FL, USA, 2002.

110. B. Cuq, N. Gontard and S. Guilbert, *Polymer*, 1997, **38**, 10, 2399.

111. B. Cuq, N. Gontard and S. Guilbert, *Polymer*, 1997, **38**, 16, 4071.

112. G. Viroben, J. Barbot, A. Gaugain and J. Gueguen in *Valorisations Non-Alimentaires des Grandes Productions Agricoles*, Les colloques No.71, Ed., J. Gueguen, INRA Editions, Paris, France, 1994, p.215.

113. F.F. Shih, *Cereal Chemistry*, 1996, **73**, 3, 406.

114. L.H. Krull and J.S. Wall, *Baker's Digest*, 1969, **43**, 4, 30.

115. R.D.B. Fraser, T.P. MacRae and G.E. Rogers, *Keratins: Their Composition, Structure and Biosynthesis*, C.C. Thomas, Springfield, IL, USA, 1972.

116. A. Veis, *The Macromolecular Chemistry of Gelatin*, Academic Press Inc., New York, NY, USA, 1964.

117. T.C. Vanaman, K. Brew and R.L. Hill, *Journal of Biological Chemistry*, 1970, **245**, 17, 4583.

118. G. Braunitzer, R. Chen, B. Schrank and A. Stangl, *Hoppe-Seyler's Zeitung fur Physiologie und Chemie*, 1972, **353**, 832.

119. T.B. Osborne, *The Vegetable Proteins*, 2nd Edition, Longman, Green and Co, London, UK, 1924.

120. I.G. Donhowe and O. Fennema in *Edible Coatings and Films to Improve Food Quality*, Ed., J.M. Krochta, E.A. Baldwin, and M.O. Nisperos-Carriedo, Technomic Publishing Company, Lancaster, PA, 1994, p.1.

121. D.H. Chou and C.V. Morr, *Journal of the American Oil Chemists' Society*, 1979, **56**, 53.

122. P.R. Shewry and B.J. Miflin in *Cereal Science and Technology*, Volume 7, Ed., Y. Pomeranz, VCH, New York, NY, USA, 1978, p.1.

123. N.K. Sian and S. Ishak, *Cereal Foods World*, 1990, **35**, 8, 748.

124. I.G. Donhowe and O. Fennema, *Journal of Food Processing and Preservation*, 1993, **17**, 231.

125. A. Gennadios, A.H. Brandenburg, C.L. Weller and R.F. Testin, *Journal of Agricultural and Food Chemistry*, 1993, **41**, 11, 1835.

126. S. Guilbert in *Food Packaging and Preservation*, Ed., M. Mathlouthi, Blackie Academic and Professional, Glasgow, UK, 1994, p.371.

127. C. Savary, P. Colonna and G. Della Valle, *Industrie des Céréales*, 1993, **10**, 17.

128. J-D. Ferry, *Viscoelastic Properties of Polymers, 3rd Edition*, John Wiley and Sons, New York, USA, 1980.

129. G. Cherian and P. Chinachoti, *Cereal Chemistry*, 1996, **73**, 618.

130. L. Slade and H. Levine in *The Glassy State in Foods*, Ed., J.M.V. Blanshard and P.J. Lillford, Nottingham University Press, Nottingham, UK, 1993, p.35.

131. G. Cherian, A. Gennadios, C.L. Weller and P. Chinachoti, *Cereal Chemistry*, 1995, **72**, 1, 1.

132. Y.I. Mateev, *Special Publication of RSC*, 1995, **156**, 552.

133. H. Madeka and J.L. Kokini, *Cereal Chemistry*, 1996, **73**, 433.

134. J. Magoshi, S. Nakamura and K-I. Murakami, *Journal of Applied Polymer Science*, 1992, **45**, 11, 2043.

135. J.W. Lawton, *Cereal Chemistry*, 1992, **69**, 351.

136. A. Redl, *Matériaux Thermoplastiques à Base de Protéines de Blé; Mise en Forme et Etude des Propriétés*, ENSA Montpellier, France, 1998. [PhD Thesis]

137. M. Peleg, *Rheological Acta*, 1993, **32**, 575.

138. R.C. Hoseney, K. Zeleznak and C.S. Lai, *Cereal Chemistry*, 1986, **63**, 3, 285.

139. M.T. Kalichevski, E.M. Jaroszkiewicz and J.M. Blanshard, *International Journal of Biological Macromolecules*, 1992, **14**, 257.

140. A.M. Cocero and J.L. Kokini, *Journal of Rheology*, 1991, **35**, 257.

141. T.R. Noel, R. Parker, S.G. Ring and A.S. Tatham, *International Journal of Biological Macromolecules*, 1995, **17**, 2, 81.

142. J.L. Kokini, A.M. Cocero and H. Madeka, *Food Technology*, 1995, 5, 74.

143. E.M. De Graaf, H. Madeka, A.M. Cocero and J.L. Kokini, *Biotechnology Progress*, 1993, **9**, 2, 210.

144. M.T. Kalichevski, J.M. Blanshard and R.D. Marsh in *The Glassy State in Foods*, Eds., J.M.V. Blanshard and P.J. Lillford, Nottingham University Press, Nottingham, UK, 1993, p.133.

145. L. Slade, H. Levine and J.W. Finley in *Protein Quality and the Effects of Processing*, Eds., R.D. Phillips and J.W. Finley, Marcel Dekker Inc., New York, NY, USA, 1989, p.9.

146. P.R. Couchman and F.E. Karasz, *Macromolecules*, 1978, **11**, 1, 117.

147. M. Gordon and J.S. Taylor, *Journal of Applied Chemistry*, 1952, **2**, 493.

148. T.K.J. Kwei, *Polymer Science*, 1984, **22**, 6, 307.

149. S. Guilbert, A. Redl, and N. Gontard in *Engineering and Food for the 21st Century*, Eds., J. Welti-Chanes, G.V. Barbosa Canovas and J.M. Aguilera, Food Preservation Technology Series, CRC Press, Boca Raton, FL, USA, 2000.

150. B. Cuq, N. Gontard and S. Guilbert, *Lebensmittel-Wissenschaft und -Technology*, 1999, **32**, 2, 107.

151. V. Micard and S. Guilbert, *International Journal of Biological Macromolecules*, 2000, **27**, 3, 229.

152. V. Micard, M.H. Morel, J. Bonicel and S. Guilbert, *Polymer*, 2001, **42**, 477.

153. B. Cuq, F. Boutrot, A. Redl and V. Lullien-Pellerin, *Journal of Agricultural and Food Chemistry*, 2000, **48**, 7, 2954.

154. A. Redl, M.H. Morel, B. Vergnes and S. Guilbert in *Wheat Gluten*, Eds., P.R. Shewry and A.S. Tatham, The Royal Society of Chemistry, Cambridge, UK, 2000, p.430.

155. V. Micard, R. Belamri, M-H. Morel and S. Guilbert, *Journal of Agricultural and Food Chemistry*, 2000, **48**, 7, 2948.

156. M-H. Morel, J. Bonicel, V. Micard and S. Guilbert, *Journal of Agricultural and Food Chemistry*, 2000, **48**, 2, 186.

157. J.R. Mitchell, J.A. Areas and S. Rasul in *La Cuisson-Extrusion*, Ed., P. Colonna and G. Della Valle, Presses de INRA, France, 1994, p.85.

158. E.R. Lieberman and S.G. Guilbert, *Journal of Polymer Science*, 1973, **41**, 1, 33.

159. I.G. Donhowe and O. Fennema, *Journal of Food Processing and Preservation*, 1993, **17**, 247.

160. A. Gennadios, C.L. Weller and R.F. Testin, *Transactions of the ASAE*, 1993, **36**, 2, 465.

161. H.J. Park, J.M. Bunn, C.L. Weller, P.J. Vergano and R.F. Testin, *Transactions of the ASAE*, 1994, **37**, 4, 1281.

162. M. Pouplin, A. Redl and N. Gontard, *Journal of Agricultural and Food Chemistry*, 1999, **47**, 2, 538.

163. M.T. Kalichevski, E.M. Jaroszkiewicz, S. Ablett, J.M. Blanshard and P.J. Lillford, *Carbohydrate Polymers*, 1992, **18**, 2, 77.

164. J. Gueguen, J. Viroben, J. Barboot and M. Subirade in *Plant Proteins from European Crops: Food and Non-Food Applications*, Eds., J. Guéguen and Y. Popineau, Springer-Verlag, Berlin, Germany, 1998, p.319.

165. J.L. Jane and S. Wang, inventors; Iowa State University Research Foundation, Inc., assignee; US 5,523,293, 1996.

166. A. Borchering and T. Luck in *Plant Proteins from European Crops, Food and Non-Food Applications*, Ed., J. Guéguen and Y. Popineau, Springer-Verlag, Berlin, Germany, 1998, p.313.

167. N. Somanathan, M.D. Naresh, V. Arumugan, T.S. Ranganathan and R. Sanjeevi, *Polymer Journal (Japan)*, 1992, **24**, 7, 603.

168. M.A. Lillie and J.M. Gosline in *The Glassy State in Foods*, Eds., J.M. Blanshard and P.J. Lillford, Nottingham University Press, Nottingham, 1993, 281.

169. C.A. Kumins, *Journal of Polymer Science – Part C*, 1965, **10**, 1.

170. R. Osawa and T.P. Walsh, *Journal of Agricultural and Food Chemistry*, 1993, **41**, 5, 704.

171. R.J. Avena-Bustillos and J.M. Krochta, *Journal of Food Science*, 1993, **58**, 4, 904.

172. J.M. Krochta, J.S. Hudson and R.J. Avena-Bustillos, *Annual Meeting of the Institute of Food Technologists*, Anaheim, CA, USA, 1990.

173. P. Kolster, J.M. Vereuken and L.A. De Graaf in *Plant Proteins from European Crops: Food and Non-Food Applications*, Eds., J. Guéguen and Y. Popineau, Springer-Verlag, Berlin, Germany, 1998, p.305.

174. C. Farnum, D.W. Stanley and J.I. Gray, *Canadian Institute of Food Science and Technology Journal*, 1976, **9**, 201.

175. H.J. Park, C.L. Weller, P.J. Vergano and R.F. Testin, *Journal of Food Science*, 1993, **58**, 6, 1361.

176. J.H. Briston, *Plastics Films*, 3rd Edition, John Wiley and Sons, New York, NY, USA, 1998.

177. L. Allen, A.I. Nelson, M.P. Steinberg and J.N. McGill, *Food Technology*, 1963, **17**, 1437.

178. H.J. Park and M.S. Chinnan in *Proceedings of the International Winter Meeting, American Society of Agricultural Engineers*, Chicago, IL, USA, 1990, Paper No. 90-6510.

179. R.D. Hagenmaier and P.E. Shaw, *Journal of Agricultural and Food Chemistry*, 1990, **38**, 9, 1799.

180. W. Landman, N.V. Lovegren and R.O. Feuge, *Journal of the American Oil Chemists' Society*, 1960, **37**, 1.

181. A.W. Myers, J.A. Meyer, C.E. Rogers, V. Stannett and M. Szwarc, *TAPPI Journal*, 1961, **44**, 58.

182. *The Wiley Encyclopedia of Packaging Technology*, Eds., M. Bakker and D. Eckroth, John Wiley and Sons, New York, NY, 1986.

183. J. Poyet in the *Symposium sur l'Alimentarité dans les Matières Plastiques et les Caoutchoucs*, Paris, France, 1993.

184. A. Gennadios, C.L. Weller and R.F. Testin in *Proceedings of the International Winter Meeting, American Society of Agricultural Engineers*, Chicago, IL, USA, 1990, Paper No. 90-6504.

185. R.J. Ashley in *Polymer Permeability*, Ed., J. Comyn, Chapman & Hall, London, UK, 1985, p.269.

186. C.C. Taylor in *The Wiley Encyclopedia of Packaging Technology*, Eds., M. Bakker and D. Eckroth, John Wiley and Sons, New York, NY, USA, 1986, p.159.

187. M. Salame in *The Wiley Encyclopedia of Packaging Technology*, Eds., M. Bakker and D. Eckroth, John Wiley and Sons, New York, NY, USA, 1986, p.48.

188. H. Mujica-Paz and N. Gontard, *Journal of Agricultural and Food Chemistry*, 1997, **45**, 10, 4101.

189. S. Guilbert, N. Gontard and L.G.M. Gorris, *Lebensmittel-Wissenschaft und -Technology*, 1996, **29**, 1-2, 10.

190. F. Debeaufort and A. Voilley, *Journal of Agricultural and Food Chemistry*, 1994, **42**, 12, 2871.

191. J.A. Torres and M. Karel, *Journal of Food Processing and Preservation*, 1985, **9**, 107.

192. J.A. Torres, M. Motoki and M. Karel, *Journal of Food Processing and Preservation*, 1985, **9**, 75.

193. B. Cuq and A. Redl in *Les Emballages Actifs*, Ed., N. Gontard, Editions Technique et Documentation, Lavoisier, Paris, France, 2000, p.43.

194. H.E. Swaisgood and G.L. Catignani, *Advances in Food and Nutrition Research*, 1991, **35**, 185.

195. C.L. Garcia-Rodenas, J-L. Cuq and C. Aymard, *Food Chemistry*, 1994, **51**, 3, 275.

12 Enzyme Catalysis in the Synthesis of Biodegradable Polymers

Amarjit Singh and David Kaplan

12.1 Introduction

Nature is responsible for the synthesis of a diverse set of polymers. These polymers function in information storage and transfer (DNA, RNA and proteins), energy storage (polyhydroxyalkanoates and polysaccharides), architectural and mechanical systems (fibrous proteins, polysaccharides) and catalysis (proteins and RNA). These polymers provide all of the requirements for survival for the organisms that synthesise them. This includes protection from environmental variables and threats, such as changes in available food supplies, desiccation, mechanical integrity and other features. Furthermore and perhaps most importantly for the present review, these polymers are 'programmed' in terms of structure and chemistry for controlled lifetimes. Biological systems cannot 'afford' to generate polymers that are not recyclable, which it can't put back into normal metabolic processes for reuse as the building blocks or elements in new structures and functions. Thus, biodegradability is an inherent feature of any biologically-derived polymer.

With this knowledge, it is logical to consider the key catalyst in these processes: enzymes, which are an important source for initiating reactions designed to generate biodegradable polymers that might have a wide range of potential uses, such as for information flow, energy storage and architectural functions. While biological systems are capable of generating a diverse set of polynucleotides, polysacccharides, polyesters, proteins and polyaromatics, isolating the catalysts responsible for these processes from the rest of the biological *milieu* should result in important control over the biosynthesis and the resulting structural and functional aspects of the polymers formed from these catalysts. The remarkable regioselective, chemoselective and enantioselective capabilities of enzymes, their ability to retain catalytic function in diverse and even non-natural environments such as organic solvents and supercritical fluids, their robust nature under some conditions and their ability to be chemically and genetically manipulated to optimise or modify functions represents the extraordinary opportunity presented by this group of catalysts. This is the focus of the present chapter. We address the impressive progress in the last ten to fifteen years in terms of *in vitro* enzyme-based polymerisation reactions to generate new biodegradable polymers.

By our previous definition for biological catalysts, any polymer generated through enzyme catalysis should be biodegradable, presuming that the monomers are naturally occurring – thus sufficient time on the evolutionary scale has been available for a degradative enzyme to develop. This would also suggest that even in the case of non-native building blocks, such as fluorinated amino acids or modified sugars, enzymes could evolve in time either in the laboratory, or naturally, to accommodate their biodegradation. Therefore, a key aspect to the field of enzyme-based polymer synthesis is that the products of such reactions should be biodegradable. This chapter will focus primarily on polyesters, polysaccharides and polyaromatics as products of *in vitro* enzymic synthesis reactions. For another recent review see Kaplan and co-workers [1]. We have neglected polyamides to maintain focus, however, recent reviews on this topic are available (for example, Gill and co-workers [2]; Döhren and co-workers [3] and Wong and co-workers [4]).

12.2 Polyester Synthesis

Microorganisms such as bacteria produce biodegradable polyhydroxyalkanoate polyesters for use as intracellular energy and carbon storage materials from a variety of different substrates such as sugars, alcohols, *n*-alkanes, *n*-alkenes, alkanoic and alkenoic acids. Isolated enzymes, mostly lipases, have been used as catalysts for the construction of polyester from various monomers, typically hydroxy acids or their esters, dicarboxylic acids or their activated derivatives with glycols, lactones, carbonates, oxirane with glycol and anhydrides with glycols. Remarkable properties of lipases like regio-, enantio-, and chemo-selectivity and mild reaction conditions in comparison to chemical processes have been exploited to produce functional polyesters, most of which are difficult to synthesise by conventional methodologies. Polyesters are particularly attractive for a variety of commodity polymer applications as well as in specialty biomedical polymer uses since their rates of degradation can be controlled through composition and processing.

12.2.1 Polycondensation of Hydroxyacids and Esters

Porcine pancreatic lipase (PPL), *Candida cylindracea* lipase (CCL), *Chromobacterium viscosum* lipase (CVL) *Candida antarctica* lipase (CAL), polyethylene glycol (PEG)-modified *Pseudomonas fluorescens* lipase (PEG-PFL) and *Candida rugosa* lipase (CRL) have been used for the construction of polyesters from the hydroxyacids [5-13] (**Table 12.1**). Usually a low molecular weight polymer is produced. Molecular sieves have been used to remove water produced during the reactions to increase the molecular weight of product. O'Hagan and co-workers [5] reported that 10-hydroxydecanoic acid was close to optimum length for CAL catalysed polymerisation as monomers with shorter and longer carbon chains (4-hydroxybutyric acid, DL-2-hydroxybutyric acid, glycolic acid and 16-hydroxyhexadecanoic

Table 12.1 Polymerisation of hydroxyacids and hydroxyesters with different enzymes

	Monomer Hydroxyacids	Enzyme	Ref.
1	HO-(CH$_3$)CH-(CH$_2$)$_8$-COOH	*C. rugosa* lipase	[7]
2	HO-CH$_2$-(CH$_2$)$_8$-COOH	*C. rugosa*, PEG-modified *P. fluorescens* lipase.	[5, 12]
3	HO-CH$_2$-(CH$_2$)$_9$-COOH	*C. cylindracea* lipase	[11]
4	HO-CH$_2$-(CH$_2$)$_{10}$-COOH	*C. rugosa, P. cepacia, P. fluorescens, C. cylindracea, C. viscosum* lipase	[6, 25]
5	HO-CH$_2$-(CH$_2$)$_{14}$-COOH	*C. rugosa, P. cepacia, P. fluorescens, C. cylindracea, C. viscosum* lipase	[6, 25]
6	HO-CH(C$_6$H$_{13}$)-(CH$_2$)$_{10}$-COOH	*C. rugosa* lipase	[6]
7	HO-CH(C$_6$H$_{13}$)-CH$_2$-CH=CH-(CH$_2$)$_7$-COOH	*C. rugosa, C.viscosum* lipase	[6]
	Hydroxyesters		
8	HO-(CH$_2$)$_3$-COOC$_2$H$_5$	*Pseudomonas sp.* lipase	[16]
9	HO-(CH$_2$)$_5$-COOC$_2$H$_5$	*Pseudomonas sp.* lipase	[16]
10	HO-(CH$_3$)-CH-CH$_2$-COOC$_2$H$_5$	*Pseudomonas sp.* lipase	[16]
11	HO-(CH$_3$)-CH-(CH$_2$)$_3$-COOC$_2$H$_5$	*Pseudomonas sp.* lipase	[16]
12	HO-(C$_{13}$H$_{27}$)CH-(CH$_2$)$_3$-COOC$_2$H$_5$	*Pseudomonas sp.* lipase	[16]
13	HO-(CH$_2$)$_{14}$-COOC$_2$H$_5$	*Pseudomonas sp.* lipase	[16]
14	HO-CH$_2$-(CH$_2$)$_3$-COOCH$_3$	Porcine pancreatic lipase	[9, 17]
15	HO-CH$_2$-(CH$_2$)$_4$-COOCH$_3$	Porcine pancreatic lipase	[9, 17]
16	HO-CH(CH$_3$)-(CH$_2$)$_4$-COOCH$_3$	Porcine pancreatic lipase	[10]
17	HO-CH(CH$_3$)-(CH$_2$)$_4$-COOCH$_2$CCl$_3$	Porcine pancreatic lipase	[10]
18	HO-CH(CH$_2$CH$_3$)-(CH$_2$)$_4$-COOCH$_3$	Porcine pancreatic lipase	[10]
19	HO-CH(C$_6$H$_5$)-(CH$_2$)$_4$-COOCH$_3$	Porcine pancreatic lipase	[10]
20	HO-CH(C$_6$H$_{13}$)-(CH$_2$)$_{10}$-COOCH$_3$	Porcine pancreatic lipase	[10]
21	HO-CH(C$_2$COOCH$_3$)-CH$_2$COOCH$_3$	Horse liver acetone powder, pig liver acetone powder, *Streptomyces griceus*	[15]
22	HO-CH(C$_6$H$_{13}$)-CH$_2$-CH=CH-(CH$_2$)$_7$-COOCH$_3$	Porcine pancreatic lipase	[10]

acid) failed to polymerise. A racemic monomer, 10-hydroxyundecanoic acid, was polymerised to produce (*S*)-enantiomer enriched polyester (60% enantiomeric excess) by CRL in dry hexane [7]. Ritter and co-workers [14] reported the formation of oligomers from cholic acid by self condensation reactions catalysed by the CAL (**Scheme 12.1**).

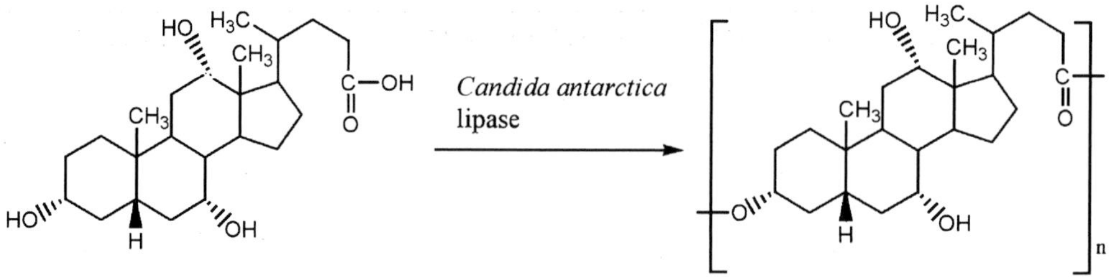

Scheme 12.1

Knani and co-workers [9-10] studied the polymerisation of hydroxyesters using PPL catalysis in hexane. Low molecular weight polymers were produced and the size of substituent at the $\in$ position influenced both reaction rate and enantioselectivity of the enzyme, the bulkier the substituent, the slower the reaction and higher the enantioselectivity. Liver acetone powders of horse and pig and the protease from *Streptomyces griceus*, were used for the asymmetric polymerisation of dimethyl β-hydroxyglutarate in dry hexane. Reactions were very slow and trimers were obtained after 10 days [15]. Dong and co-workers [16] carried out condensation polymerisation of linear hydroxyesters at 45 °C using lipase (EC 3.1.1.3) from *Pseudomonas sp.* (PSL) and formed polyesters having a number average molecular weight (Mn) in the range of 3,000 to 5,400. Gutman and co-workers reported that unsubstituted β-, δ-, ε-hydroxyesters underwent intermolecular transesterification by PPL in organic solvent to form polyesters with up to seven monomer units [17].

12.2.2 Polymerisation of Dicarboxylic Acids or Their Activated Derivatives with Glycols

Various combinations of dicarboxylic acid and their activated derivatives with glycols have been reacted enzymically to generate biodegradable polyesters under mild reaction conditions (**Table 12.2**). Okumura and co-workers [18] studied polyester formation from dicarboxylic acids (C_6~C_{14}) and diols. Those from 1,13-tridecanedioic acid and 1,3-propanediol were studied extensively. A mixture of products were separated by gel permeation chromatography (GPC) and determined by IR and MS. Trimer, pentamer and heptamer products were detected with a small amount of dimer and no tetramer and hexamer. It was assumed that pentamer was produced from dimer and trimer and heptamer from pentamer and dimmer, and dimer was the key substrate used by the lipase from *Aspergillus niger* (ANL) for the construction of the polymer. Uyama and co-workers [19-20] performed polymerisations in solvent-free systems and reported that polymer yields and molecular weights were strongly dependent on the methylene chain length of the monomers - the hydrophobicity of the monomers. The leaving groups (water or alcohol)

	Monomer	Enzyme	Refs.
	Table 12.2 Polymerisation of dicarboxylic acid and their activated derivatives with glycols		
1	HOOC-$(CH_2)_2$-COOH with HO-$(CH_2)_2$-OH	*Candida antarctica* lipase	[19]
2	HOOC-$(CH_2)_2$-COOH with HO-$(CH_2)_4$-OH	*Candida antarctica* lipase	[19, 20]
3	HOOC-$(CH_2)_4$-COOH with HO-$(CH_2)_2$-OH	*Aspergillus niger* lipase, *Candida antarctica* lipase	[18, 19, 21]
4	HOOC-$(CH_2)_4$-COOH with HO-$(CH_2)_3$-OH	*Aspergillus niger* lipase, *Candida antarctica* lipase	[18, 19]
5	HOOC-$(CH_2)_4$-COOH with HO-$(CH_2)_4$-OH	*Candida antarctica* lipase Lipozyme IM 30	[19-23]
6	HOOC-$(CH_2)_4$-COOH with HO-$(CH_2)_5$-OH	*Candida antarctica* lipase	[21]
7	HOOC-$(CH_2)_4$-COOH with HO-$(CH_2)_6$-OH	*Candida antarctica* lipase	[21]
8	HOOC-$(CH_2)_4$-COOH with HO-$(CH_2)_8$-OH	*Candida antarctica* lipase *Pseudomonas sp.* lipase	[19]
9	HOOC-$(CH_2)_4$-COOH with HO-CH_2-CH=CH-CH_2-OH	*Candida antarctica* lipase	[21]
10	HOOC-$(CH_2)_4$-COOH with HO-CH_2-CH≡C-CH_2-OH	*Candida antarctica* lipase	[21]
11	HOOC-$(CH_2)_5$-COOH with HO-$(CH_2)_2$-OH	*Aspergillus niger* lipase	[18]
12	HOOC-$(CH_2)_5$-COOH with HO-$(CH_2)_3$-OH	*Aspergillus niger* lipase	[18]
13	HOOC-$(CH_2)_6$-COOH with HO-$(CH_2)_2$-OH	*Aspergillus niger* lipase	[18]
14	HOOC-$(CH_2)_6$-COOH with HO-$(CH_2)_3$-OH	*Aspergillus niger* lipase	[18]
15	HOOC-$(CH_2)_6$-COOH with HO-$(CH_2)_4$-OH	*Candida antarctica* lipase	[18]
16	HOOC-$(CH_2)_6$-COOH with HO-$(CH_2)_6$-OH	*Pseudomonas sp.* lipase	[25]
17	HOOC-$(CH_2)_6$-COOH with HO-$(CH_2)_8$-OH	*Pseudomonas sp.* lipase	[25]
18	HOOC-$(CH_2)_6$-COOH with HO-$(CH_2)_{10}$-OH	*Pseudomonas sp.* lipase	[25]
19	HOOC-$(CH_2)_6$-COOH with HO-$(CH_2)_{12}$-OH	*Pseudomonas sp.* lipase	[25]

	Monomer	Enzyme	Refs.
20	HOOC-$(CH_2)_8$-COOH with HO-$(CH_2)_2$-OH	*Aspergillus niger* lipase, *Candida antarctica* lipase	[18-20]
21	HOOC-$(CH_2)_8$-COOH with HO-$(CH_2)_3$-OH	*Aspergillus niger* lipase, *Candida antarctica* lipase	[18-19]
22	HOOC-$(CH_2)_8$-COOH with HO-$(CH_2)_4$-OH	*Aspergillus niger* lipase, *Mucor miehei* lipase, *Pseudomonas cepacia*, *C. antarctica* lipase	[19, 20, 24]
23	HOOC-$(CH_2)_8$-COOH with HO-$(CH_2)_5$-OH	*Candida antarctica* lipase	[19]
24	HOOC-$(CH_2)_8$-COOH with HO-$(CH_2)_6$-OH	*Candida antarctica*, *P. cepacia* lipase	[19, 25]
25	HOOC-$(CH_2)_8$-COOH with HO-$(CH_2)_8$-OH	*Candida antarctica*, *P. cepacia* lipase	[19, 20, 24-26]
26	HOOC-$(CH_2)_8$-COOH with HO-$(CH_2)_{10}$-OH	*Candida antarctica*, *P. cepacia* lipase	[19, 25]
27	HOOC-$(CH_2)_8$-COOH with HO-$(CH_2)_{12}$-OH	*Candida antarctica*, *P. cepacia* lipase	[19, 20, 24, 25]
28	HOOC-$(CH_2)_6$-COOH with (benzene ring with two adjacent -CH_2OH groups, ortho)	*Pseudomonas cepacia* lipase	[25]
29	HOOC-$(CH_2)_8$-COOH with (benzene ring with two -CH_2OH groups, meta)	*Pseudomonas cepacia* lipase	[25]
30	HOOC-$(CH_2)_8$-COOH with HOH_2C-(benzene ring)-CH_2OH (para)	*Pseudomonas cepacia* lipase	[25]
31	HOOC-$(CH_2)_8$-COOH with HO-(benzene ring)-OH (para)	*Pseudomonas cepacia* lipase	[25]
32	HOOC-$(CH_2)_8$-COOH with HOH_2C-(cyclohexane ring)-CH_2OH	*Pseudomonas cepacia* lipase	[25]
33	HOOC-$(CH_2)_{10}$-COOH with HO-$(CH_2)_2$-OH	*Aspergillus niger* lipase	[18]
34	HOOC-$(CH_2)_{10}$-COOH with HO-$(CH_2)_3$-OH	*Aspergillus niger* lipase	[18]

Table 12.2 Continued...

	Monomer	Enzyme	Refs.
35	HOOC-$(CH_2)_{10}$-COOH with HO-$(CH_2)_6$-OH	*Pseudomonas cepacia* lipase	[25]
36	HOOC-$(CH_2)_{10}$-COOH with HO-$(CH_2)_8$-OH	*Pseudomonas cepacia* lipase	[25]
37	HOOC-$(CH_2)_{10}$-COOH with HO-$(CH_2)_{10}$-OH	*Pseudomonas cepacia* lipase	[25]
38	HOOC-$(CH_2)_{10}$-COOH with HO-$(CH_2)_{12}$-OH	*Pseudomonas cepacia* lipase	[25]
39	HOOC-$(CH_2)_{11}$-COOH with HO-$(CH_2)_2$-OH	*Aspergillus niger* lipase	[18]
40	HOOC-$(CH_2)_{11}$-COOH with HO-$(CH_2)_3$-OH	*Aspergillus niger* lipase	[18]
41	HOOC-$(CH_2)_{12}$-COOH with HO-$(CH_2)_2$-OH	*Aspergillus niger* lipase, *Candida antarctica* lipase	[18, 19]
42	HOOC-$(CH_2)_{12}$-COOH with HO-$(CH_2)_3$-OH	*Aspergillus niger* lipase	[18]
43	HOOC-$(CH_2)_{12}$-COOH with HO-$(CH_2)_4$-OH	*Candida antarctica* lipase	[19, 20]
44	HOOC-$(CH_2)_{12}$-COOH with HO-$(CH_2)_6$-OH	*P. cepacia* lipase	[25]
45	HOOC-$(CH_2)_{12}$-COOH with HO-$(CH_2)_8$-OH	*Candida antarctica*, *P. cepacia* lipase	[19, 24, 25]
46	HOOC-$(CH_2)_{12}$-COOH with HO-$(CH_2)_{10}$-OH	*Pseudomonas cepacia* lipase	[25]
47	HOOC-$(CH_2)_{12}$-COOH with HO-$(CH_2)_{12}$-OH	*Pseudomonas cepacia* lipase	[25]
48	HOOC-CH$\overset{cis}{=}$CH-COOH with HO-$(CH_2)_4$-OH	*Candida antarctica* lipase	[21]
49	HOOC-CH$\overset{trans}{=}$CH-COOH with HO-$(CH_2)_4$-OH	*Candida antarctica* lipase	[21]
50	HOOC-C≡C-COOH with HO-$(CH_2)_4$-OH	*Candida antarctica* lipase	[21]
51	HOOC-CH_2-C(=CH_2)-COOH with HO-$(CH_2)_4$-OH	*Candida antarctica* lipase	[21]
52	HOOC-CH_2-CH=CH-CH_2-COOH with HO-$(CH_2)_4$-OH	*Candida antarctica* lipase	[21]
53	HOOC-⟨⟩-COOH with HO-$(CH_2)_8$-OH	*Pseudomonas cepacia* lipase	[25]

Table 12.2 Continued...

	Table 12.2 Continued...		
	Monomer	Enzyme	Refs.
54	[cyclohexane structure with two COOH groups] with HO-(CH$_2$)$_8$-OH	*Pseudomonas cepacia* lipase	[25]
55	HOOC-[cyclohexane]-COOH with HO-(CH$_2$)$_4$-OH	*Pseudomonas cepacia* lipase	[25]
	Divinyl sebacate with the following diols		
56	HO-(CH$_2$)$_2$-OH	*P. cepacia*, *Candida antarctica* lipase	[28]
57	HO-(CH$_2$)$_4$-OH	*P. cepacia*, *C. antarctica*, *M. miehei*, *P. fluorenscens*, porcine pancreatic lipase	[28]
58	HO-(CH$_2$)$_6$-OH	*P. cepacia*, *Candida antarctica* lipase	[28]
59	HO-(CH$_2$)$_{10}$-OH	*P. cepacia*, *Candida antarctica* lipase	[28]
60	HO-CH$_2$-CH(OH)-CH$_2$-OH	*M. miehei*, *Candida antarctica* lipase	[30, 32-33]
61	HO-CH$_2$-CH(OH)-(CH$_2$)$_2$-OH	*M. miehei*, *Candida antarctica* lipase	[33]
62	HO-CH$_2$-CH(OH)-(CH$_2$)$_4$-OH	*M. miehei*, *Candida antarctica* lipase	[33]
63	HO-CH$_2$-CH=CH-CH$_2$-OH	*P. cepacia* lipase	[28]
64	HO-CH$_2$-C≡C-CH$_2$-OH	*P. cepacia* lipase	[28]
65	HO-CH$_2$-(CF$_2$)$_2$-CH$_2$-OH	*P. fluorensens*, *C. rugosa*, *R. oryzae*, *P. camemberti*, *M. javanicus*, *P. cepacia*, *C. antarctica*, porcine pancreatic lipase, *C. antarctica* lipase.	[42]
66	HO-CH$_2$-(CF$_2$)$_3$-CH$_2$-OH	*P. fluorensens*, *C. rugosa*, *R. oryzae*, *P. camemberti*, *M. javanicus*, *P. cepacia*, *C. antarctica*, porcine pancreatic lipase, *C. antarctica* lipase.	[42]

	Table 12.2 Continued...		
	Monomer	Enzyme	Refs.
67	HO-$(CH_2)_2$-$(CF_2)_4$-$(CH_2)_2$-OH	*P. fluorensens, C. rugosa, R. oryzae, P. camemberti, M. javanicus, P. cepacia, C. antarctica,* porcine pancreatic lipase, *C. antarctica* lipase.	[42]
68	HO-CH_2-CH-CH-CH-CH-CH_2OH 　　　OH OH OH OH	*C. antarctica* lipase	[31]
69	HO-H_2C-⬡-CH_2-OH		[29]
	Divinyl adipate with the following diols		
70	HO-$(CH_2)_2$-OH	*Pseudomonas cepacia* lipase	[27-28]
71	HO-$(CH_2)_4$-OH	*C. antarctica, M. meihei, P. cepacia, P. fluorenscens,* porcine pancreatic lipase	[28, 38-40]
72	HO-$(CH_2)_6$-OH	*P. cepacia, P. fluorescens* lipase	[27-28]
73	HO-$(CH_2)_{10}$-OH	*P. cepacia, P. fluorescens* lipase	[27-28]
74	HO-CH_2-CH=CH-CH_2-OH	*Pseudomonas cepacia* lipase	[28]
75	HO-CH_2-C≡C-CH_2-OH	*Pseudomonas cepacia* lipase	[28]
76	HO-CH_2CH(OH)-CH_2-OH	*C. antarctica, M. meihei, P. cepacia* lipase	[30, 41]
77	HO-CH_2CH(OH)-$(CH_2)_2$-OH	*C. antarctica* lipase	[41]
78	HO-CH_2CH(OH)-$(CH_2)_3$-OH	*C. antarctica, M. meihei, P. cepacia* lipase	41]

were removed from the reaction mixture, leading to a shift of the equilibrium towards polymerisation and polymer with a M_n of more than 1×10^4 was obtained in reactions under reduced pressure. Binns and co-workers [21-23] used *Aspergillus niger* NRRL 337, Lipozyme IM-20 and CAL (immobilised) for the polymerisation of unactivated diacid/diol systems for polyester synthesis. A variety of different combinations of diacids and diols have been studied. For example, condensation of 1,4-butane diol with adipic, maleic, fumaric, itaconic, (E)-hex-3-enedioic or acetylendicarboxylic acids, and condensation of adipic acid with ethane, propane, butane, pentane and hexane, (Z)-but-2-ene-1,4-diol and but-2-yne-1,4-diol. The authors analysed the enzymically synthesised reaction mixture by use of synthesised markers. For example in the reaction from a combination of adipic acid

(A) and butane-1,4-diol (B), after 4 hours a predominant amount of hydroxy terminated oligomers BAB and B(AB)$_2$ as well as butanediol and adipic acid (insoluble) were observed. The proposed oligomerisation process pathway was that adipic acid (A) acylates the enzyme: the acyl-enzyme complex is attacked by butane-1,4-diol (B) releasing AB. AB then acylates the enzyme *via* the acid terminus and the enzyme-AB complex is attacked by B to give BAB which does not react with the enzyme but, in due course, attacks an enzyme-AB complex to give B(AB)$_2$. The continuation of this cycle gives B(AB)$_n$. Kobayashi and co-workers [24-26] reported the dehydration polymerisation of diacids and diols to form polyester in water with use of lipase catalysts.

Kobayashi and co-workers [27-33] studied the polymerisation of divinyl ester of dicarboxylic acids with diols, triols and sorbitol. The polymerisation behaviour was strongly dependent on the monomer structure, enzyme origin and reaction conditions. Under appropriate conditions aliphatic polyester with a molecular weight of greater than 2×10^4 was obtained. The polymerisation of divinyl adipate with 1,4-butanediol by *Pseudomonas fluorescens* lipase (PFL) in isopropyl ether at 45 °C for 48 hours produced a polyester with a molecular weight of 6,700 with a 50% yield. The polycondensation of divinyl sebacate with triols (glycerol) gave polyesters with regioselectively incorporated 1,3-disubstituted glyceride units and a free secondary pendant hydroxyl group [34]. CAL was also used to produce crosslinkable polyesters [32]. Divinyl sebacate and glycerol were polymerised in the presence of unsaturated fatty acids oleic acid, linoleic acid and linoleinic acid. NMR analysis revealed that the reaction proceeded with regioselectivity during condensation of divinyl ester and glycerol and the pendant hydroxyl group of glycerol was acylated with fatty acids in the same reaction. Unsaturation in the fatty acid chain did not disturb the process. In the polymerisation of divinyl sebacate and polyol (sorbitol) the regioselectivity was controlled to yield sugar-containing polyesters in which the 1- and 6-positions of sorbitol were regioselectively acylated. Aromatic polyesters were produced from divinyl and dimethyl esters of isophthalic acid, terephthalic acid and *p*-phenylene diacetic acid. CAL showed high catalytic activity in the polymerisation of aromatic diesters to give polymers with molecular weights of several thousand daltons [29, 35]. Park and co-workers [36] used a protease from *Bacillus licheniformis* for the polyesterification of the diester of glutaric acid with aromatic diols.

Russell and co-workers [37-42] investigated lipase-catalysed transesterification of diesters with alkane diols, fluorinated diols and triols in organic solvents and supercritical fluids. PPL catalysed polytransesterification of bis(2,2,2-trichloroethyl) adipate by 1,4-butandiol suspended in fluoroform at 50 °C gave a polymer with molecular weight (MW) and polydispersity (PD) ranges from 739-2,189 and 1.02-1.23, respectively, by changing the pressure from 6.2-20.8 MPa. Kline and co-workers [41] polymerised divinyladipate with glycerol, 1,2,4-butanetriol and 1,2,6-trihydroxyhexane and observed that the weight average molecular weight of the resulting polyester varied according to the triol used and ranged from ~3,000 to 14,000 Da. Mesiano and co-workers [42] synthesised fluorinated polyesters from activated diesters and fluorinated diols using CAL. A maximum M_n of 5,289 was observed for solvent-free reaction between divinyl adipate and 3,3,4,4,5,5,6,6-octafluorooctan-1,8-diol.

Dimethylmaleate and dimethylfumarate differ in configuration around the double bond, the former is *cis* and the latter *trans*. Mezoul and co-workers [43-44] reported the synthesis of poly(hexamethylene maleate) and poly(hexamethylene fumerate) prepared in toluene at 60 °C in the presence of Novozyme as catalyst. The *cis* configuration (dimethylmaleate) of the double bond favoured the formation of macrolactones (24 wt% of the reaction product), whereas less than 1 wt% of lactone was formed during the polycondensation of dimethyfumarate with 1,6-hexanediol. Co-polymerisation of 1,6-hexanediol with a mixture of dimethylmaleate and dimethyfumarate also showed that cyclisation depended on the control of the maleate monomer and not on the type of catalyst. Wallace and co-workers [45] performed an enantioselective polymerisation of bis(2,2,2-trichloroethyl)*trans*-3,4-epoxyadipate with 1,4-butanediol using the enzyme PPL as a catalyst at ambient temperature in anhydrous ethyl ether (**Scheme 12.2**). End group analysis of the polymer by NMR gave an M_n of 5,300 Da, whereas GPC provided a MW of 7,900 Da. The same author used different diesters and diol combinations for the polymerisation using PPL in ether, tetrahydrofuran (THF) or hexane. Polyester molecular weights (M_n) were reported to be in the range of 3,200 to 8,200 by NMR and 4,900 to 11,800 by GPC for the polymers generated from same set of monomers [46-47]. Geresh and co-workers [48]

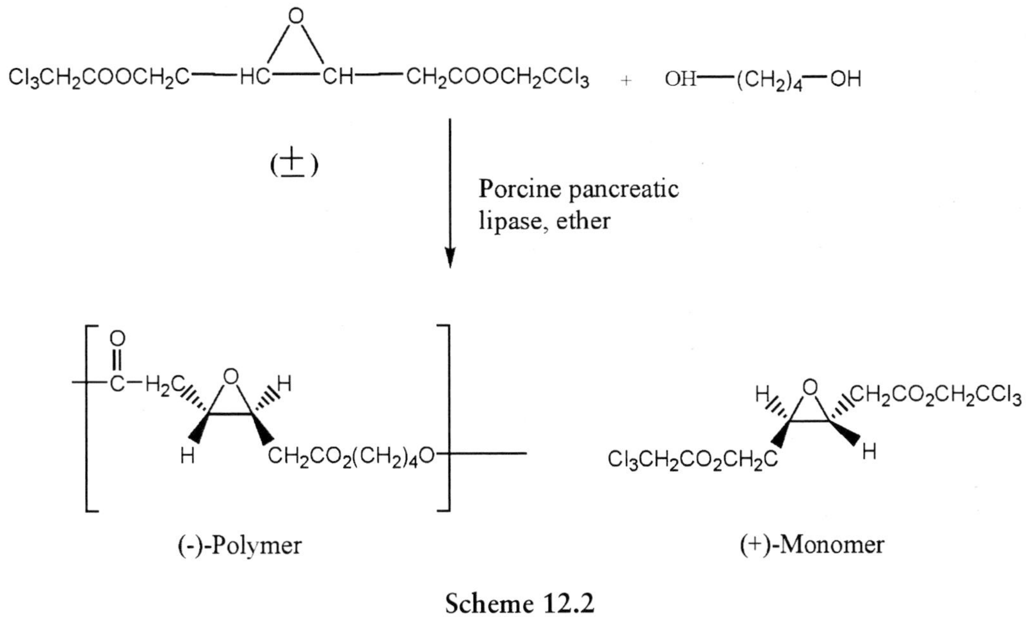

Scheme 12.2

enzymically synthesised polyester from diethyl, dipropyl, dichloroethyl, dimethoxyethyl, dicyanoethyl and ditrifluoroethyl activated esters of fumaric acid with 1,4-butanediol and the same authors [49-50] investigated the polyesterification of dichloroethylfumarate with different aliphatic and aromatic diols in THF and acetone using lipase from *Pseudomonas* and *Mucor* species.

Oxiranes and dicarboxylic anhydrides were polymerised in the presence of lipases to yield the corresponding polyesters [51-52]. Oxiranes, such as glycidyl phenyl ether and benzyl glycidate were copolymerised with succinic anhydride by lipases and preferably PFL between 60-80 °C to give biodegradable polyesters with molecular weights greater than 10,000. Succinic anhydride and glutaric anhydride polymerisation with glycols (1,6-hexanediol, 1,8-octanediol, 1,10-decanediol, 1,12-dodecanediol) by PFL catalysis at room temperature for five days yielded polyesters with M_n ranges of 800 to 2,900 depending on the diol used [53].

12.2.3 Ring Opening Polymerisation of Carbonates and Other Cyclic Monomers

Ring opening polymerisation of six membered cyclic carbonate, 1,3-dioxan-2-one has been investigated using lipases derived from *Candida antarctica*, *Candida cylindracea*, porcine pancreas, *Pseudomonas sp.* and *Mucor sp.* (**Scheme 12.3**) [54-57]. In some of these reports extraordinary differences in polymer molecular weights achieved using the same enzyme under similar experimental conditions are reported (**Table 12.3**). Matsumura and co-workers [55] reported 1,3-dioxan-2-one polymerisation yielding a MW of 84,700, PD of 3.9 with 0.5 wt% PPL and a MW of 169,000, PD 3.5 with 0.25 wt% PPL when reactions were carried out at 100 °C for 24 hours. Polymerisation occurred with PPL, CCL and PS lipases but not with Novozym-435. On the other hand, the best result in trimethylene carbonate bulk (TMC) polymerisations was at 70 °C for 120 hours using Novozym-435 from *Candida antarctica* with almost quantitative monomer conversion (97%) and the highest molecular weight (M_n = 15,000, PD = 2.2) of the seven lipases studied [56]. PPL

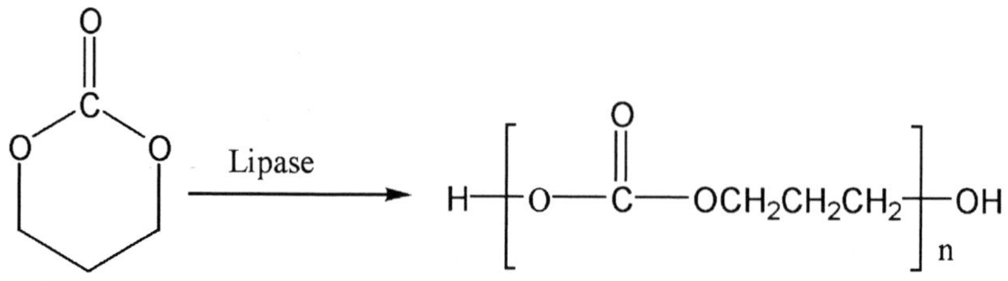

Scheme 12.3

Table 12.3 Ring opening polymerisation of carbonates and other cyclic monomers

Monomer	Enzyme/reaction conditions and polymer analysis	Ref.
1,3-dioxan-2-one	*Candida antarctica* lipase; M_n = 2500, PD = 3.4 by GPC against polystyrene standard, Reaction time 72 hours, conversion 100% at 75 °C.	[54]
	Mucor miehei lipase, M_n = 610, PD = 1.2 by GPC against polystyrene standard. Reaction time 72 hours, conversion 93% at 75 °C.	[54]
	Porcine pancreatic lipase; M_n = 800, PD = 1.4 by GPC against polystyrene standard. Reaction time 72 hours, conversion 80% at 75 °C.	[54]
	Porcine pancreatic lipase (0.25 wt%); MW = 169 000, PD = 3.5 by GPC against polystyrene standard. Reaction time 24 hours, conversion 96% at 100 °C.	[55]
	Pseudomonas sp. lipase (0.5 wt%); MW = 24000, PD = 1.9 by GPC against polystyrene standard. Reaction time 24 hours, conversion 97% at 100 °C.	[55]
	Candida cylindracea lipase (1 wt%), MW = 1000, PD = 1.2 by GPC against polystyrene standard. Reaction time 24 hours, conversion 5% at 100 °C.	[55]
1,4-dioxan-2-one	*Candida antarctica* lipase (5 wt%), MW = 28000, PD = 9.5 by GPC against polystyrene standard, Reaction time 48 hours, conversion 69% at 60 °C.	[60]
	Porcine pancreatic lipase (5 wt%), MW = 3890, PD = 2.2 by GPC against polystyrene standard. Reaction time 48 hours, conversion 66% at 100 °C.	[60]
	Bacillus thermoproteolyticus rokko; MW = 8610, PD = 6.2 by GPC against polystyrene standard, Reaction time 48 hours, conversion 36% at 100 °C.	[60]
	Candida antarctica lipase (5 wt%), MW = 41 000, by GPC against polystyrene standards. Reaction time 15 hours, conversion 77% at 60 °C.	[60]
5-methyl-5-benzyl oxycarbonyl-1,3-dioxan-2-one	*Pseudomonas fluorescens* lipase, M_n = 6100, PD = 1.6 by GPC against polystyrene standard. Reaction time 72 hours, conversion 97% at 80 °C.	[58]
	Porcine pancreatic lipase, M_n = 1300, PD = 1.3 by GPC against polystyrene standard. Reaction time 72 hours, conversion 98% at 80 °C.	[58]
	Pseudomonas cepacia lipase, M_n = 1450, PD = 1.0 by GPC against polystyrene standard. Reaction time 24 hours, conversion 50% at 80 °C.	[58]
	Candida antarctica lipase, M_n = 4400, PD = 2.1 by GPC against polystyrene standard. Reaction time 72 hours, conversion 86% at 80 °C.	[58]

Table 12.3 Continued...		
Monomer	Enzyme/reaction conditions and polymer analysis	Ref.
3(S)-isopropyl-morpholine-2,5-dione	Porcine pancreatic lipase (10 wt%), M_n = 14300, PD = 1.07 by GPC against polystyrene standard. Reaction time 72 hours, conversion 92% at 120 °C.	[62]
	Pseudomonas sp.(4.7 wt%), M_n = 12500, PD = 3.33 by GPC against polystyrene standard. Reaction time 168 hours, conversion 73.8% at 100 °C.	[61]
	Pseudomonas cepacia lipase (10 wt%), M_n = 4500, PD = 1.84 by GPC against polystyrene standard. Reaction time 72 hours, conversion 20.8% at 100 °C.	[61]
3(R)-isopropyl-morpholine-2,5-dione	Porcine pancreatic lipase (10 wt%), M_n = 12200, PD = 1.14 by GPC against polystyrene standard. Reaction time 72 hours, conversion 90% at 120 °C.	[62]
3(R,S)-isopropyl morpholine-2,5-dione	Porcine pancreatic lipase (10 wt%), M_n = 12000, PD = 1.15 by GPC against polystyrene standard. Reaction time 72 hours, conversion 90% at 120 °C.	[62]
3(S, 6R, S)-isopropyl-6-methyl-morpholine-2,5-dione	Porcine pancreatic lipase (10 wt%), M_n = 6900, PD = 1.16 by GPC against polystyrene standard. Reaction time 72 hours, conversion 9% at 120 °C.	[62]
3(S)-isobutyl morpholine-2,5-dione	Porcine pancreatic lipase (10 wt%), M_n = 9900, PD = 1.14 by GPC against polystyrene standard. Reaction time 72 hours, conversion 40% at 130 °C.	[62]
3(S)-*sec*-butyl-morpholine-2,5-dione	Porcine pancreatic lipase (10 wt%), M_n = 11500, PD = 1.09 by GPC against polystyrene standard. Reaction time 144 hours, conversion 90% at 110 °C.	[62]
6(S)-methyl-morpholine-2,5-dione	Porcine pancreatic lipase (10 wt%), M_n = 12000, PD = 1.05 by GPC against polystyrene standard. Reaction time 72 hours, conversion 76% at 100 °C.	[62]
6(R,S)-methyl-morpholine-2,5-dione	Porcine pancreatic lipase (10 wt%), M_n = 9300, PD = 1.04 by GPC against polystyrene standard. Reaction time 72 hours, conversion 34% at 120 °C.	[62]
Cyclobis(hexamethylene carbonate	*Candida antarctica* lipase, M_n = 12000, PD = 1.7 by SEC analysis agaist polystyrene standard. Reaction time 72 hours, yield 85% at 60 °C.	[63]
	Pseudomonas fluorescens lipase, M_n = 13000, PD = 2.1 by SEC against polystyrene standard. Reaction time 120 hours, yield 29% at 60 °C.	[63]
Cyclobis(diethylene glycol carbonate)	*Candida antarctica* lipase, M_n = 5300, PD = 1.8 by SEC analysis against polystyrene standard. Reaction time 72 hours, yield 72% at 60 °C.	[63]
	Pseudomonas fluorescens lipase, M_n = 9200, PD = 2.0 by SEC against polystyrene standard. Reaction time 120 hours, yield 57% at 60 °C.	[63]

exhibited high monomer conversion (>80%) over the 120 hours polymerisation time but the molecular weight (M_n = 3,500) of polymer produced was low. In contrast, Kobayashi and co-workers [54] reported the formation of low molecular weight poly (TMC) (M_n = 800, PD = 1.4) by PPL (50 wt%) catalysed polymerisation at 75 °C. Further, thermally treated CAL (heated in water at 100 °C for several hours) did not show catalysis at 75 °C but the enzyme from *Mucor miehei* and PPL (thermally inactivated) showed monomer conversion (36% and 97%, respectively). These data show that lipase from *Mucor miehei* and PPL did not loose catalytic ability even after thermal treatment or possible impurities (the enzyme contains basic and acidic groups in the side chain such as those found in lysine, glutamic acid and aspartic acid residues) in the enzyme acted as the catalysts for the polymerisation. The authors claimed that polymerisation proceeded through enzymatic catalysis as unchanged monomer was recovered in the absence of enzyme or using an inactivated enzyme. NMR spectroscopic results for polymer structural analysis have shown an absence of ether linkages and the presence of carbonate groups in the polymer chain to confirm that chain propagation proceeded without decarboxylation. It has been reported that partial decarboxylation takes place when the polymerisation of trimethylene carbonate was carried out in the absence of enzyme by cationic chemical initiators. The lipase catalysed polymerisation of the disubstituted trimethylene carbonate analogue 5-methyl-5-benzyloxycarbonyl-1,3-dioxan-2-one was also studied [58]. The bulk polymerisation, catalysed by lipase AK (from *P. fluorescens*) for 72 hours at 80 °C yielded 97% monomer conversion and a product with a M_n of 6,100. The benzyl ester protecting groups of the polymer were removed by catalytic hydrogenation (palladium/charcoal; Pd/C) in ethyl acetate to give the corresponding functional polycarbonate with pendant carboxylic acid groups in the main chain. Ring opening polymerisation of the cyclic phosphate (ethylene isopropyl phosphate), was demonstrated at 100 °C for 24 hours, M_n = 1,660 using 0.25 wt% PPL (**Scheme 12.4**) [59]. Higher polymerisation temperature and lipase concentration enhanced the polymerisation rate. Enzymic ring opening polymerisation of other cyclic monomers 1,4-dioxan-2-one, [60] 3(S)-isopropylmorpholine-2,5-dione and its derivatives have been studied [61-62]. Out of twelve enzymes (seven lipases, two esterases and three proteases) studied, 5 wt% immobilised lipase from *C. antarctica* polymerised 1,4-dioxan-2-one at 60 °C for 15 hours to the highest weight average molecular weight (MW = 41,000). Water in small amounts acted as a substrate for initiation of the process

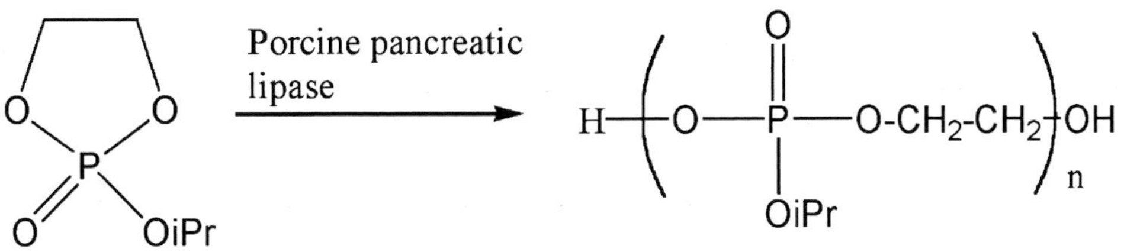

Scheme 12.4

but in excess acted as a chain cleavage agent. Enzymic ring opening polymerisations of the 6-membered cyclic depsipeptides: 3(S)-isopropyl-morpholine-2,5-dione, 3(R)-isopropyl-morpholine-2,5-dione, 3(R,S)- isopropyl-morpholine-2,5-dione, (3S, 6R, S)-3-isopropyl-morpholine-2,5-dione, 3(S)-isobutyl-morpholine-2,5-dione, 3(S)-*sec*-butyl-morpholine-2,5-dione, 6(S)-methyl-morpholine-2,5-dione and 6(R,S)-methyl-morpholine-2,5-dione was reported [62]. Cyclic dicarbonates, cyclobis(hexamethylene carbonate) and cyclobis(diethylene glycol carbonate) were polymerised by lipase from *C. antarctica* and *P. fluorescens* [63].

Poly(lactide-*co*-trimethylene carbonate) was prepared by lipase catalysed ring opening copolymerisation of different kinds of lactide (L,L-, D,D- and D,L-lactides) and trimethylene carbonate (**Table 12.4**) [64]. PPL showed the best results for both the polymerisation rate and the molecular weight attained (MW in the range of 20,000 Da) for the polylactide. The results indicated that poly(lactide-*co*-trimethylene carbonate) was a random co-polymer and the glass transition temperature (T_g) of the copolymer linearly decreased with increasing TMC content. Lipase-AK (isolated from *P. fluorescens*) catalysed the ring opening polymerisation of TMC with 5-methyl-5-benzyloxycarbonyl-1,3-dioxan-2-one (MBC) at 80 °C for 72 hours [65]. Reactivity of TMC compared to MBC was higher and the polymers produced were not ordered structures but random polymers. The benzyl ester protecting groups of poly(TMC-*co*-BMC) were removed by hydrogenolysis using H_2 over a Pd/C catalyst in ethylacetate to leave free pendant acid

Monomer + TMC	Enzyme, reaction conditions and polymer analysis	Ref.
Table 12.4 Ring opening co-polymerisation of trimethylene carbonate (TMC) with other cyclic monomers		
L,L-lactide [50:50]	Porcine pancreatic lipase, MW = 19100, PD = 1.7 by SEC against polystyrene standard. Reaction time 7 days, yield 34% at 100 °C.	[64]
D,D-lactide [50:50]	Porcine pancreatic lipase, MW = 12800, PD = 1.4 by SEC against polystyrene standard. Reaction time 7 days, yield 38% at 100 °C.	[64]
D,L-lactide [50:50]	Porcine pancreatic lipase, MW = 8100, PD = 1.4 by SEC against polystyrene standard. Reaction time 7 days, yield 25% at 100 °C.	[64]
5-methyl-5-benzyloxy carbony-1,3-dioxan-2-one [50:50]	*Pseudomonas fluorescens* lipase (4 wt%), M_n = 7500, PD = 4.4 by GPC against polystyrene standard. Reaction time 72 hours, yield 85% at 80 °C.	[65]
ω-pentadecalactone	Novozyme-435 (10 wt%), M_n = 18800, PD = 1.65 by GPC against polystyrene standard. Reaction time 24 hours, yield 90% at 70 °C.	[66]

groups. CAL (Novozyme 435) catalysed the ring opening copolymerisation of trimethylene carbonate and ω-pentadecalactone in toluene at 70 °C and gave random copolymers [66]. Changing the feed ratio of the comonomers resulted in regulation of copolymer composition. Chemical catalysts such as stannous octanoate, methylaluminoxane and aluminium isopropoxide have been used for the copolymerisation of TMC and PDL and the results showed that TMC had much greater reactivity than PDL. In contrast, for Novozyme-435 catalysed copolymerisation, PDL had a greater reactivity than TMC. Cyclic dicarbonates, cyclobis(hexamethylene carbonate) and cyclobis(diethylene glycol carbonate) have been co-polymerised with ε-CL and 12-dodecanolide using CAL in toluene at 60 °C for 48 hours (**Scheme 12.5**) [63].

A mechanism was proposed for carbonate ring opening polymerisation (**Scheme 12.6**). The mechanism was based on the identification of propanediol, a dimer of trimethylene carbonate (DTMC) and a trimer of trimethylene carbonate (TTMC) in the reaction mixture, along with the presence of symmetrical hydroxyl end group structures in the low and high molecular weight TMC polymerisation products.

1. The reaction of TMC with lipase to form the lipase-TMC enzyme-activated monomer (EAM) complex.

2. Reaction of EAM with water followed by rapid decarboxylation to form 1,3-propanediol.

3. Propagation as defined by the presence of a carbonate functionality involved in the formation of DTMC by the reaction of the EAM with 1,3-propanediol.

4. TTMC synthesis by the reaction of DTMC with the EAM.

5. Subsequent reactions to form high molecular weight chains.

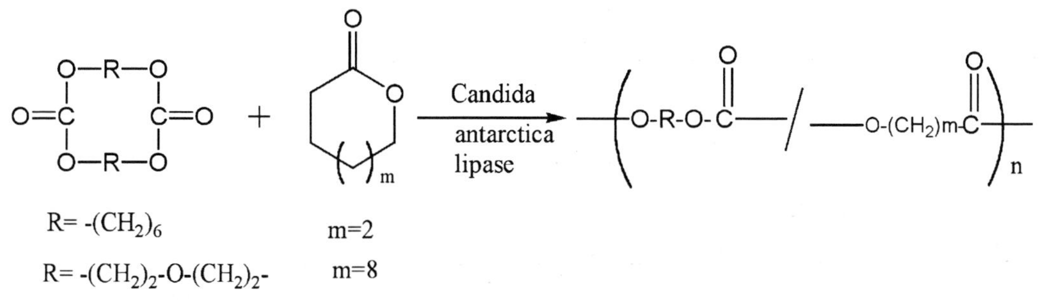

R= -(CH₂)₆   m=2
R= -(CH₂)₂-O-(CH₂)₂-   m=8

Scheme 12.5

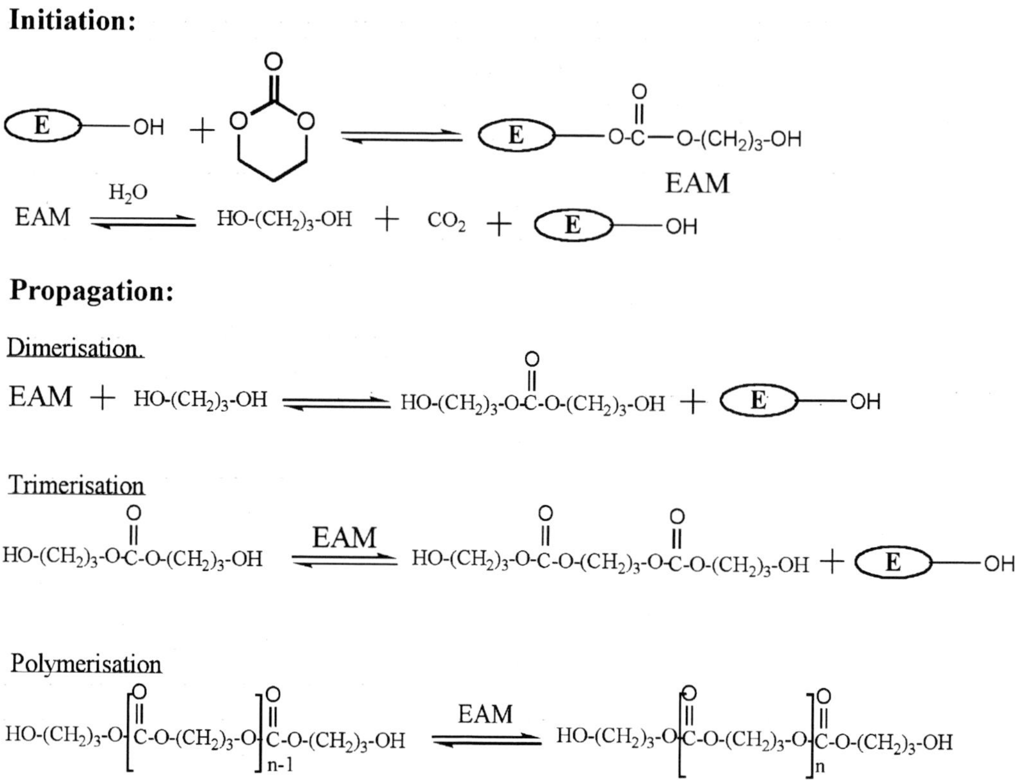

Initiation:

Propagation:

Dimerisation.

Trimerisation

Polymerisation

Scheme 12.6

12.2.4 Ring Opening Polymerisation and Copolymerisation of Lactones

Four membered ring lactones: β-propiolactone (β-PL) [26, 67-69], β-butyrolactone (β-BL) [70-75], benzyl β-malolactonate (BBM) [76] and α-methyl-β-propiolactone (MPL) [77-78] were polymerised using different lipases (**Table 12.5**). The lipase catalysed ring opening polymerisation of the four membered β-BL was first reported by Nobes and co-workers [73]. Poly(3-hydroxybutyrate) having weight average molecular weights (MW) ranging from 256 to 1,045 were prepared after several weeks of polymerisation using approximately equal weights of β-BL and lipase. An enantioselective polymerisation of four membered lactones was demonstrated. Racemic α-methyl-β-propiolactone was stereo-selectively polymerised by *Pseudomonas cepacia* lipase (PsCL) to generate optically active (S)-enriched polyester with enantiomeric excess of 50%. PHB-depolymerase (EC 3.1.1.75) was also used to polymerise the BL and the rate of polymerisation was faster compared to PPL and CCL under the same reaction conditions at 80 °C in bulk. Benzyl β-malonate was polymerised by PPL and Novozyme 435 lipase at 60 °C to yield poly(benzyl

	Lactone	Enzyme	Ref.
	4-membered ring		
1	β-PL	*C. cylindracea*, *P.cepacia*, porcine pancreatic, *P. fluorescens*, *C. antarctica*, *P. aeruginosa*, *A. niger* lipase	[26, 67-69, 73]
2	β-BL	Lipase ESL-001, *C. cylindracea*, *P. fluorescens*, porcine pancreatic, *P. cepacia*, PHB-depolymerase, *Pseudomonas sp.* lipase	[16, 70-75]
3	MPL	*P. fluorescens*, porcine pancreatic, *C. cylindracea*	[77-78]
4	BBM	Novozyme-435	[76]
	5-membered ring		
5	γ-BL	*P. cepacia*, *Pseudomonas sp.*, porcine pancreatic lipase	[16, 73]
6	γ-VL	*Pseudomonas sp.* lipase	[75]
7	γ-CL	*Pseudomonas sp.* lipase	[75]
	6-membered ring		
8	δ-VL	*C. cylindracea*, *P. fluorescens*, *P.* pancreatic, *R. japonicus* lipase	[26, 69, 79-80]
9	MVL	*C. antarctica* lipase	[81]
10	δ-DL	*Pseudomonas sp.* lipase	[16, 75]
11	δ-DODL	*Pseudomonas sp.* lipase	[16, 75]
	7-membered ring		
12	∈-CL	*C. antarctica*, *P. fluorescens*, porcine pancreatic, *P. cepacia*, *A. niger*, *C. cylindracea*, *P. delemer*, *R. japonicus*, hog liver esterase, *Pseudomonas sp.* lipase	[16, 26, 69, 73, 75, 79-80, 82-88]
13	αMCL	*C. antarctica* lipase	[81]
	9-membered ring		
14	8-OL	*C. antarctica*, *C. cylindracea*, *P. cepacia*, *P. fluorescens* lipase	[26, 69, 89]
	12-membered ring		
15	UDL	*P. fluorescens*, *C. cylindracea*, *C. antarctica* lipase	[26, 69, 84, 90-91]
	13-membered ring		
16	DDL	*C. cylindracea*, porcine pancreatic. *Pseudomonas sp.*, *P. fluorescens*, *C. antarctica*, *P. cepacia* lipase	[26, 69, 84-85, 91-92]
	16-membered ring		
17	PDL	*C. cylindracea*, *P. fluoroescens*, *Pseudomonas sp.*, *Mucor sp.*, *C. antarctica*, *M. meihei* lipase	[16, 26, 69, 90-91, 93-95]
	17-membered ring		
18	HDL	*C. antarctica*, *C. cylindracea*, *P. cepacia*, porcine pancreatic, *P. fluorescens* lipase	[69, 96]

β-maleate) having a MW greater than 7,000. The benzyl group of poly(benzyl β-maleate) was removed by catalytic hydrogenation using Pd/C to yield poly(β-D,L-malic acid).

Five membered, unsubstituted, lactone γ-butyrolactone (γ-BL) was polymerised by PPL or PCL [16, 73] into small oligomers with a degree of polymerisation (DP) of 8-11. In the *Pseudomonas sp.* lipase catalysed polymerisation of γ-VL and γ-CL, less than 10% conversion was observed at 60 °C for 480 hours [75]. Unsubstituted and substituted six membered lactone δ-valerolactone (δ-VL) [26, 69, 79-80] and α-methyl-δ-valerolactone (MVL) [81] were polymerised using *Rhizopus japonicus* lipase, CCL, PFL, PPL and CAL enzymes. For unsubstituted δ-VL, the reactions were run for 5-10 days and the highest molecular weights obtained were in the range of 2,000 Da. CAL catalysed polymerisation of α-methyl-δ-valerolactone yielded polyester with a M_n of up to 11,400 at 60 °C in 24 hours.

Several lipases (PPL, PFL, CAL, PCL) have been used in the ring opening polymerisation of ε-caprolactone [16, 26, 69, 73, 75, 79-80, 82-88]. The polymerisation of ε-caprolactone by PFL at 60 °C in bulk for 10 days generated polyesters with an average molecular weight of 7×10^3. *C. antarctica* lipase B catalysed polymerisation in bulk and produced a linear polymer with a MW of 4,701 Da and small amounts of cyclic oligomers, whereas the main product obtained in organic solvent was primarily cyclic in structure [86]. Immobilised CAL showed high catalytic activity toward the polymerisation of ε-caprolactone. A small amount of lipase (less than 1 wt%) was enough to induce the polymerisation. ε-Methyl-ε-caprolactone was polymerised in bulk (with immobilised CAL under mild reaction conditions) to aliphatic polyesters having hydroxyl groups at one end and carboxylic groups at the other end.

Lipase catalysed ring opening polymerisation of the nine membered lactone, 8-octanolide (8-OL), has been reported using various lipases in isooctane [26, 69, 89]. CAL and PFL showed high catalytic activity. In the polymerisation of 8-OL using PFL at 75 °C for 240 hours, a polymer with a M_n of 1.6×10^4 was obtained.

Anionic polymerisation of small and medium size lactones was reported to be fast (4-, 6- and 7-membered) when compared to macrolactones (12-, 13- and 16-membered) due to higher ring strain in the smaller lactones. On the other hand, four macrolactones, 11-undecanolide (UDL) [26, 69, 84, 90-91], 12-dodecanolide (DDL) [26, 69, 84-85, 91-92], 15-pentadecanolide (PDL) [16, 26, 69, 90-91, 93-95] and 16-hexadecanolide (HDL) [69, 96] showed unusual activity towards enzymic catalysis as compared to chemical polymerisations. Lipase PF-catalysed polymerisation of macrolactones proceeded much faster than that of ε-caprolactone. For the polymerisation of DDL, lipases CC, PC, PF and PPL showed high catalytic activity and the order of activity was as follows: lipase PC > lipase PF > lipase CC > PPL. The rate of UDL polymerisation using PFL was higher than that using CCL, whereas the polymerisation of UDL using CCL produced a polymer of higher molecular weight (M_n = 9,400) compared to that obtained using PFL (M_n = 8,400) [90]. Lipase PS-30 immobilised on Celite was used for bulk

PDL polymerisation and poly(PDL) with a M_n of 62,000 and a PD of 1.9 was obtained [93]. Recently, instead of bulk polymerisation, Novozyme 435 catalysed polymerisation of PDL was conducted in toluene (1:1) *w/v* Poly(PDL) with the highest molecular weight of 86,000 was obtained. Enzymic ring opening polymerisation of the 17-membered lactone, HDL, was performed at 75 °C for five days using PFL to give a polymer with a molecular weight of 5 x 10^5. Namekawa and co-workers [97] studied the lactones (ε-CL, 8-OL, UDL, DDL and PDL) in ring opening polymerisations in water. Among the various lipases used, the best results (M_n = 1,200 and 1,300 at 60 °C) were obtained with PCL and PFL, respectively. Chemoselective ring opening polymerisation of the lactone, 2-methylene-4-oxa-12-dodecanolide, was carried out using CAL yielding a polyester having the reactive exo-methylene group in the main chain [98]. According to the proposed mechanism for the enzyme catalysed polymerisation of lactones, the hydroxyl group of serine residue in the active site of lipase opens the lactone ring to form an acyl-enzyme intermediate (EAM). Polymer chain initiation is by a nucleophilic attack of water, which is probably contained in the enzyme on the acyl carbon of the intermediate to produce ω-hydroxycarboxylic acid, the shortest propagating species. In the propagation stage, the enzyme activated intermediate is nucleophilically attacked by the terminal hydroxyl group of a propagating polymer to produce an elongated polymer chain with one additional monomer unit (**Scheme 12.7**).

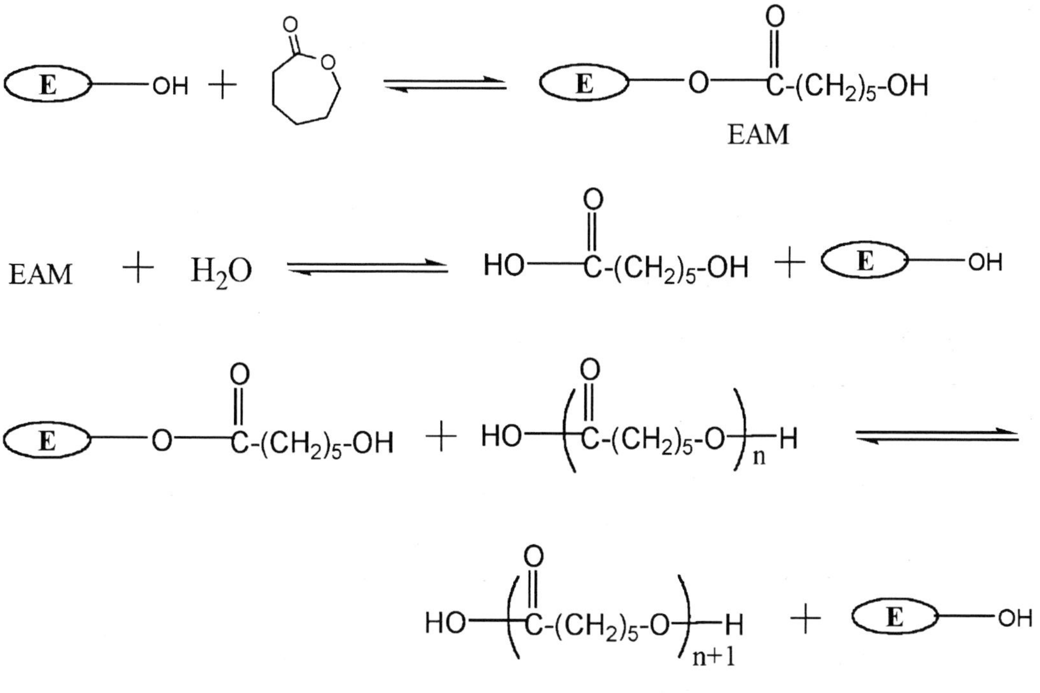

Scheme 12.7

PCL catalysed the enzymic co-polymerisation (**Table 12.6**) of β-PL with ε-caprolactone in bulk at 60 °C. Low molecular weight (M_n = 520) polyesters were produced. Ring opening co-polymerisation of another four membered lactone, benzyl malolactanate (BML), was enhanced, based on yield and molecular weight (MW), through the addition of small amounts of β-PL. The BML was polymerised in the presence of 17 mol% β-PL at 60 °C for 24 hours. The poly(BML-*co*-PL) containing 91 mol% BML units was obtained with a MW of 32,100 [99]. β-BL has been co-polymerised with ε-CL, DDL and 12-hydroxydodecanoic acid [100-101]. PPL mediated ring opening polymerisation of β-BL with 12-hydroxydodecanoic acid (HDDA) at 45 °C produced a copolymer (yield 17%, M_n = 1,800, PD = 1.11) containing hydroxy and carboxylic acid end groups. In the co-polymerisation of (±)-β-BL with PDL, the (S)-isomer was preferentially reacted to give the (S)-enriched optically active co-polymer with 69% enantiomeric excess of β-BL units [101]. PSL (*P. fluorescens*) catalysed the copolymerisation of γ-BL with ε-CL and formed a copolymer with a low molecular weight (M_n = 2.9 x 10^3) at low conversion (56%) after 20 days [16]. The six membered lactone, δ-VL has been polymerised with ε-CL and PDL [88, 94]. In the co-polymerisation of δ-VL with ε-CL in an equimolar ratio using PFL at 60 °C for 10 days, a copolymer with a molecular weight of 3.7 x 10^3 was obtained and was found to be a random copolymer structure. The molecular weight of the copolymer from PDL and δ-VL was lower (1.9 x 10^3) when compared to the co-polymer of δ-VL with ε-CL obtained under the same reaction conditions.

S.No.	lactone	lactone	Enzyme	Ref.
		Table 12.6 Lipase catalysed ring opening co-polymerisation of lactones		
1	β-PL	BML	*C. cylindracea* lipase	[99]
2	β-PL	ε-CL	*P. cepacia* lipase	[67]
3	β-BL	ε-CL	*C. antarctica* lipase	[100]
4	β-PL	DDL	*C. antarctica* lipase	[100]
5	β-PL	HDDA	Pancreatic lipase	[101]
6	γ-BL	ε-CL	*Pseudomonas sp.* lipase	[16]
7	δ-VL	ε-CL	*P. fluorescens* lipase	[88]
8	δ-VL	PDL	*P. fluorescens* lipase	[94]
9	ε-CL	δ-CL	*C. antarctica* lipase	[100]
10	ε-CL	8-OL	*C. antarctica* , *P. cepacia* lipase	[89]
11	ε-CL	PDL	*C. antarctica* lipase B, *P. cepacia*, *P. fluorescens* lipase	[94-95]
12	δ-CL	UDL	*C. antarctica* lipase	[100]
13	δ-CL	DDL	*C. antarctica* lipase	[100]
14	δ-CL	PDL	*C. antarctica* lipase	[100]
15	UDL	PDL	*P. fluorescens* lipase	[94]
16	DDL	8-OL	*C. antarctica* lipase	[89]
17	DDL	PDL	*P. fluorescens* lipase	[94]

The lipase catalysed copolymerisation of the nine membered lactone, 8-OL with ε-CL and DDL produced random copolymers [89]. In the CAL catalysed copolymerisation, 8-OL showed less reactivity than ε-CL, whereas the opposite effect was observed when PCL was used. In the co-polymerisation of 8-OL with ε-CL and DDL using lipase CA catalyst at 60 °C for 48 hours with a 50:50% feed ratio, the M_n of the copolymers were 5.4 x 10^3 and 8.6 x 10^3, respectively. Kobayashi and co-workers have explored the polymerisation of PDL with UDL and DDL in bulk at 60 °C for 240 hours using lipase PFL or PCL [94]. Copolymers with a M_n in the range of 2.0-2.1 x 10^3 were produced.

In the Novozyme-435 catalysed co-polymerisation of ε-CL and ω-PDL at 70 °C for 45 minutes in toluene (toluene to PDL 2:1 *v/w*), a copolymer with a yield of 88% and M_n of 2,000 Da was formed. Studies on monomer (ε-CL and ω-PDL) reactivity showed that ω-PDL reacted 13 times faster than ε-CL and the copolymer produced had a random sequence of the repeat units [95]. In the PSL catalysed co-polymerisation of ε-CL with hydroxyesters (ethyl lactate, ethyl 4-hydroxybutyrate and ethyl 15-hydroxypentadecanoate) low molecular weight copolymers were produced [16]. In the CAL co-polymerisation of racemic δ-CL with achiral ε-CL, UDL, DDL and PDL at 60 °C for 4 hours, a *R*-isomer enriched co-polymer with M_n of 2,000, 5,900, 7,000 and 6,100, respectively, were produced [100].

Uyama and co-workers [102-103] utilised PFL in the single step ring opening polymerisation of DDL and acylation of hydroxy termini with different vinyl esters to produce polymers having polymerisable groups only at the one terminus of the polymer chain. Kobayashi and co-workers [104] have reported that the lipase catalysed (CAL and PCL) polymerisation of lactones (12-, 13- and 16-membered), divinyl esters of adipic and sebacic acid and α,ω-glycols in one pot produced the corresponding ester copolymers in which two different type of polymerisations, ring opening polymerisation and polycondensation as well as transesterification, simultaneously occurred *via* the same enzyme intermediate to provide random copolymers. Polymerisation of macrolides (DDL, PDL) in the presence of preformed polyester (polycaprolactone) produced the corresponding copolyesters [105].

Cordova and co-workers [106] prepared macromonomers using *Candida antarctica* lipase B as catalyst. Ring opening polymerisation of ε-CL was initiated by alcohols which included 9-decenol, cinnamyl alcohol, 2-(4-hydroxyphenyl) ethanol and 2-(3-hydroxyphenyl) ethanol. In another approach acids and esters which included *n*-decanoic acid, octadecanoic acid, oleic acid, linoleic acid, 2-(3-hydroxyphenyl)acetic acid, 2-(4-hydroxyphenyl)acetic acid and 3-(4-hydroxyphenyl)propanoic acid were added to the prepolymerised ε-CL. Consequently, acid terminated PCL was formed. In the first approach 9-decenol-initiated PCL was formed (24 hours, 99% conversion of ε-CL) with an average MW of 1,980 Da. In the second approach linoleic acid terminated PCL was formed with an average MW of 2,400 Da (51 hours, 99% conversion). In an effort to simultaneously control both the hydroxyl and carboxyl end groups of macromers, esters, e.g., 9-decenyl oleate, 2-

(4-hydroxyphenyl)ethyl acrylate, 9-decenyl-2-(4-hydroxyphenyl)ethyl)acetate, methyl linoleate or a sequence of an alcohol and an acid, were added at various times during the course of *C. antarctica* lipase B-catalysed CL polymerisation. Polyesters bearing hydrophilic sugar monomers at the polymer termini were synthesised by CAL-catalysed polymerisation of ε-CL with the sugar initiators [107-108]. The enzyme selectively used the –OH group at the 6-position of the sugar to open the ring, thus no protection and deprotection for the other free hydroxyl groups was required. Similarly PPL and CLONEZYME ESL-001 catalysed graft polymerisation of ε-CL on hydroxyethylcellulose to obtain cellulose-graft-poly(ε-CL) with a degree of substitution from 0.10 to 0.32 [109]. Cordova and co-workers [110-111] reported the selective synthesis of a poly(ε-CL) monosubstituted dendrimer by using a hexahydroxy functional dendrimer.

12.3 Oxidative Polymerisation of Phenol and Derivatives of Phenol

Phenol-formaldehyde polymers, including novolaks and resoles, have a number of applications in coatings, finishes, adhesives, composites, laminates and related areas. Concerns have been raised about the continued use of phenol-formaldehyde resins due to the various toxic effects of formaldehyde. Therefore, there has been active study of alternative sources of these types of oligomers and polymers with a consideration for environmental compatibility.

Horseradish peroxidase (EC 1.11.1.7) catalyses the covalent coupling of a number of phenols and aromatic amines using hydrogen peroxide as an oxidant. This process has been successfully used for the removal of toxic aromatic pollutants from industrial wastewaters [112-114]. Reactions were not feasible for the production of polyphenols because most of the phenols are insoluble in water and the phenolic dimers and trimers formed are insoluble in water and immediately fall out of the solution, thereby preventing further polymerisation to high molecular weight polymers. Klibanov and co-workers [115] reported the polymerisation of phenols by horseradish peroxidase catalysis in water miscible solvents such as dioxane, acetone, dimethylformamide (DMF) and methylformate to produce various phenolic polymers with average molecular weights from 400 to 2.6 x 10^4 Da depending on the composition of the reaction medium and the nature of the phenol. Polyphenols and their co-polymers have been prepared from a series of phenol monomers [116-125]. Physical and chemical properties of these homo- and co-polymers such as melting point, solubility, elemental analysis, molecular weight distribution, infrared absorption, solid state ^{13}C NMR, thermal gravimetric analysis and differential scanning calorimetry were determined. The reactions were conducted in monophasic solvent, reverse micelles and air-water interfaces in a Langmuir system. In a monophasic solvent system containing 85% dioxane and 15% water, a polymer with a molecular weight as high as 4 x 10^5 was produced by the polymerisation of *p*-phenylphenol [124]. To carry out the reaction in reverse micelles (reverse micelles are organised surfactant structures that form hollow spheres usually in the nanometer length scale in organic solvent, the continuous phase, with an aqueous core within the sphere) the reaction conditions included surfactant, bis(2-ethylhexyl)sulfosuccinate, sodium salt, (AOT), (HEPES buffer and horseradish peroxidase enzyme. The concentrations of reactants were adjusted to generate a reverse micellar solution with a water to AOT molar ratio (W_o) of 15 and a final enzyme concentration of 12.5 μM. The monomer was then introduced and reactions initiated by the addition of 0.2 M of 30% H_2O_2. *p*-Ethylphenol polymerisation was extensively studied in reverse micelles to examine the feasibility of reaction in terms of kinetics, monomer conversion and the morphology of the particles generated.

In a comparative study of two systems (**Table 12.7**), (1) the monophasic organic solvent systems of dioxane plus water and (2) the reverse micellar system, the distinction in polymerisation lies in the oligomer-to-polymer ratio (soluble-to-insoluble product ratio).

Table 12.7 Quantitative aspects of polyethyl phenol synthesised in various media			
Reaction medium	Monomer conversion (after 2 hours)	Polymer yield, polymer produced/monomer converted	Comments
1. Reverse micelles W_o = 15 [AOT] = 0.15 M	≈ 95%	≈ 95%	Monomer soluble; high enzyme dispersion (solution clear); minimal oligomer formation; polymer precipitates
2. Iso-octane	< 5%	95-100%	Monomer soluble; poor enzyme dispersion (in insoluble aggregates); little oligomer formation; polymer at the air/ isooctane interface
3. Dioxane (85%) in water	95%	≈ 20%	Monomer soluble; fairly high enzyme dispersion (enzyme in suspension, cloudy solution); significant oligomer formation; polymer precipitates
4. Water	35%	≈ 55%	Poor monomer solubility; high enzyme dispersion (clearly soluble); some oligomer formation; polymer precipitates

This difference could originate because the two systems have different solvating abilities and hence may sustain the growing chain in solution to differing extents.

Oligomer products are more predominant in the dioxane/water system than in the reversed micellar system and the reactions are fast in reverse micellar system as compared to monophasic organic solvents. Polymer generated in reverse micellar systems had narrower molecular weight distributions. In reverse micellar environments, the precipitated polymer particles acquired a spherical morphology (**Figure 12.1a**) while the polymer synthesised in bulk solvent systems of dioxane (85%) and water (15%) did not show this characteristic morphology (**Figure 12.1b**). Templating effects during polymerisation in reverse micellar environments could be the reason for the generation spherical morphology particles.

The morphology of the polymer is effected by phase composition and a 3:1 surfactant (AOT) to monomer ratio has to be maintained for the generation of the spherical morphology. Observations of interest from the micrographs are the following. Comparison of **Figure 12.1c** and **12.1d** indicates that increasing AOT at constant *p*-ethylphenol concentration results in a small decrease in particle size. Secondly, a comparison of **Figure 12.1d** and **12.1a** indicates an increase in average particle size when the monomer concentration was increased at constant surfactant concentration. Further increase in monomer concentration results in a crossover from spherical to non-spherical morphology (**Figure 12.1e** and **12.1f**). The spherical particles were soluble in benzene, THF, DMF and dimethylsulfoxide. Molecular weight measurements of polyethylphenol by GPC indicated a broad distribution typically centered at about 90 kDa. The efficiency of polymer synthesis decreased dramatically if the alkyl group on the monomer was longer than 3-4 units. Poly(*p*-butylphenol) synthesis was less efficient than poly(*p*-ethyl phenol) synthesis, with only about 40% monomer conversion and negligible precipitation of polymer. No detectable conversion was observed with *p*-octylphenol, *p*-nonylphenol and *p*-dodecylphenol in reverse micelles.

In the polymerisation of alkylphenols in aqueous organic solvents [126-129], the position and chain length of the alkyl substituent, as well as the solvent type, significantly affected the polymerisation. In the polymerisation of unbranched *p*-alkylphenols, the yield of polymer increased with increasing chain length of alkyl group from 1 to 5 carbons and the yield of the polymer obtained from heptylphenol was almost the same as that from the pentyl derivative. The yield of the polymer from *p*-isopropylphenol was higher than that from the unbranched analogue at the *p*-position. No polymerisation was observed in the polymerisation of *o*- and *m*-isopropylphenols. In the case of the polymerisation of unbranched alkylphenols in aqueous 1,4-dioxane, polymer yield increased with increasing chain length of the alkyl group from 1-5 carbons and the yield of the polymer from hexyl or heptylphenol was almost the same as that of the pentyl derivative. In the reverse micellar system the highest yield was obtained from ethylphenol. The polymerisation of hexyl phenol in the reverse micellar system produced no polymeric materials. On the other hand polymer was obtained

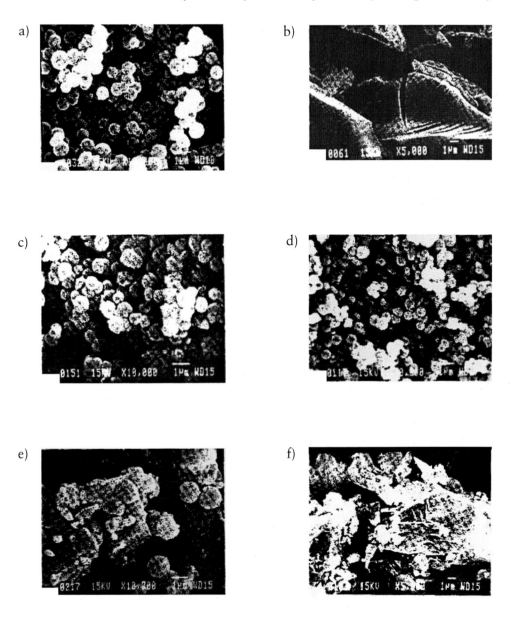

Figure 12.1 Scanning electron micrographs of polymer formed by enzymic synthesis in different synthetic conditions. (a) AOT 1.5 M, *p*-ethylphenol 0.3 M; (b) Monophasic organic solvent system of 85% dioxane and 15% water (by volume); (c) AOT 0.5 M, *p*-ethylphenol 0.15 M; (d) AOT 1.5 M, *p*-ethylphenol 0.15 M; (e) AOT 0.5 M, p-ethylphenol 0.3 M; (f) AOT 1.5 M, *p*-ethylphenol 1.5 M

in high yield in the aqueous 1,4-dioxane. The possible explanation for contrary results in reverse micelles is as follows. The enzyme is soluble in water and always present inside the reverse micelles. The phenol with shorter alkyl chain length is relatively hydrophilic in nature and prefers to stay inside the micelle leading to increased polymerisation rate. On the other hand phenols with a longer alkyl chain are hydrophobic and prefer to be inside the non-polar isooctane solution leading to a poor yield. The polymerisation behaviour of the *m*-substituted monomers greatly depends on the enzyme. Horseradish peroxidase readily polymerises monomers with small substituents, whereas for monomers with large substituents, a high yield was achieved by using soybean peroxidase as catalyst. The enzymic oxidative polymerisation of *p*-alkyl phenols by horseradish peroxidase gave a mixture of polyphenols containing phenylene and oxyphenylene units determined by NMR and IR as well as titration of the residual phenolic moiety of the polymer.

Kobayashi and co-workers [130-133] have synthesised soluble polyphenols and demonstrated that polymerisation parameters, enzyme origin, buffer pH, mixed ratio of alcohol and buffer, purity and amount of horseradish peroxidase and concentration and addition rate of hydrogen peroxide, strongly affect the molecular weight and solubility of the polymers. Of the organic solvents used in compositions with buffer, methanol content, 50% or 75% gave a completely soluble polyphenol in DMF. Polymer structure was determined by IR and NMR and found to contain a mixture of phenylene and oxyphenylene units. The number of oxyphenylene units increased with increasing methanol content, varying in the range of 32-59%. The more oxyphenylene units present, the better the solubility of the polyphenol. Polyphenol particles in the sub-micron range were prepared in a mixture of 1,4-dioxane and phosphate buffer with horseradish peroxidase catalysis by dispersion polymerisation [134].

Peroxidases (horseradish and soybean) and laccases (derived from fungus *Pycnoporus coccineous* and *Myceliophthore*) catalysed the polymerisation of 4-hydroxybenzoic acid derivatives (3,5-dimethoxy-4-hydroxy benzoic acid (syringic acid) and 3,5-dimethyl-4-hydroxy benzoic acid) in a mixture of water-miscible organic solvent and acetate buffer at room temperature under air to give poly(1,4-phenylene oxide) in good yield (**Scheme 12.8**) [135-136]. Both enzyme types and solvent composition greatly effected polymerisation results. When laccases (EC 1.10.3.2) from *P. coccineous* and *Myceliophthore* were used as catalysts no polymerisation of syringic acid was achieved when either pure acetone or the buffer were used. In 50% acetone with acetate buffer (pH = 5) the highest yield (84%) was obtained, while the highest molecular weight (7.7×10^3) was obtained when 40% acetone was used. Enzymic oxidative polymerisation of 4-hydroxybenzoic acid derivatives involves the elimination of carbon dioxide and hydrogen from the monomer to form poly(1,4-phenylene oxide) (PPO). No polymerisation was observed when unsubstituted 4-hydroxybenzoic acid was used as a substrate for peroxidases and laccases under similar experimental conditions. NMR, IR and MALDI-TOF results showed that the polymers

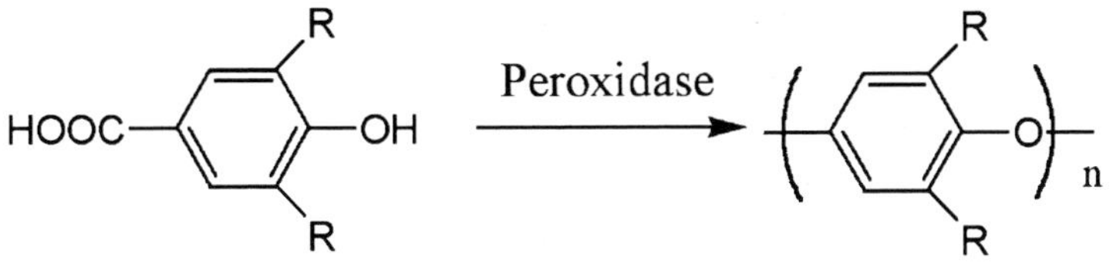

R= -O-CH₃, -CH₃

$$R= -O\text{-}CH_3, -CH_3$$

Scheme 12.8

were composed exclusively of 1,4-oxyphenylene units with a phenolic hydroxyl group at one chain terminus and a benzoic acid group at the other chain end. PPO was first prepared by oxidative coupling polymerisation of 2,6-dimethylphenol by using a copper-amine catalyst [137]. PPO is synthesised commercially by this process, which involves side reactions resulting in the incorporation of Mannich-base and 3,5,3′,5′-tetramethyl-4,4′-diphenoquinone units into the polymer. Enzymic polymerisation of 2,6-dimethylphenol using horseradish peroxidase, soybean peroxidase and laccase derived from *Pycnoporus coccineous* in aqueous organic solvent at room temperature produced polymers with molecular weights in the range of several thousand daltons. The polymerisation behaviour was dependent on the enzyme type as well as solvent composition. The resulting polymer was exclusively composed of dimethyl-1,4-oxyphenylene units according to NMR and MALDI-TOF determinations [138].

4,4′-Biphenyldiol ($HO\text{-}C_6H_4\text{-}C_6H_4\text{-}OH$) [139], bisphenol-A (2,2-bis(4-hydroxyphenyl)propane) ($HO\text{-}C_6H_4\text{-}C(CH_3)_2\text{-}C_6H_4\text{-}OH$) [140] and 4,4′-dihydroxydiphenyl ether ($HO\text{-}C_6H_4\text{-}O\text{-}C_6H_4\text{-}OH$) [141] have been polymerised using horseradish peroxidase in aqueous organic solvents. Dordick and Wang used CAL to regioselectively acylate thymidine at the 5′-hydroxyl position with a trifluoroethylester derivative of *p*-hydroxyphenylacetic acid in anhydrous acetonitrile [142]. This was followed by polymerisation of the phenolic nucleoside derivative catalysed by the peroxidase from soybean hulls in the presence of hydrogen peroxide and aqueous buffer containing 60% (v/v) acetonitrile (**Scheme 12.9**). Redox-active polymers are useful in applications such as batteries, sensors, electrical conductors and antioxidants [143]. Dordick and co-workers synthesised of poly(hydroquinone) by the enzymic oxidative polymerisation of glucose-β-D-hydroquinone and subsequent acid hydrolysis [144]. In the first step β-glucuronidase (EC 3.2.1.31) from bovine liver was used to regiospecifically attach glycoside to one of the hydroxyl groups of hydroquinone in aqueous solution to give glucose-β-D-hydroquinone (arbutin). In the second step, arbutin was polymerised by peroxidases from horseradish and soybean in aqueous buffer to form water soluble polymers (**Scheme 12.10**). Deglycosylation

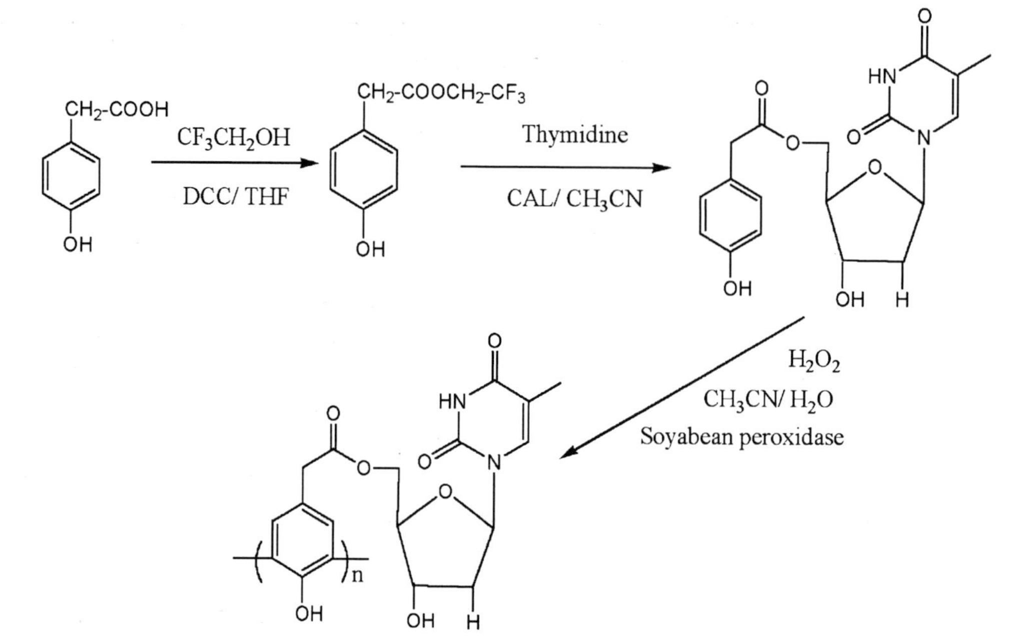

Scheme 12.9

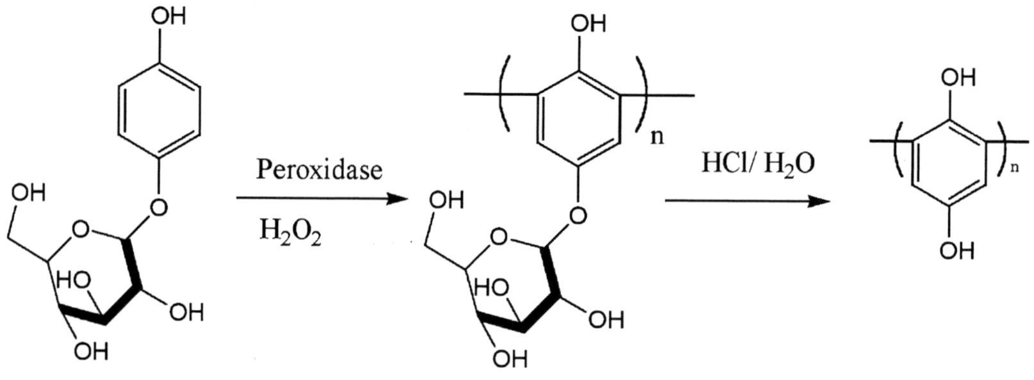

Scheme 12.10

of the poly(arbutin) gave poly(1,4-dihydroxy-2,6-phenylene). This polymer was different from electrochemically synthesised poly(hydroquinone) which is poly(1,4-dihydroxy-2,5-phenylene). Kobayashi and co-workers [145] synthesised a new kind of poly(hydroquinone) derivative with a mixture of phenylene and oxyphenylene units by using the peroxidases (horseradish and soybean) to catalyse the polymerisation of 4-hydroxyphenyl benzoate and the subsequent hydrolysis of the resulting polymer (**Scheme 12.11**). Similarly Tripathy and co-workers [146] synthesised a photoactive azopolymer, poly(4-phenylazophenol) by horseradish peroxidase catalysed polymerisation in acetone and sodium phosphate buffer. Bilirubin oxidase (EC 1.3.3.5) was shown to catalyse the regioselective polymerisation of 1,5-dihydroxynaphthalene to a polymer in a mixed solvent composed of dioxane, ethylacetate and acetate buffer [147]. Chalcones are intermediates in the biosynthesis of lignins in plants. Oligomers were produced by horseradish peroxidase-mediated polymerisation of aminochalcones in a mixture of 1,4-dioxane and phosphate buffer (**Scheme 12.12**) [148]. Acetaminophen is a widely used analgesic and antipyretic drug. Horseradish peroxidase-mediated polymerisation of acetaminophen was carried out in phosphate buffer at 25 °C [149-150]. Using NMR spectroscopy it was shown that the reaction mixture was composed of two dimers, three trimers and one tetramer. Oligomer formation was due to the formation of covalent bonds between carbons *ortho* to the hydroxyl group and to a lesser extent, between the carbons *ortho* to the hydroxyl group and the amido group of another acetaminophen molecule.

Cardanol is an analogue of phenol with a 15-carbon unsaturated chain with zero to three double bonds in the *meta*-position and is the main constituent obtained after the

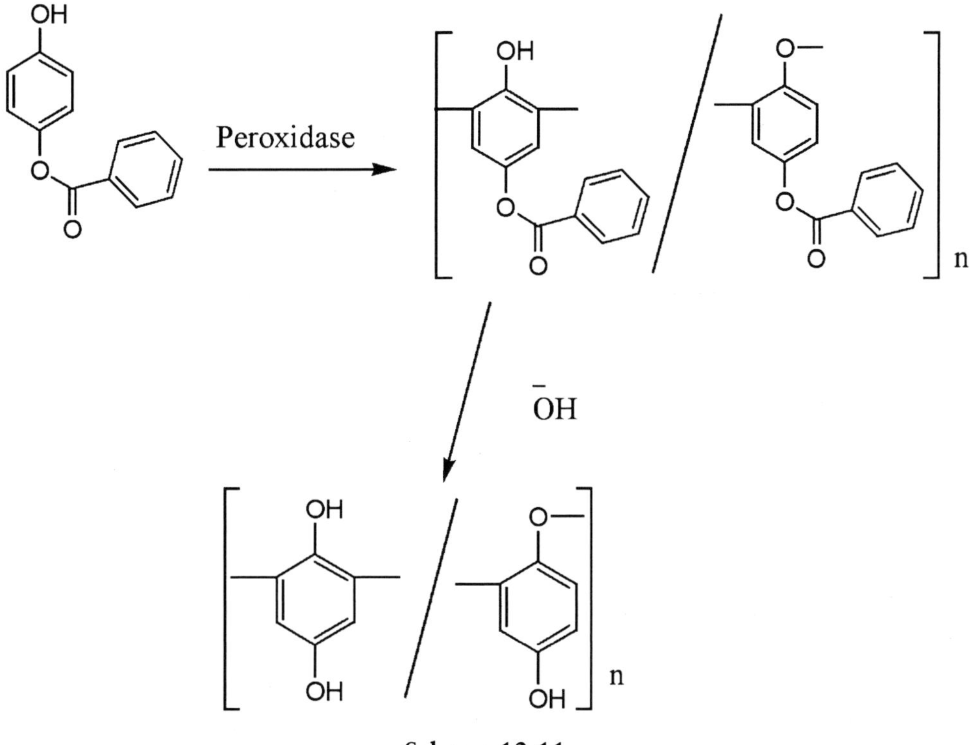

Scheme 12.11

415

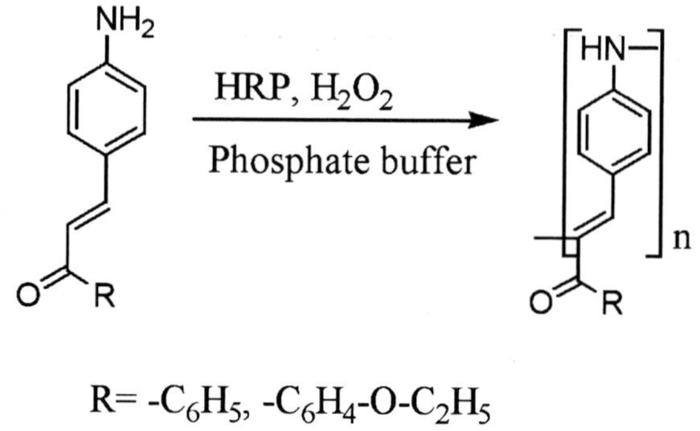

R= -C$_6$H$_5$, -C$_6$H$_4$-O-C$_2$H$_5$

Scheme 12.12

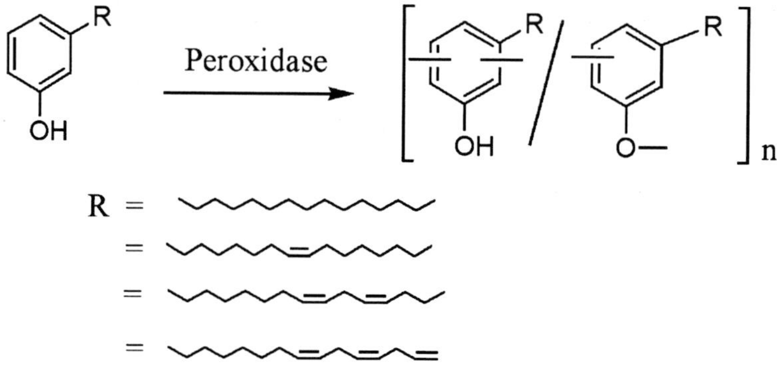

Scheme 12.13

thermal treatment of cashew nut shell liquid. Cardanol was polymerised with soybean peroxidase in a mixture of acetone:buffer (75:25), to form an oily polymer. ^{1}H NMR and FTIR analysis indicated that double bonds in the side chain were not affected by the polymerisation conditions and poly(cardanol) was a mixture of phenylene and oxyphenylene units (**Scheme 12.13**) [151]. Urushi is a Japanese traditional natural paint. The main component of urishi are urishinols, whose structure is a catechol derivative with unsaturated hydrocarbon chains consisting of a mixture of monoenes, dienes and trienes at *meta* or *para* position of catechol. Kobayashi and co-workers [152] have carried out laccase-catalysed crosslinking reactions with urishinol analogues to prepare urishi. Horseradish peroxidase catalysed polymerisation of *m*-ethynylphenol (HO-C$_6$H$_4$-C≡CH) having more than one polymerisable group showed that the phenol moiety was chemoselectively polymerised when acetylene or methacryl group were present [153].

12.4 Enzymic Polymerisation of Polysaccharides

Cellulose is the most abundant compound produced photo-chemically by plants on the earth. Kobayashi and co-workers [154-167] synthesised cellulose, chitin, xylan and non-natural derivatives of these polymers. Cellulose was produced *in vitro* by polycondensation reactions using β-cellobiose fluoride and cellulase (EC 3.2.1.4) (**Scheme 12.14**). The reaction proceeded with complete regio- and stereoselectivity, giving rise to cellulose having β (1 → 4) linkages. Polymerisation reactions generated cellulose with DP of 22 with a 54% yield when carried out in an acetonitrile/acetate buffer (5:1) mixed solvent at 30 °C with 5 wt% of cellulase. Cellulase from *Trichoderma viride* was the most effective enzyme for the synthesis of cellulose when compared with cellulases from *Aspergillus niger* or *Polyporus tulpiferae*. β-Glucosidase from almonds did not catalyse the polymerisation. Cellulose I is the native form of cellulose, with parallel glucan chains and a thermodynamically metastable form, produced by living organisms. Cellulose II with antiparallel glucan chain is the more stable form. Kobayashi and co-workers [166] reported that cellulose I and cellulose II can be selectively synthesised *in vitro* using the enzymic polymerisation of β-cellobiosyl fluoride monomer and this selectivity could be controlled by changing the purity of the enzyme and the polymerisation conditions. A new term 'choroselectivity' was therefore proposed, which is concerned with the intermolecular relationship in packing of polymers having directionality in their chains. Various β-cellobiose derivatives (methyl β-cellobioside, allyl β-cellobioside, trifluoroethyl

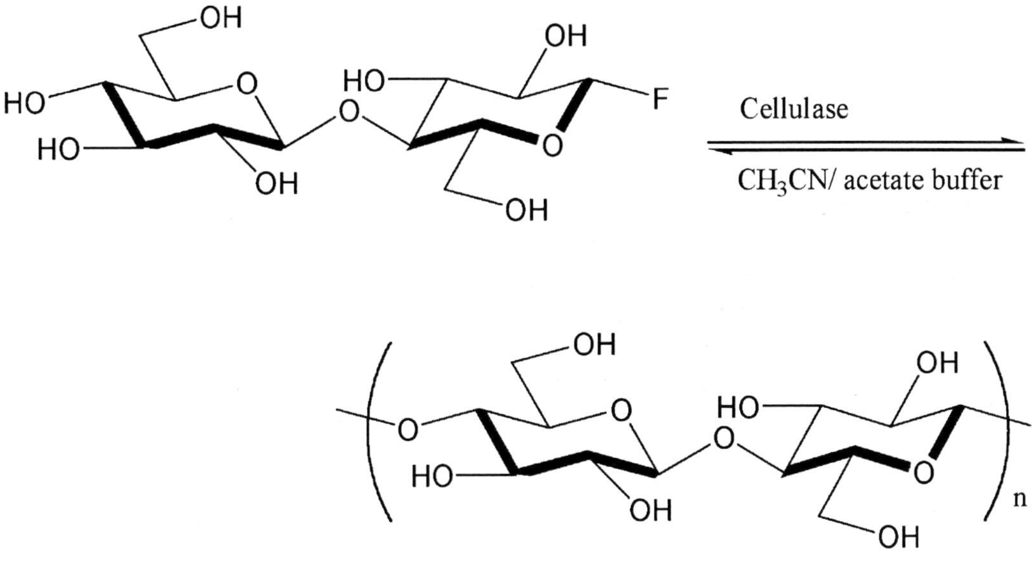

Scheme 12.14

β-cellobioside, methyl β-thiocellobioside, phenyl β-cellobiosyl sulfoxide and 1-O-acetyl β-cellobiose) have been used in enzymic polymerisations using cellulase as catalyst. Among all the activated cellobiose substrates β-cellobiosyl fluoride gave the best result in terms of DP. β-Cellotriosyl fluoride (trimer) and β-cellotetreosyl fluoride (tetramer) were found to be rapidly hydrolysed in the enzymic reactions. Non-natural 6-O-methylated cellulose was produced with high regio- and stereo-selectivity by cellulase catalysis starting from 6-O-methyl-β-cellobiosyl fluoride.

Enzymic polymerisation of α-D-maltosyl fluoride using α-amylase (EC 3.2.1.1) as the catalyst in a mixed solvent of methanol-phosphate buffer (pH 7) produced oligomeric products with α- $(1 \rightarrow 4)$-glycosidic linkages [162]. Other substrates such as D-maltose, β-D-maltosyl fluoride and α-D-glucosyl fluoride gave no condensation products. These results indicated that α-D-maltosyl fluoride with an α-configuration was essential for α-amylase catalysed polymerisation. Malto-oligosaccharides are useful substrates as food additives, medicines and enzyme substrates for clinical research. Generally they are produced by the degradation reaction of polymers such as amylose, amylopectin and glycogen. Xylan is an important component of hemicellulose in plant cell walls. Xylan was synthesised by a transglycosylation reaction catalysed by cellulase with the use of β-xylobiosyl fluoride as substrate [155]. Cellulose-xylan hybrid polymers were synthesised by the polycondensation of β-xylopyranosyl-glucipyranosyl fluoride catalysed by xylanase (EC 3.2.1.32) from *Trichoderma viridei*, in a mixed solvent of acetonitrile and acetate buffer (**Scheme 12.15**) [156].

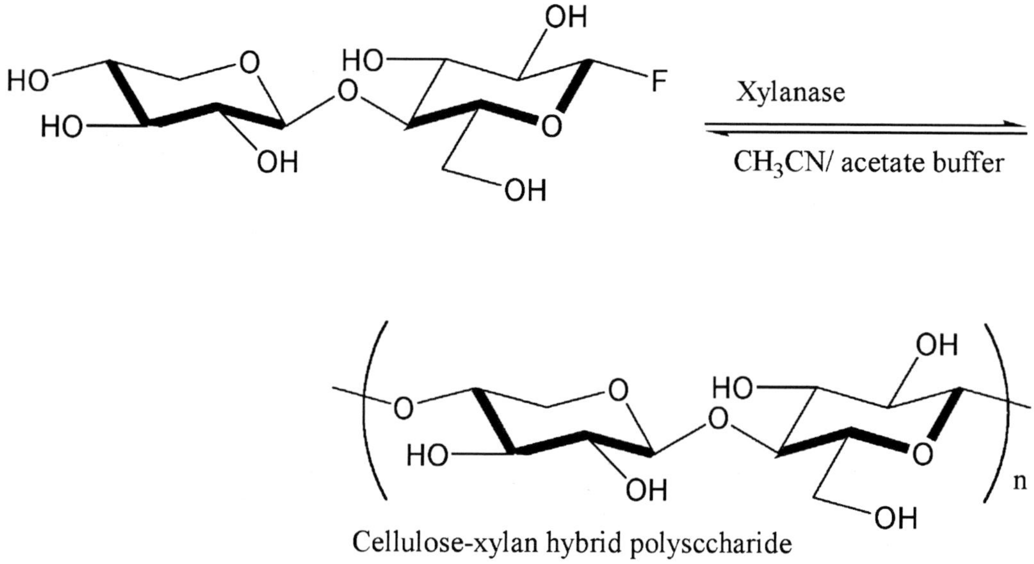

Cellulose-xylan hybrid polysccharide

Scheme 12.15

It was postulated that cellobiose (disaccharide) would be the preferred substrate for the polymerisation reactions since this is the molecular unit recognised by the binding site of enzymes such as cellulase in comparison to glucose (monosaccharide). There are two routes proposed for the chain propagation in these enzymic reactions. The first one (activated monomer mechanism) involves the formation of an active intermediate of the disaccharide unit by reaction of cellobiosyl fluoride and the cellulase, followed by an attack of the terminal 4′-hydroxy group of the propagating polymer. This interaction generates the product. In the second mechanism (active chain end mechanism), an active intermediate is formed on the chain end and the propagating process is realised by the attack of the 4′-hydroxyl group of the disaccharide unit on this intermediate.

Chitin is widely found in invertebrates and is one of the most abundant and widespread natural structural polysaccharides normally found in animals, comparable to the predominance of cellulose in plants. Kobayashi and co-workers [158-159] have produced chitin through chitinase-catalysed (EC 3.2.1.14) polymerisation of a chitobiose oxazoline derivative in phosphate buffer (**Scheme 12.16**). The product structure determined by cross-polarisation magic-angle-spinning (CP/MAS), ^{13}C-NMR spectroscopy and reported to be a β(1-4) linkage indicating regio and stereo-selective linkages between the chitobiose units and the inversion of configuration at C_1.

12.5 Conclusions

The challenges ahead are clear and do not involve the ability to make polymers *in vitro* using enzyme catalysis. As illustrated in the examples in this chapter, there are already many and varied opportunities in this field. The challenges that remain are those of scale up, optimisation and economics in order to compete with high volume commodity polymers already available as long as oil sources are reasonable in cost and supply. For biodegradable biomedical polymers, biological responses in terms of inflammation, rates of degradation *in vivo* and processing into suitable mechanically functional products are challenges that

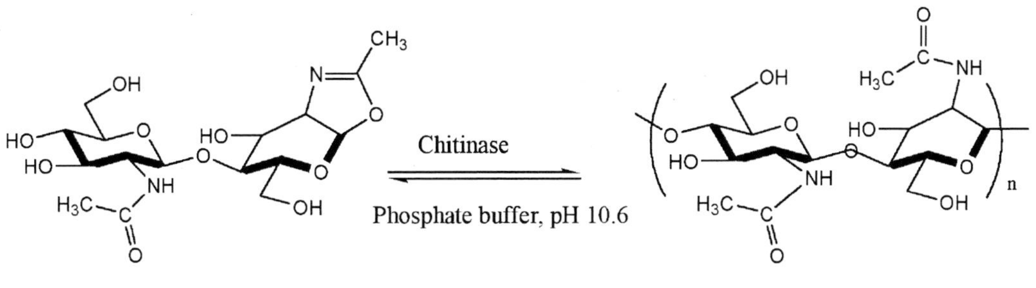

Scheme 12.16

perhaps can be met in a shorter time frame than the needs in the commodity area. In either case, the place for enzymes in the world of polymer synthesis and polymer modification is already here. This role for enzymes should gradually expand as new insights are gained into the mechanisms involved as well as the opportunities for these polymerisation reactions.

References

1. D.L. Kaplan, J. Dordick, R.A. Gross and G. Swift in *Enzymes in Polymer Synthesis,* Eds., R.A. Gross, D.L. Kaplan and G. Swift, ACS Symposium Series No. 684, ACS, Washington, DC, USA, 1998, p.2-16.

2. I. Gill, R. Lopez-Fandino, X. Jorba and E.N. Vulfson, *Enzyme and Microbial Technology*, 1996, **18**, 3, 162.

3. H.V. Dohren, U. Keller, J. Vater and R. Zocher, *Chemical Reviews,* 1997, **97**, 7, 2675.

4. C-H. Wong and G.M. Whitesides, *Enzymes in Synthetic Organic Chemistry, Tetrahedron Organic Chemistry Series,* Volume 12, Pergamon Press, Oxford, UK, 1994.

5. D. O'Hagan and N.A. Zaidi, *Journal of the Chemical Society, Perkin Transactions 1,* 1993, **20**, 2389.

6. S. Matsumura and J. Takahashi, *Macromolecular Rapid Communications,* 1986, **7**, 6, 369.

7. D. O'Hagan and A.H. Parker, *Polymer Bulletin,* 1998, **41**, 5, 519.

8. A.L. Gutman, K. Zuobi and A. Boltansky, *Tetrahedron Letters,* 1987, **28**, 33, 3861.

9. D. Knani, A.L. Gutman and D.H. Kohn, *Journal of Polymer Science, Part A: Polymer Chemistry,* 1993, **31**, 5, 1221.

10. D. Knani and D.H. Kohn, *Journal of Polymer Science, Part A: Polymer Chemistry,* 1993, **31**, 12, 2887.

11. D. O'Hagan and N.A. Zaidi, *Polymer,* 1994, **35**, 16, 3576.

12. A. Ajima, T. Yoshimoto, K. Takahashi, Y. Tamaura, Y. Saito and Y. Inada, *Biotechnology Letters,* 1985, **7**, 5, 303.

13. K. Pavel and H. Ritter, *Die Makromolekulare Chemie,* 1991, **192**, 9, 1941.

14. O. Noll and H. Ritter, *Macromolecular Rapid Communications,* 1996, **17**, 8, 553.

15. A.L. Gutman and T. Bravdo, *Journal of Organic Chemistry,* 1989, **54**, 24, 5645.

16. H. Dong, H. Wang, S. Cao and J. Shen, *Biotechnology Letters,* 1998, **20**, 10, 905.

17. A.L. Gutman, D. Oren, A. Boltanski and T. Bravdo, *Tetrahedron Letters,* 1987, **28**, 44, 5367.

18. S. Okumura, M. Iwai and Y. Tominaga, *Agricultural and Biological Chemistry,* 1984, **48**, 11, 2805.

19. H. Uyama, K. Inada and S. Kobayashi, *Polymer Journal,* 2000, **32**, 5, 440.

20. H. Uyama, K. Inada and S. Kobayashi, *Chemistry Letters,* 1998, **27**, 12, 1285.

21. F. Binns, P Harffey, S.M. Roberts and A. Taylor, *Journal of the Chemical Society, Perkin Transactions 1,* 1999, **19**, 2671.

22. F. Binns, S.M. Roberts, A. Taylor and C.F. Williams, *Journal of the Chemical Society, Perkin Transactions 1,* 1993, 8, 899.

23. F. Binns, P. Harffey, S.M. Roberts and A. Taylor, *Journal of Polymer Science, Part A: Polymer Chemistry,* 1998, **36**, 12, 2069.

24. S. Kobayashi, H. Uyama, S. Suda and S. Namekawa, *Chemistry Letters,* 1997, **26**, 1, 105.

25. S. Suda, H. Uyama and S. Kobayashi, *Proceedings of the Japan Academy Series B: Physical and Biological Sciences,* 1999, **75**, 201.

26. S. Kobayashi, H. Uyama and S. Namekawa, *Polymer Degradation and Stability,* 1998, **59**, 1-3, 195.

27. H. Uyama and S. Kobayashi, *Chemistry Letters,* 1994, **23**, 9, 1687.

28. H. Uyama, S. Yaguchi and S. Kobayashi, *Journal of Polymer Science, Part.A: Polymer Chemistry,* 1999, **37**, 15, 2737.

29. H. Uyama, S. Yaguchi and S. Kobayashi, *Polymer Journal,* 1999, **31**, 4, 380.

30. H. Uyama, K. Inada and S. Kobayashi, *Macromolecular Rapid Communications,* 1999, **20**, 4, 171.

31. H. Uyama, E. Klegraf, S. Wada and S. Kobayashi, *Chemistry Letters*, 2000, **29**, 7, 800.

32. T. Tsujimoto, H. Uyama and S. Kobayashi, *Biomacromolecules*, 2001, **2**, 1, 29.

33. H. Uyama, K. Inada and S. Kobayashi, *Macromolecular Bioscience*, 2001, **1**, 1, 40.

34. A.K. Chaudhary, B.J. Kline, E.J. Beckman and A.J. Russell, *Polymer Preprints*, 1997, **38**, 2, 396.

35. G. Mezoul, T. Lalot, M. Brigodiot and E. Marechal, *Polymer Bulletin*, 1996, **36**, 5, 541.

36. H.G. Park, H.N. Chang and J.S. Dordick, *Biotechnology Letters*, 1995, **17**, 10, 1085.

37. A.K. Chaudhary, E.J. Beckman and A.J. Russell, *Journal of the American Chemical Society*, 1995, **117**, 13, 3728.

38. A.K. Chaudhary, E.J. Beckman and A.J. Russell, *Biotechnology and Bioengineering*, 1997, **55**, 1, 227.

39. A.K. Chaudhary, J. Lopez, E.J. Beckman and A.J. Russell, *Biotechnology Progress*, 1997, **13**, 3, 318.

40. A.K. Chaudhary, E.J. Beckman and A.J. Russell, *Biotechnology and Bioengineering*, 1998, **59**, 4, 428.

41. B.J. Kline, E.J. Beckman and A.J. Russell, *Journal of the American Chemical Society*, 1998, **120**, 37, 9475.

42. A.J. Mesiano, E.J. Beckman and A.J. Russell, *Biotechnology Progress*, 2000, **16**, 1, 64.

43. G. Mezoul, T. Lalot, M. Brigodiot and E. Marechal, *Macromolecular Rapid Communications*, 1995, **16**, 8, 613.

44. G. Mezoul, T. Lalot, M. Brigodiot and E. Marechal, *Macromolecular Chemistry & Physics*, 1996, **197**, 11, 3581.

45. J.S. Wallace and C.J. Morrow, *Journal of Polymer Science, Part.A: Polymer Chemistry*, 1989, **27**, 8, 2553.

46. J.S. Wallace and C.J. Morrow, *Journal of Polymer Science, Part A: Polymer Chemistry*, 1989, **27**, 10, 3271.

47. E.M. Brazwell, D.Y. Filos and C.J. Morrow, *Journal of Polymer Science, Part A: Polymer Chemistry*, 1995, **33**, 1, 89.

48. S. Geresh and Y. Gilboa, *Biotechnology and Bioengineering*, 1990, **36**, 3, 270.

49. S. Geresh and Y. Gilboa, *Biotechnology and Bioengineering*, 1991, **37**, 9, 883.

50. S. Geresh, Y. Gilboa, S. Abrahami and A. Bershadsky, *Polymer Engineering and Science*, 1993, **33**, 5, 311.

51. S. Matsumura, T. Okamoto, K. Tsukada and K. Toshima, *Macromolecular Rapid Communications*, 1998, **19**, 6, 295.

52. S. Matsumura, T. Okamoto, K. Tsukada, N. Mizutani and K. Toshima, *Macromolecular Symposia*, 1999, **144**, 219.

53. S. Kobayashi and H. Uyama, *Macromolecular Rapid Communications*, 1993, **14**, 12, 841.

54. S. Kobayashi, H. Kikuchi and H. Uyama, *Macromolecular Rapid Communications*, 1997, **18**, 7, 575.

55. S. Matsumura, K. Tsukada and K. Toshima, *Macromolecules*, 1997, **30**, 10, 3122.

56. K.S. Bisht, Y.Y. Svirkin, L.A. Henderson, R.A. Gross, D.L. Kaplan and G. Swift, *Macromolecules*, 1997, **30**, 25, 7735.

57. F. Deng and R.A. Gross, *International Journal of Biological Macromolecules*, 1999, **25**, 1-3, 153.

58. T.F. Al-Azemi and K.S. Bisht, *Macromolecules*, 1999, **32**, 20, 6536.

59. J. Wen and R. Zhuo, *Macromolecular Rapid Communications*, 1998, **19**, 12, 641.

60. H. Nishida, M. Yamashita, M. Nagashima, T. Endo and Y. Tokiwa, *Journal of Polymer Science, Part A: Polymer Chemistry*, 2000, **38**, 9, 1560.

61. Y. Feng, J. Knufermann, D. Klee and H. Hocker, *Macromolecular Rapid Communications*, 1999, **20**, 2, 88.

62. Y. Feng, D. Klee, H. Keul and H. Hocker, *Macromolecular Chemistry and Physics*, 2000, **201**, 18, 2670.

63. S. Namekawa, H. Uyama, S. Kobayashi and H.R. Kricheldorf, *Macromolecular Chemistry & Physics*, 2000, **201**, 2, 261.

64. S. Matsumura, K. Tsukada and K. Toshima, *International Journal of Biological Macromolecules,* 1999, **25**, 1-3, 161.

65. T.F. Al-Azemi, J.P. Harmon and K.S. Bisht, *Biomacromolecules,* 2000, **1**, 3, 493.

66. A. Kumar, K. Garg and R.A. Gross, *Macromolecules,* 2001, **34**, 11, 3527.

67. S. Matsumura, H. Beppu, K. Tsukada and K. Toshima, *Biotechnology Letters,* 1996, **18**, 9, 1041.

68. S. Namekawa, H. Uyama and S. Kobayashi, *Polymer Journal (Japan),* 1996, **28**, 8, 730.

69. S. Namekawa, S. Suda, H. Uyama and S. Kobayashi, *International Journal of Biological Macromolecules,* 1999, **25**, 1-3, 145.

70. W. Xie, J. Li, D. Chen and P.G. Wang, *Macromolecules,* 1997, **30**, 22, 6997.

71. S. Matsumura, Y. Suzuki, K. Tsukada, K. Toshima, Y. Doi and K. Kasuya, *Macromolecules,* 1998, **31**, 19, 6444.

72. Y. Suzuki, T. Ohura, K-I. Kasuya, K. Toshima, Y. Doi and S. Matsumura, *Chemical Letters,* 2000, **29**, 4, 318.

73. G.A.R. Nobes, R.J. Kazlauskas and R.H. Marchessault, *Macromolecules,* 1996, **29**, 14, 4829.

74. J. Xu, R.A. Gross, D.L. Kaplan and G. Swift, *Macromolecules,* 1996, **29**, 11, 3857.

75. H. Dong, S-G. Cao, Z-Q. Li, S-P. Han, D-L. You and J-C. Shen, *Journal of Polymer Science, Part A: Polymer Chemistry,* 1999, **37**, 9, 1265.

76. S. Matsumura, H. Beppu, K. Nakamura, S. Osanai and K. Toshima. *Chemistry Letters,* 1996, 25, 9, 795.

77. J. Xu, R.A. Gross, D.L. Kaplan and G. Swift, *Macromolecules,* 1996, **29**, 13, 4582.

78. Y.Y. Svirkin, J. Xu, R.A. Gross, D.L. Kaplan and G. Swift, *Macromolecules,* 1996, **29**, 13, 4591.

79. S. Kobayashi, K. Takeya, S. Suda and H. Uyama, *Macromolecular Chemistry & Physics,* 1998, **199**, 8, 1729.

80. H. Uyama and S. Kobayashi, *Chemistry Letters,* 1993, **22**, 7, 1149.

81. K. Küllmer, H. Kikuchi, H. Uyama and S. Kobayashi, *Macromolecular Rapid Communications,* 1998, **19**, 2, 127.

82. R.T. MacDonald, S.K. Pulapura, Y.Y. Svirkin, R.A. Gross, D.L. Kaplan, J. Akkara, G. Swift and S. Wolk, *Macromolecules,* 1995, **28**, 1, 73.

83. L.A. Henderson, Y.Y. Svirkin, R.A. Gross, D.L. Kaplan and G. Swift, *Macromolecules,* 1996, **29**, 24, 7759.

84. H. Uyama, S. Suda, H. Kikuchi and S. Kobayashi, *Chemistry Letters,* 1997, **26**, 11, 1109.

85. H. Uyama, S. Namekawa and S. Kobayashi, *Polymer Journal (Japan),* 1997, **29**, 3, 299.

86. A. Cordova, T. Iversen, K. Hult and M. Martinelle, *Polymer,* 1998, **39**, 25, 6519.

87. A. Kumar and R.A. Gross, *Biomacromolecules,* 2000, **1**, 1, 133.

88. H. Uyama, K. Takeya and S. Kobayashi, *Proceedings of the Japan Academy,* 1993, **69B**, 8, 203.

89. S. Kobayashi, H. Uyama, S. Namekawa and H. Hayakawa, *Macromolecules,* 1998, **31**, 17, 5655.

90. H. Uyama, K. Takeya and S. Kobayashi, *Bulletin of the Chemical Society of Japan,* 1995, **68**, 1, 56.

91. H. Uyama, H. Kikuchi, K. Takeya, N. Hoshi and S. Kobayashi, *Chemistry Letters,* 1996, **25**, 2, 107.

92. H. Uyama, K. Takeya, N. Hoshi and S. Kobayashi, *Macromolecules,* 1995, **28**, 21, 7046.

93. K.S. Bisht, L.A. Henderson, R.A. Gross, D.L. Kaplan and G. Swift, *Macromolecules,* 1997, **30**, 9, 2705.

94. H. Uyama, H. Kikuchi, K. Takeya and S. Kobayashi, *Acta Polymerica,* 1996, **47**, 8, 357.

95. A. Kumar, B. Kalra, A. Dekhterman and R.A. Gross, *Macromolecules,* 2000, **33**, 17, 6303.

96. S. Namekawa, H. Uyama and S. Kobayashi, *Proceedings of the Japan Academy*, 1998, **74B**, 4, 65.

97. S. Namekawa, H. Uyama and S. Kobayashi, *Polymer Journal*, 1998, **30**, 3, 269.

98. H. Uyama, S. Kobayashi, M. Morita, S. Habaue and Y. Okamoto, *Macromolecules*, 2001, **34**, 19, 6554.

99. S. Matsumura, H. Beppu and K. Toshima, *Chemistry Letters*, 1999, **28**, 3, 249.

100. H. Kikuchi, H. Uyama and S. Kobayashi, *Macromolecules*, 2000, **33**, 24, 8971.

101. Z. Jedlinski, M. Kowalczuk, G. Adamus, W. Sikorska and J. Rydz., *International Journal of Biological Macromolecules*, 1999, **25**, 1-3, 247.

102. H. Uyama, H. Kikuchi and S. Kobayashi, *Chemistry Letters*, 1995, **24**, 11, 1047.

103. H. Uyama, H. Kikuchi and S. Kobayashi, *Bulletin of the Chemical Society of Japan*, 1997, **70**, 7, 1691.

104. S. Namekawa, H. Uyama and S. Kobayashi, *Biomacromolecules*, 2000, **1**, 3, 335.

105. S. Namekawa, H. Uyama and S. Kobayashi, *Macromolecular Chemistry & Physics*, 2001, **202**, 6, 801.

106. A. Cordova, T. Iversen and K. Hult, *Polymer*, 1999, **40**, 24, 6709.

107. K.S. Bisht, F. Deng, R.A. Gross, D.L. Kaplan and G. Swift, *Journal of the American Chemical Society*, 1998, **120**, 7, 1363.

108. A. Cordova, T. Iversen and K. Hult, *Macromolecules*, 1998, **31**, 4, 1040.

109. J. Li, W. Xie, H.N. Cheng, R.G. Nickol and P.G. Wang, *Macromolecules*, 1999, **32**, 8, 2789.

110. A. Cordova, A. Hult, K. Hult, H. Ihre, T. Iversen and E. Malmstrom, *Journal of the American Chemical Society*, 1998, **120**, 51, 13521.

111. A. Cordova, *Biomacromolecules*, 2001, **2**, 4, 1347.

112. I.D. Buchanan and J.A. Nicell, *Biotechnology and Biotechnology*, 1997, **54**, 3, 251.

113. J.A. Nicell, J.K. Bewtra, N. Biswas, C.C. St. Pierre and K.E. Taylor, *Canadian Journal of Civil Engineering*, 1993, **20**, 725.

114. J.A. Nicell, J.K. Bewtra, K.E. Taylor, N. Biswas and C.St. Pierre, *Water Science and Technology*, 1992, **25**, 3, 157.

115. J.S. Dordick, M.A. Marletta and A.M. Klibanov, *Biotechnology and Bioengineering*, 1987, **30**, 1, 31.

116. A.M. Rao, V.T. John, R.D. Gonzalez, J.A. Akkara and D.L. Kaplan, *Biotechnology and Bioengineering*, 1993, **41**, 5, 531.

117. M.S. Ayyagari, K.A. Marx, S.K. Tripathy, J.A. Akkara and D.L. Kaplan, *Macromolecules*, 1995, **28**, 15, 5192.

118. R.S. Premachandran, S. Banerjee, X.K. Wu, V.T. John, G.L. McPherson, J. Akkara, M. Ayaggari and D.L. Kaplan, *Macromolecules*, 1996, **29**, 20, 6452.

119. R. Premachandran, S. Banerjee, V.T. John, G.L. McPherson, J.A. Akkara and D.L. Kaplan, *Chemistry of Materials*, 1997, **9**, 6, 1342.

120. S. Banerjee, V.T. John, G.L. McPherson, C.J. O'Conor, Y.S.L. Buisson, J.A. Akkara and D.L. Kaplan, *Colloid & Polymer Science*, 1997, **275**, 10, 930.

121. M. Ayyagari, J.A. Akkara and D.L. Kaplan, *Materials Science and Engineering C*, 1996, **4**, 3, 169.

122. J.A. Akkara, M. Ayyagari, F Bruno, L. Samuelson, V.T. John, C. Karayigitoglu, S. Tripathy, K.A. Marx, D.V.G.L.N. Rao and D.L. Kaplan, *Biomimetics*, 1994, **2**, 4, 331.

123. J.A. Akkara, D.L. Kaplan, V.T. John and S.K. Tripathy in *Polymeric Materials Encyclopedia*, Ed., J.C. Salamone, CRC Press, Boca Raton, FL, USA, 1996, **3**, p.2115.

124. J.A. Akkara, K.J. Senecal and D.L. Kaplan, *Journal of Polymer Science, Part A: Polymer Chemistry*, 1991, **29**, 11, 1561.

125. C.F. Karayigitoglu, N. Kommareddi, R.D. Gonzalez, V.T. John. G.L. McPherson, J.A. Akkara and D.L. Kaplan, *Materials Science and Engineering C*, 1995, **2**, 3, 165.

126. H. Tonami, H. Uyama, S. Kobayashi and M. Kubota, *Macromolecular Chemistry & Physics*, 1999, **200**, 10, 2365.

127. H. Uyama, H. Kurioka, J. Sugihara, I. Komatsu and S. Kobayashi, *Bulletin of the Chemical Society of Japan*, 1995, **68**, 11, 3209.

128. H. Uyama, H. Kurioka, J. Sugihara, I. Komatsu and S. Kobayashi, *Journal of Polymer Science, Part.A: Polymer Chemistry,* 1997, **35**, 8, 1453.

129. H. Kurioka, I. Komatsu, H. Uyama and S. Kobayashi, *Macromolecular Rapid Communications,* 1994, **15**, 6, 507.

130. T. Oguchi, S. Tawaki, H. Uyama and S. Kobayashi, *Bulletin of the Chemical Society of Japan,* 2000, **73**, 6, 1389.

131. H. Uyama, H. Kurioka, J. Sugihara and S. Kobayashi, *Bulletin of the Chemical Society of Japan,* 1996, **69**, 1, 189.

132. T. Oguchi, S. Tawaki, H. Uyama and S. Kobayashi, *Macromolecular Rapid Communications,* 1999, **20**, 7, 401.

133. N. Mita, T. Oguchi, S. Tawaki, H. Uyama and S. Kobayashi, *Polymer Preprints,* 2000, **41**, 1, 223.

134. H. Uyama, H. Kurioka and S. Kobayashi, *Chemistry Letters,* 1995, **24**, 9, 795.

135. R. Ikeda, H. Uyama and S. Kobayashi, *Macromolecules,* 1996, **29**, 8, 3053.

136. R. Ikeda, J. Sugihara, H. Uyama and S. Kobayashi, *Polymer International,* 1998, **47**, 3, 295.

137. A.S. Hay, *Journal of Polymer Science, Part A: Polymer Chemistry,* 1998, **36**, 4, 505.

138. R. Ikeda, J. Sugihara, H. Uyama and S. Kobayashi, *Macromolecules,* 1996, **29**, 27, 8702.

139. S. Kobayashi, H. Kurioka and H. Uyama, *Macromolecular Rapid Communications,* 1996, **17**, 8, 503.

140. S. Kobayashi, H. Uyama, T. Ushiwata, T. Uchiyama, J. Sugihara and H. Kurioka, *Macromolecular Chemistry & Physics,* 1998, **199**, 5, 777.

141. T. Fukuoka, H. Tonami, N. Maruichi, H. Uyama, S. Kobayashi and H. Higashimura, *Macromolecules,* 2000, **33**, 24, 9152.

142. P. Wang and J.S. Dordick, *Macromolecules,* 1998, **31**, 3, 941.

143. P. Novak, K. Muller, K.S.V. Santhanam and O. Haas, *Chemical Reviews,* 1997, **97**, 1, 207.

144. P. Wang, B.D. Martin, S. Parida, D.G. Rethwisch and J.S. Dordick, *Journal of the American Chemical Society,* 1995, **117**, 51, 12885.

145. H. Tonami, H. Uyama, S. Kobayashi, K. Rettig and H. Ritter, *Macromolecular Chemistry & Physics,* 1999, **200**, 9, 1998.

146. W. Liu, S. Bian, L. Li, L. Samuelson, J. Kumar and S. Tripathy, *Chemistry of Materials,* 2000, **12**, 6, 1577.

147. L. Wang, E. Kobatake, Y. Ikariyama and M. Aizawa, *Journal of Polymer Science, Part A: Polymer Chemistry,* 1993, **31**, 11, 2855.

148. C. Goretzki and H. Ritter, *Macromolecular Chemistry & Physics,* 1998, **199**, 6, 1019.

149. D.W. Potter, D.W. Miller and J.A. Hinson, *Journal of Biological Chemistry,* 1985, **260**, 22, 12174.

150. D.W. Potter, D.W. Miller and J.A. Hinson, *Molecular Pharmacology,* 1986, **29**, 2, 155.

151. R. Ikeda, H. Tanaka, H. Uyama and S. Kobayashi, *Polymer Preprints,* 2000, **41**, 1, 83.

152. S. Kobayashi, R. Ikeda, H. Oyabu, H. Tanaka and H. Uyama, *Chemistry Letters,* 2000, **29**, 10, 1214.

153. H. Tonami, H. Uyama, S. Kobayashi, T. Fujita, Y. Taguchi and K. Osada, *Biomacromolecules,* 2000, **1**, 2, 149.

154. S-I. Shoda, T. Kawasaki, K. Obata and S. Kobayashi, *Carbohydrate Research,* 1993, **249**, 1, 127.

155. S. Kobayashi, X. Wen and S. Shoda, *Macromolecules,* 1996, **29**, 7, 2698.

156. M. Fujita, S. Shoda and S. Kobayashi, *Journal of the American Chemical Society,* 1998, **120**, 25, 6411.

157. S. Kobayashi, L.J. Hobson, J. Sakamoto, S. Kimura, J. Sugiyama, T. Imai and T. Itoh, *Biomacromolecules,* 2000, **1**, 2, 168.

158. S. Kobayashi, T. Kiyosada and S. Shoda, *Journal of the American Chemical Society,* 1996, **118**, 51, 13113.

159. J. Sakamoto, J. Sugiyama, S. Kimura, T. Imai, T. Itoh, T. Watanabe and S. Kobayashi, *Macromolecules*, 2000, **33**, 11, 4155.

160. S. Kobayashi, K. Kashiwa, T. Kawasaki and S. Shoda, *Journal of the American Chemical Society*, 1991, **113**, 8, 3079.

161. S-I. Shoda, K. Obata, O. Karthaus and S. Kobayashi, *Journal of the Chemical Society, Chemical Communications*, 1993, 1402.

162. S. Kobayashi, J. Shimada, K. Kashiwa and S. Shoda, *Macromolecules*, 1992, **25**, 12, 3237.

163. S-I. Shoda, E. Okamoto, T. Kiyosada and S. Kobayashi, *Macromolecular Rapid Communications*, 1994, **15**, 10, 751.

164. S. Kobayashi, S. Shoda, J. Lee, K. Okuda, R.M. Brown and S. Kuga, *Macromolecular Chemistry & Physics*, 1994, **195**, 4, 1319.

165. E. Okamoto, T. Kiyosada, S-I. Shoda and S. Kobayashi, *Cellulose*, 1997, **4**, 2, 161.

166. S. Kobayashi, E. Okamoto, X. Wen and S. Shoda, *Journal of Macromolecular Science - Pure and Applied Chemistry*, 1996, **A33**, 10, 1375.

167. S. Kobayashi, S. Shoda, M.J. Donnelly and S.P. Church in *Carbohydrate Biotechnology Protocols, Methods in Biotechnology, Volume 10*, Ed., C. Bucke, Humana Press, Totowa, NJ, USA, 1999, p.57.

13 Environmental Life Cycle Comparisons of Biodegradable Plastics

Martin Patel

13.1 Introduction

In Europe, biodegradable polymers were originally developed and introduced to the market for two main reasons. Firstly, the limited volume of landfill capacity became more and more of a threat and secondly, the bad image of plastics held by the public prompted the call for more environmentally friendly products. While the first issue has largely disappeared from the top of the agenda due to the introduction of plastics recycling schemes and due to newly built incineration plants, the environmental performance is currently the main argument for biodegradable polymers. Against this background, this comparative review of publicly available life cycle studies may provide useful information to manufacturers, processors, consumers and policy makers.

In this chapter all major biodegradable plastics are discussed, i.e., starch polymers, polyhydroxyalkanoates (PHA), polylactic acids (PLA) and a small group of other materials. All of these polymers are manufactured by use of renewable resources. Only materials which are generally acknowledged as being *completely* biodegradable according to established standards are included in this chapter. For this reason, starch-polyolefin blends and polyolefins containing ferric salts have not been taken into account. The types of *end products* covered are pellets, loose-fill packaging material (packaging chips), films, bags and mulch films. These are compared with products made from petrochemical polymers, usually polyethylene (PE), polypropylene (PP) or polystyrene (PS).

All major European producers of biodegradable polymers and several European research organisations known for environmental assessments in this area were approached to ensure that the coverage of this chapter would be as comprehensive as possible. Several experts in the field also commented on earlier versions of this chapter.

13.2 Methodology of LCA

A life cycle assessment (LCA) consists of four independent elements (ISO [1-4]; CML [5]):

i) the definition of goal and scope,

ii) the life cycle inventory analysis,

iii) the life cycle impact assessment and

iv) the life cycle interpretation.

The **definition of the goal and scope** includes a decision about the functional unit which forms the basis of comparison, the product system to be studied, system boundaries, allocation procedures, assumptions made and limitations. The functional unit can either be a certain service or a product, with the latter being the usual choice for the type of studies reviewed here, e.g., comparison of 1 m³ loose-fill packaging material made of starch polymer versus PS. Critical LCA issues regarding biodegradable polymers are, among others, the share of renewable *versus* fossil fuel raw materials, the way of growing the agricultural raw materials (intensive *versus* extensive cultivation), the type of conventional polymer that is chosen as a reference and the mix of waste management processes assumed for both the biodegradable and the non-degradable polymer (landfilling, incineration, recycling, composting and digestion). It is generally assumed that the carbon dioxide originating from the biomass is equivalent to the amount which was previously withdrawn from the atmosphere during growth and that it therefore does not contribute to global warming (fossil fuels required for transport, processing the crops and producing auxiliaries, e.g., fertilisers, are accounted for separately).

The **life cycle inventory analysis** involves data collection and calculation procedures to quantify the total system's inputs and outputs that are relevant from an environmental point of view, i.e., mainly resource use, atmospheric emissions, aqueous emissions, solid waste and land use.

The **life cycle impact assessment** aims at evaluating the significance of potential environmental impacts using the results of the life cycle inventory analysis. One important goal of the life cycle impact assessment is to aggregate outputs with comparable effects, (e.g., all greenhouse gases or all acidifying components), by use of so-called characterisation factors (characterisation factors hence serve for aggregation within the impact categories, e.g., to determine the total global warming potential of a gas mixture containing CO_2, CH_4 and N_2O. Characterisation factors are sometimes also referred to as equivalence factors). This leads to a limited number of parameters, called *impact categories*. As an optional step, the results by impact categories can be divided by a reference value, (e.g., total greenhouse gas emissions of a country), in order to understand better the relative importance of the various impacts; this step is referred to as *normalisation*. Finally it is, in principle, possible to aggregate the results determined for the various impact categories. However, this *valuation step* is based not only on scientific facts but also on subjective choices and societal values. So far, there is no generally accepted methodology to translate life cycle inventory data to highly aggregated, let alone, single-score, indicators (according to the draft ISO standard for the life cycle impact assessment 'weighting, as described in [paragraph] 6.4, shall not be used for comparative assertions, disclosed to public' (EN ISO 14 042 [3], draft 1998, paragraph 9)). In some of the LCA reviewed in this chapter, two single-score aggregation methods have been applied [6]; for example see CARBOTECH [7]). Given the missing general acceptance of these approaches, the results will, however, not be discussed here.

The **life cycle interpretation** is the final step of the LCA where conclusions are drawn from both the life cycle inventory analysis and the life cycle impact assessment or, in the case of life cycle inventory studies, from the inventory analysis only. As an outcome of the interpretation stage, recommendations can be formulated which, for example, may be directed to producers or policy makers.

The main objective of this chapter is to review full-sized LCA. However, a few more studies were taken into account that are much more limited in scope, e.g., by carrying out an assessment only for non-renewable energy and CO_2 emissions. It was felt that these studies nevertheless contribute to a better understanding of the environmental aspects because they address materials that have not been studied from this angle so far or because they provide an indication about how certain or uncertain the results are.

13.3 Presentation of Comparative Data

In total, thirteen publications were reviewed six of which deal with starch polymers, four with PHA, two with PLA and one with other biodegradable polymers. The dominance and the size of the studies analysing starch polymers reflect the current economic importance of this type of material among the biodegradable polymers. Appendix 13.1 provides an overview of the reviewed publications. They were all prepared by Swiss or German organisations and several of them are available in German only.

Regarding the choice of the functional unit, some of the studies only analyse the production and waste management of materials in the form of pellets without referring to a specific application of use while other studies refer to a certain type of end use. The first type of study has the advantage that it provides a first impression about the opportunities of clean production. For example, if the environmental performance is not attractive at the material level, there is a good chance that this will also be true at the product level. On the other hand studies that exclusively analyse the material level have the disadvantage of not taking decisive parameters at the end use level into account, for example:

- materials processing, where the amount of material required to manufacture a certain end product might be higher or lower than for petrochemical polymers.

- transportation, which can be substantial for end products with a low density such as loose fill packaging material.

- the use phase, where consumer behaviour can play a role, e.g., in the case of compost bins without an inliner where the way of cleaning the bin has a large influence on the overall environmental impact.

- the waste stage where logistics and recycling processes can be tailored to a specific product or product group.

For these reasons both approaches, the analysis at the material level and the end product level, provide valuable insight. They are therefore both included in this review.

The regional scope (compare Appendix 13.1) is relevant due primarily to national differences in CO_2 emissions from electricity generation (power is used for plastics production) and the type of waste management infrastructure in place, e.g., treatment of municipal solid waste by incineration, landfilling, recycling and/or composting. Due to its relevance for biodegradable polymers, most studies include composting as a waste management option (Appendix 1). The output of the composting process, i.e., compost, can be used for soil amelioration. Pathogenic microorganisms are eliminated. Organic nitrogen and phosphorus compounds are converted into inorganic compounds which can be utilised by plants [8]. The option of waste management by digestion is taken into consideration only by one of the studies reviewed [9].

The number of impact categories varies greatly (see Appendix 1), giving an indication of the differences in depth of analysis among the various studies. The studies also differ considerably in the amount of published background data and the degree of detail regarding explanations about the methodology and results. In some cases the results are given in natural units while in others, they are expressed as an index relative to the reference case, which makes it more difficult for the user of this information to draw comparisons with other sources.

13.3.1 Starch Polymers

All studies discussed in this chapter deal with thermoplastic starch (TPS) which is manufactured through destructurisation in presence of specific amounts of plasticisers and under certain extrusion conditions. Depending on the type of application either pure starch polymers or various types of blends with different ratios of petrochemical copolymers are used.

13.3.1.1 Starch Polymer Pellets

• CARBOTECH (1996)

The Swiss Federal Agency for the Environment, Forests and Landscape (BUWAL) commissioned CARBOTECH to prepare the first detailed and publicly available LCA for biodegradable polymers [7]. This study analyses starch polymers and compares them to polyolefins. The report also contains data for blends of starch polymers and polyolefins (not included in this review for the reasons given previously). The LCA refers to Switzerland. The thermoplastic starch polymers are based on two starch sources, i.e., potatoes (85% of input) and maize (15%). The system studied covers the entire production process (cradle-to-factory gate) and the waste management stage (compare Appendix 13.1). Two types of waste management have been distinguished, i.e., firstly a combination of combustion in Municipal Solid Waste Incineration (MSWI) plants and landfilling and secondly, composting. For most impact categories, this difference has only little impact on the data (**Table 13.1**). The largest difference is found for the indicator 'deposited waste' which is clearly lower for composting. The use of compost contributes to soil amelioration and it may replace synthetic fertilisers to some extent. If incinerated in waste-to-energy facilities, starch polymers yield certain amounts of electricity and/or steam. In contrast to other LCA (see below) the CARBOTECH Study does not ascribe any credits to these benefits. Hence the environmental impacts tend to be overestimated.

In the CARBOTECH study, thermoplastic starch is compared to virgin low-density polyethylene (LDPE; see **Table 13.1**). These data originate from an earlier study commissioned by BUWAL [10]. The waste management assumed for LDPE consists of 80% incineration and 20% landfilling. Based on these assumptions the CARBOTECH study comes to the conclusion that thermoplastic starch performs better than LDPE in all impact categories except for eutrophication (see row 4). The use of TPS is particularly advantageous for energy resources, greenhouse gas (GHG) emissions, human toxicity and salinisation. GHG emissions are reported to be dominated by CO_2 while N_2O emissions (from agriculture) and methane emissions (from energy supply) are of minor importance (5% and 1-2%, respectively, of the total GHG emission potential [11]). The impacts on biodiversity and soil quality were assessed in qualitative terms. Here, no additional negative impacts were determined if starch crops are grown on areas which are currently used for agricultural purposes. In contrast, the effects are clearly negative if natural areas are displaced [12].

It is concluded in the CARBOTECH study [7] that the preferences among the environmental targets determine whether starch polymers are found to be environmentally attractive. If the reduction of eutrophication was the prime objective then starch polymers would not represent an attractive option. Regarding biodiversity, the type of land used plays an important role. For all the other parameters the results are in favour of starch polymers.

Table 13.1 LCA results for TPS and LDPE (Functional unit = 100 kg of plastic material [13, 14]

All data refer to 100 kg plastic	Energy resources (MJ)	GHG emissions (kg CO_2 eq.)	Ozone precursors (kg ethylene eq.)	Human toxicity (y x m³)	Acidification (kg SO_2 eq.)	Eutrophication (kg PO_4 eq.)	Ecotoxicity (d x l)	Salinisation (H⁺/mol)	Deposited waste (10^{-3} EPSY)
(1) TPS (80% MSWI, 20% landfilling)	2550 ±5%	120 ±15%	0.47 ±20%	20 ±40%	1.09 ±5%	0.47 ±40%	2.8 ±55%	180 ±15%	5.1 ±10%
(2) TPS (100% composting)	2540 ±5%	114 ±15%	0.50 ±20%	20 ±35%	1.06 ±5%	0.47 ±40%	2.8 ±55%	180 ±15%	0.72 ±10%
(3) LDPE (80% MSWI, 20% landfilling)	9170 ±5%	520 ±20%	1.3 ±15%	70 ±60%	1.74 ±5%	0.11 ±55%	4.6 ±25%	860 ±10%	5.5 ±5%
(4) Ratio (1)/(3)	28%	23%	36%	29%	63%	427%	61%	21%	93%

EPSY: Environmental Point System

436

- **Fraunhofer ISI (1999)**

As for the LCA just discussed, the Fraunhofer ISI Study [15] restricts itself to starch polymer pellets and compares them with pellets made of PE. The main difference to the other studies discussed in this chapter is that the Fraunhofer Study compares various blends with different ratios of petrochemical copolymers. Information about the composition of the blends were provided by starch polymer manufacturers (Novamont, Biotec). It was assumed that both the starch polymers and PE are incinerated in MSWI plants after their useful life. No credits have been assigned to steam and/or electricity generated in waste-to-energy facilities. The results of this analysis that is restricted to energy and CO_2 are shown in **Table 13.2**. According to this comparison starch polymers offer saving potentials relative to PE in the range of 28-55 GJ/t plastic and 1.4-3.9 t CO_2/t plastic depending on the share of petrochemical co-polymers (it must be borne in mind that there are still considerable uncertainties even for conventional, fossil fuel-based polymers (see Section 13.5).

Table 13.2 Energy requirements and CO_2 emissions for different types of starch polymers and for LDPE [15]			
Type of plastic	Share of petrochemical compounds	Cradle-to-factory gate energy use [1]	Fossil CO_2 emissions throughout life cycle (production and waste incineration)
	% (wt)	GJ/t product	kg CO_2/t product
TPS [2]	0	25.4	1140
TPS/polyvinyl alcohol [3]	15	24.9	1730
TPS/polycaprolacton [3]	52.5	48.3	3360
TPS/polycaprolacton [3]	60	52.3	3600
LDPE [4]	100	80.6	4840

[1] *Non-renewable energy (fossil and nuclear)*
[2] *Source of data in this row: CARBOTECH [14] (without waste management). The CO_2 energy ratio according to this dataset is very low (45 kg CO_2/GJ). The reason might be co-firing of biomass waste*
[3] *Fraunhofer ISI [15]*
[4] *Embodied carbon: 3140 kg CO_2/t PE [16]*

13.3.1.2 Starch Polymer Loose Fills

- COMPOSTO (July 2000)

The Italian starch polymer manufacturer Novamont commissioned COMPOSTO to conduct an LCA for loose fill packaging material made from Novamont's product Mater-Bi

PE01U and from expanded polystyrene (EPS) [17]. Mater-Bi PE01U is a starch polymer containing about 15% of polyvinylalcohol (PVOH). The study evaluates the use of Mater-Bi loose fill in Switzerland. The entire production chain and waste management is included in the system boundaries while the use phase is excluded (Appendix 13.1). The transportation of loose fill is also included in the LCA. This is of general importance for products with low density because of the relatively high energy requirements needed for transportation. It is, moreover, of particular importance for the products studied here since the density of starch polymer loose fill is about twice as high than that of EPS loose fill. Regarding waste management, incineration has been assumed for EPS loose fill while composting has been assumed for Mater-Bi loose fill (more than 90% of all organic waste delivered to commercial plants in Switzerland is processed by composting, the rest is fed to digestion plants). The calculations are based on composting in open stacks since more than 80% of the organic waste in Switzerland is composted in this way while the remainder is treated in boxes located in buildings [8]. An important assumption made here is that 60% of the carbon absorbed in the vegetable material is released to the atmosphere during composting (97% as CO_2, 3% as CH_4) and that the rest (40%) is sequestered in the compost (these assumptions here have been described in the COMPOSTO Study for bags [23]). The authors of the LCA consider these data to be particularly uncertain [18, 19].

The PVOH required for the manufacture of Mater-Bi PE01U is of petrochemical origin. The production of PVOH is reported to be the largest consumer of energy resources throughout the life cycle of starch polymer loose fill [20]. The authors consider the data used for PVOH to be another major source of uncertainty [21].

To account for these data uncertainties when comparing the results for starch and EPS loose fill, threshold values were determined. For example, the ecological damage is considered 'significantly higher' if the impact is at least double as high in the case of energy and waste and if there is a difference of at least a factor of five for all the other impact categories. Similarly, threshold values are determined to quantify the relationships 'higher', 'comparable', 'lower' and 'significantly lower' [22].

Two different approaches were used to generate aggregated values for the various impact categories (compare Appendix 1). One of them is based on a CARBOTECH report [7] which uses the characterisation factors generated by Heijungs [24] and various other sources (nine impact categories). In the other approach characterisation factors according to Eco-indicator '95 [6] are used (eight impact categories). Since there is some overlap between the two there are 13 impact categories in total [25]. In eight of these 13 impact categories the production and disposal of Mater-Bi loose fill causes less environmental damage than EPS loose fill. The environmental impact of Mater-Bi loose fill is reported to be *significantly lower* for the categories of winter smog, air toxicity and carcinogeneity. The impact of Mater-Bi loose fills is *lower* than EPS loose fills for energy use, global warming,

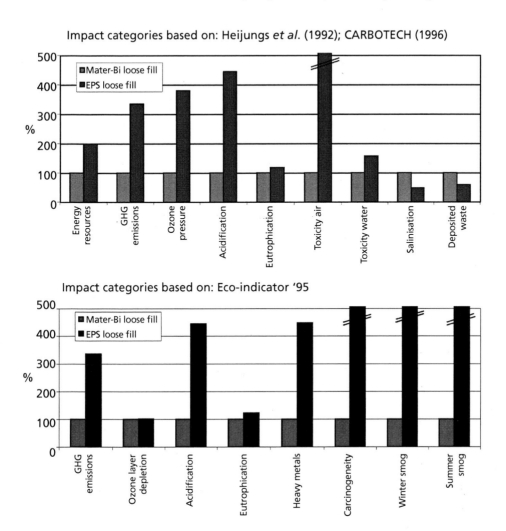

Figure 13.1 LCA for 1 m³ loose fills: Environmental performance using two different methodologies in the impact assessment stage [17]

acidification, ozone creation/summer smog and heavy metals. In two categories, Mater-Bi loose fill has a *larger* environmental impact than EPS loose fill (salinisation and deposited waste) while the effects are *comparable* for the three remaining categories (eutrophication, toxicity water and ozone layer depletion). The overall conclusion of this LCA is that Mater-Bi loose fills are ecologically less damaging than EPS loose fills [26].

• BIFA (2001)

Together with the LCA on starch polymers prepared by CARBOTECH [7], the LCA prepared by BIFA [9] is the most detailed study that is publicly available. Two types of loose fill produced by the German company FloPak are compared: one based on starch polymers and the other on EPS. The study follows a 2-step-approach:

• In the first step, various options of production, use and waste management are evaluated sequentially: first, the effects of variations in production are studied while the assumptions made for the use phase and for waste management remain unchanged. This is followed by similar sensitivity analyses for the use phase and for waste management. All these analyses are prepared both for starch polymer (**Table 13.3**) and EPS loose fill (not reported here).

• In the second step, various schemes are formed by combining selected options in the production, use and waste management stage (**Table 13.4**).

For starch polymer loose fill, the first column of **Table 13.3** provides a ranking of the various options studied in production, in the use phase and in waste management. Twenty enviromental parameters were determined. The ranking shown in **Table 13.3** has been determined by comparing for how many impact categories the environmental damage is lower in one case compared to the other without normalisation and valuation. Similar comparisons were also made for EPS loose fill (results not given).

The results of the schemes studied in the second step (**Table 13.4**) are shown in **Table 13.5**. The main results can be summarised as:

• Among the various schemes for starch polymer loose fill, the environmental impacts are lowest for scheme Starch IV, followed by Starch III, Starch II and Starch I (see **Table 13.4**).

• Among the schemes for EPS loose fills, the environmental effects are lowest for scheme EPS IV, followed EPS III, EPS II and EPS I.

• As shown in **Table 13.5** loose-fill production from maize (Starch I) and from virgin EPS (EPS I) assuming no recycling of the products in either of the two cases and waste management according to current practices are roughly comparable in environmental terms (nine impact categories in favour of starch, 11 impact categories in favour of EPS).

• EPS loose fills exclusively produced from post-consumer waste score better than starch polymer loose fills in most cases (**Table 13.5**). Exceptions are the comparisons Starch III – EPS II, Starch IV – EPS II and Starch IV – EPS III.

Table 13.3 Overall environmental ranking for starch polymer loose fills according to step 1 of the BIFA study [9]			
	Phases throughout the life cycle		
Environmental performance	Production	Use	Waste Management
Best	Wheat, extensive (13/1/2/2) *	4 cycles (within company) (20/0/0/0)	Optimised MSWI plant (12/3/0/3)
2nd	Potato, with effluent use (6/7/4/1)	4 cycles (recycling station) (9/10/0/0)	Standard MSWI plant (2/4/0/3)
3rd	Corn (maize) (4/3/6/3)	2 cycles (within company) (0/1/0/0)	- Current MSW management practice (3/2/3/7)
4th	Potato (0/3/0/8)	1 cycle (0/0/12/8)	- Digestion (4/2/3/8) - Blast furnace (2/1/6/5)
Worst	Wheat, intensive (2/2/3/11)	1 cycle (w/o allocation) (0/0/0/20)	Composting (3/0/3/8)

In step 1 of the BIFA study, the options of production, use and waste management listed in this table are studied separately and sequentially in a ceteris paribus approach; ranking of the environmental performance has been determined by comparing for how many of the 20 impact categories the environmental damage is lower in one case compared to the other (comparison without weighting)

The following 20 impact assessments have been taken into account: GHG emissions, carcinogenicity, eutrophication, acidification, diesel particles, ozone precursors, ozone precursors N-corrected, use of natural land, ozone depletion, eutrophication, cumulative fossil energy demand, cumulative nuclear energy demand, oil equivalents, lead, sulfur dioxide, fluorinated hydrocarbons, ammonia, nitrous oxide, adsorbable halogenated organic compounds, biocide use

*Example: Among all the production schemes studied, the option marked with * scores best in 13 impact categories, it obtains the second-best score for one impact category, the second-to-worst position in two cases and the worst score also for two impact categories. This result is abbreviated by (13/1/2/2). The number of scores does not add up to a total of 20 in may cases because partly 5 and partly 6 grades have been distinguished in the BIFA study while only 4 grades are reported here.*

The assumption that all EPS loose fills can be manufactured exclusively from post-consumer waste might not be realistic at the national scale and over longer periods of time due to logistics and cost restrictions. From this point of view the choice of the production process might be too optimistic in the schemes EPS II, EPS III and EPS IV (see **Table 13.4**). However, the comparison of these results with those for the starch polymer schemes show that important tradeoffs exist between recycling and biodegradables. According

Table 13.4 Definition of four starch polymer schemes and four EPS schemes in step 2 of the BIFA study on loose fill packaging material [9]

	Starch I	Starch II	Starch III	Starch IV	EPS I	EPS II	ESP III	EPS IV
Production	Maize	Wheat, intensive	Potato, with effluent use	Wheat, extensive	Virgin polystyrene (PS)	Recycled PS [1]	Recycled PS [1]	Recycled PS [2]
Use [3]	1 cycle	2 cycles (within company)	2 cycles (within company)	2 cycles (within company)	1 cycle	2 cycles (within company)	2 cycles (within company)	2 cycles (within company)
Waste management	Current MSW management practice [4]	Composting	Digestion	Optimised MSWI plant	Current MSW management practice [4]	DSD (blast furnace)	Open-loop-recycling	Optimised MSWI plant

[1] 1/3 EPS packaging, 1/3 music cassette/compact disc covers (MC/CD), 1/3 pre-consumer waste
[2] MC/CD covers
[3] 1 cycle = single use of loose fill packaging material; 2 cycles = re-use of loose-fill packaging materials
[4] Refers to the average situation in Germany: 30% incineration (including waste-to-energy facilities) and 70% landfilling
DSD: Duales System Deutshland

Table 13.5 Comparison of scores between the four starch polymer schemes and the four EPS schemes (based on [9])			
	Number of impact categories with lower damage caused by:		Overall judgement
	Starch polymers	EPS	
Starch I – EPS I	9	11	Comparable
Starch II – EPS II	5	15	EPS better
Starch III – EPS III	7	12	EPS better
Starch IV – EPS IV	7	12	EPS better
Starch II – EPS III	3	17	EPS better
Starch III – EPS II	11	9	Comparable
Starch II – EPS IV	3	17	EPS better
Starch III – EPS IV	4	16	EPS better
Starch IV – EPS II	16	4	Starch better
Starch IV – EPS III	9	10	Comparable

The abbreviation of the schemes used in the first columns are described in Table 13.4
The number of scores to not add up to a total of 20 if there is at least one impact category
for which the results are identical for starch polymers and EPS

to the BIFA study [9], starch polymers can hardly compete with petrochemical polymers on environmental grounds if the latter are recycled. It can be concluded that the use of renewable raw materials for the production of recyclable materials offers unexploited potentials which deserve further R&D.

13.3.1.3 Starch Polymer Films and Bags

• CARBOTECH (1996)

The CARBOTECH study discussed in Section 13.3.1.1 also contains a comparative LCA for films made from TPS and LDPE. The study refers to Switzerland and assumes that 80% of the waste is incinerated and the remaining 20% is landfilled. The results are again lower for TPS compared to LDPE for most impact categories with eutrophication and deposited waste being the main exceptions; for acidification and exotoxicity the impacts are practically identical. The results for human toxicity are reported to be subject to major uncertainties [27]. Compared to CARBOTECH's analysis for materials (Section 13.3.1.1) the starch polymer's advantage compared to LDPE is smaller. This is due to the assumption that the polymer input required to manufacture a given area of film is about 60% larger for starch polymers compared to LDPE (22.1 kg TPS compared to

Table 13.6 LCA results for films (100 m², 150 µm) made from TPS compared to LDPE; assumed waste management: 80% incineration, 20% landfilling [7]

	Energy resources (MJ)	GHG emissions (kg CO₂ eq.)	Ozone precursors (kg ethylene eq.)	Human toxicity (y m³)	Acidification (kg SO₂ eq.)	Eutrophication (kg PO₄ eq.)	Ecotoxicity (d l)	Salinisation (H⁺/mol)	Deposited waste (10⁻³ EPS)
(1) TPS film	649 ± 5%	25 ± 15%	0.10 ± 20%	4.3 ± 40%	0.24 ± 5%	0.13 ± 40%	0.62 ± 75%	40 ± 15%	1.1 ± 10%
(2) LDPE film	1340 ± 5%	67 ± 20%	0.18 ± 15%	9.7 ± 60%	0.24 ± 5%	0.02 ± 50%	0.65 ± 40%	120 ± %	0.8 ± 5%
(3) Ratio (1)/(2)	48%	38%	56%	44%	100%	687%	95%	33%	138%

13.8 kg LDPE for 100 m^2 of film of thickness 150 μm). In the meantime, the raw material requirements for starch polymer films have decreased and now exceed that for LDPE by only 30% (personal communication Novamont, 2001).

- COMPOSTO (1998)

In this study prepared by COMPOSTO [23] for Novamont, biodegradable waste bags made from Mater-Bi ZF03U/A material are compared to high density polyethylene (HDPE) bags and Kraft paper bags. Mater-Bi ZF03U/A is a blend of TPS and poly(ε-caprolactone) (PCL). It is assumed that these bags are used as liners for compost bins. The comparison is made for the smallest bags which were commercially available in Switzerland in 1998 and which could be used for a 10 litre bin. The considerable difference in the size of the bags (Mater-Bi bag: 16.6 l; paper bag: 13.6 l; HDPE bag: 35.6 l) reflects the standard products available on the Swiss market. Most energy resources for the production of Mater-Bi bags are required to manufacture PCL. Mater-Bi bags and paper bags are assumed to be composted while PE bags are incinerated. Data for composting are considered to be particularly uncertain (see Section 13.3.1.2 and [18]).

To account for the uncertainty of the data when comparing the results for the three materials, threshold values were determined by analogy to the COMPOSTO study for loose fill (Section 13.3.1.2). The impact categories distinguished here are also identical with those for loose fill (Section 13.3.1.2). In 11 of these 13 impact categories Mater-Bi compost bags cause less environmental damage than paper compost bags (for energy resources, GHG emissions, ozone precursors/summer smog, acidification, eutrophication, toxicity air, toxicity water, deposited waste, heavy metals, carcinogeneity, winter smog; **Figure 13.2**). In the two remaining impact categories, Mater-Bi bags cause a comparable or a greater degree of environmental damage (salinisation; ozone layer depletion). The Mater-Bi compost bags and the HDPE multipurpose bags are equivalent in seven impact categories (energy resources, acidification, eutrophication, toxicity water, ozone layer depletion, carcinogeneity, winter smog). The Mater-Bi bag achieves better scores in four categories (GHG emissions, toxicity air, ozone precursors, heavy metals) but worse results in the two remaining categories (salinisation and deposited waste). However, Mater-Bi bags have a smaller environmental impact than HDPE multipurpose bags in ten categories if one considers that the waste adhering to the bags is incinerated together with the bags (ozone layer depletion and carcinogeneity). It is not specified in the LCA whether the organic waste is considered as neutral in CO_2 terms (this would be expected due to its predominantly biogenic origin) and whether an energy yield according to its heating value has also been taken into account.

A sensitivity analysis was carried out to study whether the production of maize in France instead of Switzerland changes the final results. It is concluded that this is not the case since maize production has a relatively small influence on the total life cycle of Mater-Bi

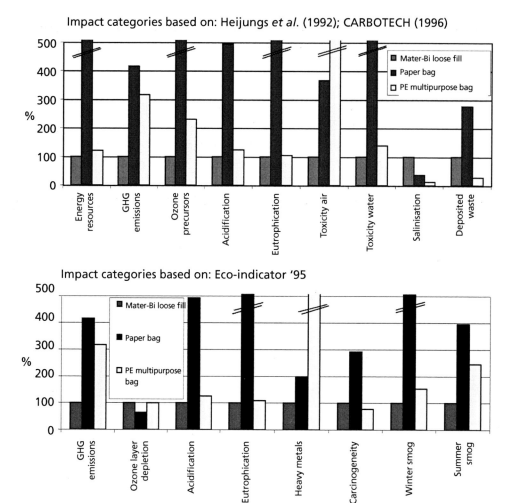

Figure 13.2 LCA results for bags made from TPS compared to HDPE multipurpose bags and compost paper bags [23]

bags and since there are only slight differences between maize production in France and Switzerland. In another sensitivity analysis it was taken into account that the organic waste adhering to the PE bags is co-combusted in MSWI plants. These calculations show a clearer environmental advantage for Mater-Bi bags ('significantly better'). However, as previously, it is unclear whether the CO_2 neutrality of natural organic waste and the additional energy yield have been taken into account.

No sensitivity analysis was conducted to determine how the availability of PE bags of the same size as Mater-Bi and paper bags would influence the results.

- COMPOSTO (1998)

In this study prepared by COMPOSTO [8] for the Kompostforum Schweiz, various types of options for the collection of organic household waste were compared, among them three bags made from biodegradable materials, one PE bag and finally cleaning of the compost bin instead of the use of a bag. The five options studied are:

1. CompoBag 9 l, made from PCL and polyester amide (these raw materials are both made from petrochemical resources).

2. COMPOSAC 14 l, made from Mater-Bi Z, i.e., a blend of maize starch, PCL and additives.

3. Ecosac 6.5 l, also made from Mater-Bi Z.

4. PE bag, 30 x 45 cm.

5. No bag (instead the compost bin is cleaned).

The collection of organic kitchen waste in a compost bin without bag results in the lowest environmental impact of all the options, if the bin is cleaned after use with cold water or with washing-up water. Since, however, 54% of the surveyed Swiss households use hot fresh water and 38% use, in addition, detergents for this purpose, the average environmental impact is highest for those households that do not use an inner liner. Compared to these impacts in the use phase, the environmental damage originating from the production and waste management of the bags is relatively small. The production of biodegradable bags has a larger environmental impact than PE bags. On the other hand the use of PE inner liners results in environmental impacts especially due to incineration of the bag and adhering compostable waste (toxicity air and toxicity water). Interestingly, one of the compostable bags (CompoBag) performs very well compared to the other biodegradable products in spite of being produced exclusively from petrochemical raw materials (the impact categories salinisation and deposited waste are exceptions, see **Figure 13.3**). It is also interesting to note that the results for air and water toxicity differ considerably for the COMPOSAC and ecosac although both are produced from the same material (this difference is explained only partly by the difference in size). It is unclear whether any credits have been allocated to the co-production of electricity and/or steam when incinerating the PE compost bags. It is also unclear whether any credits have been ascribed to composting due to carbon sequestration in the compost. To summarise, the authors of the LCA recommend cleaning the compost bin with cold water or with washing-up water. For consumers with higher standards of cleanliness they recommend biodegradable inner liners.

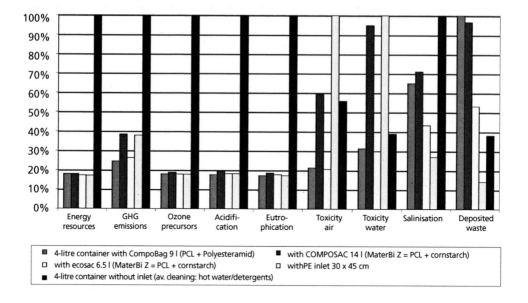

Figure 13.3 LCA for household composting bags (including average consumer behaviour in the use phase [8]

Similar calculations were carried out for container bags with a volume of 240 l (CompoBag 240 l, MaterBi bag 240 l, PE bag 240 l). According to the LCA results the use of these bags is comparable in environmental terms with the generally practised cleaning with cold water. In six out of nine impact categories, the biodegradable inner liners score better than the PE inner liner.

13.3.2 Polyhydroxyalkanoates

• GERNGROSS AND SLATER

The main representatives of PHA are polyhydroxybutyrate (PHB) and polyhydroxyvalerate (PHV). For this family of polymers, no full LCA are available but only studies and estimates comparing the energy requirements and CO_2 or greenhouse gas emissions, among them the papers by Gerngross and Slater [28] and Gerngross [29]. In **Table 13.7** their data are compared to LCA data for petrochemical polymers according to APME (as mentioned earlier there are still considerable uncertainties even for conventional, fossil fuel-based polymers, see Chapter 13.5). The table shows that the total cradle-to-factory gate fossil energy requirements of PHA can compete with HDPE depending on the type of the PHA

Table 13.7 Energy requirements for plastics production [16, 28]			
	Cradle-to-factory gate fossil energy requirements, in GJ/tonne plastic		
	Process energy	Feedstock energy	Total
PHA grown in corn plants	90	0	90
PHA by bacterial fermentation	81	0	81
HDPE	31	49	80
PET (bottle grade)	38	39	77
PS (general purpose)	39	48	87
Data for PHA from [28] *Data for petrochemical polymers from [16]*			

production process. Compared to polyethylene terephthalate (PET), the minimum total energy input for PHA production (fermentation) is somewhat higher while it is lower compared to PS. In contrast, the *process* energy requirements of PHA are two to three times higher than that for petrochemical polymers (**Table 13.7**). Limiting the discussion to these *process energy* data Gerngross and Slater drew the conclusion that PHA do not offer *any* opportunities for *emission* reduction. This finding is valid for certain system boundaries, e.g., for the system 'cradle-to-factory gate', the output of which are plastics pellets. The conclusion is also correct if all plastic waste is deposited in landfills in which the conditions are such that no biodegradation takes place (neither CO_2 nor CH_4 emissions). In contrast, the finding is not correct if other types of waste management processes are assumed within the '*cradle-to-grave*' concept. As the last column of **Table 13.7** shows, the total fossil energy requirements are practically identical for PE and PHA manufactured by bacterial fermentation. Hence, if combusted in a waste incinerator, both plastics result in comparable CO_2 emissions throughout the life cycle.

It must also be taken into consideration that PHA production by bacterial fermentation is in an early stage of development compared to the manufacture of polyolefins and that efficiency improvements are likely to accrue from upscaling of the production process. In the medium term, this can result in a better environmental performance of PHA throughout its life cycle compared to PE and PET.

• HEYDE AND LUCK

Heyde [30] compared the energy requirements of PHB production by bacterial fermentation using various feedstocks and processes to those of HDPE and PS. The PHB options studied include substrate supply from sugar beet, starch, fossil methane and fossil-based methanol and moreover, in the processing stage, the options of enzymic treatment and solvent extraction.

As **Figure 13.4** shows the energy requirements for biotechnological PHB production can substantially exceed the requirements for conventional plastics, but on the other hand there is also scope to outpace fossil-based polymers in terms of energy requirements (PHB Best Case). However, an earlier publication by Luck [31] shows that the choice in the waste management process has a decisive influence on the results. For example, PHB manufactured in an efficient way and disposed of with municipal solid waste (MSW) (German average) requires more energy resources and leads to higher GHG emissions than HDPE if this is recycled according to the German 1995 Packaging Ordinance (64% material recycling). If, on the other hand, the plastics waste is fed to average MSWI plants in both cases then the results are comparable for energy and GHG emissions.

It can be concluded that energy use and CO_2 emissions are nowadays often larger for PHB than for conventional polymers but that there is also scope to avoid this disadvantage if the entire system covering all stages of the life cycle is carefully optimised. In spite of these prospects, Monsanto, a front runner in PHA technology, decided in 1999 to postpone further research and to close down their production line based on fermentation. The original strategy had been to produce PHA as an interim step on the way towards PHA production in genetically modified plants. The goal of PHA production by fermentation was hence to gain experience with this product and to develop the market. The overall strategy was given up when it turned out that it would not be possible to reach in the short-term the target of increasing PHA yields in genetically modified plants from around 3% of dry weight to at least 15% [32].

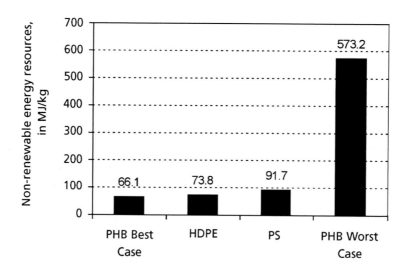

Figure 13.4 Cradle-to-factory gate requirements of non-renewable energy for the production of various polymers [30]

13.3.3 Polylactides (PLA)

• CARGILL DOW

LCA data for PLA are very scarce. Cargill Dow Polymers, the major manufacturer of this type of biodegradable polymer, has only published energy and CO_2 data but no comprehensive dataset so far [33, 34]. As shown in **Table 13.8** *total* fossil energy requirements of PLA are clearly below the respective figures for the petrochemical polymers while the *process* energy requirements are higher for the first commercial PLA plant (PLA-Year 1 in **Table 13.8**). Additional fossil energy savings in the short and long-term are, for example, possible by use of renewable energy as a fuel source (for power and heat), by energy integration of the PLA unit with the lactic acid facility and by optimised product separation. Further options are changes in feedstocks and in the production processes, such as the direct use of agricultural waste and biomass without the intermediate step of isolating dextrose and the use of improved biocatalysts [33].

• VTT

In cooperation with the Neste company, the Technical Research Centre of Finland (VTT) [37] prepared a comparative LCA for two diaper systems one of which is based on polyolefins (PP, PE) while the other uses PLA (made from maize, wheat and sugar beet). For most of the parameters studied, the polyolefin-based diaper shows better results than its biodegradable

	Process energy, fossil (GJ/t plastic)	Feedstock energy, fossil (GJ/t plastic)	Total fossil energy (GJ/t plastic)	Fossil CO_2 from process energy (kg/t plastic)	CO_2 absorption, plant growth (kg/t plastic)	Net fossil CO_2 (kg/t PLA)
PLA – Year 1	57	0	57	3840*	-2020	1820
PLA – Year 5	34	0	34	1939*	-2020**	-81
PLA – Long-term	5	0	5	520*	-2020**	-1500
HDPE	31	49	80	1700	0	1700
PET (bottle grade)	38	39	77	4300	0	4300
Nylon 6	81	39	120	5500	0	5500

Table 13.8 Cradle-to-factory gate energy requirements and CO_2 emissions for plastics production [29, 34-36]

* *Determined by deducting CO_2 absorption from totals (Net fossil CO_2)*
** *Refers to feedstock carbon only. Data for 'Year 5' and 'long-term' assumed to be identical with 'Year 1'*
Data for PLA from CARGILL DOW [34] and personal communication with E. Vink [35]
Data for petrochemical polymers from APME [16, 36]

counterpart. However, the differences in environmental impacts between the two systems are small. Moreover, the lactide and PLA production process were still under development when the LCA was prepared [38] which gives the study a more preliminary, indicative quality.

Based on the information that is currently available one can expect the environmental performance of PLA according to current production methods to be less advantageous than most starch polymers but clearly more beneficial than for PHA. Substantial improvements are expected for PLA for the future.

13.3.4 Other Biodegradable Polymers

• FAT/CARBOTECH [39]

This study was prepared by the Swiss Research Institute for Agriculture (Eidgenössische Forschungsanstalt für Agrarwirtschaft und Landtechnik, FAT) and the environmental consultancy CARBOTECH for the Swiss Federal Office of Agriculture. It contains LCA results for three products (in the FAT/CARBOTECH Study (1997), these three products are referred to as No.3, No.10 and No.12).

(i) Mulch films made of kenaf could be used as an alternative to PE mulch films.

(ii) Loose-fill chips made of miscanthus represent a potential substitute for EPS chips.

(iii) Shredded and ground miscanthus, combined with a matrix based on renewable resources, could be used as a substitute for PE in injection moulding. This material may not fully comply with the latest requirements for biodegradability. The material is not being produced any more.

For kenaf mulch films (i), the LCA shows large advantages compared to the PE alternatives for all indicators except for eutrophication. Disadvantages have, however, been determined for biodiversity, soil fertility and economics (these parameters were not studied as part of the LCA). The two products made of miscanthus (ii, iii) clearly score better than or at least as good as their counterparts based on fossil fuels. In these cases, unresolved technical issues and the forthcoming commercialisation represent the main challenges for the future.

13.4 Summarising Comparison

A comparison of the life cycle inventory results for pellets, loose fills and films/bags is given in **Table 13.9**. Only those impact categories are listed for which at least one dataset was available by type of product.

Table 13.9 provides data for **PCL and PVOH** which are both used as co-polymers for starch plastics. Life cycle practitioners consider these data to be subject to major uncertainties. This is supported by the considerable range of values for energy use in the case of PVOH and for CO_2 emissions for both PCL and PVOH.

For **starch polymer pellets** energy requirements are mostly 35%-70% below those for PE and GHG emissions are 30%-80% lower (disregarding differences that result from differences in waste management). Starch polymers also score better than PE for all the other indicators listed in the table with eutrophication being the sole exception. The lower the share of petrochemical copolymers, the smaller the environmental impact of starch polymers generally is. However, the application areas for pure starch polymers and blends with small amounts of copolymers are limited due to inferior material properties. Hence, blending can extend the applicability of starch polymers and thus lower the overall environmental impact at the macroeconomic level.

For **starch polymer loose fill**, the results differ decisively depending on the source. Much of these differences can be explained by different assumptions regarding the bulk density of the loose fills (see second column in **Table 13.9**) and different approaches for the quantification of the ozone depletion potential (inclusion *versus* exclusion of NO_x; personal communication, E. Würdinger [40]). It therefore seems more useful to compare the results of each study separately. One can conclude from both the Composto and the BIFA study that starch polymer loose fills generally score better than their equivalents made of virgin EPS. GHG emissions represent an exception where the release of CH_4 emissions from biodegradable compounds in landfills results in a disadvantage for starch polymers compared to virgin EPS (only according to BIFA). Loose fill produced from recycled PS may represent a serious option compared to starch polymers according to these calculations.

By comparison to loose fills, the range of results for **starch polymer films and bags** is to a large extent understandable from the differences in film thickness. Taking this factor into account, the environmental impacts of the starch films/bags are lower with regard to energy, GHG emissions and ozone precursors. The situation is less clear for acidification. For eutrophication, PE films tend to score better.

The cradle-to-factory-gate energy requirements for **PLA** are 30%-40% below those for PE, while GHG emissions are about 25% lower. The results for **PHA** vary greatly (only energy data are available). Cradle-to-gate energy requirements in the best case (66.1 GJ/t) are 20%-30% lower than those for PE. For more energy intensive production processes PHA does not compare well with petrochemical polymers. As mentioned earlier, PHA also compares less favourably for process energy (see Section 13.3.2).

It must be noted that all data in **Table 13.9** refer to the current state-of-the-art. Technological progress, improved process integration and various other possibilities for optimisation are likely to result in more favourable results for biodegradable polymers in the future.

Table 13.9 Summary of key indicators from the LCA studies reviewed (state-of-the-art technologies only)

Type of plastic	Functional unit	Cradle-to-grave non-renewable energy use [1] (MJ/functional unit)	Type of waste treatment assumed for calculation of emissions	GHG emissions (kg CO_2 eq./functional unit)	Ozone precursors (g ethylene eq.)	Acidification (g SO_2 eq.)	Eutrophication (g PO_4 eq.)	Refs.
Petrochemical polymers								
HDPE	1 kg	80	Incineration	4.84 [2]	n/a	n/a	n/a	[16]
LDPE	1 kg	80.6	Incineration	5.04 [2]	n/a	n/a	n/a	[16]
LDPE	1 kg	91.7	80% incineration + 20% landfilling	5.20	13.0	17.4	1.1	[7]
Nylon 6	1 kg	120	Incineration	7.64 [2]	n/a	n/a	n/a	[16]
PET (bottle grade)	1 kg	77	Incineration	4.93 [2]	n/a	n/a	n/a	[16]
PS (general purpose)	1 kg	87	Incineration	5.98 [2]	n/a	n/a	n/a	[16]
EPS	1 kg	84	Incineration	5.88 [2]	n/a	n/a	n/a	[16]
EPS	1 kg	88	None (cradle-to-factory gate)	2.80	43.0	170.0	5.8	[17]
EPS (PS + 2% SBR + Pentan + Butan)	1 kg	87	None (cradle-to-factory gate)	2.72	1.2	18.5	1.5	[9]
Petrochemical copolymers								
Polycaprolactone (PCL)	1 kg	83	Incineration	3.1 [2]	6.1	5.5	0.5	[8]
Polycaprolactone (PCL)	1 kg	77	Incineration	5.0–5.7 [2]	n/a	n/a	n/a	[16]
Polyvinyl alcohol (PVOH)	1 kg	102	Incineration	2.7 [2]	8.9	8.0	0.9	[17]
Polyvinyl alcohol (PVOH)	1 kg	58	Incineration	4.1–4.3 [2]	n/a	n/a	n/a	[17]
Biodegradable polymers (pellets)								
TPS	1 kg	25.4	Incineration	1.14	n/a	n/a	n/a	[15]
TPS	1 kg	25.5	80% incineration + 20% composting	1.20	4.7	10.9	4.7	[7]
TPS	1 kg	25.4	100% composting	1.14	5.0	10.6	4.7	[7]

Table 13.9 Continued...

Type of plastic	Functional unit	Cradle-to-grave non-renewable energy use [1] (MJ/functional unit)	Type of waste treatment assumed for calculation of emissions	GHG emissions (kg CO_2 eq./ functional unit) emissions	Ozone precursors (g ethylene eq.)	Acidification (g SO_2 eq.)	Eutrophication (g PO_4 eq.)	Refs.
TPS (maize starch + 5.4 maize grit + 12.7% PVOH	1 kg	18.9	None (cradle-to-factory gate) [3]	1.10 [3]	0.2	4.6	0.5	[9]
TPS + 15% PVOH	1 kg	24.9	Incineration	1.73	n/a	n/a	n/a	[15]
TPS + 52.5% PCL	1 kg	48.3	Incineration	3.36	n/a	n/a	n/a	[15]
TPS + 60% PCL	1 kg	52.3	Incineration	3.60	n/a	n/a	n/a	[15]
Mater-Bi foam grade	1 kg	32.4	Composting	0.89	5.5	20.8	2.8	[17]
Mater-Bi foam grade	1 kg	36.5	Waste water treatment plant	1.43	5.8	20.7	3.1	[17]
Mater-Bi film grade	1 kg	53.5	Composting	1.21	5.3	10.4	1.1	[8]
PLA	1 kg	57	Incineration [3]	3.84 [3]	n/a	n/a	n/a	[34]
PHA by fermentation	1 kg	81	n/a	n/a	n/a	n/a	n/a	[28]
PHB, various processes	1 kg	66-573	n/a	n/a	n/a	n/a	n/a	[30]
Loose fills								
Mater-Bi starch loose fills	1 m³ (10 kg)	492	Waste water treatment plant	21.0	115	276	39.0	[17]
FloPak starch loose fill	1 m³ (12 kg)	277	30% incineration, 70% landfilling	33.5	10	83	9.9	[9]
EPS loose fill	1 m³ (4.5 kg)	680	Incineration	56.0	1200	325	42.0	[17]
FloPak EPS loose fill	1 m³ (4 kg)	453	30% incineration, 70% landfilling	22.5	57	85	8.0	[9]
EPS loose fill (by recycling of PS waste)	1 m³ (4 kg)	361	30% incineration, 70% landfilling	18.6	55	107	9.9	[9]
Films and bags								
TPS film	100 m², 150 µm [4]	649	80% incineration + 20% landfilling	25.30	100	239	103.0	[7]

Table 13.9 Continued...

Type of plastic	Functional unit	Cradle-to-grave non-renewable energy use [1] (MJ/functional unit)	Type of waste treatment assumed for calculation of emissions	GHG emissions (kg CO_2 eq./ functional unit)	Ozone precursors (g ethylene eq.)	Acidification (g SO_2 eq.)	Eutrophication (g PO_4 eq.)	Refs.
Mater-Bi starch film	100 m², 200 µm [4]	133	Composting	2.98	14.0	26.5	2.8	[8]
PE film	100 m², 150 µm [4]	1340	80% incineration + 20% landfilling	66.70	180	238	15.0	[7]

[1] Total of process energy and feedstock energy. Non-renewable energy only, i.e., total of fossil and nuclear energy. In the 'cradle-to-factory gate' concept the downstream system boundry coincides with the output of the polymer or the end product. Hence, no credits are ascribed to valuable by-products from waste management (steam, electricity, secondary materials)

[2] Only CO_2. Embodied carbon: 3.14 kg CO_2/kg PE, 2.34 kg CO_2/kg Nylon 6, 2.29 kg CO_2/t PET, 3.38 kg CO_2/t PS, 2.32 kg CO_2/t PCL, 2.00 kg CO_2/t PVOH

[3] No credit for carbon uptake by plants

[4] An important explanation for the large difference between the values reported is that Carbotech assumes a film thickness of 150 µm while it is only 20 µm in the case of Composto

13.5 Discussion

The comparison of the main assumptions made in the various studies and the comparison with the current state-of-the-art reveals a number of uncertainties with the most important being:

- LCA data for PCL and PVOH are generally considered to be subject to major uncertainties. In view of the widespread use of these compounds in biodegradable materials and given the strong impact on the final results especially for some starch polymers, reliable LCA data need to be generated.

- The data used for composting are are subject to major uncertainties. This is partly explicitly stated by the authors [19], partly it becomes obvious by comparing the assumptions made in the various studies (wherever these are described in detail). According to COMPOSTO [23] 40%-60% of the carbon absorbed in the vegetable material is released to the atmosphere during composting. To avoid the underestimation of GHG emissions, the COMPOSTO Studies [17, 23] assume that 60% of the absorbed carbon is released. The assumption can be considered as safe if compared to Schleiss and Chardonnens [41] who state that the average carbon dissipation in the form of CO_2 amounts to 40% (average of all composting plants in Switzerland). While these data refer to the average of all inputs and outputs of a composting plant, the question arises whether it also holds true for the materials discussed here, i.e., for biodegradable polymers. Since biodegradable polymers decompose to a large extent within a short period of time the question arises as to whether the approach chosen by BIFA [9] might be more accurate where it was assumed that the buildup of organic matter, and hence, the effect of carbon-carbon sequestration is negligible. According to biodegradation tests conducted by several institutes, the degradation of starch polymers during composting (59 °C, 45 days) amounts to about 80% to 90% (the test refers to a mixture of 15% starch polymers and 85% pure cellulose). Since biodegradation in the subsequent maturation phase is negligible, Novamont draws the conclusion that an average conversion rate of 80% is realistic (personal communication, L. Marini, [42]). The specific characteristics of the starch polymer considered and the type of composting technology applied may influence the biodegradation fraction.

- The various studies differ in the accounting method for waste incineration of biodegradable polymers. Even though the detailed assumptions are hardly ever spelled out it is quite obvious that the chosen approaches are not comparable. For example, the BIFA study assumes that incineration takes place in waste-to-energy facilities, resulting in a net output of electricity and/or heat. Credits are assigned to these useful products. In contrast, the COMPOSTO studies do not account for co-produced electricity/steam. It is unlikely that this reflects the differences in the share of energy recovery (waste-to-energy facilities *versus* simple incineration without energy recovery) among the countries studied; it rather represents a methodological difference.

- The environmental assessment of the incineration of mulch films with adhering organic waste (soil) raises particular questions. In one of the sensitivity analyses, the COMPOSTO study [23] introduces a CO_2 penalty in order to account for the emissions resulting from the incineration of this adhering organic waste. This may be justified if the moisture of the organic waste is so high that the vapourisation of the water contained requires more energy than the calorific value of the organic waste. In this case the incineration of the adhering waste represents a net energy sink. In practice, this is typically compensated by co-firing of fossil fuels or of other high-calorific combustible waste leading to CO_2 and other environmental impacts. On the other hand it is also possible that the moisture content of the adhering waste is low, resulting in a net energy yield in the incineration process. Moreover, if the organic waste is of biogenic origin, its incineration is neutral in CO_2 terms (due to extraction of CO_2 from the atmosphere during plant growth). These considerations show that specific circumstances determine whether the co-combustion of adhering organic waste – be it soil, organic kitchen waste or any other type of biogeneous waste – results in net environmental benefits or disadvantages.

- In the case of landfilling some studies account for methane (CH_4) emissions due to anaerobic emissions while others do not take this into consideration. This can have a considerable impact on the results due to the relatively strong greenhouse gas effect of CH_4. As a consequence the overall global warming potential (GWP) of biodegradable polymers manufactured from renewable raw materials may be higher than for petrochemical plastics depending on the waste management system chosen for the latter [9].

- The characterisation factors for global warming potentials used in most of the studies reviewed are outdated in the meantime (GWP_{100} for methane and nitrous oxide) The GWP equivalence factors used in the various studies are 11 or 21 for CH_4 and 270 or 310 for N_2O, while - according to the current state of research - more accurate figures are 23 (CH_4) and 296 (N_2O) [44]. Since the contribution of CO_2 dominates the overall GHG effect, this uncertainty is considered to be less important.

- When making comparisons with conventional fossil fuel-based polymers it must be remembered that LCA data for these products are also uncertain and continue to be corrected. This is in spite of the fact that petrochemical polymers are manufactured by use of mature technologies that are applied globally with only limited variations. For example, energy data for PE production range between approximately 65 GJ/t and 80 GJ/t according to a comparison of various sources [43] while the CARBOTECH study assumes about 92 GJ/t. While this does not change the overall conclusions for energy use and CO_2 emissions, the implications are unclear for the other environmental parameters covered by the CARBOTECH study.

The problem related to these uncertainties can be resolved to some extent by taking into account the significance of the difference in values for the systems compared (thresholds for the categories 'significantly higher', 'higher', 'comparable' etc., see [17], Section 13.3.1.2). In addition, it is an important goal of future research to reduce further the existing uncertainties.

In all of the studies reviewed ecological ranking was determined by comparing for how many indicators the environmental impact is lower for biodegradable polymers compared to the petrochemical polymers (in addition, single-score parameters were used; as indicated earlier these results are not described here since the method applied is not generally accepted). The disadvantage of this approach is that the selection of the indicators compared can have an influence on the final conclusions. Together with the fact that the relative difference in the results for the various impact categories (a few per cent *versus* a few hundred per cent) is hardly ever accounted for; this shows the urgent need for the further development of the LCA methodology, e.g., by introduction of significance thresholds. Finally, when interpreting the results, it must be remembered that the studies reviewed partly differ in regional scope. Since the results are to some extent subject to country specific circumstances, (e.g., GHG emissions from national power production), care must be taken when drawing more general conclusions. On the other hand, the uncertainties related to conclusions can be reduced if several independent analyses for different countries arrive at similar conclusions. A summary of the aspects to be considered in future LCA studies for biodegradable polymers is given in the checklist in Appendix 13.2.

13.6 Conclusions

The number of published LCA for bioplastics is very limited. This seems to be in contrast to the general public interest for this issue and the more recent interest by policy makers. For example, within the European Commission's 'European Climate Change Programme' (ECCP) a subgroup to the Working Group 'Industry' deals with Renewable Raw Materials (ECCP, 2001, pp.78-91). Since most biodegradable polymers are made from renewable raw materials, a substantial share of biodegradable polymers are covered. For example, no comprehensive LCA have been published so far for PLA (plant-based), cellulose polymers (plant-based) and some fossil fuel-based biodegradable polymers, such as BASF's product Ecoflex.

All of the biodegradable polymers covered by this review are manufactured from renewable resources. It is important to keep this in mind since some of the findings would probably differ for biodegradable polymers based on petrochemical polymers.

The existing LCA contain uncertainties which should be addressed by future research and analysis. A prominent example is the environmental assessment of the composting process for biodegradable polymers. In some studies further sensitivity analyses would be required to ensure that the final findings are well founded, e.g. for smaller PE bags in [23]. Moreover, many of the environmental analyses choose a cradle-to-factory-gate perspective, i.e., the analysis ends with the product under consideration. While this approach provides valuable results, additional analyses taking a *cradle-to-grave* perspective by inclusion of the waste management stage should also be carried out. Due to their strong impact on the final results several alternatives in the waste management stage should be evaluated.

In spite of these uncertainties and the information gaps it is safe to conclude that biodegradable polymers offer important environmental benefits today and for the future. Of all biodegradable polymers studied, starch polymers are considered to perform best in environmental terms under the current state of the art – with some differences among the various types of starch polymers. Compared to starch polymers the environmental benefits seem to be smaller for PLA (LCA results only available for energy and CO_2). For PHA, the environmental advantage currently seems to be very small compared to conventional polymers (LCA results are only available for energy use). For PLA, a comprehensive life cycle assessment would provide valuable, additional insight. For both PLA and PHA, the production method, the scale of production and the type of waste management treatment can influence decisively the ultimate conclusion about the overall environmental balance. When making comparisons of this type it must be taken into account that the material properties of the various polymers differ which has important consequences for their suitability for certain types of application.

Starch polymers are currently the only type of biodegradable polymer for which several comprehensive LCA studies are available. According to these assessments starch polymers do not perform better than their fossil fuel-based counterparts in all environmental categories, including biodiversity and soil quality, which are generally outside the scope of LCA. However, most studies come to the conclusion that starch polymers (pellets and end products) are more beneficial in environmental terms than their petrochemical counterparts; this conclusion is drawn without weighting and in most cases without significance thresholds. The preferences among the environmental targets determine whether biodegradable polymers are considered to be environmentally attractive.

For the time being, it is not possible to make a *general judgement* about whether biodegradable plastics should be preferred to petrochemical polymers from an environmental point of view. As a prerequisite for drawing such a conclusion, full-sized LCA studies would be needed for PHA, PLA and other important biodegradable polymers including those manufactured from fossil feedstocks, e.g., BASF's product 'Ecoflex' and Eastman's 'Eastar'. But even if those were available one would be left with considerable uncertainties, e.g., because it will never be feasible to cover all possible products and all possible impact categories (compare [46] and Section 7 in Appendix 13.2). In spite of these limitations one can conclude that the results for the use of fossil energy resources and GHG emissions are already more favourable for most biodegradable polymers today. As an exception, landfilling of biodegradable polymers can result in methane emissions (unless landfill gas is captured) which makes the system unattractive in terms of reducing greenhouse gas emissions.

The CARBOTECH study [7] reaches the conclusion that polymers based on starch, kenaf and miscanthus offer larger opportunities for energy saving and GHG mitigation than *bioenergy* (CARBOTECH, 1996, 92; partly based on [39]). On this basis the authors of the CARBOTECH study draw the conclusion that the use of biomass as a chemical feedstock *generally* offers larger potentials for energy saving and GHG emission reduction (per km^2 of cultivated land) than biomass use for energy purposes. It is, however, an important shortcoming that only a small selection of bioenergy technologies was taken into account (rapeseed oil methyl ester; incineration of miscanthus). To keep track of competition and synergies between bioenergy and biomaterials, comparative assessments will therefore continue to be needed. This is also necessary to account for innovations in both areas. It would ease such comparisons and the usefulness for decision-makers if future studies dealing with bioenergy and biomaterials always also studied the land use requirements of the various options.

To improve the 'environmental competitiveness' of biodegradable polymers further R&D is required to optimise the production by increasing the efficiencies of the various unit processes involved, (e.g., separation processes), and by process integration. Substantial scope for improvement can be expected here given the fact that all biodegradable polymers are still in their infancy while the manufacture of petrochemical polymers has been optimised for decades. Some of the LCA are already outdated due to the substantial progress made in manufacturing and processing of biodegradable polymers, e.g., for films see Section 13.3.1.3. This means that the real environmental impacts caused by biodegradable polymers tend to be lower than established in the LCA studies reviewed.

A promising line for future R&D - even though somewhat outside the scope of the biodegradability concept - could be the development of biomass-derived polymers that can be recycled mechanically, preferably also in blends with petrochemical polymers. Such recyclable polymers made from renewable raw materials are likely to be unrivalled in environmental terms provided that their manufacture is not too resource-intensive in the first place. This may offer longer term prospects to PHA, PLA and other bio-based polymers while post-consumer recycling of starch polymers seems hardly viable due to the sensitivity of these products to water.

To summarise, the existing LCA studies and environmental assessments support the further development of biodegradable polymers. Careful monitoring of as many environmental

461

impacts as possible continues to be necessary both for decision makers in companies and in policy. This requires that comprehensive LCA studies are conducted on a regular basis to account for changes in the production process and in the infrastructure (power generation, waste management). For some materials the environmental benefits achieved are substantial already today and in most cases the prospects are very promising.

Acknowledgements

The preparation of this chapter was only possible due to the large support from various experts in the field. Special thanks are directed to Mrs. Catia Bastioli and Mr. Luigi Marini, both Novamont (Italy), for providing valuable data from background reports and for their very constructive comments. I am also very grateful to Mr. Eduard Würdinger, Bayrisches Amt für Abfallwirtschaft (BIFA, Germany) who provided unpublished results from a large study on loose fill packaging materials. Mr. Erwin Vink, Cargill Dow (Netherlands), kindly provided the latest available data for polylactides. Many other experts in the field, including Mrs. Bea Schwarzwälder, Composto (Switzerland), Mr. Fredy Dinkel, Carbotech (Switzerland) and Mr. Urs Hänggi, Biomer (Germany) helped to clarify important details. Last but not least I would like to thank Ms. Frances Powers, RAPRA (United Kingdom) who provided numerous articles on biodegradable materials, and thus ensured the comprehensiveness of this chapter.

References

1.   ISO 14040 (DIN EN ISO 14040), *Environmental Management – Life Cycle Assessment – Principles and Framework*, 1997.

2.   EN ISO 14041, *Environmental Management – Life Cycle Assessment – Goal and scope definition and inventory analysis*, 1998.

3.   EN ISO 14042, Environmental *Management – Life Cycle Assessment – Life Cycle Impact Assessment*, 1998.

4.   EN ISO 14043, *Environmental Management – Life Cycle Assessment – Life Cycle Interpretation*, 1999.

5.   J.B. Guinée, M. Gorrée, R. Heijungs, G. Huppes, R. Kleijn, A. de Koning, L. van Oers, A.W. Sleeswijk, S. Suh, H.A. Udo de Haes, H. de Bruin, R. van Duin and M.A.J. Huijbregts, *Life Cycle Assessment – An Operational Guide to the ISO Standards*, Parts 1-3, Kluwer Academic Publishers, Dordrecht, The Netherlands, 2001.

6. *Eco-indicator 95 - Weighting Method for Environmental Effects that Damage Ecosystems or Human Health on a European Scale*, Pré Consultants and DUIF Consultancy bV, Amersfoort, The Netherlands, 1995.

7. F. Dinkel, C. Pohl, M. Ros and B. Waldeck, *Life Cycle Assessment (LCA) of Starch Polymers (Ökobilanz stärkehaltiger Kunststoffe)*, Report No. 271, 2 volumes, BUWAL, Berne, Switzerland, 1996. [In German]

8. R. Estermann, *Test of Biodegradable Bags for Green Waste (Test von Säcken aus biologisch abbaubaren Werkstoffen für die Grünabfallsammlung)*, Composto, Olten for the Kompostforum Schweiz, Uerikon, Switzerland, 1998. [In German]

9. *Polymers from Renewable Resources – Comparative Life Cycle Assessment for Loose-Fill Packaging Material from Starch and Polystyrene (Kunststoffe aus Nachwachsenden Rohstoffen - Vergleichende Ökobilanz für Loose-Fill-Packmittel aus Stärke bzw aus Polystyrol; interim report)*. Bayrisches Amt für Abfallwirtschaft, Würzburg (BIFA), Institut für Energie- und Umweltforschung Heidelberg (IFEU), Flo-Pak GmbH, Germany. [In German]

10. *Life Cycle Assessment of Packaging Material (Ökobilanzen von Packstoffen)*, Environmental Reports Series (Schriftenreihe Umweltschutz) No.132, Bundesamt für Umwelt und Landschaft (BUWAL), Bern, Switzerland. [In German]

11. F. Dinkel, C. Pohl, M. Ros and B. Waldeck, *Life Cycle Assessment (LCA) of Starch Polymers (Ökobilanz stärkehaltiger Kunststoffe)*, Report No. 271, 2 volumes, BUWAL, Berne, Switzerland, 1996, p.58. [In German]

12. F. Dinkel, C. Pohl, M. Ros and B. Waldeck, *Life Cycle Assessment (LCA) of Starch Polymers (Ökobilanz stärkehaltiger Kunststoffe)*, Report No. 271, 2 volumes, BUWAL, Berne, Switzerland, 1996, p.12. [In German]

13. F. Dinkel, C. Pohl, M. Ros and B. Waldeck, *Life Cycle Assessment (LCA) of Starch Polymers (Ökobilanz stärkehaltiger Kunststoffe)*, Report No. 271, 2 volumes, BUWAL, Berne, Switzerland, 1996, p.13. [In German]

14. F. Dinkel, C. Pohl, M. Ros and B. Waldeck, *Life Cycle Assessment (LCA) of Starch Polymers (Ökobilanz stärkehaltiger Kunststoffe)*, Report No. 271, 2 volumes, BUWAL, Berne, Switzerland, 1996, p.51. [In German]

15. M. Patel, E. Jochem, F. Marscheider-Weidemann, P. Radgen and N. von Thienen, *C-STREAMS - Estimation of Material, Energy and CO_2 Flows for Model Systems in the Context of Non-Energy Use, From a Life Cycle Perspective – Status and Scenarios, Volume I: Estimates for the Total System*, Fraunhofer Institute for

Systems and Innovation Research (FhG-ISI), Karlsruhe, Germany, 1999. [In German, English abstract].

16. *Eco-profiles of plastics and related intermediates (about 55 products)*, Association of Plastics Manufacturers in Europe (APME), Brussels, Belgium, 1999. Downloadable from *http://lca.apme.org*

17. R. Estermann, B. Schwarzwälder and B. Gysin, *Life Cycle Assessment of Mater-Bi and EPS Loose Fills*, Composto, Olten, Switzerland, 2000.

18. R. Estermann and B. Schwarzwälder, *Life Cycle Assessment of Mater-Bi bags for the Collection of Compostable Waste*, Composto, Olten, Switzerland, 1998, p.19.

19. R. Estermann, B. Schwarzwälder and B. Gysin, *Life Cycle Assessment of Mater-Bi and EPS Loose Fills*, Composto, Olten, Switzerland, 2000, p.16.

20. R. Estermann, B. Schwarzwälder and B. Gysin, *Life Cycle Assessment of Mater-Bi and EPS Loose Fills*, Composto, Olten, Switzerland, 2000, p.19.

21. R. Estermann, B. Schwarzwälder and B. Gysin, *Life Cycle Assessment of Mater-Bi and EPS Loose Fills*, Composto, Olten, Switzerland, 2000, p.28.

22. R. Estermann, B. Schwarzwälder and B. Gysin, *Life Cycle Assessment of Mater-Bi and EPS Loose Fills*, Composto, Olten, Switzerland, 2000, p.15.

23. R. Estermann and B. Schwarzwälder, *Life Cycle Assessment of Mater-Bi bags for the Collection of Compostable Waste*, Composto, Olten, Switzerland, 1998.

24. R. Heijungs, J.B. Guinee, G. Huppes, R.M. Lankreijer, H.A. Udo de Haes, A. Wegener Sleeswijk, *Environmental Life-Cycle Assessment of Products – Guide and Background*, CML (Centre for Environmental Science), Leiden, The Netherlands, Netherlands Organisation for Applied Scientific Research (TNO), Fuels and Raw Materials Bureau (B&G), Leiden, The Netherlands, 1992.

25. R. Estermann, B. Schwarzwälder and B. Gysin, *Life Cycle Assessment of Mater-Bi and EPS Loose Fills*, Composto, Olten, Switzerland, 2000, p.26.

26. R. Estermann, B. Schwarzwälder and B. Gysin, *Life Cycle Assessment of Mater-Bi and EPS Loose Fills*, Composto, Olten, Switzerland, 2000, p.27.

27. F. Dinkel, C. Pohl, M. Ros and B. Waldeck, *Life Cycle Assessment (LCA) of Starch Polymers (Ökobilanz stärkehaltiger Kunststoffe)*, Report No. 271, 2 volumes, BUWAL, Berne, Switzerland, 1996, p.60. [In German]

28. T.U. Gerngross and S. Slater, *Scientific American*, 2000, August, 36.

29. T.U. Gerngross, *Nature Biotechnology*, 1999, **17**, 541.

30. M. Heyde, *Polymer Degradation and Stability*, 1998, **59**, 1-3, 3.

31. T. Luck, *Feasibility Study Polyhydroxyalkanoates – Summary Report of the Studies 'Feasibility Study for Eestimation of the Market Potential of New Polyhydroxy Fatty Acids (PHF)' and 'Shortcut LCAs for Polymeric Materials Made from Biologically Produced Polyhydroxy Fatty Acids'*, (Feasability-Studie Polyhydroxyfettsäuren – Kurzdarstellung der Ergebnisse der Studien 'Feasibility-Studie zur Abschätzung des Marktpotentials neuer Polyhydroxyfettsäuren (PHF)' und 'Abschätzende Ökobilanzen zu Polymerwerkstoffen auf der Basis biologisch erzeugter Polyhydroxyfettsäuren'), Fraunhofer Institut für Lebensmitteltechnologie und Verpackung (ILV), Munich, Germany, 1996. [In German]

32. *Chemistry & Industry*, 1999, October, 729.

33. R. E. Conn, *Life Cycle Inventory as a Process Development Tool*, Presentation at the Süddeutsches Kunststoffzentrum (SKZ), Würzburg, Germany.

34. E.T.H. Vink, *NatureWorks – A new generation of biopolymers*, Presentation on 29 March 2001, Birmingham, UK.

35. E.T.H. Vink, personal communication, Cargill Dow.

36. *Eco-profiles of Plastics and Related Intermediates (about 55 Products)*, Association of Plastics Manufacturers in Europe (APME), Brussels, Belgium. Downloadable from the internet (*http://lca.apme.org*).

37. S. Hakala, Y. Virtanen, K. Meinander and T. Tanner, *Life-cycle Assessment, Comparison of Biopolymer and Traditional Diaper Systems*, Technical Research Centre of Finland (VTT), Jyväskylä, Finland, 1997.

38. S. Hakala, Y. Virtanen, K. Meinander and T. Tanner, *Life-cycle Assessment, Comparison of Biopolymer and Traditional Diaper Systems*, Technical Research Centre of Finland (VTT), Jyväskylä, Finland, 1997, p.28.

39. U. Wolfensberger and F. Dinkel, *Assessment of Renewable Resources in Switzerland in the Years 1993-1996 – Comparative Assessment of Products made from Selected Renewable Resources and Equivalent Conventional Products with Regard to Environmental Effects and Economics* (Beurteilung Nachwachsender Rohstoffe in der Schweiz in den Jahren 1993-1996 – Vergleichende Betrachtung

von Produkten aus Ausgewählten Nachwachsenden Rohstoffen und Entsprechenden Konventionellen Produkten Bezüglich Umweltwirkungen und Wirtschaftlichkeit.), Eidgenössische Forschungsanstalt für Agrarwirtschaft und Landtechnik (FAT) and CARBOTECH for the Federal Office of Agriculture, Bern, Switzerland, 1997. [In German]

40. E. Würdinger, personal communication, BIFA.

41. K. Schleiss and M. Chardonnens, *Status and Development of Composting in Switzerland 1993*, (Stand und Entwicklung der Kompostierung in der Schweiz 1993), BUWAL Umwelt-Mterialien No.21, Bundesamt für Umwelt, Wald und Landschaft (BUWAL), Bern, Switzerland, 1994. [In German]

42. L. Marini, personal communication.

43. M. Patel, *Energy*, 2003, **28**, 7, 721.

44. *Climate Change 2001 – The Scientific Basis*, Eds., J.T. Houghton, Y. Ding, D.J. Griggs, M. Noguer, P.J. van der Linden, D. Xiaosu, K. Maskell and C.A. Johnson, Intergovernmental Panel on Climate Change (IPCC), Geneva, Switzerland, 2001. Full text downloadable from *www.ipcc.ch*.

45. *European Climate Change Programme (ECCP)*, Long Report, European Commission, Brussels, Belgium 2001.

46. G. Finnveden, *International Journal of Life Cycle Assessment*, 2000, **5**, 4, 229.

47. H. A. Udo de Haes, *Towards a Methodology for Life Cycle Impact Assessment*, Society of Environmental Toxicology and Chemistry (SETAC), Brussels, Belgium, 1996.

48. L-G. Lindfors, K. Christiansen, L. Hoffman, Y. Virtanen, V. Juntilla, O.J. Hanssen, A. Ronning, T. Ekvali and G. Finnveden, *Nordic Guidelines on Life-Cycle Assessment*, Nordic Council of Ministers, Copenhagen, Denmark.

Appendix 13.1 Overview of environmental life cycle comparisons for biodegradable polymers included in this review

Table 13.10 LCA studies for starch polymer pellets and films		
	CARBOTECH, 1996 (pellets and films)	Fraunhofer ISI, 1999 (pellets)
Biodegradable product *Reference*	- Starch polymer pellets - Starch polymer film - LDPE pellets - LDPE film	- Starch polymer pellets (different blends) - PE pellets
Region/time	Switzerland 1990s Exceptions: - Germany for the production of petrochemical plastics - Partly Europe for electricity generation (for manufacturing processes outside Switzerland)	Germany, mid 1990s
System boundaries *Production* *Use phase* *Waste management*	 All process steps included Use of pesticides taken into account Excluded Included Specific aspects: - MSWI plants: No credits for co-production electricity/heat - Composting plants: No credits for composting - Pre-consumer recycling: Taken into account - Post-consumer recycling: Not taken into account	 All process steps included Excluded Included Specific aspects: - MSWI plants: No credits for electricity/heat

Table 13.10 Continued ...		
	CARBOTECH, 1996 (pellets and films)	Fraunhofer ISI, 1999 (pellets)
Parameters		
Quantative analysis	*Impact category* *Unit* 1. Energy resources (MJ) 2. GHG emissions (kg CO_2 eq.) 3. Ozone precursors (kg ethylene eq.) 4. Human toxicity (a m^2) 5. Acidification (kg SO_2 eq.) 6. Eutrophication (kg PO_4 eq.) 7. Ecotoxicity (d x l) 8. Salinisation (H^+/mol) 9. Deposited waste (10^{-3} EPS) Impact factors [1] From Heijungs [24] [2] and additional estimates based on various sources	*Impact category* *Unit* 1. Energy resource (MJ) 2. CO_2 (kg CO_2) Impact factors [1] -
Functional unit	100 kg pellets 100 m^2 film, thickness 150 micrometers	1000 kg pellets
Qualitative analysis	Quality of soil Biodiversity	
[1] For aggregation to the various impact categories. In addition, aggregation to one single parameter using two different methods. Eco-indicator 95 and Environmental Pollution Score (EPS; see CARBOTECH, 1996) [2] *The values used for the global warming potential of individual gases are outdated (e.g., for CH_4: 11 kg CO_2 eq./kg CH_4; for N_2O: 270 kg CO_2 eq./kg N_2O*		

Table 13.11 LCA studies for starch polymer loose fill packaging material

	COMPOSTO, 2000 (loose fill)	BIFA, 2001 (loose fill)
Biodegradable product	- Starch polymer loose fills (Mater-Bi) - EPS loose fills	- Starch polymer loose fills (Flo-Pak) - EPS loose fills (Flo-Pak)
Reference		
Region/time	Switzerland 1990s	Germany, 1990s schemes to account for possible future developments.
System boundaries		
Production	All process steps included, except for packaging and distribution. Sensitivity analyses for (i) production at customer's site and (ii) direct production from starch.	All process steps included. The following options are distinguished for starch polymers and EPS: *Starch polymers* - Corn (maize) - Potato - Potato with effluent use - Wheat, intensive - Wheat, extensive *EPS* - Virgin PS - PS pre-consumer waste (industrial waste) - PS post-consumer waste from MC/CD covers - PS post-consumer waste from EPS packaging - PS post-consumer waste from DSD (cups)
Use phase	Excluded	Five options are distinguished both for starch polymers and for EPS: - Single use - 2 cycles (within company) - 4 cycles (within company) - 4 cycles (recycling station) - Single use (without allocation)

Table 13.11 Continued ...

	COMPOSTO, 2000 (loose fill)	BIFA, 2001 (loose fill)
Waste management	Composting of starch polymers, incineration of EPS. Collection after use is excluded. Sensitivity analyses for disposal via waste water.	The following options are distinguished for starch polymers and EPS: *Starch polymers* - Current MSW management practice (landfill, incineration) - Average MSW incineration plant - Separate collection & optimised MSWI plant - Composting - Fermentation - DSD collection, blast furnace *EPS* - DSD collection system and blast furnace - Closed-loop mechanical recycling - Open-loop mechanical recycling

Table 13.11 Continued ...

Parameters	COMPOSTO, 2000 (loose fill)	Unit	BIFA, 2001 (loose fill)	Unit
Quantative analysis	*Impact category*	*Unit*	*Impact category*	*Unit*
	(a) See column for CARBOTECH (1996)		1. GHG emissions	(kg CO_2 eq.)
	(b) 1. GHG emissions	index	2. Carcinogeneity	(kg arsenic eq.)
	2. Ozone precursors	index	3. Eutrophication (terrestrial)	(kg PO_4 eq.)
	3. Acidification	index	4. Acidification	(kg SO_2 eq.)
	4. Eutrophication	index	5. Diesel particles	(kg)
	5. Heavy metals	index	6. Ozone precursors	(kg ethylene eq.)
	6. Carcinogeneity	index	7. Ozone precursors, N-corrected	(kg NCPOCP)
	7. Winter smog	index	8. Use of natural land (class VI)	(m^2)
	8. Summer smog	index	9. Ozone depletion	(kg N_2O)
			10. Eutrophication (aquatic)	(kg PO_4 eq.)
			11. Cumulative Energy Demand fossil	(GJ)
			12. Cumulative Energy Demand nuclear	(GJ)
			13. Oil equivalents	(kg oe)
			14. Lead	(g)
			15. Sulfur dioxide	(SO_2)
			16. Fluorinated hydrocarbons	(kg)
			17. Ammonia	(kg)
			18. Nitrous oxides	(kg)
			19. Halogenated organic hydrocarbons	(mg)
			20. Biocide use	(kg)
	Impact factors [1]		*Impact factors* [1]	
	(a) see column for CARBOTECH (1996)		Various sources including Heijungs [24]	
	(b) Eco-indicator 95			

Table 13.11 Continued ...

	COMPOSTO, 2000 (loose fill)	BIFA, 2001 (loose fill)
Functional unit	1 m³ loose fill	100 m³ loose fill
Qualitative analysis		

1) For aggregation to the various impact categories. In addition, aggregation to one single parameter using two different methods. Eco-indicator 95 and Environmental Pollution Score (EPS; see [27])

2) The values used for the global warming potential of individual gases are outdated (e.g., for CH_4: 11 kg CO_2 eq./kg CH_4; for N_2O: 270 kg CO_2 eq./kg N_2O

Table 13.12 LCA studies for compost bags

	COMPOSTO, September 1998 (waste bags)	COMPOSTO, January 1998 (waste bags)
Biodegradable product	- Starch polymer bag (Mater-Bi Z) - Compost paper bag - PE multi-purpose bag	Bag for 4 l and 10 l compost bins (also for 240 l): - CompoBag 9 l (PCL + polyester amide) - COMPOSAC 14 l (Mater-Bi Z = starch + PCL) - ecosac 6.5 l (Mater-Bi Z = starch + PCL) - PE bag 30 x 45 cm - No bag
Reference		
Region/time	Switzerland 1990s (sensitivity analysis: France)	Switzerland, 1990s (Raw materials from USA/Europe)
System boundaries		
Production	All process steps included, except for packaging and distribution	All process steps included
Use phase	Excluded	Included
Waste management	Composting of starch polymer bags and of kraft paper bags, incineration of HDPE bags. Collection after use is excluded. Sensitivity analyses for agricultural production of maize in France and for compostable waste incineration.	Included Specific aspects: - Composting of bag if biodegradable - Incineration of bag if not biodegradable (PE) - Unclear whether credits are allocated to electricity/heat production in MSWI plants - Unclear whether composting is considered to sequester carbon and if so, whether credits are allocated.

Table 13.12 Continued ...

Parameters		COMPOSTO, September 1998 (waste bags)		COMPOSTO, January 1998 (waste bags)	
Quantitative analysis		*Impact category*	*Unit*	*Impact category*	*Unit*
		a) See column CARBOTECH (1996)		1. Energy resource	(MJ)
		b) 1. GHG emissions	index	2. GHG emissions	(kg CO_2 eq.)
		2. Ozone precursors	index	3. Ozone precursors	(kg ethylene eq.)
		3. Acidification	index	4. Acidification	(kg SO_2 eq.)
		4. Eutrophication	index	5. Eutrophication	(kg PO_4 eq.)
		5. Heavy metals	index	6. Toxicity air	(a m^3)
		6. Carcinogeneity	index	7. Toxicity water	(d x l)
		7. Winter smog	index	8. Salinisation	(H^+/mol)
		8. Summer smog	index	9. Deposited waste	(10^{-3} EPS)
		Impact factors		*Impact factors*	
		a) See column for CARBOTECH (1996)		See column for CARBOTECH (1996)	
		b) Eco-indicator 95			
Functional unit		Bags for compost bins with a volumetric content of 5-17 litres volumetric content		Bags for compost bins with a volumetric content of 5-15 litres volumetric content	
Qualitative analysis					

Table 13.13 LCA studies for PHA, PLA and other biodegradable polymers

	Gerngross & Slater, 2000	VTT, 1997	FAT/CARBOTECH, 1997
Biodegradable product	- PHA pellets - PLA pellets	- PLA diaper (1000 pieces)	(i) Mulch film made of kenaf (ii) Loose-fill chips made of miscanthus (iii) 'Miscanthus polymer' for injection moulding
Reference	- PE pellets - PLA pellets - Nylon	- PP/PE diaper (1000 pieces)	(i) PE mulch film (ii) EPS loose-fill chips (iii) PE pellets
Region/time	USA, end 1990s Exceptions:	Western Europe, mid 1990s	(Switzerland, mid 1990s)
System boundaries *Production*	All process steps included	All process steps included	All process steps included. To ensure comparability in land use, fallow land or extensive land use was assumed in the reference cases where fossil fuel based products are produced.
Use phase	Excluded	Excluded	Excluded
Waste management	Excluded (Landfilling possibly included implicitly)	Included Specific aspects: - MSWI plants: No credits for electricity/ heat	Included (i) Degradation of mulch film on the field (ii) Composting of miscanthus loose-fill chips (iii) (Composting of 'Miscanthus polymer') For products based on petrochemical feedstocks: incineration in MSWI plants.

Table 13.13 Continued ...

Parameters	Gerngross & Slater, 2000		VTT, 1997		FAT/CARBOTECH, 1997	
Quantitative analysis	*Impact category*	*Unit*	*Impact category*	*Unit*	*Impact category*	*Unit*
	1. Energy resource	(MJ)	1. Primary energy	(MJ)	1. Energy resources	index
			2. GHG emissions	(kg CO_2 eq.)	2. Land use	index
			3. Acidification	(mol H^+ eq.)	3. GHG emissions	index
			3. Eutrophication		4. Ozone precursors	index
			4. Photooxidant formation	(g O_2 eq.)	5. Acidification	index
			5. Toxicity air	(kg NO_X, CH_4, CO and VOC))	6. Eutrophication	index
			8a. Toxicity water	(kg Critical Body Weight, Air)	7. Toxicity air	index
			8b. Toxicity water	(kg Critical Body Weight, Water)	8. Toxicity water	index
			8. Use of natural land (class VI)	(m^3 units polluted water)	9. Toxicity soil	index
					10. Waste	index
Functional unit	1 kg pellets		1000 kg pellets		(i) 1 kg mulch film	
					(ii) 1 m^3 loose-fill chips	
					(iii) 1 m^3 pellets	
Qualitative analysis	Emissions				- Biodiversity	
					- Soil fertility	
					- Economics	

VOC: volatile organic compounds

Appendix 13.2 Checklist for the preparation of an LCA for biodegradable plastics

The following aspects should be taken into account when preparing a life cycle assessment for biodegradable plastics. All the methodological decisions, assumptions and key data should be specified in the text. In addition to the aspects listed in this checklist the life cycle assessment must comply with the requirements specified in the ISO standards 14040 to 14043 [1-4].

1. **Biomass production**

1.1 Country of origin: Where is the biomass used grown?

1.2 Type of cultivation: Is the biomass grown by intensive or extensive cultivation?

1.3 Fertilisers: Have the effects related to the production of fertilisers been taken into account?

1.4 Carbon balance plant growth: Is carbon uptake during plant growth:

a) considered as a separate process which is therefore reflected in the LCA calculations as negative CO_2 emissions or is it
b) combined with the process of decomposition (after the use of the product) resulting in overall net zero emissions?

Note: Both concepts are possible and the aggregated results throughout the life cycle are identical; however, differences in approaches result in difficulties when comparing disaggregated results of studies (results for subsystems). It is therefore recommended to apply approach a) since this is the more differentiated approach by breaking down the entire activity into more subprocesses.

2. **Plastics production and use**

2.1 Country: In which country is the biodegradable plastic (or the end product) manufactured?

2.2 Power generation: Are the assumptions regarding the average efficiency of power generation and the specific emissions, (e.g., in kg per MWh), stated?

2.3 By-products: If any by-products are produced (materials or energy), how are these taken into account (by means of credits or by extension of the system)?

2.4 System boundaries: Does the LCA refers to the system 'cradle-to-factory-gate' or to the system 'cradle-to-grave'?

Note: 'Cradle-to-factory-gate' refers to the entire production system from the extraction of the required resources to the production of the product under consideration. The system 'Cradle-to-grave' includes, moreover, the waste management after the useful life of the product.

2.5 Functional unit: Is the functional unit a certain amount of polymer (in mass terms or in terms of volume?), a semi-finished product, (e.g., 1 m² of film), or an end product, (e.g., 100 plastic bags)? A further option is to choose the product service as the functional unit, e.g. 100 m³ of packed goods in the case of packaging materials. Has it been taken into account that the amount of material required (in kg) for a given functional unit might differ for biodegradable polymers and their potential alternatives, (e.g., non-degradable polymers, paper)?

2.6 Use phase (for end products only): Is it clearly specified whether the use phase is or is not included in the system boundaries? If so, have the assumptions been specified?

Note: For example, the inclusion of the use phase for compost bags means that the functional unit is the collection of biodegradable household waste. In this case, comparisons are generally made with the collection in a compost bin without a bag; the compost bin is therefore cleaned after use; this implies the use of water and detergents which are included in the system boundaries while this would, for example, not be the case for the system 'cradle-to-factory-gate'.

3. Plastics waste management

The following aspects are only relevant if the system 'cradle-to-grave' has been chosen, otherwise they are irrelevant.

3.1 Waste management system: Which waste treatment process/es has/have been assumed? i.e., what are the shares of waste landfilled, recycled, incinerated without energy recovery, fed to waste-to-energy facilities, composted, digested and/or disposed of via sewage treatment?

Note: Sewage treatment is a practical option, e.g., for loose-fill polymers.

3.2 Landfill emissions: Have emissions from biodegradation in landfills been taken into account (especially: methane which orginates from inaerobic processes in landfills)? How high are the emissions?

3.3 Composting: Has any sequestration of carbon in the compost been assumed and if so, to which extent?

3.4 Waste-to-energy: In the case of waste-to-energy facilities, what are the yields of power and/or heat? Have these useful outputs been accounted for by credits (for input-related impact categories like energy resources and output-related impact categories such as greenhouse gas emissions)? And if so, which assumptions have been made when establishing these credits?

3.5 Recycling processes: In the case of recycling, have the types of technologies been specified?

- Mechanical or feedstock recycling?
- Which type(s) of feedstock recycling?

3.6 Mechanical recycling: Which substitution factor has been assumed in the case of mechanical recycling?

Note: It may be necessary to blend the recycled plastics with virgin material in order to obtain the desired material properties; it may also be necessary to use more recyclates than virgin polymers for the same functional unit. In both cases the substitution factors is less than 100%, i.e., each kilogram of recycled plastics substitutes less than one kilogram virgin material.

3.7 Waste schemes: Have separate schemes been developed for waste management, e.g., in order to account for different practices depending on the country or to account for (future) changes in waste policy?

4. Transportation

4.1 Particularly light products: Has transportation been taken into account for products with a particularly high volume/mass ratio?

Note: Transportation energy generally does not play any major role, neither for the production of plastics (including biodegradable ones) nor for final products made thereof. Particularly light products, e.g., loose fill packaging material, are exempted from this general rule.

4.2 Assumptions: Have all the assumptions been made clear in these cases (transportation mode, transportation distances, load factors, fuel efficiencies)?

5. Overall assessment

5.1 Impact categories and impact subcategories: The choice of the impact *sub*categories specified in addition to the impact categories can have a major impact on the final conclusions. Has the selection of the impact *sub*categories been throughly reflected and is the choice justified in the text?

Note: Background information is given in Section 7. In Table 13.4, a set of impact categories is listed that are considered to be particularly relevant for biodegradable polymers.

5.2 Significance thresholds: In comparative LCA, the uncertainty of the results and the importance of the differences in the results can be taken into account by distinguishing between significance thresholds. Has any approach of this type been applied?

*Note: **Table 13.15** gives an example of how this can be done.*

5.3 Characterisation factors: Have updated characterisation factors been used for aggregation within the impact categories? Characterisation factors are sometimes also referred to as equivalence factors.

Note: The characterisation factors for climate change are referred to as GWP values. The most recent figures can be found in Houghton and co-workers [44].

5.4 Weighting: Have the aggregated scores of the various impact categories been weighted?

Note: If not - i.e., if the conclusions are drawn without an weighting procedure - even more attention must be paid to the selection of the impact categories (see Section 5.1).

5.5 Non-biodegradable polymer: Has the non-biodegradable (conventional) polymer been named (in the case of comparative LCA)? Are the approach and the assumptions consistent with those used for biodegradable polymers and are all these assumptions clearly described?

6. Further aspects (consider only if relevant)

6.1 Alternative use of biomass: Has the alternative use of renewable raw materials for other material purposes or as an energy source been studied in order to put the results for biodegradable polymers into perspective?

Note: The background of this question is that it is already known that biodegradable polymers based on renewable raw materials generally score better than petrochemical polymers with regard to fossil energy use and greenhouse gas emissions while they score worse with regard to land use, ecotoxicity and eutrophication. Given this knowledge and considering the limited availability of biomass-derived raw materials it might be of interest to study the environmental impacts of other options of using the same renewable raw materials as used for the manufacture of biodegradable polymers.

6.2 National level: Have the results been translated to the national level?

Note: For materials that are or that could be used in bulk quantities this can be relevant for strategy development in companies and governmental policy.

This section reviews seven LCA results for biodegradable polymers - consequences of the choice of impact categories and impact subcategories and limitations of LCA

The choice of the impact categories can have an important impact on the final conclusions drawn from an LCA study. So far, only suggestions for the choice of impact categories to be included have been made [47, 48] but no obligatory set has been decided upon by the Society of Environmental Toxicology and Chemistry (SETAC). Neither does a minimum or a maximum list exist. The following criteria have been put forward for the choice of the impact categories [47]:

- Completeness: The list should include all relevant environmental problems.

- Independence: The categories should be as independent of each other as possible in order to avoid double counting of indirect effects.

- Practicality: For practical reasons the list should be as concise as possible.

A distinction can be made between input related and output related impact categories. Input related impact categories refer to resource depletion or competition while output related categories are metrics for emission and pollution.

Apart from the main impact categories, subcategories can be defined. This is usually done in those cases where the overall impact within one main category is caused by two or more factors that differ decisively, e.g., input of materials and energy as two subcategories of the main category 'resources'. The definition of subcategories raises the question how to aggregate this information in the further course of an LCA. There are two options [47]:

- Impact subcategories may be aggregated in the 'characterisation step', (e.g., aggregation of the subcategories 'energy' and 'materials' to the main category 'resources'), if the distinction of subcategories was mainly made due to lack of information for the impact category as a whole.

- The impact subcategories should be aggregated as a part of the valuation step if the effects and the underlying mechanisms are so diverse that the aggregation is primarily determined by the value system (this means that *sub*categories are upgraded to categories, for example: ecotoxicity *versus* human toxicity).

Table 13.14 provides an overview of firstly, the *main* impact categories proposed by Udo de Haes [47] and secondly, a selection of *sub*categories (and other impact indicators) encountered in LCA for biodegradable polymers. The crosses in the columns give a rough indication for which of the categories either polymers based on renewable resources or on petrochemical raw materials score better and in which cases the difference tends to be insignificant (column 'Neutral'). The table points out once more the potential importance of the selection of impact categories for the findings of an LCA. If, for example, the selection is such that it includes many impact categories that are relatively insignificant (column 'Neutral'), comparative LCAs may come to the conclusion that the differences between the options are relatively small. This type of misinterpretation can be avoided to some extent by introducing 'Significance thresholds' (see N° 5.2 in checklist) and by conducting a full-sized LCA including normalisation and valuation.

Finally, it is important to note the limitations of an LCA as a tool for decision support (**Tables 13.15** and **13.16**). Finnveden [46] points out that it is, strictly speaking, impossible to show by means of an LCA that one product is environmentally preferable to another. This has to do with the fact that universal statements are logically impossible to prove. Let us, for example, assume that a product A is (objectively) preferable to product B in environmental terms. Even if there is an LCA showing this, it is likely to contain some methodological and empirical choices that are uncertain to some extent. For example, it will probably be impossible to show that *all* relevant impact categories have been taken into account. It will therefore not be possible to prove the general environmental superiority of product A. If such a proof must be provided as a precondition for a decision at the company or governmental level, it is very likely that no action will ever take place. If, on the other hand, society wants to be able to act then it is inevitable to make decisions on a less rigid basis [46].

		Advantage renewables	Advantage petrochemicals	Neutral
	Table 13.14 Main impact categories and subcategories in LCA studies for biodegradable polymers (developed on the basis of Udo de Haes [47])			
I)	Impact categories			
	Input related categories			
I.A)	- Abiotic resources	X		
	- Biotic resources		X	
	- Land use		X	
I.B)	Output related categories			
	- Global warming	X		
	- Depletion of stratospheric ozone			X
	- Human toxicity			X
	- Ecotoxicity		X [1]	
	- Photo-oxidant formation			X
	- Acidification			X
	- Eutrophication		X	
	- Odour			X
	- Noise			X
	- Radiation			X
	- Casualties			X
II)	Impact subcategories and other impact indicators [2]			
	- Carbon resources (renewable & non-renewable)		X	
	- Non-renewable resources	X		
	- Non-renewable energy (fossil & nuclear)	X		
	- Nuclear energy			X
	- Adsorbable organic halogens (AOX)			X
	- Lead			X
	- Carcinogeneity			X
	- Diesel particulates			X
	- Total waste		X	
	- Hazardous waste	X		

[1] *Mainly due to biocide and pesticide use*
[2] *Selection of indicators used in LCA studies on biodegradable polymers: some of these indicators are categorised by Udo de Haes [47] as 'Pro Memoria Categories'. Udo de Haes defines these as 'truncated flows' that cannot be allocated to the categories extraction or emissions. Examples named are energy and waste*

Table 13.15 List of LCA impact categories of particular relevance for biodegradable polymers
1. Non-renewable energy (fossil & nuclear)
2. Land use [*]
3. Inorganic resources
4. Global warming
5. Depletion of stratospheric ozone
6. Human toxicity
7. Ecotoxicity
8. Photo-oxidant formation
9. Acidification
10. Eutrophication
11. Hazardous waste
[*]: *The subcategories biodiversity and soil quality may be accounted for in qualitative terms*

Table 13.16 Significance thresholds used in an LCA study on biodegradable polymers [17]		
The environmental impact is ...	Environmental categories	
	Energy, waste	All others
... much higher	> 200%	> 500%
... higher	125-200%	167-500%
... comparable	80-125%	60-167%
... lower	50-80%	20-80%
... much lower	< 50%	< 20%

14 Biodegradable Polymers and the Optimisation of Models for Source Separation and Composting of Municipal Solid Waste

Enzo Favoino

14.1 Introduction

As source separation, recycling and integrated waste management strategies grow up, there are more and more important suggestions about waste streams on which efforts have to be concentrated in order to reach high recycling rates and an overall improvement of environmental performance of waste management.

Traditionally, source separation systems have tackled only dry recyclables and most times were simply *added* to a collection of mixed municipal solid waste (MSW). Collection of paper, glass and plastics by means of road containers did not imply structural changes in the MSW collection. With such systems, separation rates range between 1% and 15%, depending on the distribution of road containers.

More recently, *integrated* source separation systems have been introduced. 'Integrated' source separation implies higher recycling rates by means of the segregation of the compostable fractions; in turn, this makes it possible to change also the features of collection systems for Restwaste (residual waste). In such respect, a central role is played by source separation of food waste.

From a *quantitative* point of view, fermentable material (food waste) accounts for a major percentage of MSW; and this is particularly true in Southern Europe and former Eastern European Countries. For instance, in Northern Italy percentages of food waste range between 25% and 40% of the total MSW; whereas in Southern regions they range between 35% and 50%, mainly due to the lower presence of packaging in a less wealthy, mainly rural economy where the habit is to have meals at home where pre-cooked and/or frozen products (which produce less food waste) are used less.

From a *qualitative* point of view, the more fermentable material gets sorted and recycled, the less production of biogas and leachate is to be expected in landfilling and the better thermal valorisation of 'restwaste' can be envisaged.

485

14.1.1 The Development of Composting and Schemes for Source Separation of Biowaste in Europe: A Matter of Quality

Since the late 1980s, composting has been undergoing an impressive growth across Europe, and above all in many countries belonging to the European Union (EU).

Even before that time, composting had been adopted as a disposal route for MSW, by attempting to sort the putrescible fraction mechanically. Such a strategy proved to be unsuccessful due mainly to the following reasons:

- the increasing presence of contaminants inside municipal waste

- the lack of suitable refining technologies that could effectively clean up the end product so that it was accepted by end users

- the consequent lack of confidence among farmers and other potential users

- the increasing awareness, among scientific bodies and institutions, of the importance to keep soils unpolluted – with specific reference to potentially toxic elements such as heavy metals.

As a consequence, the recent and effective growth of composting programmes started in parallel to the growth of schemes for source segregation of biowaste. These were increasingly adopted as the proper answer to the need to have quality products suitable for a profitable use in farmlands and other cropping conditions (forestry, nursery, gardening, cultivation of plants in pots, etc.).

With reference to activities in the field of source separation and composting of biowaste, European countries can be grouped into four categories (**Figure 14.1**). In Austria, Belgium (Flanders in particular), Germany, Switzerland, Luxembourg and the Netherlands strategies and policies are already fully implemented nationwide. The contribution of these countries – and Germany in particular – to the overall recovery of biowaste in the EU is fundamental and was around 80% in 1999. In the second category we find Denmark, Sweden, Italy, Spain (Catalunya) and Norway. In these countries policies are fully outlined but there is still an ongoing development of schemes, of the composting capacity needed and of the marketing framework.

Finland, France, the United Kingdom and Wallonie (Belgium) belong to a third category, where programmes are at the starting point though policies have been sometimes fully laid out.

In the fourth category are countries where there is no effort towards the composting of source separated organic waste just yet; these include most regions in Spain, as well as Greece,

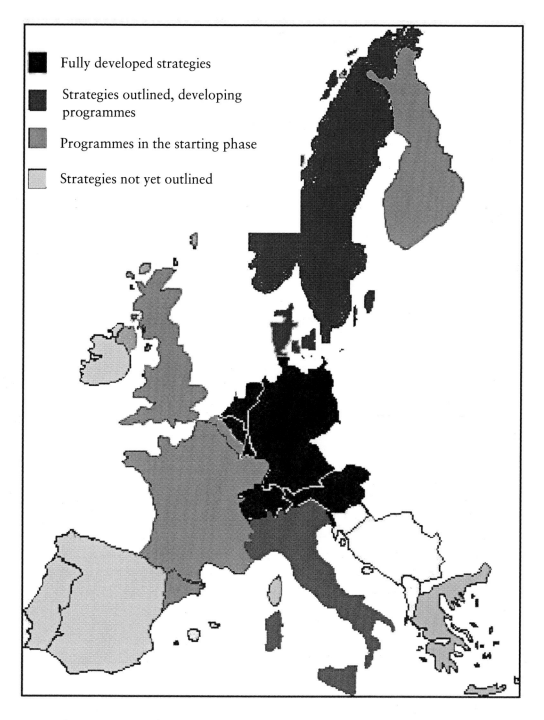

Figure 14.1 Development of source separation and composting in Europe

Ireland and Portugal. In these countries composting from mixed urban waste is still found, this sometimes plays an important role, e.g., many local strategies in Spain and Portugal.

14.2 The Driving Forces for Composting in the EU

There is a diffused awareness among technicians and decision-makers that composting will still play a most important role in forthcoming European strategies for waste management. In this section the most important driving forces at EU level for that are described.

14.2.1 The Directive on the Landfill of Waste (99/31/CE)

The Directive on Landfill of Waste basically provides for the landfilled biowaste to be sharply reduced within next years. This is aimed at effectively reducing the production of biogas at landfilling sites (one of highest contribution to the global warming potential from waste management) and to improve the conditions at which landfills get operated, (e.g., lower chemical strength of leachates, less settlings in the shape of the site after the landfill gets shut down).

Biowaste to be landfilled should be reduced by:

- 25% (with reference to 1995) within 5 years

- 50% within 8 years

- 65% within 15 years

Though this could also be achieved through thermal treatment, biological treatment and composting is likely to play a major role. In the end, composting is the most '*natural*' way to manage biowaste, and its cost is generally lower than that of incineration – above all incineration eventually has to comply with the provisions of the recent Directive on Incineration which mandates much tighter limit values for emissions from incinerators.

14.2.2 The Proposed Directive on Biological Treatment of Biodegradable Waste

The European Commission (EC) recently took the initiative to propose a Directive on Biological Treatment of Biodegradable Waste, in order to:

- ensure a balanced approach to the commitments on reduction of landfilled biowaste outlined in Directive 99/31/CE, i.e., to state that recycling of organic matter is a better option than thermal recovery (energetic exploitation of putrescible waste is made most difficult by its high moisture content).

- fix some recycling targets for biowaste, so as to ensure an even development of composting across Europe.

- define common limit values and conditions for use and marketing of composted products across Europe.

- give a further boost for the production of high-quality composted soil improvers to be used in organic farming and as a tool to fight desertification processes in southern European countries.

- pay attention also to those processes, usually described as mechanical-biological treatment (former MSW composting) that are at present experiencing a wide development above all to treat the residual waste. The Directive could in order to define their role in integrated waste management strategies and rule the conditions of use, (in land reclamation, landfilling, etc.) of their end product.

One of most important provisions included in current proposed Draft is that *source separation of biowaste should be mandated by all EU Countries. According to the current draft Directive, source separation ought to be developed in big cities (with possible exceptions only in inner cities) as well as in rural areas and little municipalities.* Such a provision could be disputed, as in general, it is argued that purity of sorted food waste inevitably tends to get much lower in highly populated areas. Actually, on the contrary, *the quality of collected biowaste seems to be much more dependent on the system adopted for collection than on the size of towns,* and reportedly in many situations, schemes also prove to be successful in big towns and inner cities. Considering Italian schemes, for example, the outcome of purity of separated biowaste (percentage of compostable materials) reported in various sorting analysis performed across Italy (**Figure 14.2**). What turns out is that no relationship can actually be detected between the size of the population covered and the purity of the waste.

This means that other factors are affecting the purity more than the population covered by the scheme, namely the *type* of the scheme put in place; schemes with collection at the doorstep generally perform much better than schemes run through containers on the road. Similar outcomes are reported in Catalunya, where similarly both types of scheme are currently run with a progressive switch to collection at the doorstep.

Statistical treatment of numbers yields a very low relationship ($R^2 = 0.0015$), and this is in itself a demonstration of a low dependence of purity of waste on the size of towns running the scheme for source separation. Even at a first glance, it is easy to see that high

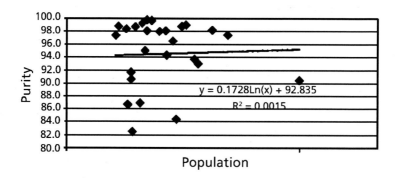

Figure 14.2 Purity of food waste *versus* population [1]

purity waste can as easily come from small villages as from medium to big towns, and the reverse is also true: in certain situations low purity of separated biowaste is recorded in tiny villages.

14.3 Source Separation of Organic Waste in Mediterranean Countries: An Overview

As a consequence of a growing number of provisions in national or local legislation, and/or mandatory programmes, a growing number of districts also in EU Southern Member States have lately adopted those strategies already well developed in Central and Northern Europe, aimed at source segregation of the organic fraction of municipal waste. During the last few years, the development has been particularly noticeable in Northern Italy and Catalunya.

Italy has recently seen a huge growth of source separation of food waste mainly due to the issuing of the National Waste Management Act (Decree 22/97).

This decree clearly states that:

• waste reduction and material recovery, re-use and recycling must be preferred to energy recovery and landfilling (which is seen as last resort);

• specific recycling targets (for each Province) are set at:

15% by March 1999

25% by March 2001

35% by March 2003, and

landfilling is allowed only for non-recyclable or treated materials (since July 2001).

Although source separation of organic waste (kitchen and garden waste) is not compulsory, it is becoming fundamental in the waste management system, in order to achieve the recycling targets. In fact it yields (particularly when operated with door-to-door systems) recycling rates as high as 20-40% on its own; this puts the overall recycling rate (including for example, paper, glass, etc.), at more than 50% even in towns, while hundreds of municipalities among the medium to small ones reach more than 60% (some as high as 70-75%). **Table 14.1** reports on best performing municipalities in 1999, and the contribution of food and yard waste to the overall quantity of recycled waste.

In general, the intensive collection of dry recyclables alone (paper, glass, metal and plastic) will not allow municipalities to meet the 35% recycling goal for 2003. Accordingly, most regions and provinces now plan to promote food waste source separation from households and major producers (restaurants, canteens, greengrocers, etc).

Table 14.1 Italian municipalities with highest recycling rates in 1999 and the contribution of compostable fractions				
Municipality	Inhabitants	% Recycling rate	Garden waste kg/per person/year	Food waste kg/per person/year
Masate	2296	79.6	196	55
Villa di Serio	5742	76.2	87	68
Presezzo	4512	71.7	77	57
Mesero	3430	70.8	106	66
Fara Gera d'Adda	6533	70.1	41	43
Gambellara	3146	69.0	-	42
Albairate	4062	68.8	44	69
Cassago Brianza	3936	67.7	65	41
Arcore	16,495	67.3	43	66
Usmate Velate	8252	67.3	70	62
Aicurzio	1947	66.9	116	70
Fumane	3736	66.1	37	52
Bariano	3923	66.0	45	55
Trezzo sull'Adda	11,425	66.0	55	82
Guido Visconti	1307	65.5	87	56
Azzano San Paolo	6786	65.4	37	58

The number of municipalities which are running schemes for source separation of food waste is steadily growing and it is likely to be far beyond the 1500 municipalities at present (the overall number of municipalities being about 8000). Though mostly concentrated in Northern Italy, with Lombardia having led its development since the early 1990s, during the last 2 years the strategy has also been developing in the Southern regions. After the first pilot schemes in Abruzzo, which led some municipalities to pass 50% recycling rate, during the early months in 2001 some 50 municipalities have implemented the schemes in Campania – including some medium to big towns. A wide development is expected during next years in Southern regions, as composting has been steadily included in Waste Management Plans drawn by the Governmental Task Forces committed to outline a sustainable waste management strategy in those Regions (Campania, Sicily, Calabria and Puglia).

Thanks to the wide diffusion of schemes, it is now possible to assess the effectiveness of these systems, in terms of:

- Quantitative effectiveness: This feature is expressed as specific collection capacity (in grams per person per day or kilograms per person per year); captures of food waste on its own are most often reported in Northern Italy at some 200 g/inhabitant/day, while pilot schemes in Southern regions often reach 250-300 g/inhabitant/day, as a consequence of the much more diffused habit to cook and have meals at home and of the higher percentage of vegetables and fish in the daily diet. On the contrary, it will be seen in Section 14.5 that schemes run through road containers allow much lower captures of food waste, while a high percentage of yard waste gets delivered inside the containers.

- Purity of the fraction collected: as **Table 14.2** clearly shows, random analyses of food waste, indicate the excellent quality of organic material collected. In fact usually, where schemes with collection at the doorstep are being run, the purity (percentage of compostable materials inside collected food waste) ranges between 97% and 99%. This result is to be compared to the 95% purity meant to be the 'excellence' level to have high quality composted products without affording expensive pre-sorting and final refining technologies in the composting plants.

Table 14.2 Purity (at sorting analysis) of collected food waste		
Municipality/Area	Inhabitants	Compostable materials (% weight)
Milan Province 17 municipalities	493,673	97.28
Municipality Monza	119,187	97.4
District 'Padova 1' 26 municipalities	203,429	98.7
Sources: [2, 3]		

Composting is also developing fast also in Spain. The start up of pilot schemes for source segregation of *'basura orgánica'* (also known as FORM or FORSU, organic fraction of municipal waste) dates back some time and has been developed in many Spanish districts, both rural and urban. Among the latter, an outstanding scheme – if the population covered is considered – has already been run in Cordoba for a long time (some 300,000 inhabitants).

Nonetheless, if we consider schemes for source segregation, Catalunya is undoubtedly in the lead, in Spain. The Catalan development takes as its source a Regional Law (Law 6/93) setting out compulsory programmes for the source segregation of organic waste in all municipalities with a population of over 5000 inhabitants. This mandate affects 158 municipalities with a population of 5.3 million inhabitants, or nearly 90% of Catalunya's population. The remaining municipalities, those with populations of under 5,000 inhabitants, are not required to comply, although they may participate – and many are doing so – on a voluntary basis.

From November 2000, 72 municipalities in Catalunya were reported to source separate biowaste, for an overall population of some 640,000 inhabitants. In the Barcelona metropolitan area itself, there were 21 out of 33, covering 150,000 inhabitants.

Catalan schemes were based, in the beginning, on a collection of organic waste by means of road containers, as had been previously done in other Spanish districts. Lately – after effective outcomes were reported in Northern Italy – doorstep schemes have been introduced and developed in various municipalities (Tona, Tiana, Riudecanyes being the first ones) with sharply different and better outcomes, which gives new perspectives in the growth and optimisation of strategies for composting.

As for recycling rates, these are shown to be impressively higher where doorstep schemes are used rather than in traditional schemes (**Figure 14.3**).

Also specific captures (directly related to recycling rates) and purity show sharply different and positive trends in doorstep schemes (**Table 14.3**).

These numbers are showing once again, as has already been shown for a long time on a broader scale in Italy, the different and much better outcomes that doorstep collection of food waste can yield. Having stressed the higher contribution of food waste to meet recycling targets met in doorstep schemes, the implications of its higher captures on the side of collection methods for restwaste, its simplified features and cost-optimisation still need to be considered. This can actually lead to optimised and cost-competitive schemes, as shown in Section 14.7.

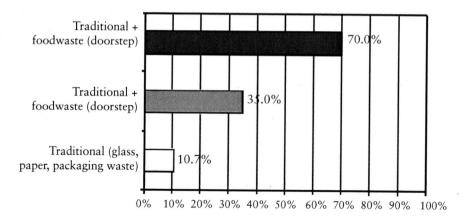

Figure 14.3 Recycling rates under different source schemes in Catalan municipalities

14.4 'Biowaste', 'VGF' and 'Food Waste': Relevance of a Definition on Performances of the Waste Management System

In Germany and Austria, the fraction targeted by the source separation system is referred to as 'Bioabfall' (biowaste), that means a mixture of food scraps and yard waste. In the Netherlands, in Belgium (Flanders) and in many sites in Germany and Austria, the definition 'VGF' (vegetable, garden, fruit) is used, addressing a mixture of yard waste and the food waste before cooking (not including, for example, meat and fish scraps). This choice is due to the troublesome, highly fermentable nature of cooked food residues.

On the other hand, we have to underline that *the recycling of dry fractions and packaging materials (paper, glass, plastics, etc), produces – as an undesired side-effect – the concentration of the fermentable material inside 'restwaste'*. This is to say that a higher percentage of fermentable material is found in residual waste whereas the dry fractions and packaging materials (paper, glass, plastics, etc.), are diverted with another recycling scheme and where the food waste is not effectively captured by an intensive collection scheme. This is what actually occurs in those countries (Germany, Holland, Austria, etc.) even though source separation of biowaste has already gone a long way, there. This means, in those countries separation of dry recyclables is likely to be more effective than that of food waste. For example, in the Netherlands and Germany, the percentage of food waste inside 'restwaste' is often reported at 30-50% [4, 5]. When transferred to warmer climates – as in the Mediterranean area – this system would need an increased frequency of collections of residual waste in order to escape odour problems (or nuisance effects).

494

Municipalities/schemes	Performances of schemes for food waste	
	Quantity (g/per person/day)	Quality (% impurities *w/w*)
Torrelles de Llobregat	139	1.8
Molins de Rei	116	2.1
Baix Camp	175	5.2
Igualada	125	3.8
Castelldefels	292	7.2
Castelldefels (March 2000)		4.5
Gavà	223	4.7
Viladecans	128	2.8
Viladecans		3.6
Castellbisbal	254	2.1
Vilanova i la Geltrú	239	—
Sant Cugat del Vallès (April 2000)	213	2.6
Barcelona (Major de Gràcia)	52	18.7
Barcelona (Gracia Comercial) (January 2000)		5.7
Barcelona (38 markets) (January 2000)		3.7
AVERAGE road container	177	4.9
Tona (October 2000)	265	0.9
Tiana (August 2000)	285	4.0
Riudecanyes (October 2000)	298	1.9
AVERAGE doorstep	283	2.3

Table 14.3 Specific capture and purity in schemes for source segregation of food waste in Catalunya. Schemes where a doorstep collection is in place are highlighted

Source: [6], updated

Moreover, in central Europe, in the 'biobin' (the bin supplied to households to separate biowaste) a large proportion of garden waste can be found (up to 80-90%, wet weight basis, out of the total bin content) in addition to food waste. The delivery of garden waste is much increased as households – even in detached houses with gardens – are provided with large-volume bins that allow the delivery of bulky materials as yard waste. We focus on the possibility of adopting a different scheme for the collection of compostable organics, by means of which the collection of food waste and that of yard waste are kept separated. This means, one collection route has to target only 'food waste' as a whole (including cooked stuffs such as meat and fish), by means of small volume bins and buckets; a different system targets yard waste only. This distinction between the two collection schemes takes into account:

- *The troublesome features of food scraps* (high putrescence and moisture). This needs the adoption of *specific tools, systems and collection frequencies in order to have the system clean and 'user-friend'*. When people feel comfortable using a system, the overall participation is enhanced. This leads to better quality and a higher quantity collected; lowers the percentage of food stuffs inside the restwaste, making it possible to collect it less frequently. In effect, analytical measurements – where a door-to-door collection is adopted – report the content of food stuffs inside restwaste at an average 15-20% and even less [2], which is much lower than in previous source separation programmes across Europe (see numbers reported above).

- *The different biochemical and seasonal feature of the food scraps as compared to yard waste.* In Italy – where a door-to-door collection for food waste is adopted, and in contrast with what is generally being done in Central Europe – the collection of the garden waste, that does not stink, doesn't attract flies and rodents and does not produce leachate, uses different schemes and tools compared to those for food waste. This in turn makes it possible an overall optimisation of the scheme, as 'intensive' features of the collection of food waste (high frequencies due to climatic conditions, watertight bags) do not apply to yard waste, which doesn't need such intensive, expensive collection patterns. Splitting the collection into food waste and yard waste separately makes it possible to build up a scheme where the total bin/vehicles volume fits the specific production of food waste, as it does not show huge seasonal fluctuations. On the contrary doorstep systems used for yard waste need to be more elastic and can be seasonally adapted varying the frequency of collection or using vehicles with different volumes according to the specific production.

- *The different bulk density of yard and food waste.* This forces the use of compacting vehicles (packer trucks) to collect yard waste, for food waste (which shows a much higher density) compacting vehicles can be replaced by small bulk lorries that are much cheaper at an equivalent working capacity. This is one of the most powerful ways to optimise the operational features and costs related to systems for source separation of compostable waste.

A system that does not set any difference between food and yard waste is a system where a huge delivery of garden waste is to be expected. It is noteworthy that in Central Europe – where a door-to-door bin collection for compostables is in place – it has often recorded an overall organic waste collection of some 200-250 kg per person per year and more. This is due, above all, to the ease of delivering yard waste to the collection service (households are allowed to deliver it in the same bins adopted for food waste collection, even in detached houses with private gardens, that makes the percentage of yard waste out of total compostable waste collected much higher). The general outcome is a high recycling rate, but the overall MSW production figure gets much higher, as well. In such situations, it is common to record an overall MSW production of some 500-600 kg per person per year. The same has already been reported in a few situations in Italy where similar collection systems have been adopted [7, 8]. Such a situation makes recycling rates rise, but also increases the overall quantity of waste to be collected and treated. One should for example mention the case of Forte dei Marmi (Tuscany), which after having implemented a curbside collection for yard waste reached 462 kg per person per year yard waste collected in 1998, though it led to an awful 850 kg per person per year figure for total waste [8]. Deliveries are much lower where the collection of yard waste is performed through collections at Civic Amenity Sites, or by means of collection at the doorstep, but with much lower frequencies (once monthly). Such systems keep among households a certain attitude to participate in home composting programmmes, as delivery is not made extremely easy – as it would be, on the contrary, with bins at the doorstep of detached houses with gardens.

14.5 The Importance of Biobags

Running source separation for food waste by households, means that it is necessary to find out the best way to face the specific troublesome features of such a material: *its fermentable nature and its high moisture content*. A good feature of the service, where households are provided with tools to avoid nuisance (buckets, biobins, biobags, besides 'intensive' collection schedules and a comfortable collection system, namely the door-to-door one), will result in an enhanced participation and will thus determine higher collection quantity/quality [3].

The 'Italian answer' to this issue – above all where a 'door-to-door' collection system is adopted – has been, typically:

- a relatively 'intensive' collection schedule (once to three times a week; it has to be noted that in Southern Italy, as in Spain, Portugal, etc., collection is scheduled up to 6-7 times a week; in Northern Italy the collection for MSW is usually 3 times/week);

- the use of 'door to door' collection systems so that they are more 'user-friend' and enhance participation;

- the use of watertight, transparent tools to hold the waste ('Biobags').

To allow people to feel comfortable with the biowaste collection service, municipalities usually provide them with small watertight bags for food waste. The use of the bags:

- prevents insect proliferation and leachate production and keeps the bins clean; therefore, it makes it possible to lower the frequency for washing rounds. Actually, in many cases, bins are washed by households themselves, other times the public Cleansing Service provides bin-washing a much lower number of times than would be possible without using the bags;

- Avoids nuisances generally related to delivery of 'loose' material inside the bin, makes it possible to collect even meat and fish scraps along with vegetable and fruit residues;

- Increases capture of foodstuffs, which in turn allow a significant reduction in collection frequency for 'restwaste';

- the small bag size prevents bulky materials, (e.g., bottles, cans), from being included into the collection, ensuring higher purity;

- the transparency of the bags allows an easy quality control of the material captured and defines the need for further information to be forwarded to households, (e.g., in a particular neighbourhood).

The 'bio-bag' is then placed:

- directly on the roadside on the collection day, either as it is or – most frequently – inside the family small bin (6.5 litres); this system is often used in small towns and villages to reduce the pick-up time for each dwelling and to prevent households from delivering garden waste inside the bins;

- in a bigger bin whose capacity usually ranges from 80 to 240 litres for 10 to 20 families depending on the collection frequency; this system is used where dwellings are in high-rise buildings.

14.5.1 Features of 'Biobags': The Importance of Biodegradability and its Cost-Efficiency

In general, it is possible to use polyethylene (PE) bags or biodegradable ones, as a matter of fact, both are used nowadays. Nevertheless, the biodegradable bag does not interfere with the composting process as it degrades during the composting cycle; whereas PE bags can be used only if the composting plant where the biowaste has to be delivered is provided with:

- *a pre-treatment section* (in general, a bag opener plus primary screen) and

- *an aeraulic facility* (separates by air blowing the plastic fragments from the compostable fraction) or equivalent in order to separate non-biodegradable plastic fragments.

The separation itself of course is not 100% effective, and often compels plant managers to shrink the sieving size so as to get rid of little (PE) fragments; very often screening holes are kept in such a case at 10 mm or less, whereas with biodegradable bags it is possible to screen at 12-15 mm and more (depending on the targeted use). But *the shrinking of the screen holes leads, in turn, to a dramatic product loss*, as many composted material particles are rejected and go to disposal. As for the rejects themselves, (which actually are – on a weight basis – mainly wooden materials not degraded yet) recycling gets much more difficult, as they are contaminated with plastics that would get more and more concentrated.

This is why *composting plants only accept PE bags, if ever, with much higher tipping fees*. The average additional cost related to the use of biodegradable bags (for the time being, in Italy, mainly corn starch based materials with 6.5 or 10 litres unit volumes) is about 10 Euro per ton collected; this has to be compared with additional operational costs and fees (about 15-20 Euro per ton or so) applied by composting plants when biowaste is delivered in PE bags.

In both cases, a *transparent* bag allows an easy quality check. As a consequence, the waste hauler might reject the bag, if it does not meet the quality criteria demanded by the composting plant. Furthermore, as already stated, it is possible to define which are the troublesome issues to be addressed in further information campaigns to improve collection quality.

It has to be mentioned that *nowadays in Milan Province and Northern Italy more than 95% of municipalities adopt biodegradable bags*. Furthermore, many more are abandoning previous PE bags and providing households with biodegradable ones.

The use of biobags is a very effective means to enhance participation and cut collection costs down. It has to be mentioned that in some municipalities, *watertight biodegradable liners have been adopted for use in bins* to further prevent them from getting dirty; thus the goal of stopping the expensive washing rounds is fully achieved. The liner is then collected with the biowaste itself as the collection truck empties the bin. In such situations, an average cost figure for the liner to be placed is about 0.5 Euro/bin for each collection round – including the manpower; but this makes it possible to save the much higher costs related to bin washing (about 1.5-2.5 Euro/bin).

14.6 Cost Assessment of Optimised Schemes

One of the major waste management concerns across Europe is the lack of cost-competitiveness of source separation systems aiming at reaching high recycling rates, as compared to the traditional mixed MSW collection. Operators in general think that sorting food waste leads to higher costs of the overall collection scheme.

Cost analyses carried out so far across Europe have traditionally focused on costs per kilogram (or per ton) for a single waste material collected. However, there is evidence that this biases the picture, because *the more the waste collected, the lower the costs of the collection service per kilogram*. This distortion obscures some important outcomes of integrated source separation and waste management:

- the reduction of total waste delivered as a consequence of effective waste reduction policies;

- the much lower delivery of industrial waste to the MSW collection route where large-volume road containers get substituted by curbside low-volume bins and bags;

- the contribution of home composting programmes to the overall reduction of organic waste collected.

Furthermore, the evaluation of the cost for a single waste flow, does not allow one to compare its costs to the likely advantages on collection costs for other materials, flowing from 'operational integration'. In effect, the collection of food waste – above all when it shows high captures – allows important changes in the collection scheme, by reducing, for instance, frequencies of collection for residual waste ('restwaste').

It is therefore incorrect to express the cost of the service per kilogram collected, rather it should be expressed as cost per person. Once an overall cost of a certain scheme is given, the municipality could only be happy with lower deliveries – that would on the contrary make the cost per kilogram higher! This is why we shall focus on costs per person.

To allow a comparison among different collection systems, the Research Group on Composting and Integrated Waste Management at Scuola Agraria del Parco di Monza has led some surveys on the costs of different collection systems run in Italy [9], grouped by their main features and above all according to the way food waste gets separated (or not).

The three system groups might be described as follows:

- traditional source separation, based on the use of plastic bags or road containers (up to 3.3 m^3) for mixed MSW and source separation through road containers only for dry recyclables (paper, glass, plastics). The food waste is not sorted and it's delivered along with the mixed waste; this holds fermentable waste (actually, food waste gets concentrated in it due to the withdrawal of paper, board, glass, plastics, etc.), and has to be collected frequently.

- intensive source separation, including that of food waste, based on road containers (120-240 litres, up to 3.3 m^3) both for food waste and dry recyclables; residual waste is delivered by road containers. This is usually referred to as the 'double container' collection (beside that for residual waste, households find the one for food waste).

It's pretty diffused in Central Italy (Emilia, Tuscany) and has been the most diffused, so far, also in Spain.

- intensive source separation, including that of food waste, with door-to-door (DtD, also known as 'collection at the doorstep') collection for food waste and residual waste. In general, some high-yield dry recyclables are also collected with a collection at the doorstep (usually paper and board, due to the much higher capture per person than with road containers). It's the most diffused system in those municipalities and provinces where highest recycling rates have been met (up to 70% in single municipalities). The system is well diffused in central Europe, as well, though the 'Italian version' uses buckets in place of bins for detached houses with gardens (to prevent deliveries of yard waste, as already mentioned). Also Spain (Catalunya, in particular) is now recording some first successful attempts to introduce this scheme, (e.g., municipalities of Tiana, Tona and Ruydecanes started in the year 2000).

Surveys have led to some unexpected outcomes, e.g., data from district 'Venezia 4', close to Venice (**Figure 14.4**), show that source segregation of food waste with doorstep schemes can be run with no substantial increase in overall cost, and sometimes costs are even lower than with traditional collections (no segregation of food waste) or with food waste segregation by means of road containers (in order to have a lower number of pick-up points).

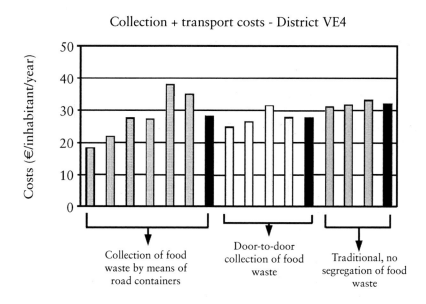

Figure 14.4 Cost comparison (€ per person per year) for different collection schemes in a single district

Such a paradox requires of course detailed insight into it in order to understand which tools are best suited to optimise operational and cost features. Actually, we have consciously developed such tools since the very beginning to ensure a steady development of source separation as a cost-effective strategy across Italy. Features of optimised schemes have also been adopted in national guidelines such as those reported in the National Handbook on Source Separation issued by the Italian National Environmental Protection Agency (ANPA) [10].

14.6.1 Tools to Optimise the Schemes and their Suitability in Different Situations

To understand such unexpected outcomes, it should be emphasised that if source separation of food waste is added to that of commingled municipal waste, with no modification in the previous scheme for MSW collection, total costs are bound to rise. This actually happens with the segregation of food waste by means of road containers. But this does not happen when food collection is integrated into the overall collection scheme: namely, when schemes for collection at the doorstep are implemented.

The trick is that intensive doorstep schemes for food waste – when made 'comfortable' for households – yield high captures. This in turn cuts the percentage of food waste inside the residual waste, which can then be collected less frequently. Furthermore, food waste on its own needs no compaction – letting operators use cheaper collection vehicles; this holds true in those schemes where the delivery of yard waste with food waste (that would be pretty high in areas with detached housing and private gardens) is being prevented by means of low-volume buckets which allow households to deliver only their food waste.

14.6.1.1 Collection Frequency for Residual Waste

Obviously collection frequencies for residual waste can be cut only when an effective separation of foodstuffs, yielding high captures, is run. Under such a viewpoint we have to mention (**Table 14.4**) that doorstep schemes enable much higher performances. Some 170-250 grams per person per day have been reported for food waste; outcomes tend to be higher in Southern Regions, thanks to a higher presence of food scraps in municipal waste (also first numbers reported in Catalunya, Spain, are confirming such high captures). Large road containers yield much lower quantities; their capture is actually sometimes similar, but a high percentage of yard waste contributes, and actual capture of food waste is low. We could therefore assume that 'collection using road containers results in a lower participation rate'; which is quite obvious due to the higher average distance between households and the container.

Table 14.4 Performances of different collection schemes for biowaste in Italy			
System	Overall yield (typical) g per person per day	Yard waste %	Actual capture of food waste g per person per day
Door-to-door (DtD)	170-250	0% (where delivery is banned) to 10% (maximum, due to low available volumes)	160-250
Road containers	150-200	40-70% (seasonal)	60-120
Sources: [3, 11]			

Cutting down collection frequencies for residual waste is one of the most important tools for optimising schemes for source segregation of food waste. Its use is particularly effective in those areas where high collection frequencies are in place for traditional, mixed MSW collection (above all Southern Europe).

Table 14.5 shows typical collection frequencies for mixed MSW and for 'integrated' collection systems whereby food waste is being segregated. Frequencies applied in Southern Italy do perfectly work in many Mediterranean situations, as well, where mixed collection is traditionally run six times weekly.

Table 14.5 Frequencies for the collection of:				
Area	Mixed MSW (with no segregation of food waste)	Food waste (both with DtD schemes and road containers)	Restwaste in DtD schemes (frequencies cut down, thanks to high capture of food waste)	Restwaste in road container schemes (no difference from previous mixed collection)
Northern Italy	3 times weekly	2 times weekly (sometimes once weekly during wintertime)	1-2 times weekly	3 times weekly
Southern Italy	6 times weekly	3-4 times weekly	2-3 times weekly	6 times weekly

14.6.1.2 Diversifying the Fleet of Collection Vehicles

In the doorstep system for food waste (split from green waste) small bags are delivered in bins (for high-rise buildings) or small buckets (for single families in houses with gardens).

The material collected shows a high bulk density (0.5-0.7 kg/l) so that it can be transported using open tank lorries (bulk lorries) instead of packer trucks.

These are suitable only when schemes effectively prevent the delivery of yard waste with food waste. So it is advisable to limit the size of containers supplied to households where gardens are available (6-10 litres, up to 30 litres); bins (80-240 litres) have to be supplied only to high-rise buildings to serve many families with a single pick-up. The use of buckets in detached houses also requires much less time per pick-up point (20 to 40 seconds on average, while bins require 2 to 3 minutes) as bins have to be hung up to the loading device, then unloaded and put back on their place: too time-consuming – for a single household – as compared to the simple, quick action of picking up and emptying a bucket manually. Assessment of course leads to a different outcome if we consider high-rise buildings, where a single bin can serve up to 10-20 families, thus making much more time-effective the single pick-up.

Households can manage yard waste through:

- home composting, promoted effectively by the municipality

- delivery to local recycling centres ('Déchetteries' in France, 'Civic Amenity Sites' in the UK, 'Recyclinghöfe' in German-speaking countries, 'Piattaforme Ecologiche' or 'Ecocentri' in Italy);

- specific collection of yard waste at the doorstep with low frequencies, (e.g., once monthly, only in the growing season, in general April-October).

We could therefore say that *collection rounds for food waste will have costs reduced through the use of low-tech vehicles and time-saving containers*. In our surveys, it was calculated and found out that a two-shift scheme for food waste collection using bulk lorries tends to equal the cost of a single-shift collection for residual waste with packer trucks (**Table 14.6**). This is partly due to the higher cost of a packer truck itself, partly to the much higher time spent on each pick-up point.

14.7 Conclusions

According to the numbers shown, it is clear that the main mistake made when planning sorting schemes, is the added feature of the scheme. Which means, a new collection scheme is run in addition to the previous mixed MSW collection, and cannot therefore yield savings to fund a new scheme. It is vital – on the contrary – that the new separate collection is integrated into the established waste management system, e.g., changing frequencies and volumes to collect residual waste, provide the collection of food waste

Table 14.6 Costs of collection routes (€ per person per year) for food waste and restwaste in door-to-door schemes			
Municipality (Province)	Population	Cost for collection of Restwaste (once weekly, with packer trucks)	Cost for collection of food waste (twice weekly, with bulk lorries)
Calcio (Bergamo)	4,765	5.14	4.21
Caravaggio (Bergamo)	14,181	5.46	6.01
Sommacampagna, Sona (Verona)	26,036	7.28	8.88

yields high captures through a comfortable scheme. Furthermore, 'integration' has to take into account the features of the area where the scheme has to be put in place; above all considering the need to find specifically suited systems for food and yard waste, where a big amount of yard waste is to be expected (areas with many gardens).

It must be remembered that collection frequencies of Restwaste can be cut only where a high capture of food waste reduces the fermentability of Restwaste. From such a standpoint, the use of comfortable tools such as watertight, biodegradable bags has proven to be very effective. This is why an 'intensive' collection, run through door-to-door schemes, notwithstanding a much higher number of pick-up points, has shown to be suitable for cost-optimisation, thanks to the integration of the system and much lower collection costs for restwaste.

Collection of food waste at the doorstep allows municipalities to perform much higher recycling rates (greater even than 60-70% and more in small municipalities, 50% in Monza, with a population around 120,000) and a much better quality of collected food waste [12, 13, 14].

A further tool to optimise the scheme is the use of suitable vehicles to collect food waste, due to its high bulk density when yard waste is kept away from the collection scheme for food waste.

References

1. *Analisi merceologiche dei rifiuti organici: 1999, [Report on Sorting Analysis of Biodegradable Waste]*, Azienda Municipale di Igiene Ambientale di Torino: Torino, Italy, 1999. [In Italian]

2. *Exercise Book – Municipal Solid Waste Management 1998 – Programmatic Trends and In-Depth Analysis*, Provincia di Milano, Milan, Italy, 1998, 47-74. [In Italian]

3. E. Favoino in *Proceedings of the Biodegradable Plastics 99 Conference*, Frankfurt, Germany, 1999.

4. K. Wiemer and M. Kern in *Abfall-Wirtschaft: Neues aus Forschung und Praxis*, Ed., Witzenhausen, Institute for Waste, Environment and Energy, MIC Balza - Verlag, Germany, 1995.

5. Department of the Environment, Baden-Baden Municipality, *Management of the Residual Waste: Findings of the Research Programme*, personal communication, 1996.

6. F. Girò in *Proceedings of RICICLA 2000, 2nd National Conference on Composting*, Rimini, Italy, 2000, p.53.

7. *Comuni Ricicloni 1997. Standings of the National Award to Highest Municipal Recycling Rates*, Legambiente, Rome, Italy, 1998.

8. *Comuni Ricicloni 1998. Standings of the National Award to Highest Municipal Recycling Rates*, Legambiente, Rome, Italy, 1999.

9. Scuola Agraria del Parco di Monza, *Proceedings of Comuni Ricicloni 1999*, National Award to best performing Municipalities in source separation, Rome, Italy, 1999. [In Italian]

10. *ANPA Handbook*, ANPA - Osservatorio Nazionale Rifiuti, Rome, Italy, 1999. [In Italian]

11. *Production, Disposal, Separate Collection in 1996/1997*, Provincia di Milano, Milan, Italy, 1998. [In Italian]

12. *Municipal and Assimulated Waste Management in 1997*, Consorzio Provinciale della Brianza Milanese, Seregno, Italy, 1997. (In Italian)

13. E. Favoino in *Proceedings of Jornadas Sobre Compostaje*, La Rioja, Spain, 2000.

14. *Municipal Solid Waste Production and Separate Collection Trend*, Provincia di Lecco, Lecco, Italy, 1997.

Abbreviations

β-BL	β-Butyrolactone
ε-CL	ε-Caprolactone
αMCL	α-Methyl-ε-caprolactone
β-PL	β-Propiolactone
γ-VL	γ-Valerolactone
3HB	3-Hydroxybutyrate
3HB-*co*-3HV	3-Hydroxybutyrate-*co*-3-hydroxyvalerate
3HB-*co*-4HB	3-Hydroxybutyrate-*co*-4-hydroxybutyrate
3HD	3-Hydroxydecanoate
3HHX	3-Hydroxyhexanoate
3HP	3-Hydroxypropionate
^{3}HPE	Tritium labelled polyethylene
3HV	3-Hydroxyvalerate
4HB	4-Hydroxybutyrate
4HV	4-Hydroxyvalerate
8-OL	8-Octanolide
AATCC	American Association of Textile Chemists and Colorists
ABS	Acrylonitrile-butadiene-styrene copolymer
ACP	Acyl carrier protein
ADM	Archer Daniels Midland Co
Ala	Alanine
ANL	*Aspergillus niger* lipase
AOT	Bis(2-ethylhexyl)sulfosuccinate, sodium salt
AOX	Adsorbable organic halogens
Arg	Arginine

Asn	Asparagine
Asp	Aspartic acid
ASTM	American Society for Testing and Materials
AVI	AIB-Vinçotte International
BBM	Benzyl β-malolactonate
BML	Benzyl malolactonate
BOD	Biological oxygen demand
BPI	International Biodegradable Products Institute
BPS	Biodegradable Plastics Society
BTA	Butanediol, terephthalic acid and adipic acid
BUWAL	Bundesamt für Umwelt Wald und Landschaft, Switzerland
CAL	*Candida antarctica* lipase
CAS	Continuous activated sludge
CCL	*Candida cylindracea* lipase
CEC	Cation exchange capacity
CEN	Comité Européen de Normalisation – European Committee for Standardisation
COD	Chemical oxygen demand
CP/MAS	Cross-polarisation magic-angle-spinning
cP[3HB]	Complexed P[3HB]
CRL	*Candida rugosa* lipase
CVL	*Chromobacterium viscosum* lipase
Cys	Cysteine
DAM	Draft amendment
DCW	Dry cell weight
DDL	12-Dodecanolide
DELTA-VL	δ-Valerolactone
DEV	German Standard Procedures for Investigation of Water, Wastewater and Sludge
DIC	Dissolved inorganic carbon
DIN	Deutsches Institut für Normung
DIS	Draft international standard
DMA	Dynamic mechanical analysis
DMF	Dimethylformamide

DMT	Dimethyl terephthalate
DMTA	Dynamic mechanical thermal analysis
DNA	Deoxyribonucleic acid
DOC	Dissolved organic carbon
DP	Degree of polymerisation
DS	Degrees of substitution
DSC	Differential scanning calorimetry
DSD	Duales System Deutschland
DtD	Door-to-door
DTMC	Dimer of trimethylene carbonate
EAA	Ethylene-acrylic acid copolymer
EAM	Enzyme-activated monomer
EC	European Commission
EC50	Concentration at which half the maximal effect is observed
ECCP	European Climate Change programme
EFTA	European Fair Trade Association - Iceland, Norway, Switzerland
EKI	Essem Kashoggi Industries
EN	European Norm
EPG	Environmental Polymers Group
EPS	Expanded polystyrene
EPSY	Environmental Point System
EU	European Union
EVOH	Ethylene vinyl alcohol
FabD	Malonyl-CoA-ACP transacylase
FDA	Food and Drug Administration
FTIR	Fourier-transform infra red spectroscopy
g/d	Grams/denier
GAMMA-BL	Lactone γ-butyrolactone
GAMMA-CL	γ-Caprolactone
GHG	Greenhouse gas emissions
Gln	Glutamine
Glu	Glutamic acid
Gly	Glycine

GPC	Gel permeation chromatography
gr30	gr is per gram and 30 is the thickness of the film
GWP	Global warming potential
HB	Hydroxybutyrate
HDDA	12-Hydroxydodecanoic acid
HDL	16-Hexadecanolide
HDPE	High-density polyethylene
HEPES	(*N*-[2-Hydroxyethyl]piperazine-*N*´-[2-ethanesulfonic acid])
HFCS	High-fructose corn syrup
His	Histidine
HPMC	Hydroxy propyl methylcellulose
HRP	Horseradish peroxidase
HV	Hydroxyvalerate
IBAW	Interessengemeinschaft Biologisch Abbaubare Werkstoffe (Industry association of bioplastic producers)
ICI	Imperial Chemical Industries
Ile	Isoleucine
iPP	Isotactic polypropylene
IR	Infra-red
ISO	International Organisation for Standardisation
JIS	Japanese Institute for Standards Organisation
LCA	Life cycle assessment(s)
LDPE	Low-density polyethylene
Leu	Leucine
Liapse CC	Lipase from *Candida cylindrica*
Lipase PC	Lipase from *Pseudomonas cepacia*
Lipase PF	Lipase from *Pseudomonas fluorescens*
LLDPE	Linear low-density polyethylene
Lys	Lysine
MALDI-TOF	Matrix assisted laser desorption ionisation time-of-flight
MARPOL	International Convention for the Prevention of Pollution from Ships
MBC	5-Methyl-5-benzyloxycarbonyl-1,3-dioxan-2-one
MCL	Medium chain length

Met	Methionine
MFI	Melt flow index
MFR	Melt flow rate
MITI	The Japanese Ministry of International Trade and Industry
MJL	*Mucor javanicus* lipase
ML	Mass loss
MML	*Mucor miehei* lipase
M_n	Number average molecular weight
MPL	α-Methyl-β-propiolactone
MS	Mass spectrometry
MSW	Municipal solid waste
MSWI	Municipal solid waste incineration plant
MVL	α-Methyl-δ-valerolactone
MW	Molecular weight(s)
NADPH	Nicotinamide adenine dinucleotide phosphate (reduced form)
NCIB	National Collections of Industrial and Marine Bacteria Ltd., Aberdeen
NMR	Nuclear magnetic resonance
NOEC	No Effect Concentration Level
OECD	Organisation for Economic Co-operation and Development
P[3HB]	Poly (R-3-hydroxybutyrate)
P[3HB-*co*-3HV]	Poly [3-hydroxybutyrate-*co*-3-hydroxyvalerate]
P[3HB-*co*-3MP	Poly [3-hydroxybutyrate-*co*-3-mercaptopropionate]
P[3HB-*co*-4HB]	Poly [3-hydroxybutyrate-*co*-4-hydroxybutyrate
P[3HV]	Poly[3-hydroxyvalerate]
P[4HB]	Poly[4-hydroxybutyrate]
PBSU	Polybutylene succinate
PBT	Polybutylene terephthalate
PCB	Polychlorinated phenols
PCL	Poly(ε-caprolactone)
PsCL	*Pseudomonas cepacia* lipase
PD	Polydispersity
Pd/C	Palladium/charcoal

PDL	15-Pentadecanolide
PDLA	Poly(D, L-lactide)
PDLLA	DL polylactic acid
PE	Polyethylene(s)
PEG	Polyethylene glycol
PEIP	Poly(ethylene isophthalate)
PESU	Polyethylene succinate
PET	Polyethylene terephthalate
PFL	*Pseudomonas fluorescens* lipase
PGA	Polyglycolic acid
PHA	Polyhydroxyalkanoate(s)
PhaB	Acetyl-CoA reductase
PhaG	3-Hydroxyl-ACP:CoA transferase
PhaJ1	(R)–Enol-CoA hydratase
PHA_{MCL}	Medium chain length PHA
PHA_{SCL}	Short chain length PHA
PHB	Polyhydroxybutyrate(s)
PHBV	Poly(hydroxybutyrate-*co*-hydroxyvalerate)
Phe	Phenylalanine
PHS	Poly(ethylene succinate)
PHT	Poly(hexamethylene terephthalate)
PHV	Polyhydroxyvalerate
PLA	Polylactic acid(s)
PLA/CL	Polylactide-*co*-caprolactone(s)
PLA/GA	Poly(lactide-*co*-gylcolide)(s)
PLLA	Poly(L-lactic acid)
Poly(BML-*co*-PL)	Poly(benzyl malolactonate-*co*-propiolactone)
PP	Polypropylene
PPL	Porcine pancreatic lipase
ppm	*Parts per million*
PPO	Poly(1,4-phenylene oxide)
PPP	Poly(1,2-propanediyl phthalate)
PPT	Polypropylene terephthalate
Pro	Proline

PS	Polystyrene
PSL	Lipase from *Pseudomonas sp.*
PU	Polyurethane(s)
PVC	Polyvinyl chloride
PVDC	Poly(vinylidene chloride)
PVOH	Polyvinyl alcohol
R&D	Research & Development
RAL	Reichs-Ausschuss für Lieferbedingungen
RH	Relative humidity
RNA	Ribonucleic acid
ROL	*Rhizopus oryzae* lipase
SCAS	Semi-continuous activated sludge
SCL	Short chain length
SEC	Size exclusion chromatography
SEM	Scanning electron microscopy
Ser	Serine
SETAC	The Society of Environmental Toxicology and Chemistry
SOM	Soil organic matter
Tc	Crystallisation temperature
TCA	Tricarboxylic acid
T_d	Decomposition temperature
TEM	Transmission electron microscopy
T_g	Glass transition temperature
THF	Tetrahydrofuran
ThOD	Theoretical oxygen demand
Thr	Threonine
T_m	Melting temperature
TMC	Trimethylene carbonate
TPA	Terephthalic acid
TPPS	Thermoplastically processable starch
TPS	Thermoplastic starch
Trp	Tryptophan
TTMC	Trimer of trimethylene carbonate
Tyr	Tyrosine

UDL	11-Undecanolide
UHMW	Ultra-high molecular weight
USCC	US Composting Council
USP	US Pharmacopeia
UTS	Ultimate tensile strength
UV	Ultraviolet
Val	Valine
VGF	Vegetable, garden, fruit
WO	Water to AOT molar ratio
WR	The rate of sub-unit weight (%) in raw materials
WVP	Water vapour permeability

Contributors

Catia Bastioli
Novamont SpA, Via G. Fauser, 8, 28100 Novara, Italy

Gregory Bohlmann
SRI Consulting, 4300 Bohannon Drive, Suite 200, Menlo Park, CA 94025, USA

Bernard Cuq
Ecole Nationale Supérieure Agronomique de Montpellier/INRA, Unité Technologie des Céréales et des Agro-Polymères, 2 place Viala, 34060 Montpellier, Cedex 1, France

Yoshiharu Doi
Polymer Chemistry Laboratory, Riken Institute, Hirosawa 2-1, Wako-shi, Saitama 351-0198, Japan

Enzo Favoino
Working Group on Composting and Integrated Waste Management, Scuola Agraria del Parco di Monza, V.le Cavriga 3, I-20052 Monza, Italy

Johann Fritz
Universität für Bodenkultur Wien, Department IFA-Tulln, Abteilung Umweltbiotechnologie, Konrad Lorenz Strasse 20, A-3430 Tulln, Austria

Stéphane Guilbert
Ecole Nationale Supérieure Agronomique de Montpellier/INRA, Unité Technologie des Céréales et des Agro-Polymères, 2 place Viala, 34060 Montpellier, Cedex 1, France

515

Samuel J Huang

Institute of Materials Science, University of Connecticut, Storrs, CT 06269, USA

Francesco Degi Innocenti

NOVAMONT SpA, via Fauser 8, I-28100 Novara, Italy

David Kaplan

Tufts University, Department of Chemical & Biological Engineering, Bioengineering Center, 4 Colby Street, Medford, MA 02155, USA

Rolf-Joachim Müller

Gesellschaft für Biotechnologische Forschung, Department TUBCE, Mascheroder Weg 1, D-38124 Braunschweig, Germany

Martin Patel

Utrecht University, Department of Science, Technology and Society, Heidelberglaan 2, NL-3584 CS Utrecht, Netherlands

Amarjit Singh

Tufts University, Department of Chemical & Biological Engineering, Bioengineering Center, 4 Colby Street, Medford, MA 02155, USA

Kumar Sudesh

School of Biological Sciences, Universiti Sains Malaysia, 11800 Penang, Malaysia

Maarten Van der Zee

Agrotechnology & Food Innovations, Business Unit Biobased Products, PO Box 17, NL-6700 AA Wageningen, The Netherlands

Bruno De Wilde

OWS nv, Dok Noord, 4, B-9000 Ghent, Belgium

Index

Page numbers in italic, e.g. *12*, refer to figures. Page numbers in bold, e.g. **346**, signify entries in tables.

Z